APPLIED ELECTROMAGNETICS AND ELECTROMAGNETIC COMPATIBILITY

APPLIED ELECTROMAGNETICS AND ELECTROMAGNETIC COMPATIBILITY

Dipak L. Sengupta
The University of Michigan
and The University of Detroit Mercy

Valdis V. Liepa
The University of Michigan

A JOHN WILEY & SONS, INC., PUBLICATION

Published by John Wiley & Sons, Inc., Hoboken, New Jersey.
Published simultaneously in Canada.

Library of Congress Cataloging-in-Publication is available.

ISBN-13 978-0-471-16549-1
ISBN-10 0-471-16549-2

Printed in the United States of America.

10 9 8 7 6 5 4 3 2 1

This book is dedicated to
Sujata Basu Sengupta
and
Austra Liepa

CONTENTS

PREFACE

Over the past two decades the electromagnetic compatibility (EMC) considerations in the design of digital electronic devices or components have grown in importance throughout the world. This is because the United States and other industrial nations do not allow electronic devices to be marketed in their countries unless their electromagnetic noise emissions have been tested and certified to meet certain limits. As a result the electronic industries are showing increasing interest in electrical engineering (EE) graduates with an EMC background. Currently, except for a handful of schools, the undergraduate EE programs in the United States do not address the EMC issues directly, although most of them require at least one 3- or 4-credit course in electromagnetics.

Students specializing in fields and waves are probably equipped to investigate certain fundamental problems of EMC. A few well-known schools with strong programs in electromagnetics often express the opinion that a good training in fields and waves prepares the students sufficiently to meet the challenges of EMC in their professional life. Nevertheless, to meet the growing demand from industries, the IEEE is actively encouraging schools to include EMC as a course topic in their curricula.

During the summer of 1994 the first named author (DLS) introduced a senior/graduate level course in EMC at the Electrical Engineering and Physics Department of the University of Detroit Mercy. It was taken mostly by practicing engineers from the Detroit metropolitan area's Big Three automobile manufacturing and other related industries. Our own attending EE students had varied levels of background in electromagnetics but none had the expected familiarity with Maxwell's equations and plane electromagnetic waves. Because of this we faced difficulties in the planning of the course. In addition at that time we had a very limited selection of textbooks on EMC. We chose *Introduction to Electromagnetic Compatibility* by C. R. Paul supplemented by *Noise Reduction Techniques in Electronic Systems* by H. R. Ott, but found them not completely suitable for our students' needs. It was therefore necessary to develop lecture notes specialized to the class. The initial lectures were devoted to bringing the students' background level in electromagnetics up to a uniform level of familiarity with Maxwell's equations. After this, plane waves and related topics, transmission lines, antennas, and radiation were introduced. Overall, knowledge of these topics was deemed to be a necessary minimum background for an EMC course. The rest of the lectures were on selected topics in EMC. The course was so well received that it was repeated the next year (1995). Because of continued demand it is still being offered every alternate year.

Our experience motivated us to write a textbook combining the fundamentals of fields and waves, a few selected topics of applied electromagnetics, and a variety of topics typical of EMC. The descriptions of electromagnetics are placed in the context of EMC, and those of EMC are presented where they help in the analysis of EMC phenomena as well as in planning the measurements needed for compliance with EMC specifications. The book is also an outgrowth of classroom lecture notes for a number of undergraduate/graduate level courses in electromagnetic theory and applied electromagnetics given by the first author over many years at the electrical engineering departments of the University of Michigan, Ann Arbor and the University of Detroit Mercy.

A brief outline of the book follows. Chapter 1 introduces electromagnetic interference in general, and describes the evolution of EMC in the digital electronics era. It also defines various acronyms that are used alternatively, and often erroneously, to describe interference effects. The electromagnetic environment consists of a variety of natural and human-made noise sources in which electronic devices are expected to operate. These noise sources are described in Chapter 2. Chapter 3 is about the fundamental concepts and relations of electromagnetic fields and waves. Basic laws of electricity and magnetism, their generalizations, and their mathematical descriptions by Maxwell are described. Boundary conditions, the Poynting theorem, and energy transfer are then discussed. The time harmonic formulation of Maxwell's equations are introduced next and their applications to general problems are described. Finally, uniform plane waves in lossless and lossy media, skin effect phenomena, and reflection and refraction of plane waves are discussed. Chapter 4 describes the frequency spectra of known electromagnetic sources to the extent necessary for the characterization of their electromagnetic emissions as functions of frequency from the viewpoint of EMC. Basic characteristics and applications of TEM transmission lines and, in particular, the two-wire, coaxial, microstrip, stripline, and parallel plate lines are briefly described in Chapter 5. The time dependent or the transient solutions for a two-wire line are also briefly mentioned here. Chapter 6 discusses the fundamentals of antennas and radiation, including the equivalent circuits for receiving and transmitting antennas. The radiation from basic antennas, such as the electric and magnetic dipoles, is described in detail; these descriptions are then utilized in the discussion of certain general characteristics of radiation. In addition the half-wave dipole and the biconical antenna are described.

The behavior of the lumped circuit parameters R, L and C are described in Chapter 7. The field theory definitions of these parameters are introduced at first; they are then used to analyze performance as functions of frequency. Chapter 8 gives analytical descriptions of the radiated emissions from certain components of an electronic device and their susceptibility to outside noise. Simple wire and transmission-line models are used to estimate these emissions and susceptibility of the components when illuminated by incident plane waves from outside sources. Principles of electromagnetic shielding are briefly described in Chapter 9. The inductive and capacitive coupling effects in selected circuit configurations are outlined in Chapter 10. Chapter 11 deals with the electrostatic discharge (ESD) phenomenon and its impact on the design of electronic systems from the viewpoint of EMC. Chapter 12 gives the typical standards for EMC prescribed by the FCC for both Class A and Class B types of electronic devices. Some European standards are also mentioned. Chapter 13 describes briefly the measurement procedures that are followed to test the compliance of a device to the emission limits required by the enforcing agency. Appendix A gives a rather complete description of the vectors and vector calculus that are essential background knowledge for any course in electromagnetics. Problems, and answers to many of them, are provided at the end of some chapters.

The book is intended to serve as a textbook for courses on applied electromagnetics and electromagnetic compatibility at the senior/graduate level in EE. The prerequisites for such a course are completion of basic undergraduate EE and physics courses in electricity and magnetism, analog and digital electronic circuits, and advanced calculus.

The description of fields and waves starts at the basic level and then proceeds to a fairly high level. Topics in EMC are described such that the electromagnetic interference effects associated with them can be better understood.

Depending on the electromagnetic background of the class, the instructor may apply his/her discretion to adjust the emphasis on specific course materials.

For a class with a sufficient background in electromagnetic fields, the book can be used by the instructor to delve into the discussed EMC topics in more detail and also to put forward additional EMC topics. For example, one could include designs for EMC that are not considered here and extend the discussion of EMC measurements. The appropriate materials for taking this direction are in Chapters 1, 2, and 7 through 13.

This book is also designed to serve as a textbook for coursework on applied electromagnetics. The appropriate chapters are 3 through 6 (and perhaps 7) and Appendix A. The instructor may choose to include more discussion of these topics as well as more materials on antennas, for example. Such a course might even be followed by the course on EMC described earlier.

Finally, practicing engineers in industry interested in exploring EMC may find the book useful for self-study. The topics and descriptions are such that engineers involved in the design of electronic devices for EMC will find the book useful as a reference tool.

D. L. SENGUPTA

V. V. LIEPA

Ann Arbor, Michigan

ACKNOWLEDGMENTS

We gratefully acknowledge the support received from the Department of Electrical Engineering and Computer Science of the University of Michigan and the Department of Electrical Engineering and Physics of the University of Detroit Mercy, where virtually all the material presented in this book was taught over many years. The suggestions and comments received from our students helped us in the organization and presentation of the material. We thank our students for their critical comments. The final preparation of the manuscript was accomplished at the University of Michigan Radiation Laboratory. We thank the Director of the Radiation Laboratory for generously providing the Laboratory facilities for this purpose. The tedious task of transforming the handwritten manuscript to its final form was accomplished by a team of people. We are especially grateful to Joseph Brunett who supervised and actively carried out the electronic formatting and graphic design. He was assisted by Richard Carnes, Bradley Koski, and Sanita Liopa. Our grateful thanks to all of them for performing an excellent job. We are grateful to the staff of John Wiley & Sons, Inc., especially to George Telecki, Associate Publisher, Wiley-Interscience, for his interest, support, cooperation, and production of the book; Danielle Lacourciere, Senior Production Editor, for the production of the book; and Rachel Witmer, Editorial Program Coordinator, for managing the production schedule and the cover design.

The writing of this book has been a long and arduous task. It would not have been completed without the patience and continuous support of our wives and children.

CHAPTER 1

GENERAL CONSIDERATIONS

1.1 INTRODUCTION

Modern design of an electronic device or electric system requires it to be compatible with its electromagnetic environment, which may contain a number of sources emitting electromagnetic disturbances or noises. The design should be such that these disturbances cause minimum impact on the system performance. Also it is required that the system-emitted noise(s) in the environment cause minimum impact on the performance of other electronic systems in its vicinity. The entire class of such events can be classified as electromagnetic interference (EMI).

In this chapter we define a few popular terms that are used to describe EMI phenomena and briefly discuss some selected mechanisms by which EMI effects can manifest in an electronic device. This way we can place in proper perspective the interrelationship of applied electromagnetics and electromagnetic compatibility (EMC), which, together, form the subject matter of the present book.

Applied Electromagnetics and Electromagnetic Compatibility. By D. L. Sengupta, V. V. Liepa
ISBN 0-471-16549-2

1.2 DEFINITIONS

The *IEEE Standard Dictionary of Electrical and Electronics Terms* [1] defines electromagnetic interference as "impairment of the reception of a wanted electromagnetic signal caused by an electromagnetic disturbance." Electromagnetic disturbances can be in the form of any unwanted electromagnetic signal, including any multipath form of the desired signal. The disturbances can be continuous or discontinuous and repetitive or nonrepetitive in time. In general, any unwanted electromagnetic signal or disturbance is frequently referred to as *noise*.

Since the birth of radio communication, the term *radio frequency interference* (RFI) has been extensively and often erroneously used interchangeably with EMI to describe the interference phenomena. To bring out the subtle difference between the two terms we quote the IEEE definition of RFI: "the impairment of the reception of a wanted radio signal caused by an unwanted radio signal, i.e., a radio disturbance" [1]. We will assume that the range of frequencies of radio signals (or radio frequency) extends from 9 kHz to 3000 GHz, as defined by the Federal Communications Commission (FCC) [2]. Thus RFI can be described as the impairment of the reception of wanted signal caused by a radio frequency disturbance.

The existence of ambient noise in the environment makes it necessary that proper considerations be given during the initial design phase of an electronic device so that the device is immune (not susceptible) to performance degradation due to interaction with a pre-assigned minimum level of such electromagnetic noise. At the same time, it is also important to ensure that the system does not emit electromagnetic noise above some pre-assigned level so as not to cause performance degradation of other electronic systems in its vicinity. In addition considerations must be given so that system generated internal noise does not interfere with itself, thereby degrading the system's performance.

With the proliferation of a large variety of electronic systems (particularly digital devices, which are efficient radiators of electromagnetic energy), the design of such systems in modern time attempts to fulfill these requirements so that the system designed is compatible with the ambient electromagnetic environment.

We define *electromagnetic compatibility* (EMC) as "the capability of electromagnetic equipments or systems to be operated in the intended operational electromagnetic environment at designated levels of efficiency" [1]. In the United States these levels are generally assigned by the FCC. They will be described in a later chapter.

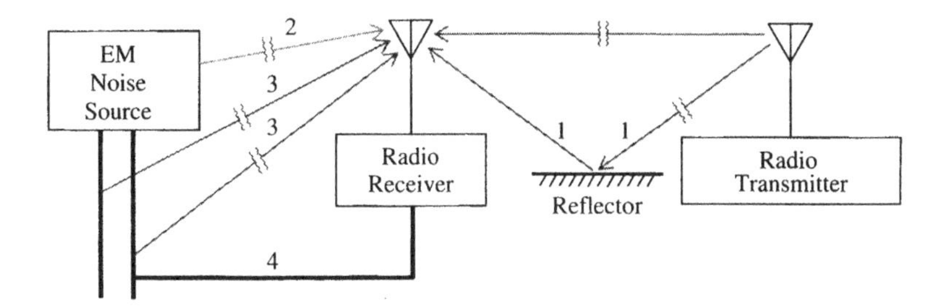

Figure 1.1 A radio receiver in free space receiving signals from a distant transmitter. Identifying selected noise paths to the receiver.

1.3 INTERFERENCE MECHANISMS

A system may suffer performance degradation due to EMI in a variety of ways. For illustration, we consider a simple case of a radio receiver receiving the desired signal from a distant transmitter (source) in free space shown in Figure 1.1. For simplicity we have assumed only one isolated electromagnetic noise source, and a simple reflector (of the desired signal) acting as a multipath source in the vicinity of the radio receiver.

Ideally, only the direct signal should be received by the radio receiver. Unwanted electromagnetic disturbances can also reach the receiver by a selected number of paths, as shown in Figure 1.1. The four paths indicate the following five electromagnetic disturbances received by the receiver in addition to the desired signal:

1. The multipath signal reaches the receiver through path 1. The signal is similar to the desired signal but reaches the receiver after suffering a reflection off the reflector. Often the amplitude is approximately the same as that of the desired signal but the phase is different at the receiving antenna. The magnitude of this phase difference generally determines the amount of corruption in the reception.

2. Disturbances radiated from the noise source reach the receiver by path 2.

3. A variety of electrical disturbances exist in the power line. Also the EM noise source may introduce noise by conduction into the power line. Path 3 shows the total noise in the power line reaching the receiver by radiation.

4. The EM noise source can conductively couple noise into its signal or control cable. This noise can reach the receiver by radiation as shown by path 3.

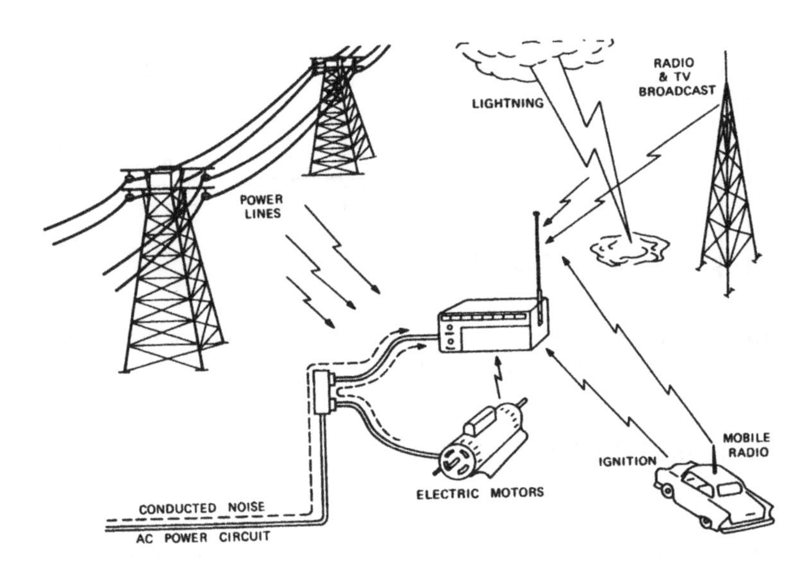

Figure 1.2 Outside of the laboratory, electronic equipment such as this radio is subject to a variety of electromagnetic noise sources. Careful design is required to guarantee compatibility with the environment. (Source: From [3], p. 3.)

5. Noises in the power line can reach the receiver by conductive coupling, path 4. It is assumed that the radio receiver and the EM noise source have a common power supply.

A more realistic illustration, taken from [3], is shown in Figure 1.2. From the two cases given in Figures 1.1 and 1.2 it is observed that electromagnetic disturbances can reach a device under consideration through the following three mechanisms: (1) radiation, (2) conduction, and (3) combination of radiation and conduction.

In the examples above we considered the noise existing outside the system. Within electronic equipment, such as the radio receiver, individual circuit components can interfere with each other in several ways. An example is shown in Figure 1.3, taken from [3].

It is also important that the system, in this case the radio receiver, not behave as a source of noise capable of interfering with other equipment in its vicinity, as depicted in Figure 1.4, taken from [3]. Here again, radiative and conductive coupling as well as a combination of both can cause undesirable interference.

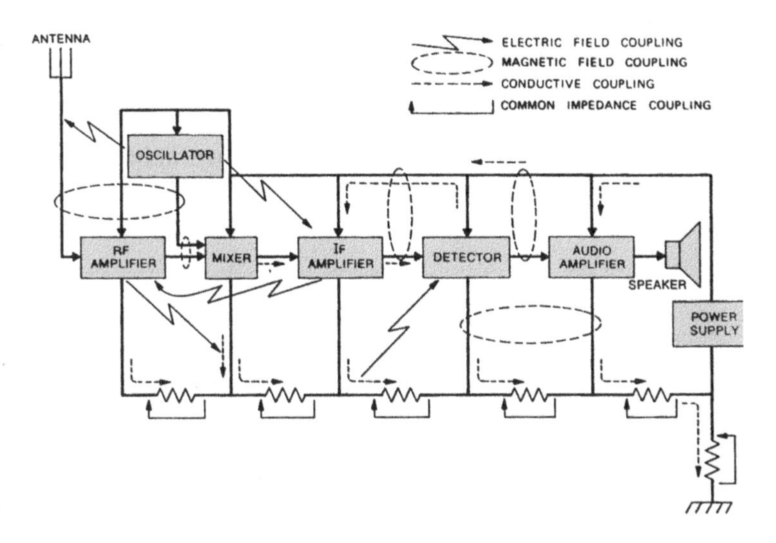

Figure 1.3 Within equipment, such as this radio receiver, individual circuit elements can interfere with one another in several ways. (Source: From [3], p. 2.)

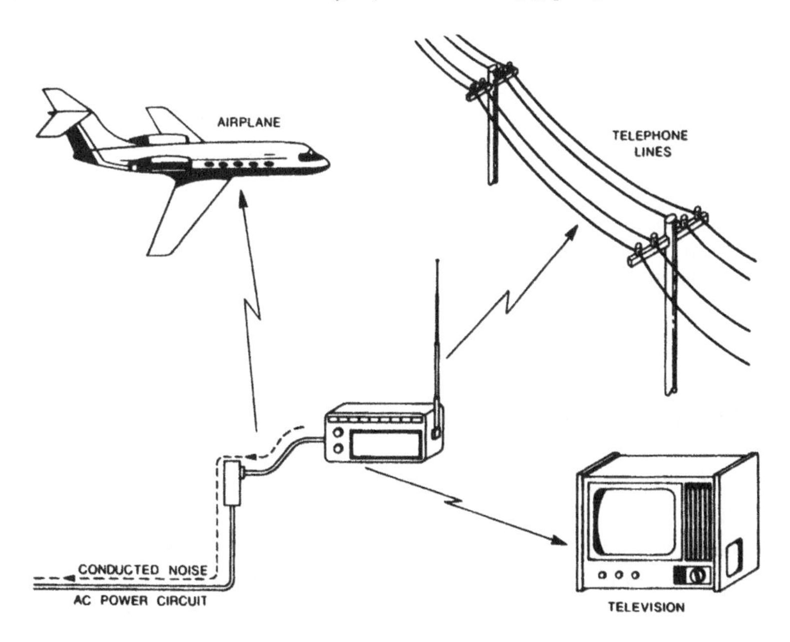

Figure 1.4 Electronic equipment such as this radio can emit noise that may interfere with their circuits. Consideration of noise during equipment design can avoid these emissions. (Source: From [3], p. 4.)

Table 1.1 Examples of Suspected PED-Caused EMI Events

System	Suspected Device	Interference
Autopilot	Walkman, computer	Aircraft abruptly banked right
CDI	Laptop computer and portable radio	EFIS screens blanked suddenly, then indicated missed approach fail along with loss of all auto navigation functions
Compass	Phone, laptop computer	Compasses lost synchronization and moved off course
HSI	Phone	Discrepancy between captain's HSI and first officer's HSI
ILS	Stereo	ILS signal was disrupted
Omega	Tape player, phone, Nintendo, computer	Omega vector was off course
EFIS HIS	Cellular phone	Discrepancy in captain's and first officer's HSI
EFIS	Cellular phone	Captain versus first officer's headings indicated approximately 10° difference
VOR	Walkman, computer, TV	Erroneous VOR signal caused aircraft to vector off course

Source: From [7] p.31.
Notes: CDI: control/display indicator or course deviation indicator (Honeywell); HSI: horizontal situation indicator; ILS: instrument landing system; EFIS: electronic flight system; VOR: very high frequency omni-directional range.

1.4 EXAMPLES

Examples of malfunction suffered by electronic devices of various kinds due to EMI effects caused by a variety of other electronic systems have been reported and discussed in technical journals, trade journals, and even in newspapers. Such events are also briefly discussed in technical books [5–7].

Here we give a sampling of such events suffered by some aircraft navigation equipment suspected to be caused by airborne operation of portable electronic devices (PEDs) [7]. Typically PEDs operate from very low frequencies to thousands of MHz. The active electronic/electric components of these mostly digital PEDs emit frequencies, usually with harmonics, that can overlap with

Table 1.2 Navigation-Frequency Operating Bands/Sensitivities of Avionics, and PED Emission Limitations

Equipment	Frequency	Minimum Level in Service	FCC Part 15 Limit from PED at 3 m
VOR	108–118 MHz	90 μV/m	150 μV/m
GLS	328–335 MHz	400 μV/m	200 μV/m
DME	978–1213 MHz	1375 μV/m	500 μV/m
GPS	1500 MHz	48 μV/m	500 μV/m
MLS	5031–5091 MHz	237 μV/m	500 μV/m

Source [7], p. 34.
Note: (VOR) Very high frequency omni-directional range, (GLS) glide slope system, (DME) distance measuring equipment, (GPS) global positioning system, (MLS) microwave landing system.

the operating frequencies of the communication and navigation systems of the airplane. Thus navigation and other electronic systems of the airplane may be susceptible to performance degradation due to the EMI effects of passenger-carried PEDs if they radiate emission levels above the sensitivity levels of the airplane systems. Table 1.1, taken from [7], gives a sampling of suspected cases of PED interference, along with the systems affected and the suspect device causing the malfunction.

Table 1.2 shows that the emission levels of some FCC-certified PEDs exceed the sensitivities of some operating avionic systems, thereby causing malfunction of the latter. Further discussion of the EMI effects of PEDs is given in [7].

1.5 DISCUSSION

An electronic system is compatible with its electromagnetic environment if it satisfies the following two criteria: it does not emit (unintentional) electromagnetic energy above a certain minimum level, and it is not susceptible to malfunction if unintentional electromagnetic energy below a certain level is incident on it. Design of electronic devices from the EMC considerations above requires understanding of electromagnetic fields and waves and their application to a variety of problems.

REFERENCES

1. IEEE Std. 100–1972, *IEEE Standard Dictionary of Electrical and Electronics Terms,* Institute of Electrical and Electronics Engineers/Wiley-Interscience, New York, 1972.

2. Code of Federal Regulations, Title 47 (47CFR), Part 15, Subpart B: "Unintentional Radiators."

3. H. W. Ott, *Noise Reduction Techniques in Electronic Systems,* 2nd ed., Wiley, New York, 1988.

4. C. R. Paul, *Introduction to Electromagnetic Compatibility,* Wiley, New York, 1992, pp. 12–13.

5. V. P. Kodali, *Engineering Electromagnetic Compatibility,* IEEE Press, New York, 1991, pp. 8–13.

6. J. D. Kraus and D. A. Fleisch, *Electromagnetics,* 5th ed., McGraw-Hill, New York, 1999, pp. 521–545.

7. L. Li, J. Xie, O. M. Ramahi, M. Pecht, and B. Donhaw, "Airborne Operation of Portable Electronic Devices," *IEEE Ant. Prop. Mag.,* 44 (August 2002): 30–39.

CHAPTER 2

THE ELECTROMAGNETIC ENVIRONMENT

2.1 INTRODUCTION

In this chapter we describe a variety of possible EMI sources (or noise sources) existing in the electromagnetic environment; these sources can potentially degrade the performance of an electronic system operating in their vicinity. It is difficult to classify the noise sources because of their variety. However, on the basis of their origin, they can be classified into two broad categories: as natural noise and as man-made noise. The literature on both types of noise is vast. Brief discussions of such noises in the context of EMC are given in [1–4], and more detailed discussions are given in [5–10]. In the following sections we present listings of some noise sources selected on the basis of their relevance to EMC. Where appropriate, brief comments and discussions are given about their general nature and behavior.

ISBN 0-471-16549-2

2.2 NATURAL NOISE

Principal natural phenomena that can cause disturbances for an electronic device are listed in Table 2.1.

Atmospherics are caused by electric discharge in the atmosphere; strong sources of atmospheric noise are lightning and electrostatic discharge (ESD) [5–7]. ESD is a natural phenomenon in which accumulated static charges are discharged and in turn cause EMI [3]. We will discuss ESD in a later chapter.

Table 2.1 Classification of Natural EMI Sources

Terrestrial	Extraterrestrial
Atmospherics (noise from lightning around the world)	Cosmic/galactic radio noise
Nearby and medium distant lightning	Solar noise (whistles, solar-disturbed and quiet radio noise)
Electrostatic Discharge (ESD)	

Source: From [7], p. 12.

The Milky Way galaxy and the sun are the primary sources of extraterrestrial or cosmic radio noise; these sources can be of two types: spatially extensive and spatially discrete. Spatially extensive sources emit broadband noise while emissions from the discrete sources may be either narrowband or broadband [9].

2.3 MAN-MADE NOISE

Man-made noise sources are so varied that it is difficult to list them extensively. Comprehensive data on the amplitude and statistics of man-made radio noise sources are given in [7–10]. Table 2.2 gives a list of man-made sources due to different causes [8].

Man-made noise is generally broadband in nature and can arise from many sources as indicated in Table 2.2.

2.4 CW AND TRANSIENT SOURCES

For analytical purposes and for performing design of electronic systems for EMC, it is convenient to classify the EMI sources on the basis of the time domain behavior of their emissions into continuous wave (CW) and transient sources. Table 2.3 gives a listing of various sources on this basis [8].

Table 2.2 Classification of Man-made EMI Sources

EMI sources due to the power network and its equipment:	
Switching operations	Static and rotary connectors
Power faults	Rectifiers
Electric motors	Contractors
EMI sources due to industrial and commercial equipment:	
Arc furnaces	Fluorescent lamps
Induction furnaces	Neon displaces
Air conditioning	Medical equipment
EMI sources due to machines and tools:	
Workshop machines	Rotary saws
Rolling mills	Compressors
Cotton mills	Ultrasonic cleaners
Welders	
EMI sources due to communication systems:	
Radio broadcast stations	Citizens-band
TV stations	Mobile telephones
Radar	Remote control door-opening transmitters
EMI sources due to consumer devices:	
Microwave ovens	Vacuum cleaners
Refrigeratiors/freezers	Hair dryers
Thermostats	Shavers
Mixers	Light dimmers
Washing machines	Personal computers

Source: From [7], p. 13.

Table 2.3 Continuous and Transient EMI Sources

Sources of Continuous EMI (Sources with Fixed Frequencies)	Sources of Transient EMI / Sources with a Large Frequency Spectrum
Broadcast stations	Lightning
Highpower radar	Nuclear EMP
Electric motor noise	Powerline faults (sparking)
Fixed and mobile communications	Switches and relays
Computers, visual display units, pointers	Electric welding equipment
High-repetition ignition noise	Low-repetition ignition noise
AC/multiphase power rectifiers	Electric train power pickup arcing
Solar and cosmic radio noise	Human electromotor discharge

Source: From [11], p. 14.

Important characteristics of some typical CW and transient EMI sources are discussed in [8], and are given in Table 2.4.

2.5 CHARACTERISTIC PARAMETERS OF AUTHORIZED RADIATORS

Radio, television, and radar are authorized sources of radiation. They can also be potential sources of interference to certain electronic devices not intended to receive their signal but operating in their vicinity. Table 2.5 gives the effective radiated power, estimated field strength, frequency, and so forth, for a few selected authorized systems that may be found useful to estimate their interference effect on the performance of certain electronic devices operating in their vicinity.

2.6 NOISE EMISSION INTENSITY

Frequently an electronic device is vulnerable to the EMI effects caused by man-made noise sources in its vicinity if proper EMC considerations are not given during the design of the device. Noise emission intensities of man-made noise sources vary with frequency [9]. Despite the wide variations of frequency and intensity of these emissions, an approximate ranking of intensity exists for these sources [9], and it can be used as a guide for estimating their potential impact. The following list gives the ranking of unintentional man-made noise sources:

Table 2.4 Frequencies and Noise Levels from Typical Interference Sources

Source Type	Comments
Power mains disturbances	Double exponential transients with rise times of about 1 ms, fall times of tens of ms, and peak value of about 10 kV, 100 KHz-ringing waveform with 0.5 ms rising edge. Power dips up to 100 ms long. Power frequency harmonics up to 2kHz.
Unintended radiations, switches and relays	Transients with rise times of a few ns and levels up to 3 kV, producing frequencies into the VHF band.
Commutation motions	Frequencies up to 300 MHz at repetition rates up to 10 kHz.
Human electrostatic discharge	Rise time of 1 to 10 ns.
Switching semiconductors	Rise time from 10 to 100 ns at repetition rate of 1 kHz to 10 MHz for voltages up to 300 V.
Switched-mode power supplies	Continuous noise from 1 kHz to 100 MHz.
Digital logic	Continuous noise from 1 kHz to 500 MHz.
Industrial and medical equipment	Metals heating in the range 1 to 199 kHz; medical equipment operates form 13 to 40 MHz at a high power of hundreds of watts.
Intended radiations: Broadcast stations, other RF transmitters, including radar	See Table 2.5

Source: From [11], p.14.

Table 2.5 Effective Radiated Power and Field Strength from Authorized Services

Service	Frequency Range (MHz)	Effective Radiated Power (dBW)	Usual Range of Separation Distances (km)	Estimated Range of Field Strength (V/m)
Low frequency (LF) Communication and Navigation aids	0.014–0.5	5.4	5–10	0.25–1.0
AM broadcast	0.5–1.6	47–50	0.5–2	0.2–6.0
HF amateur	18–30	30	20–100	3–15
HF communications	1.6–30	40	4–20	0.001–0.25
Citizens' band	27–27.5	11	10–100	0.3–3.0
Fixed and mobile communications	30–470 900–1000	17–21	0.04–0.2	0.2–1.5
Television (VHF)	54–216	50–55	0.5–2.0	1.0–7.3
FM broadcast	88–108	50	0.25–11.0	2.0–8.3
Television (UHF)	470–890	67	0.5–3.0	1–3
Radar	1000–10,000	90–97	2–20	0.1–90

Source: From [11], p. 15.

Unintentional Man-made Noise Sources ([9], p. 17)

1. Automotive Sources
   - Ignition circuitry
   - Alternators, generators, and electric motors
   - Buzzers, switches, regulators, and horns
2. Power Transport and Generating Facilities
   - Distribution lines
   - ac Transformer substations
   - dc Rectifier stations
   - Generating stations
3. Industrial Equipment
   - RF stabilized arc welders
   - Electric discharge machines

- Induction heating equipment
- RF soldering machines
- Dielectric welder and cutting machines
- Dielectric and plastic preheater
- Wood gluing equipment
- Silicon control rectifiers (SCR)
- Electric motors and inverters
- Circuit breakers
- Circuit switches
- Microwave heaters
- Electric calculators and office machines
- Cargo-loading cranes

4. Consumer Products
    - Appliance motors
    - Fluorescent, sodium vapor, and mercury vapor lights
    - Vibrators
    - Citizen-band AM transmitter spurious emissions
    - Electric door openers
    - Television local oscillator radiation
5. Lighting systems
    - Neon, mercury, argon, and sodium
    - Vapor lights
    - Fluorescent light fixtures
6. Medical Equipment
    - Diathermy
7. Electric Trains and Buses
    - dc-drive motors
    - Pantograph and third-rail contacts

Table 2.6 Intensity of Electric Field Levels in a Typical American Home

Location	Electric field intensity (V/m)
Laundry room	0.8
Dining room	0.9
Bathroom	1.2–1.5
Kitchen	2.6
Bedroom	2.4–7.8
Living room	3.3
Hallway	13.0

Source: From [3], p. 39; [10, 13].

2.7 HOME ENVIRONMENT

A typical American home environment is filled with electrical and magnetic emissions from a variety of electrical, electromechanical and electronic systems that are in use in the home. Brief discussion of such noises and how to model them electrically for analysis are given in [3]. Electromagnetic fields originating from various sources in the home have been probed and reported in [12–14].

Intensities of electric fields in various rooms of a typical American home are given in Table 2.6 and the measured electric field intensity levels at 30 cm from 115 V home electrical appliances are given in Table 2.7. Magnetic flux densities measured at different distances from various 115 V appliances are given in Table 2.8.

2.8 DISCUSSION OF NOISE SOURCES

Comprehensive data on the amplitude and time statistics of man-made radio noise are given in [10]. A brief description of such noises is given in [4], while detailed treatment of the major sources of man-made noise and their measurement and physical description are given in [9].

Automobile ignition systems and power distribution and transmission lines are the two unintentional noise sources of primary importance for which ample data exist [9]. The causes of radiated EMI noise arising from vehicular ignition systems are known to be gap discharges produced by the distributor and breaker points and radiation from various leads, each enhanced by the presence of circuit and engine cavity resonance [9].

Table 2.7 Electric Field Intensity Levels at 30 cm from 115 V Home Electrical Appliances

Appliance	Electric Field Intensity (V/m)
Electric blanket	250
Boiler	130
Stereo	90
Refrigerator	60
Electric iron	60
Hand mixer	50
Toaster	40
Hair dryer	40
Color TV	30
Coffee pot	30
Vacuum cleaner	16
Incandescent lamp	2

Source: From [3], p. 42; [13].

Table 2.8 Magnetic Flux Densities Measured at Different Distances from 115/200 V Appliances

Appliance	Magnetic Flux Density (mT)		
	Distance 3 cm	Distance 30 cm	Distance 1 m
Electric ranges (over 10 kW)	6–200	0.35–4	0.01–01
Electric ovens	1–50	0.15–0.5	0.01–0.04
Microwave ovens	75–200	4–8	0.25–0.6
Garbage disposals	80–250	1–2	0.03–0.1
Coffee-makers	1.8–25	0.08–0.15	< 0.01
Can openers	1000–2000	3.5–30	0.07–1
Vacuum cleaners	200–800	2–20	0.13–2
Hair dryers	6–2000	< 0.01–7	< 0.01–0.3
Electric shavers	15–1500	0.08–9	< 0.01–0.3
Televisions	25–50	1.04–2	< 0.01–0.15
Fluorescent fixtures	5–200	0.2–4	0.01–0.3
Saber and circular saws	250–1000	1–25	0.01–1

Source: From [3], p. 42, [13].

Power line EMI noise signals are produced by one or more of the following causes: (1) gap discharge or insulating film breakdown, (2) high-voltage corona discharge from conducting points with high potential gradient, and (3) line radiation of interference signal injected into the transmission system by equipment loads. Also open wire lines and electric power transmission lines can easily pick up EM noise from lightning and thunderstorms [3, 5]. Main power supply lines in industrial and home environments also carry transients, resulting from switches, circuit-breakers, heavy load switching, and the like [3]. These disturbances are strong enough to impair the operation of computers and many IT products (especially digital).

The source of atmospheric radio noise is lightning discharges produced during thunderstorms occurring worldwide [5, 6]; a short discussion is given in [9].

2.9 SUBJECT MATTER OF THE BOOK

It is apparent from the discussions given so far that knowledge of basic properties of electromagnetic fields and waves is necessary for the understanding of EMI effects and for the design of electronic systems from the viewpoint of EMC. The present book, at first, discusses the fundamentals of electromagnetics and selected topics of electromagnetic fields and waves deemed to be a necessary minimum for an understanding of various phenomena associated with EMC. Then selected topics in EMC are discussed.

REFERENCES

1. C. R. Paul, *Introduction to Electromagnetic Compatibility,* Wiley, New York, 1992.
2. H. W. Ott, *Noise Reduction Techniques in Electronic Systems,* 2nd ed., Wiley, New York, 1988.
3. V. P. Kodali, *Engineering Electromagnetic Compatibility,* IEEE Press, New York, 1996.
4. A. A. Smith, Jr., *Radio Frequency Principles and Applications,* IEEE Press, The Institute of Electrical and Electronics Engineers, New York, 1998.
5. M. A. Uman, *Understanding Lightning,* Bek Publications, Carnegie, PA, 1971.
6. J. Molan, *The Physics of Lightning,* English Universities Press, London, 1963.
7. E. N. Skomal and A. A. Smith Jr., *Measuring the Radio Frequency Environment,* Van Nostrand Reinhold, New York, 1977.
8. D. A. Morgan, *A Handbook for EMC Testing and Measurement Series 8,* Peter Peregrinus, London, 1994.
9. E. N. Skomal, *Man-made Radio Noise,* Van Nostrand Reinhold, New York, 1978.

10. A. D. Spaulding and R. T. Disney, *Man-made Radio Noise, Part I, Estimates for Business, Residential, and Rural Areas,* US Dept. of Commerce, Office of Telecommunications, Institute for Telecommunications Sciences, OT Report 74–38, June 1974.

11. F. M. Tesche, M. V. Ianuz and T. Karlsson, *EMC Analysis Methods and Computational Models,* Wiley, New York, 1997.

12. F. S. Barnes, "Typical electric and magnetic field exposure at power line frequencies and their coupling to biological systems," in *Biological Effects of Environmental Electromagnetic Fields,* ed. M. Bland, ACS Books, Washington DC, 1995.

13. EPRI Project 19955–07, Final Report TR100580, June 1992.

14. ITT Research Institute Technical Report E06549–3.

CHAPTER 3

FUNDAMENTALS OF FIELDS AND WAVES

3.1 INTRODUCTION

Basic concepts and relations of electromagnetic fields and waves are considered in the present chapter. Fundamental laws of electricity and magnetism, their generalization, and mathematical descriptions by Maxwell are discussed. Boundary conditions, characterizations of electromagnetic media, the Poynting theorem, and energy transfer are then described. Time harmonic (i.e., sinusoidal time dependence) formulation of Maxwell's equations and their applications to general electromagnetic problems are discussed. Finally, uniform plane electromagnetic waves in lossless and lossy media, the skin effect phenomena, reflection and refraction of plane waves, and the rudiments of electromagnetic shielding are described. Knowledge of these topics, although discussed briefly, is considered to be a minimum background necessary for conducting investigations of the EMC problems that constitute the subject matter of this book. For more in depth discussion of the above mentioned

Applied Electromagnetics and Electromagnetic Compatibility. By D. L. Sengupta, V. V. Liepa
ISBN 0-471-16549-2

topics and other useful concepts the reader is referred to standard books on electromagnetics [1–10].

It is assumed that the reader is familiar with vectors and vector analysis, which are used extensively in electromagnetic fields and waves and other fields of engineering and physics. For the benefit of those who are not familiar, a sufficiently detailed discussion of vector analysis is given in (Appendix A) of this book.

3.2 BASIC PARAMETERS

The six basic parameters used to describe electromagnetic quantities, the symbols used to represent them, and their units in SI (Systeme Internationale) units [4] are given below.

Field Quantities

Electric field intensity	$\mathcal{E} \equiv \mathcal{E}(\mathbf{r}, t)$	$\frac{\text{Volts}}{\text{meter}}$	V/m
Electric flux density	$\mathcal{D} \equiv \mathcal{D}(\mathbf{r}, t)$	$\frac{\text{Coulombs}}{\text{meter}^2}$	C/m^2
Magnetic field intensity	$\mathcal{H} \equiv \mathcal{H}(\mathbf{r}, t)$	$\frac{\text{Ampères}}{\text{meter}}$	A/m
Magnetic flux density	$\mathcal{B} \equiv \mathcal{B}(\mathbf{r}, t)$	$\frac{\text{Webers}}{\text{meter}^2}$	$\text{Wb/m}^2 = \text{T (Teslas)}$

Source Quantities

Volume electric current density	$\mathcal{J} \equiv \mathcal{J}(\mathbf{r}, t)$	$\frac{\text{Ampères}}{\text{meter}^2}$	A/m^2
Volume electric charge density	$\varrho \equiv \varrho(\mathbf{r}, t)$	$\frac{\text{Coulombs}}{\text{meter}^3}$	C/m^3

As the names imply, in any situation the given source quantities produce the field quantities. Although only two source quantities are given above, in certain applications it is found convenient to use the following two more source quantities:

Volume magnetic current density	$\mathcal{J}_{\rm m}(\mathbf{r}, t)$	$\dfrac{\text{Volts}}{\text{meter}^2}$	V/m^2
Volume magnetic charge density	$\varrho_{\rm m}(\mathbf{r}, t)$	$\dfrac{\text{Webers}}{\text{meter}^3}$	Wb/m^3

However, the two quantities above are fictitious, and we will not need them in the present book. The six field and source quantities given earlier are space and time dependent in general; since they all are physical quantities, they must be real quantities. Of the six parameters the charge density is a scalar quantity and the remaining five are vectors quantities. Thus the five vectors and one scalar quantities together constitute a maximum of $5 \cdot 3 + 1 = 16$ variables in a general electromagnetic problem. It is important to be familiar with the definitions and physical implications of the source and field quantities that are discussed in standard books on electromagnetics; for example in [1–3].

3.3 TIME DEPENDENT RELATIONS

3.3.1 Continuity of Current and Conservation of Charge

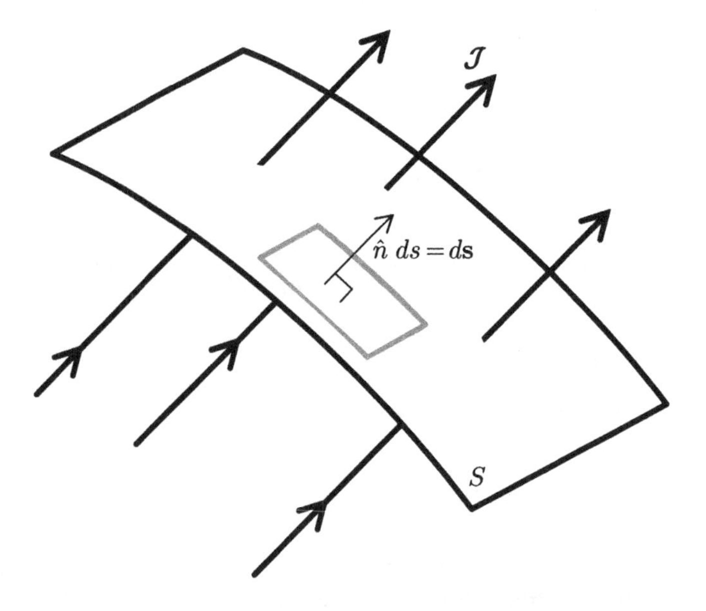

Figure 3.1 A current distribution $\mathcal{J}(\mathbf{r}, t) = \mathcal{J}$ in space at any time.

A current distribution $\mathcal{J}$ in space at any time is represented by Figure 3.1, where the various notations used are as explained in Appendix A. By definition, the current I in amperes through the given surface S is

$$I = \int_S \mathcal{J} \cdot d\mathbf{s} = \int_S \mathcal{J} \cdot \hat{n}\, ds, \tag{3.1}$$

where $d\mathbf{s}$ is the vector surface element and $\hat{n}$ is the unit positive normal to the surface element (see Appendix A for vectors, vector notations, etc.).

For a closed surface S,

$$\oint_S \mathcal{J} \cdot d\mathbf{s} = \text{total outward flux } \mathcal{J} \text{ (or current } I\text{) through closed surface} S,$$

which then represents the total outward flow of current. If current is defined as the flow of change across a surface, then it follows that the above must be a measure of the loss of charge from the volume enclosed by S, that is,

$$\oint_S \mathcal{J} \cdot ds \equiv \text{rate of decrease in the total charge } Q \text{ within volume } V \text{ bounded by } S.$$

Therefore

$$\oint_S \mathcal{J} \cdot ds = -\frac{d}{dt}\int_V \varrho\, dv = -\frac{dQ}{dt},$$

where ϱ is the volume density of electric charge within V. Assuming the volume and the surface to be stationary in time, we obtain

$$\oint_S \mathcal{J} \cdot d\mathbf{s} = -\int_V \frac{\partial \varrho}{\partial t}\, dv. \tag{3.2}$$

Equation (3.2) represents the conservation of charge principle, which states that under ordinary conditions charges are conserved. That is, charge is neither created nor destroyed. On differential basis, assuming a small volume enclosed by surface S, we can take the following limit in (3.2):

$$\lim_{\Delta v \to 0} \oint_S \frac{\mathcal{J} \cdot d\mathbf{s}}{\Delta v} = -\frac{\partial \varrho}{\partial t}.$$

From the definition of $\nabla \cdot \mathcal{J}$ (Appendix A) the expression above can be rewritten as

$$\nabla \cdot \mathcal{J} = -\frac{\partial \varrho}{\partial t}, \tag{3.3}$$

at any point in space where $\mathcal{J}$ and ϱ exist. Equation (3.3) is called the equation of continuity of current, and it indicates that the divergence of the current at any point equals the rate of decrease of the electric charge density at that point.

3.3.2 Faraday's Law

Faraday's law states that the induced electromotive force (emf) along a closed path C is equal to the negative of the time rate of change of magnetic flux linkage through the surface S enclosed by the path. The direction of the magnetic flux density, the closed path, and the orientation of the surface are as shown in Figure 3.2*a*.

Thus

$$\text{induced emf along } C = -\frac{d}{dt}\,(\text{mag. flux linkage } \psi \text{ through } S \text{ bounded by } C). \tag{3.4}$$

The flux linkage through S is

$$\psi = \int_S \boldsymbol{\mathcal{B}} \cdot d\mathbf{s}. \tag{3.5}$$

By definition, in the loop

$$\text{induced emf} = \oint_C \boldsymbol{\mathcal{E}} \cdot d\boldsymbol{\ell} = V_{\text{induced}}, \tag{3.6}$$

where

$\boldsymbol{\mathcal{E}}$ is the emf generating induced electric field intensity,

V_{induced} is defined as the open-circuit voltage across a small gap in a conducting loop C, see Figure 3.2*b*.

Thus Faraday's law can be mathematically defined as

$$\oint_C \boldsymbol{\mathcal{E}} \cdot d\boldsymbol{\ell} = -\frac{d\psi}{dt} = -\frac{d}{dt}\int_S \boldsymbol{\mathcal{B}} \cdot d\mathbf{s}. \tag{3.7}$$

If the contour is stationary and $\boldsymbol{\mathcal{B}}$ is time varying, we obtain

$$\oint_C \boldsymbol{\mathcal{E}} \cdot d\boldsymbol{\ell} = -\int_S \frac{\partial \boldsymbol{\mathcal{B}}}{\partial t} \cdot d\mathbf{s}. \tag{3.8}$$

Note in (3.7) and (3.8) that the direction of integration along the path C and the positive direction of the normal bounded by it are related by the right-hand rule. The negative signs in (3.7) and (3.8) imply that the induced voltage must oppose the inducing agent so that Lenz's law is satisfied.

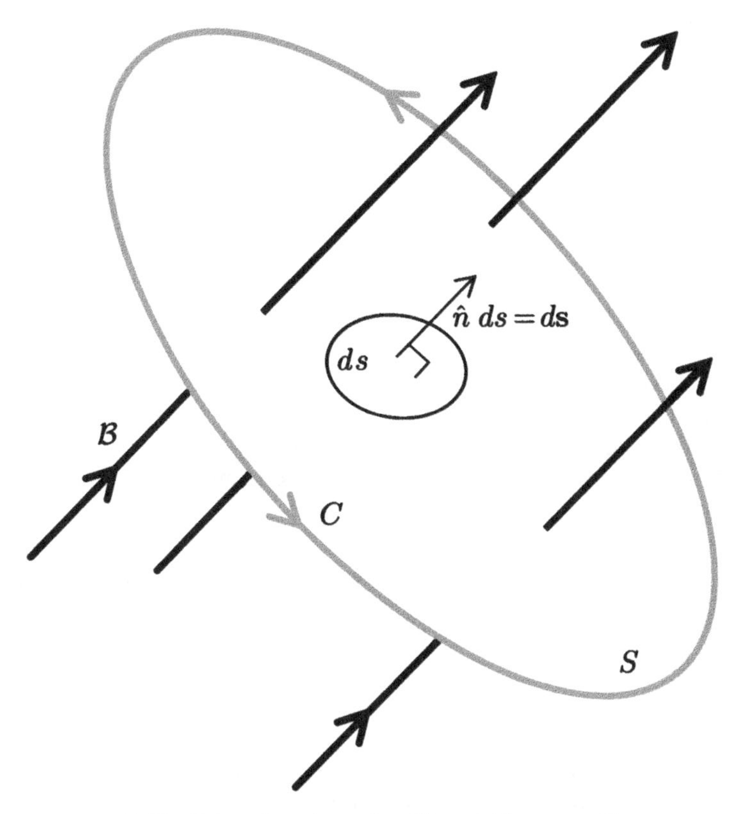

(a) Flux linkage through a surface S bounded by contour C.

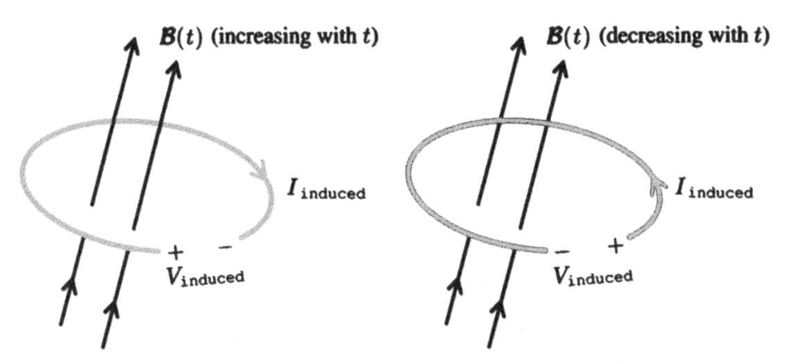

(b) Flux induced voltage and current due to magnetic field

Figure 3.2 Flux linkage through a surface S bounded by contour C. Note the direction of integration along C and of the B are related by the right hand rule.

By the definition of $\nabla \times \mathcal{E}$ (Appendix A) it now follows from (3.8) that

$$\lim_{\Delta S \to 0} \oint_C \frac{\mathcal{E} \cdot d\boldsymbol{\ell}}{\Delta S} = \hat{n} \cdot \nabla \times \mathcal{E} = -\frac{\partial \mathcal{B}}{\partial t} \cdot \hat{n}, \tag{3.9}$$

where we have assumed that C is the contour of a differential surface ΔS and the other notations are as given in Figure 3.2. Since $\hat{n}$, meaning the orientation of the surface, is arbitrary, we obtain from (3.9)

$$\nabla \times \mathcal{E} = -\frac{\partial \mathcal{B}}{\partial t} \tag{3.10}$$

which is Maxwell's first curl equation. It should be mentioned here that the integral and differential forms of Faraday's law given by (3.8) and (3.10), respectively, were first described by Maxwell.

Under static conditions $\mathcal{E}$ and $\mathcal{B}$ are independent of time and $\partial/\partial t \equiv 0$. Therefore (3.8) and (3.10) reduce to

$$\oint_C \mathcal{E} \cdot d\boldsymbol{\ell} = 0 \tag{3.11a}$$

$$\nabla \times \mathcal{E} = 0. \tag{3.11b}$$

Equation (3.11a) means that static electric field is conservative, and (3.11b) means that it is irrotational.

Considering that all physical fields are generated at a specific time t_0, they must have been zero at time $t < t_0$ in the past. Now it can be shown from (3.10) that

$$\nabla \cdot \mathcal{B} = 0, \tag{3.12}$$

which also implies that

$$\oint_S \mathcal{B} \cdot ds = 0. \tag{3.13}$$

Equations (3.12) and (3.13) may be considered as the differential and integral forms of Gauss's law for magnetic field.Gauss's Law, for magnetic field Equation (3.12) is more commonly referred to as one of Maxwell's two divergence equations. The divergence equation for magnetic field implies that the magnetic flux lines are closed, which also indicates that there is no isolated magnetic charge. It is important to note that (3.12) is obtained from (3.10), and hence it is not an independent equation. Similar comments apply to (3.13) and (3.8).

3.3.3 Ampère's Circuital Law

Ampère's circuital law states that the line integral of static magnetic field intensity taken about any closed path (i.e., the circulation of static magnetic field intensity about any closed path) must equal the total current enclosed by the of path. The mathematical statement for the law is

$$\oint_C \mathcal{H} \cdot d\boldsymbol{\ell} = \int_S \mathcal{J} \cdot d\mathbf{s}$$
$$= \text{total current } I \text{ enclosed by the path } C, \tag{3.14}$$

where S is the surface enclosed by the path C as shown in Figure 3.3.

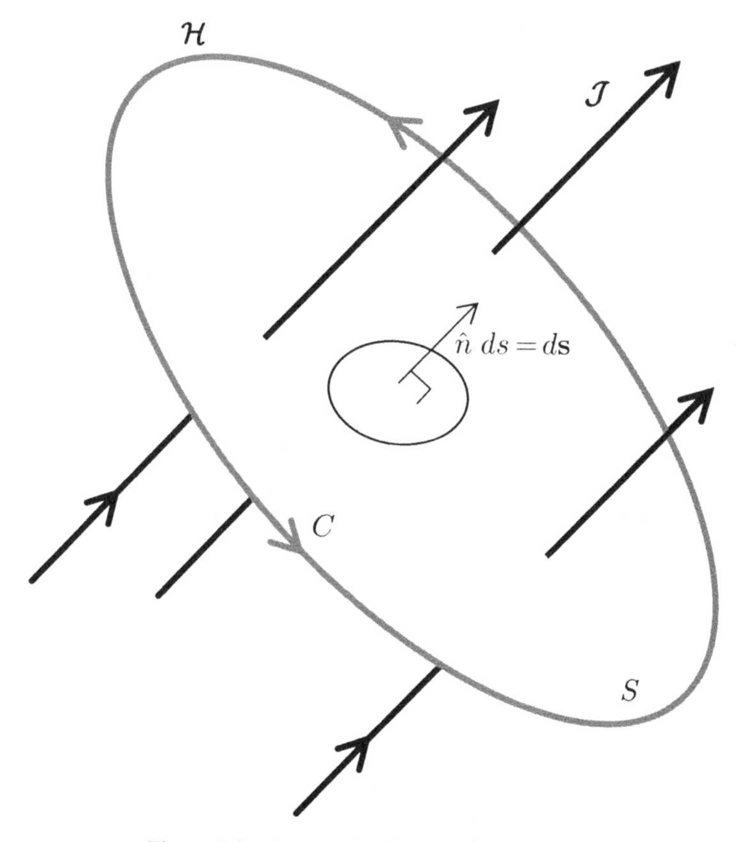

Figure 3.3 Current distribution $\mathcal{J}$ generating $\mathcal{H}$.

The sign convention for the current distribution for the current on the right-hand side of (3.14) is that it is positive if it has the sense of advance of right hand screw rotated in the directions of circulations chosen for the line integrations as in Figure 3.3 As was done for (3.8), to obtain (3.10), it can be shown that the differential form of (3.14) is

$$\nabla \times \mathcal{H} = \mathcal{J}. \tag{3.15a}$$

Taking the divergence of (3.15a) and using the well-known vectors identity (Appendix A) obtains

$$\nabla \cdot \nabla \times \mathcal{H} \equiv 0 = \nabla \cdot \mathcal{J}, \tag{3.15b}$$

which in view of (3.3) indicates that in the time-varying case Ampère's law violates the equation of continuity of current. To overcome this difficulty, Maxwell postulated the existence of the displacement current density

$$\mathcal{J}_{\mathrm{D}} = \frac{\partial \mathcal{D}}{\partial t}, \qquad \mathrm{C/m^2} \tag{3.16}$$

which can exist even in nonconducting and free space medium. After adding the displacement current to the right-hand side of (3.15a), Maxwell obtained

$$\nabla \times \mathcal{H} = \mathcal{J} + \mathcal{J}_{\mathrm{D}} = \mathcal{J} + \frac{\partial \mathcal{D}}{\partial t} \tag{3.17}$$

which now leads to the Ampère's circuital law generalized for the time-varying case

$$\oint_C \mathcal{H} \cdot d\ell = \int_S \left(\mathcal{J} + \frac{\partial \mathcal{D}}{\partial t} \right) \cdot ds. \tag{3.18}$$

Equation (3.17) is referred to as Maxwell's second curl equation, and (3.18) is its equivalent integral form. After taking the divergence of (3.17) and applying the physical reasoning that the field quantities were generated at some time in the past, and by using the equation of continuity (3.3), it can be shown that

$$\nabla \cdot \mathcal{D} = \varrho, \tag{3.19a}$$

which can be represented in the following integral form

$$\oint_S \mathcal{D} \cdot ds = \int_V \rho \, dv = Q, \tag{3.19b}$$

where it is implied that the volume is enclosed by the surface S. Equations (3.19a) and (3.19b) are the differential and integral form of generalized Gauss's law.Gauss's Law, for electric field Equation (3.19a) is commonly referred to as another of Maxwell's divergence equations. It should be noted that neither (3.19a) nor (3.19b) is an independent equation. For example, (3.19a) can be obtained from (3.17) and (3.3), and (3.19b) can be obtained from (3.10) and (3.2).

3.3.4 Lorentz Force Law

The Lorentz force law relates the force experienced by a positive point charge q moving at a velocity $\overrightarrow{v}$ through fields $\boldsymbol{\mathcal{E}}$ and $\boldsymbol{\mathcal{B}}$ given by [1–3]:

$$\boldsymbol{\mathcal{F}} = \varrho \left[\boldsymbol{\mathcal{E}} + \mathbf{v} \times \boldsymbol{\mathcal{B}}\right]. \tag{3.20}$$

If the charge is not discrete but instead given by a volume charge density ρ moving with velocity $\mathbf{v}$, then the Lorentz force law relates the force density, meaning the force per unit volume experienced by ϱ and $\boldsymbol{\mathcal{J}}$. Let the total charge in a differential volume be $\Delta q = \varrho\, \Delta v$. Then using (3.20), we obtain the total electromagnetic force on the charge Δq is

$$\Delta\boldsymbol{\mathcal{F}} = \varrho\, \Delta v(\boldsymbol{\mathcal{E}} + \mathbf{v} \times \boldsymbol{\mathcal{B}}). \tag{3.21}$$

Since a charge density ϱ moving with velocity $\mathbf{v}$ constitute a current density $\boldsymbol{\mathcal{J}} = \varrho\mathbf{v}$, we obtain the force per unit volume from (3.21) as

$$\begin{aligned}\mathcal{F}_{\Delta V} &= \lim_{\Delta\to 0} \frac{\Delta\mathcal{F}}{\Delta V} \\ &= \varrho\boldsymbol{\mathcal{E}} + \varrho(\mathbf{v} \times \boldsymbol{\mathcal{B}}) \\ &= \varrho\boldsymbol{\mathcal{E}} + \boldsymbol{\mathcal{J}} \times \boldsymbol{\mathcal{B}}, \qquad \frac{\mathrm{N}}{\mathrm{m}^3}.\end{aligned} \tag{3.22}$$

The total force on a macroscopic volume ΔV containing a charge density ϱ and current density $\boldsymbol{\mathcal{J}}$, in the presence of the fields $\boldsymbol{\mathcal{E}}$ and $\boldsymbol{\mathcal{B}}$, is

$$\boldsymbol{\mathcal{F}} = \int_V \mathcal{F}_{\Delta V}\, dv = \int_V \left[\varrho\boldsymbol{\mathcal{E}} + (\boldsymbol{\mathcal{J}} \times \mathcal{B})\right] dv. \tag{3.23}$$

Lorentz force law finds applications to many macroscopic problems: for example, to find the interaction of electric and magnetic fields with moving charges and ionized media, or the frequency response of dielectric and conducting media. However, we are not interested in such problems in this text. Maxwell's equations along with the continuity equation and the constitutive relations (to be discussed later) are sufficient for analyzing macroscopic electromagnetic problems of our interest.

3.3.5 Maxwell's Equations

The relationships between the source and field quantities at any point in a stationary medium are described by Maxwell's equations discussed earlier. For future reference we regroup here the differential and integral forms of these equations as follows:

$$\nabla \times \mathcal{E} = -\frac{\partial \mathcal{B}}{\partial t} : \quad \text{Faraday's law} \tag{3.24a}$$

$$\nabla \times \mathcal{H} = \mathcal{J} + \frac{\partial \mathcal{D}}{\partial t} : \quad \text{Generalized Ampère's law} \tag{3.25a}$$

$$\nabla \cdot \mathcal{D} = \rho : \quad \text{Generalized Gauss's law for electric fields} \tag{3.26a}$$

$$\nabla \cdot \mathcal{B} = 0 : \quad \text{Gauss's law for magnetic fields} \tag{3.27a}$$

$$\nabla \cdot \mathcal{J} + \frac{\partial \varrho}{\partial t} = 0 : \quad \text{Equation of continuity} \tag{3.28a}$$

The first four equations are Maxwell's equations and the equation of continuity (3.28a) is included here to complete the set of equations required for solving a general electromagnetic problem. For future reference, we include below the corresponding integral form of the equations above:

$$\oint_C \mathcal{E} \cdot d\boldsymbol{\ell} = -\int_S \frac{\partial \mathcal{B}}{\partial t} \cdot d\mathrm{s} \tag{3.24b}$$

$$\oint_C \mathcal{H} \cdot d\boldsymbol{\ell} = \int_S \left(\mathcal{J} + \frac{\partial \mathcal{D}}{\partial t} \right) \cdot d\mathrm{s} \tag{3.25b}$$

$$\oint_S \mathcal{D} \cdot d\mathrm{s} = \int_V \varrho \, dv \tag{3.26b}$$

$$\oint_S \mathcal{B} \cdot d\mathrm{s} = 0 \tag{3.27b}$$

$$\oint_S \mathcal{J} \cdot d\mathrm{s} + \int_V \frac{\partial \varrho}{\partial t} = 0 \tag{3.28b}$$

It is appropriate here to examine whether equations (3.24a)-(3.28a) are sufficient for solving a general electromagnetic problem. In the present case, the field and source quantities $\mathcal{E}, \mathcal{B}, \mathcal{D}, \mathcal{H}, \mathcal{J}$, and ϱ together compose $3 \cdot 5 + 1 = 16$ unknown scalar variables. It is important to note that of the five equations above, only (3.24a), (3.25a), and (3.28a) are independent, while (3.26a) and (3.27a) can be obtained from the previous three equations. Consequently, only the three independent equations providing $3 + 3 + 1 = 7$ independent scalar equations are available for sixteen unknown variables. Thus, the mathematical description of the problems provided by Maxwell's equations and the continuity equation is not complete. We need nine $(16 - 7)$ more scalar relations for a complete description of the problem. These are provided by the three constitutive vector relations that characterize the medium under consideration. Media considerations will be discussed in a later section. For the present the constitutive relations for a linear and isotropic medium are given as follows:

$$\mathcal{D}(\mathbf{r},t) = \epsilon(r)\,\mathcal{E}(\mathbf{r},t) \tag{3.29}$$
$$\mathcal{B}(\mathbf{r},t) = \mu(r)\,\mathcal{H}(\mathbf{r},t) \tag{3.30}$$
$$\mathcal{J}(\mathbf{r},t) = \sigma(r)\,\mathcal{E}(\mathbf{r},t) \tag{3.31}$$

where

$\epsilon(r)$ is the permittivity of the medium

$$\text{in } \frac{\text{Farads}}{\text{meter}} = \text{F/m} \tag{3.32a}$$

$\mu(r)$ is the permeability of the medium

$$\text{in } \frac{\text{Henries}}{\text{meter}} = \text{H/m} \tag{3.32b}$$

$\sigma(r)$ is the conductivity of the medium

$$\text{in } \frac{\text{mhos}}{\text{meter}} = \frac{\mho}{m} \text{ or } \frac{\text{Siemens}}{\text{meter}} = \text{S/m}. \tag{3.32c}$$

In the above, the linearity of the medium implies that $\epsilon(r)$, $\mu(r)$, and $\sigma(r)$ are independent of the field quantities involved. Note that ϵ, μ, and σ are assumed to be time independent for our purpose, although they can be time dependent in general. In a linear medium Maxwell's equations are linear, and therefore the superposition theorem applies. In problem-solving terms it means that with a certain linear combination of sources, the resultant field can be obtained by the similar linear (vectors) combinations of the individual fields produced by the individual sources [1,2]. Superposition theorem is extremely useful and has many practical applications.

3.3.6 Historical Comments on Maxwell's Equations

Maxwell's equations play such a fundamental role in electromagnetics that it is appropriate to make some brief comments here on certain historical aspects of the genesis of these equations, as are given elsewhere [11]. Interestingly, Maxwell's equations as we know them now (in the form given earlier) are not the same as those presented by Maxwell himself. In the year 1864, at the Royal Society (London), James Clerk Maxwell (1832–1879) read a paper entitled "A Dynamical Theory Electromagnetic Field [12]," where he proposed his electromagnetic theory in the form of 20 scalar equations involving 20 variables. Maxwell's 20 variables consisted of the 16 variables mentioned earlier and 4 more variables provided by the scalar electric potential and the vector magnetic potential, which he used as unknown variables. Through these 20 equations

Maxwell essentially summarized the observations of Charles A. de Coulomb (1736–1806), Karl F. Gauss (1777–1855), Hans C. Orsted (1777–1851), Andre M. Ampère (1775–1836), Michael Faraday (1791–1867) and others, and introduced his own radical concept of "Displacement Current" to complete his "Theory of Electromagnetic Field." In the same paper Maxwell made the theoretical observation that electromagnetic disturbances travel in free space as transverse waves with the velocity of light. He then conjectured that light is an electromagnetic wave. It is important to realize that Maxwell was silent about the generation and reception of electromagnetic waves. Maxwell's proposed theory, his equations along with his revolutionary prediction, did not receive general acceptance by the scientific community during his lifetime. It took almost a quarter century before Heinrich Rudolph Hertz (1857–1894) generated and observed electromagnetic waves around 1888; the results of his epoch-making experiments confirmed Maxwell's prediction of light as electromagnetic waves. Hertz and Oliver Heaviside (1850–1975) almost simultaneously modified Maxwell's equations by eliminating electric and magnetic potentials thereby reducing the number of variable to 16. They also discarded Maxwell's original assumption about the existence of a medium to sustain the displacement current. Hertz was the first to introduce the modified 16 scalar component equations involving 16 variables mentioned earlier. Almost immediately afterward, Heaviside formulated them using vector notations. The equations in component form due to Hertz and in vector form due to Heaviside are now universally accepted as Maxwell's equations. It should be noted that Maxwell's equations mathematically describe fundamental observations in electric and magnetic phenomena. As such, they are not derived equations. However, justification for the validity of Maxwell's equation is their enormous success in explaining macroscopic physical and experimental findings. No macroscopic electromagnetic observation has been made yet that violates the validity of these equations.

3.3.7 Media Considerations

In Section 3.3.5 we discussed the constitutive relationships between the various field vectors in a linear and isotropic medium, and thereby introduced the parameters ϵ, μ and σ characterizing the medium. In this section we describe the characteristic parameters appropriate for more general media. In an isotropic medium the constitutive parameters are defined as follows:

$$\text{permittivity } \epsilon = \frac{\mathcal{D}}{\mathcal{E}} \qquad \text{F/m} \tag{3.33a}$$

$$\text{permeability } \mu = \frac{\mathcal{B}}{\mathcal{H}} \qquad \text{H/m} \tag{3.33b}$$

$$\text{conductivity } \sigma = \frac{\mathcal{J}}{\mathcal{E}} \qquad \text{S/m.} \tag{3.33c}$$

In general, ϵ, μ, and σ may be functions of field amplitudes and frequency, and they may also be functions of time and space. We will exclude the case of time dependence for these parameters.

Classifications of Media

Classification of a medium is governed by the constitutive relationships and the parameters appropriate for the medium. The commonly used definitions for a medium are discussed as follows:

1. If $\mathcal{D}, \mathcal{B}, \mathcal{J}$ in (3.33a) through (3.33c) vary linearly with $\mathcal{E}, \mathcal{H}, \mathcal{E}$, respectively, then ϵ, μ and σ are independent of the field amplitudes. Under these conditions the medium is called **linear** otherwise it is **nonlinear**.
2. If ϵ, μ, and σ do not depend on the spatial coordinates, the medium is **homogeneous**; otherwise, it is **inhomogeneous**.
3. If $\mathcal{D}$ is parallel to $\mathcal{E}$, $\mathcal{B}$ is parallel to $\mathcal{H}$, and $\mathcal{J}$ is parallel to $\mathcal{E}$, then the medium is **isotropic**; otherwise, it is **anisotropic**.

A medium can be linear, isotropic, and homogeneous, and its characteristic parameters ϵ, μ, and σ are constant in time and space. Such a medium is called a **simple medium** when it is lossless, meaning when it has $\sigma = 0$. An example of a simple medium is the free space. The constitutive parameters for free space are:

$$\text{permittivity } \epsilon = \epsilon_0 = \frac{1}{36\pi} \times 10^{-9} = 8.854 \times 10^{-12} \quad \text{F/m} \tag{3.34}$$

$$\text{permeability } \mu = \mu_0 = \frac{1}{8\pi} \times 10^{-7} \quad \text{H/m} \tag{3.35}$$

$$\text{conductivity } \sigma \equiv 0. \tag{3.36}$$

For any linear, isotropic, and homogeneous medium the constitutive relations are

$$\mathcal{D} = \epsilon \mathcal{E} \tag{3.37a}$$

$$\mathcal{B} = \mu \mathcal{H} \tag{3.37b}$$

$$\mathcal{J} = \sigma \mathcal{E}, \tag{3.37c}$$

where ϵ, μ and σ are constant parameters.

It is customary to express the permittivity in terms of those for free space. Thus, the permittivity of a given medium is expressed by

$$\epsilon = \epsilon_0 \epsilon_r \quad \text{F/m}. \tag{3.38}$$

Frequently, ϵ is expressed alternatively as

$$\epsilon = \epsilon_0(1 + \chi_e) \tag{3.39a}$$

where the dimensionless parameters χ_e is called the electric susceptibility of the medium. Comparing (3.38) and (3.39a), we obtain an alternative expression for the dielectric constant as

$$\epsilon_r = 1 + \chi_e. \tag{3.39b}$$

Similarly, the permeability of a medium is expressed by

$$\mu = \mu_0 \mu_r \tag{3.40}$$

where μ_r is the relative permeability of the medium. An alternate expression for μ is

$$\mu = \mu_0(1 + \chi_m) \tag{3.41a}$$

where the dimensionless parameter χ_m is called the magnetic susceptibility of the medium. Again, comparing (3.38) with (3.39a) we obtain

$$\mu_r = 1 + \chi_m. \tag{3.41b}$$

Note that with the free space being lossless, it is characterized by $\sigma = 0$, $\epsilon_r = 1(\chi_e = 0)$, $\mu_r = 1(\chi_m = 0)$; with $\epsilon = \epsilon_0$, $\mu = \mu_0$ as given earlier.

We now classify linear, isotropic, and homogeneous media in the following manner:

1. **Perfect Conductor:** It has $\sigma = \infty$, $\epsilon = \epsilon_0$, and $\mu = \mu_0$. Perfect conduction is not realistic, but the assumption of infinite conductivity simplifies the theoretical analysis.

2. **Conductor:** If the conductivity σ is large but not infinite, then the material is referred to as a conductor (i.e., most metals). It should be noted that under steady state conditions, a conducting medium (including the case $\sigma = \infty$) cannot sustain a free volume density of charge; in fact, for the general media under consideration, the volume density of charge is zero if $\sigma > 0$.

3. **Dielectric Medium:** Materials with $\epsilon_r > 1$, $\mu = \mu_0$, and $\sigma = 0$ are called perfect dielectric, and materials with small values of σ are called lossy dielectric media.

4. **Magnetic Medium:** Materials for which $\mu \simeq \mu_0$ are called nonmagnetic; otherwise, they are called magnetic. The medium is lossless if $\sigma = 0$ and lossy if $\sigma > 0$.

Free space is a dielectric medium having $\chi_e = 0$ and $\epsilon_0 = 1$. Ordinarily, for most other dielectric media $\chi_e \geq 0$ and $\epsilon_0 \geq 1$. However, for an ionized medium like plasma it is possible that $\chi_e = -1$ and even $|x_e| \gg 1$, which indicates that the dielectric constant of a plasma medium can not only be zero but it can also be a large negative number. These have significant implications in the area of communication blackout during the reentry phase of a satellite into the earth's atmosphere.

Magnetic materials are classified according to their values of susceptibility or relative permeability. For example, the material is

diamagnetic if $\mu_r \lesssim 1,\ \chi_m \simeq -10^{-15}$

paramagnetic if $\mu_r \gtrsim 1,\ \chi_m \simeq +10^{-5}$

ferromagnetic if $\mu_r \gg 1,\ \chi_m$ is very large (typically, $\mu_r \simeq 50$–50,000 or more).

Ferromagnetic materials are usually non-linear. Further discussion of material characteristics may be found in [1–9].

For linear, anisotropic media the appropriate field vectors are not parallel to each other. As a result the constitutive parameters become direction dependent. In such cases matrix notations are found convenient to represent the constitutive parameters. For example, for an anisotropic dielectric medium

$$\mathcal{D} = \epsilon_0 \tilde{\epsilon} \mathcal{E} \tag{3.42a}$$

with

$$\tilde{\epsilon} = \begin{pmatrix} \epsilon_{xx} & \epsilon_{xy} & \epsilon_{xz} \\ \epsilon_{yx} & \epsilon_{yy} & \epsilon_{yz} \\ \epsilon_{zx} & \epsilon_{zy} & \epsilon_{zz} \end{pmatrix}, \tag{3.42b}$$

where $\tilde{\epsilon}$ is the dielectric constant matrix, and similarly for medium with anisotropic permeability and conductivity. When the constitutive parameters are independent of field strengths, all relation relationships between field vectors, including Maxwell's equations, become linear. Under such conditions the superposition principle applies, and frequently its application simplifies the solution of certain electromagnetic problems. We will not be interested in anisotropic and nonlinear media, and hence we have not discussed them here.

3.3.8 Boundary Conditions

Solutions of Maxwell's equations provide relations between the fields in a continuous medium. The region of interest to most practical problems generally consist of more than one medium. Thus the medium properties in the region change discontinuously across the boundary or the interface between any two media involved. Therefore, to obtain a solution appropriate for a specific problem, it is necessary to know the relations that exist between the fields at the two sides of an interface. These relations are provided by the boundary conditions that must be satisfied by the field quantities so that they are valid over the entire region of interest. The required boundary conditions are obtained by using the integral form of Maxwell's equations and the continuity equations (3.24b) through (3.28b). These conditions play such a fundamental role in the solution of practical electromagnetic problems that frequently they are referred to as boundary value problems.

General Case

We consider the interface S between two semi-infinite stationary media shown in Figure 3.4 where various electromagnetic quantities existing in the media and on the interface are identified. To make the situation general, we have arbitrarily assumed that a surface current density $\mathcal{J}_S$ in A/m and a surface charge density ϱ_s in C/m^2 exist on the interface. Observe that the unit positive normal $\hat{n}$ to the interface is directed from medium 2 to medium 1.

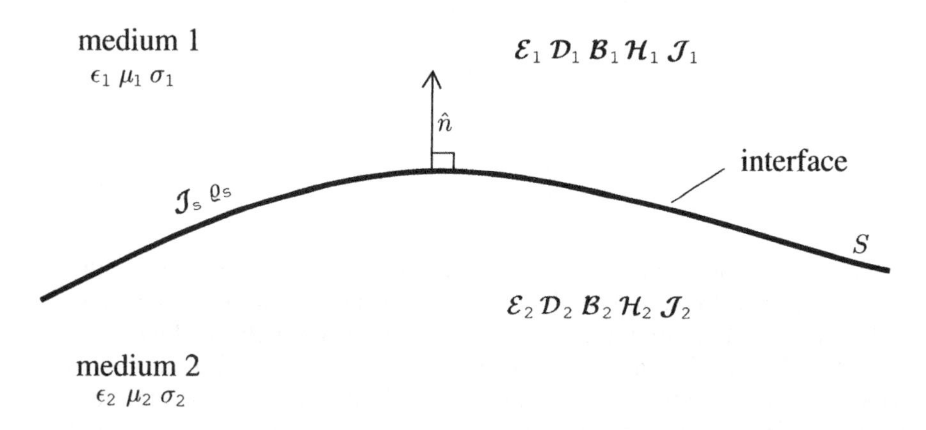

Figure 3.4 Interface S between two semi-infinite stationary media.

We construct a small rectangular contour Figure 3.5a and a small "pill box" (Figure 3.5*b*), symmetrically oriented with respect to the interface S; they provide the closed contour and the closed surface required for the derivation of the conditions. We rewrite the field equation (3.25b) using the contour C_0

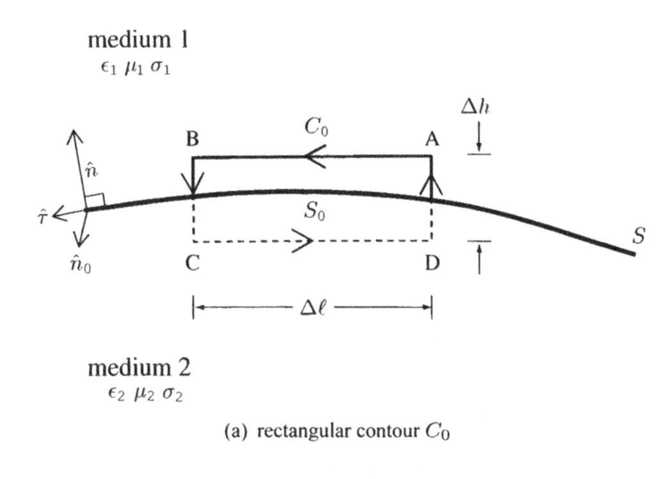

(a) rectangular contour C_0

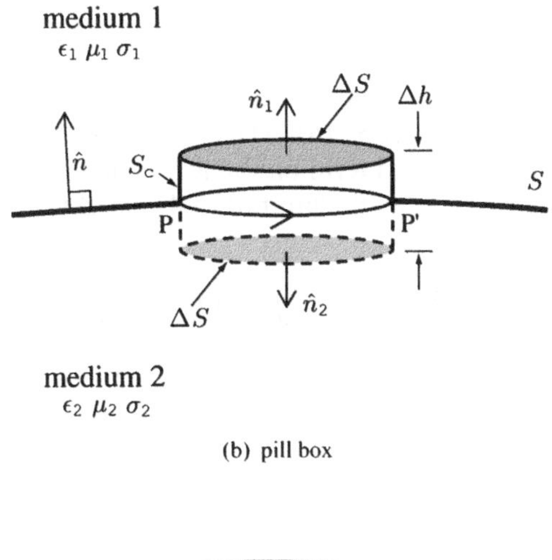

(b) pill box

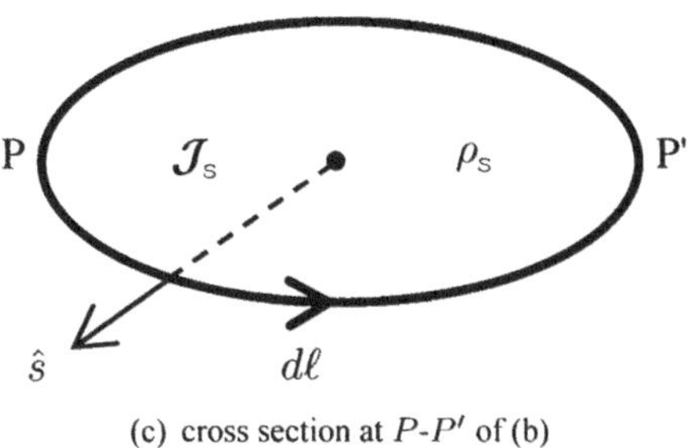

(c) cross section at P-P' of (b)

Figure 3.5 Configurations used to derive the boundary conditions at the interface S between two semi-infinite stationary media.

enclosing surface S_0, as in Figure 3.5a, to obtain

$$\oint_{C_0} \boldsymbol{\mathcal{H}} \cdot d\boldsymbol{\ell} = \int_{S_0} \left(\boldsymbol{\mathcal{J}} + \frac{\partial \boldsymbol{\mathcal{D}}}{\partial t} \right) \cdot d\mathbf{s}$$
$$= \int_{S_0} \left(\boldsymbol{\mathcal{J}} + \frac{\partial \boldsymbol{\mathcal{D}}}{\partial t} \right) \cdot \hat{n}_0 \, ds \tag{3.43}$$

where $d\boldsymbol{\ell}$ and $d\mathbf{s} = \hat{n}_0 \, ds$ are the elementary vectors lengths along C_0 and vector surface on S_0 , respectively. It is assumed that the field vectors $\boldsymbol{\mathcal{H}}$ and $\boldsymbol{\mathcal{D}}$ and their time derivatives are bounded (i.e., finite quantities in the region of integration). For sufficiently small S_0 the field vectors can be assumed constant on C_0 and over S_0, so that (3.43) can be approximated by

$$\hat{\tau} \cdot (\boldsymbol{\mathcal{H}}_1 - \boldsymbol{\mathcal{H}}_2) \, \Delta\ell + (\text{contributions from the sides } BC \text{ and } DA)$$
$$= \hat{n}_0 \cdot \left(\boldsymbol{\mathcal{J}} + \frac{\partial \boldsymbol{\mathcal{D}}}{\partial t} \right) \Delta h \, \Delta\ell, \tag{3.44}$$

where $\hat{\tau}$ is a unit vector tangential to the interface and $\hat{n}_0$ is the positive unit vector normal to S_0 as in Figure 3.5(a). The contributions from the sides being proportional to Δh, the second term on the left-hand side of (3.44) vanishes as $\Delta h \to 0$. Thus we can write the limit of (3.44) as $\Delta h \to 0$

$$\hat{\tau} \cdot (\boldsymbol{\mathcal{H}}_1 - \boldsymbol{\mathcal{H}}_2) = \hat{n}_0 \cdot \lim_{\Delta h \to 0} \left(\boldsymbol{\mathcal{J}} + \frac{\partial \boldsymbol{\mathcal{D}}}{\partial t} \right) \Delta h. \tag{3.45}$$

Now the assumption of surface current density $\boldsymbol{\mathcal{J}}_S$ on S means, by definition, that

$$\boldsymbol{\mathcal{J}}_S = \lim_{\substack{\Delta h \to 0 \\ \boldsymbol{\mathcal{J}} = \infty}} \boldsymbol{\mathcal{J}} \, \Delta h \tag{3.46a}$$

and the finiteness of $\partial \boldsymbol{\mathcal{D}} / \partial t$ means that

$$\lim_{\Delta h \to 0} \frac{\partial \boldsymbol{\mathcal{D}}}{\partial t} \Delta h = 0. \tag{3.46b}$$

Using (3.46b) in (3.45), we obtain

$$\hat{\tau} \cdot (\boldsymbol{\mathcal{H}}_1 - \boldsymbol{\mathcal{H}}_2) = \hat{n}_0 \cdot \boldsymbol{\mathcal{J}}_S. \tag{3.47}$$

In view of $\hat{\tau} = \hat{n}_0 \times \hat{n}$ and the vector identity $\mathbf{a} \times \mathbf{b} \cdot \mathbf{c} = \mathbf{a} \cdot \mathbf{b} \times \mathbf{c}$ for three arbitrary vectors (Appendix A), we obtain from (3.47),

$$\hat{n}_0 \cdot [\hat{n} \times (\boldsymbol{\mathcal{H}}_1 - \boldsymbol{\mathcal{H}}_2)] = \hat{n}_0 \cdot \boldsymbol{\mathcal{J}}_S. \tag{3.48}$$

Since the orientation of the rectangular loop C in Figure 3.5a is arbitrary, so is the orientation of $\hat{n}_0$. Hence it follows from (3.48) that

$$\hat{n} \times (\mathcal{H}_1 - \mathcal{H}_2) = \mathcal{J}_S. \tag{3.49}$$

Equation (3.49) indicates that the transition of the tangential component of the vector $\mathcal{H}$ across the interface between two different general media is discontinuous by the amount of assumed surface current density $\mathcal{J}_S$ measured in amperes per meter on the interface.

To derive the condition on the tangential component of $\mathcal{E}$, we rewrite the field equation (3.24b) as

$$\oint_{C_0} \mathcal{E} \cdot d\boldsymbol{\ell} = -\int_{S_0} \frac{\partial \mathcal{B}}{\partial t} \cdot ds$$
$$= -\int_{S_0} \frac{\partial \mathcal{B}}{\partial t} \cdot \hat{n}_0 \, ds \tag{3.50}$$

where C_0 and S_0 are as in Figure 3.5(a). Approximating (3.50) in a manner similar to that made with (3.43), we obtain the following approximation to (3.50):

$$\hat{\tau} \cdot (\mathcal{E}_1 - \mathcal{E}_2) \, \Delta\ell + (\text{contributions from ends } BC \text{ and } DA)$$
$$= -\hat{n}_0 \cdot \frac{\partial \mathcal{B}}{\partial t} \, \Delta h \, \Delta\ell. \tag{3.51}$$

As $\Delta h \to 0$, the second term on the left-hand side of (3.51) vanishes, and because of the finiteness of $\partial \mathcal{B}/\partial t$ the term on the right-hand side also vanishes. Thus, it can be shown that as $\Delta h \to 0$, $\Delta \ell \to 0$ (3.51) reduces to

$$\hat{n}_0 \cdot [\hat{n} \times (\mathcal{E}_1 - \mathcal{E}_2)] = 0. \tag{3.52}$$

Since $\hat{n}_0$ is arbitrary, we obtain from (3.52),

$$\hat{n} \times (\mathcal{E}_1 - \mathcal{E}_2) = 0, \tag{3.53}$$

which indicates that the transition of the tangential components of the vector $\mathcal{E}$ across the interface between two different general media is continuous.

To derive the conditions on the $\mathcal{B}$, we rewrite (3.27b)

$$\oint_{S_p} \mathcal{B} \cdot ds = 0, \tag{3.54}$$

where S_p is the total surface enclosing "pill box" in Figure 3.5b: $S_p = \Delta S + \Delta S + S_c$, where S_c the cylindrical side surfaces shown in Figure 3.5b. We now approximate (3.54) as

$$(\hat{n}_1 \cdot \mathcal{B}_1 + \hat{n}_2 \cdot \mathcal{B}_2)\,\Delta S \\ + (\text{contribution from } S_\text{c}) = 0. \tag{3.55}$$

Boundedness of $\mathcal{B}$ makes the cylindrical surface contribution in (3.54) vanish as $\Delta h \to 0$, and in view of $\hat{n}_1 = \hat{n},\ \hat{n}_2 = -\hat{n}_1 = -\hat{n}$ we have

$$\hat{n} \cdot (\mathcal{B}_1 - \mathcal{B}_2) = 0, \tag{3.56}$$

which indicates that the transition of the normal component of the vector $\mathcal{B}$ across the interface between two different general media is continuous.

To derive the condition on $\mathcal{D}$, we rewrite (3.26b) as

$$\oint_{S_\text{p}} \mathcal{D} \cdot d\text{s} = Q = \int_V \varrho \, dv, \tag{3.57}$$

where Q is the total charge assumed to be distributed with density $varrh$ within the volume V enclosed by S_p as in Figure 3.5b. We approximate (3.57) as

$$(\hat{n}_1 \cdot \mathcal{D}_1 + \hat{n}_2 \cdot \mathcal{D}_2)\,\Delta S = Q = \varrho\,\Delta h\,\Delta S. \tag{3.58}$$

To keep the total charge Q constant in the limiting case of $\Delta h \to 0$, it is convenient to define a surface charge density ϱ_s per unit area as follows:

$$\varrho_s = \lim_{\substack{\varrho \to \infty \\ \Delta h \to 0}} \varrho\,\Delta h, \quad \text{C/m}^2 \tag{3.59a}$$

such that

$$Q = \varrho_s\,\Delta S. \tag{3.59b}$$

Taking the limit of (3.58) as $\Delta h \to 0$ in conjunction with (3.59a), we obtain

$$\hat{n} \cdot (\mathcal{D}_1 - \mathcal{D}_2) = \varrho_s, \tag{3.60}$$

where we have used $\hat{n}_1 = -\hat{n}_2 = \hat{n}$. Equations (3.59a) and (3.59b) indicate that the transition of the normal component of $\mathcal{D}$ across the interface between two different general media is discontinuous by the amount of surface charge density on the interface.

To obtain the condition on $\mathcal{J}$, we rewrite the equation of continuity (3.28b) as

$$\oint_{S_\text{p}} \mathcal{J} \cdot d\text{s} = -\int_V \frac{\partial \varrho}{\partial t}\, dv, \tag{3.61}$$

where S_p is the total surface (end surfaces + the cylindrical surface S_c) of the pill-box enclosing the volume V (Figure 3.5b). As before, we appropriate (3.61) as

$$\hat{n}_1 \cdot (\mathcal{J}_1 - \mathcal{J}_2)\, \Delta S + (\text{contribution by } S_c \text{ excluding the effect of } \mathcal{J}_s = \mathcal{J}\, \Delta h,\ \mathcal{J} \to \infty) + \oint_{L_0} \hat{s} \cdot \mathcal{J}\, \Delta h\, d\ell = -\frac{\partial \varrho}{\partial t}\, \Delta h\, \Delta s, \tag{3.62}$$

where L_0 is the contour of the section of the interface containing $\mathcal{J}_s$, ϱ_s and $\hat{s}$ is a positive outward drawn normal as shown in Figure 3.5c. The second term on the left-hand side of (3.62) vanishes in the limit $\Delta h \to 0$ because of the boundedness of $\mathcal{J}_1$ and $\mathcal{J}_2$. As before, we define $\mathcal{J}_s = \mathcal{J}\, \Delta h$, as $\mathcal{J} \to \infty$, $\Delta h \to 0$, and $\varrho_s = \varrho_s\, \Delta h$, as $\varrho_s \to \infty$, $\Delta h \to 0$. Under these conditions we write (3.62), as $\Delta h \to 0$,

$$\hat{n} \cdot (\mathcal{J} - \mathcal{J}_2) + \oint_{L_0} \frac{\hat{s} \cdot \mathcal{J}_s\, d\ell}{\Delta S} = -\frac{\partial \varrho_s}{\partial t}. \tag{3.63}$$

Now by definition (Appendix A) of two-dimensional divergence of $\mathcal{J}_s$,

$$\lim_{\Delta\rho \to 0} \oint_{L_0} \frac{\hat{s} \cdot \mathcal{J}_s\, d\ell}{\Delta S} = \nabla_s \cdot \mathcal{J}_s. \tag{3.64}$$

If the interface L_0 is in the xy plane (Figure 3.5c), then

$$\nabla_s \cdot \mathcal{J}_{sx} = \frac{\partial \mathcal{J}_{sx}}{\partial x} + \frac{\partial \mathcal{J}_{sy}}{\partial y},$$

where $\mathcal{J}_{sx}$ and $\mathcal{J}_{sy}$ are, respectively, the x and y components of $\mathcal{J}$. We now write (3.63) as

$$\hat{n} \cdot (\mathcal{J}_1 - \mathcal{J}_2) + \nabla \cdot \mathcal{J}_s = -\frac{\partial \varrho_s}{\partial t} \tag{3.65}$$

which gives the most general condition on surface current and charge densities as a consequence of the equation of continuity. It should be mentioned that the fictitious quantity $\mathcal{J}_s$ can exist on the interface only when one of the media has infinite conductivity and is zero in all other cases. The surface charge density ϱ_s exists when one of the media has infinite conductivity, and it can also exist when both media are imperfectly dielectric, meaning they are also lossy.

For future reference we summarize below the most general boundary conditions to be satisfied by the field quantities at an interface between two arbitrary media, and also the condition on the assumed surface current and charge densities required by the equation of continuity:

$$\hat{n}_1 \times (\boldsymbol{\mathcal{E}}_1 - \boldsymbol{\mathcal{E}}_2) = 0 \tag{3.66a}$$

$$\hat{n}_1 \times (\boldsymbol{\mathcal{H}}_1 - \boldsymbol{\mathcal{H}}_2) = \boldsymbol{\mathcal{J}}_s \tag{3.66b}$$

$$\hat{n}_1 \cdot (\boldsymbol{\mathcal{D}}_1 - \boldsymbol{\mathcal{D}}_2) = \varrho_s \tag{3.66c}$$

$$\hat{n}_1 \cdot (\boldsymbol{\mathcal{B}}_1 - \boldsymbol{\mathcal{B}}_2) = 0 \tag{3.66d}$$

$$\hat{n}_1 \cdot (\boldsymbol{\mathcal{J}}_1 - \boldsymbol{\mathcal{J}}_2) + \boldsymbol{\nabla}_s \cdot \boldsymbol{\mathcal{J}}_s = -\frac{\partial \varrho_s}{\partial t}, \tag{3.66e}$$

where the notations are as explained earlier. Only the conditions (3.66a), (3.66b), and (3.66e) are independent relations; (3.66c) and (3.66d) follow from (3.66a) and (3.66b), respectively. This can be seen from the discussion of the appropriate Maxwell's equations given in Section 3.3.5.

Some Special Cases

Case 1. **Perfect Conductor–Perfect Dielectric Interface**

Assume an interface between free space and a perfect conductor. For example, $\epsilon_1 = \epsilon_0$, $\mu_1 = \mu_0$, $\sigma_1 = \infty$, and $\epsilon_2 = \epsilon_0$, $\mu_2 = \mu_0$, $\sigma_2 = \infty$. For such an interface we can assume $\boldsymbol{\mathcal{J}}_1 = 0$, and $\boldsymbol{\mathcal{E}}_1, \boldsymbol{\mathcal{E}}_2, \boldsymbol{\mathcal{D}}_2$, $\boldsymbol{\mathcal{H}}_2$, and $\boldsymbol{\mathcal{J}}_2$ to vanish within a perfect conductor. However, $\boldsymbol{\mathcal{J}}_s$ and ϱ_s are nonzero at the interface. It is clear that we can ignore the subscript 1 on the quantities and write the conditions (3.66a), (3.66b) as

$$\hat{n} \times \boldsymbol{\mathcal{E}} = 0 \tag{3.67a}$$

$$\hat{n} \times \boldsymbol{\mathcal{H}} = \boldsymbol{\mathcal{J}}_s \tag{3.67b}$$

$$\hat{n} \cdot \boldsymbol{\mathcal{D}} = \varrho_s \tag{3.67c}$$

$$\hat{n} \cdot \boldsymbol{\mathcal{B}} = 0 \tag{3.67d}$$

$$\boldsymbol{\nabla}_s \cdot \boldsymbol{\mathcal{J}}_s + \frac{\partial \varrho_s}{\partial t} = 0, \tag{3.67e}$$

where $\hat{n}$ now represents the unit positive normal to the interface S, drawn in the directions from medium 2 to medium 1.

Case 2. **Perfect Dielectric–Perfect Dielectric Interface**

We assume $\sigma_1 = \sigma_2 = 0$ Under this assumption $\varrho_s = 0$, $\boldsymbol{\mathcal{J}}_s = 0$, meaning there cannot be any surface current and surface electric charge density on the interface. We then have

$$\hat{n}_1 \times (\boldsymbol{\mathcal{E}}_1 - \boldsymbol{\mathcal{E}}_2) = 0 \tag{3.68a}$$

$$\hat{n}_1 \times (\boldsymbol{\mathcal{H}}_1 - \boldsymbol{\mathcal{H}}_2) = 0 \tag{3.68b}$$

$$\hat{n}_1 \cdot (\boldsymbol{\mathcal{D}}_1 - \boldsymbol{\mathcal{D}}_2) = 0 \tag{3.68c}$$

$$\hat{n}_1 \cdot (\boldsymbol{\mathcal{B}}_1 - \boldsymbol{\mathcal{B}}_2) = 0. \tag{3.68d}$$

The expressions above indicate that the tangential components of $\boldsymbol{\mathcal{E}}, \boldsymbol{\mathcal{H}}$ and the normal components of $\boldsymbol{\mathcal{D}}, \boldsymbol{\mathcal{B}}$ are continuous across an interface between two perfectly dielectric media.

Case 3. **Imperfect Dielectric–Imperfect Dielectric Interface**

We assume here that σ_1 and σ_2 are finite and not infinite. Here, in general, $\boldsymbol{\mathcal{J}}_s = 0$ on S, but ϱ_s may or may not be zero. The appropriate conditions then follow from (3.66a), (3.66b):

$$\hat{n}_1 \times (\boldsymbol{\mathcal{E}}_1 - \boldsymbol{\mathcal{E}}_2) = 0 \tag{3.69a}$$

$$\hat{n}_1 \times (\boldsymbol{\mathcal{H}}_1 - \boldsymbol{\mathcal{H}}_2) = 0 \tag{3.69b}$$

$$\hat{n}_1 \cdot (\boldsymbol{\mathcal{D}}_1 - \boldsymbol{\mathcal{D}}_2) = \varrho_s \tag{3.69c}$$

$$\hat{n}_1 \cdot (\boldsymbol{\mathcal{B}}_1 - \boldsymbol{\mathcal{B}}_2) = 0 \tag{3.69d}$$

$$\hat{n}_1 \cdot (\boldsymbol{\mathcal{J}}_1 - \boldsymbol{\mathcal{J}}_2) = -\frac{\partial \varrho_s}{\partial t}\,. \tag{3.69e}$$

Let us further examine (3.69c) and (3.69e) and rewrite them as

$$\epsilon_1 \mathcal{E}_{1n} - \epsilon_2 \mathcal{E}_{2n} = \varrho_s \tag{3.70}$$

$$\sigma_1 \mathcal{E}_{1n} - \sigma_2 \mathcal{E}_{2n} = -\frac{\partial \varrho_s}{\partial t}\,, \tag{3.71}$$

where $\mathcal{E}_{1n}, \mathcal{E}_{2n}$ are the normal components of the electric field; we have used the relations $\boldsymbol{\mathcal{J}}_1 = \sigma_1 \boldsymbol{\mathcal{E}}_1$ and $\boldsymbol{\mathcal{J}}_2 = \sigma_2 \boldsymbol{\mathcal{E}}_2$ to obtain (3.71). It is interesting to examine under what conditions ϱ_s can be zero in (3.70) and (3.71). If we assume $\varrho_s = 0$, then (3.70) and (3.71) can have a nontrivial solution (for $\mathcal{E}_{1n}$ and $\mathcal{E}_{2n}$) provided that the determinant of the coefficients (3.70) and (3.71) vanishes:

$$\begin{vmatrix} \epsilon_1 & -\epsilon_2 \\ \sigma_1 & -\sigma_2 \end{vmatrix} = 0 \tag{3.72}$$

which also means that

$$\frac{\epsilon_1}{\sigma_1} = \frac{\epsilon_2}{\sigma_2}\,. \tag{3.73}$$

Equation (3.73) shows that the relaxation times in the two media are equal. If they are not the same, then ϱ_s cannot be zero. This may have significant implications if a capacitor contains an interface between two lossy dielectrics having different relaxation times [6].

3.3.9 Energy Flow and Poynting's Theorem

Poynting's Theorem The Poynting theorem provides a representation for en-

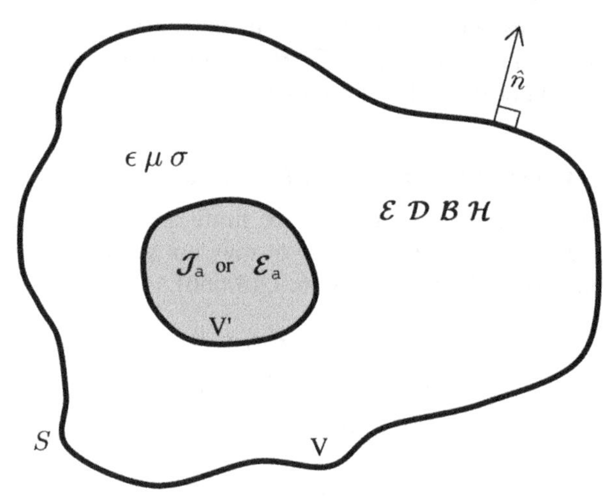

Figure 3.6 A source region V' within a closed volume V, bounded by surface S. $\hat{n}$ is outward (away from the bounded region) drawn normal to S.

ergy flow in a time-varying electromagnetic field. Consider a closed volume V bounded by a closed surface S, the medium in V is identified by ϵ, μ, and σ, as shown in Figure 3.6. Assume a source region V' within V where a known current $\mathcal{J}_a$ or field $\mathcal{E}_a$ is maintained by some applied emf and produces the fields of interest in the region. Thus the fields $\mathcal{E}, \mathcal{D}, \mathcal{B}, \mathcal{H}$ in V are due to $\mathcal{J}_a$ or $\mathcal{E}_a$ in V'. We will assume for the time being that $\mathcal{E}_a \equiv 0,\ \mathcal{J}_a \neq 0$. The field quantities above are solutions of the following equations:

$$\nabla \times \mathcal{E} = -\frac{\partial \mathcal{B}}{\partial t} \tag{3.74}$$

$$\nabla \times \mathcal{H} = \mathcal{J}_a + \mathcal{J} + \frac{\partial \mathcal{D}}{\partial t} \tag{3.75}$$

subject to the constitutive relations (3.29) through (3.31). Observe that in (3.75) we have separated the current density term on the right-hand side into the source current density $\mathcal{J}_a$ existing in V' and the induced current density $\mathcal{J}$—compare with (3.25a). To obtain the desired relation for Poynting's theorem, we start with the vector identity

$$\nabla \cdot \mathcal{E} \times \mathcal{H} = \mathcal{H} \cdot \nabla \times \mathcal{E} - \mathcal{E} \cdot \nabla \times \mathcal{H}. \tag{3.76}$$

We now use (3.74) and (3.75) to obtain

$$\nabla \cdot \mathcal{E} \times \mathcal{H} = -\mathcal{E} \cdot \mathcal{J}_{\mathrm{a}} - \mathcal{E} \cdot \mathcal{J} - \frac{\partial}{\partial t}\left(\frac{\mathcal{H} \cdot \mathcal{B}}{2} + \frac{\mathcal{E} \cdot \mathcal{D}}{2}\right). \tag{3.77}$$

Next we integrate (3.77) over the volume V, apply the divergence theorem (Appendix A) to transform the volume integral on the left-hand side into a surface integral over the closed surface S, and finally after some rearrangement of terms we obtain the following:

$$-\int_{V'} \mathcal{E} \cdot \mathcal{J}_{\mathrm{a}}\, dv = \oint_S (\mathcal{E} \times \mathcal{H}) \cdot \hat{n}\, ds \\ \frac{\partial}{\partial t} \int_V \left(\frac{\mathcal{H} \cdot \mathcal{B}}{2} + \frac{\mathcal{E} \cdot \mathcal{D}}{2}\right) dv \\ + \int_V \mathcal{E} \cdot \mathcal{J}\, dv, \tag{3.78}$$

which should be examined in the context of Figure 3.6. The source free form ($\mathcal{J}_{\mathrm{a}} \equiv 0$) of (3.78) was first given by Poynting in 1884, and again in the same year by Heaviside. To interpret the implication of (3.78), we define the following instantaneous (i.e., at any time) quantities:

$$w_{\mathrm{e}}(t) = \int_V \frac{\mathcal{E} \cdot \mathcal{D}}{2} = \int_V \frac{1}{2} \epsilon |\mathcal{E}|^2\, dv$$

$$= \text{stored electric energy in } V \tag{3.79a}$$

$$w_{\mathrm{m}}(t) = \int_V \frac{\mathcal{H} \cdot \mathcal{B}}{2} = \int_V \frac{1}{2} \mu |\mathcal{H}|^2\, dv$$

$$= \text{stored magnetic energy in } V \tag{3.79b}$$

$$p_{\mathrm{L}}(t) = \int_V \mathcal{E} \cdot \mathcal{J}\, dv = \int_V \sigma |\mathcal{E}|^2\, dv$$

$$= \text{power loss in } V \text{ (Joule or heat loss)} \tag{3.79c}$$

$$p_{\mathrm{in}}(t) = -\int_{V'} \mathcal{E} \cdot \mathcal{J}_{\mathrm{a}}\, dv$$

$$= \text{power input in } V \tag{3.79d}$$

$$p_{\mathrm{r}}(t) = \oint_S \mathcal{E} \times \mathcal{H} \cdot ds$$

$$= \text{no name yet, but it has the unit of power.} \tag{3.79e}$$

With the definitions above, we rewrite (3.78) as

$$p_{\text{in}}(t) = p_{\text{r}}(t) + \frac{\partial}{\partial t}[w_{\text{m}}(t) + w_{\text{e}}(t)] + p_{\text{L}}(t). \tag{3.80}$$

In terms of various power or energy (note power = energy / time dimensionally) within volume V, we can write (3.80) in words, except for the term $p_{\text{r}}(t)$,

instantaneous power input $= p_{\text{r}}(t) +$ (rate of increase of instantaneous stored energy) or increase of instantaneous stored power + instantaneous power lost. (3.81)

Inspection of (3.80) or (3.81) indicates that the term $p_{\text{r}}(t)$ must be interpreted as some instantaneous power, and its form given by (3.79e) indicates that it expresses the power lost across the bounding surface S. Hence, it expresses the power lost out of the volume V. Thus (3.80) expresses the power balance relation in the sense that the instantaneous power input to the volume V accounts for the instantaneous power (gain and loss) in the volume V, and it may be called the Poynting theorem with source present. In its original form the Poynting theorem applies to the source free case. With $p_{\text{in}}(t) = 0$, the right-hand side of (3.80) now expresses the power balance in the region V. The instantaneous power flow out of the volume V given by (3.79e) is rewritten in the following form:

$$p_{\text{r}}(t) = \oint_S (\boldsymbol{\mathcal{E}} \times \boldsymbol{\mathcal{H}}) \cdot \hat{n}\, ds \qquad \text{W}, \tag{3.82}$$

which indicates the power flow is across the surface S bounding the volume V, and it takes place in the direction $\hat{n}$ (out of the volume). We will see later that if S is at a large distance from the source (in this case V'), then (3.82) represents the total power radiated by the source. We now define the time-dependent Poynting vector.

$$\boldsymbol{\mathcal{S}}(t) = \boldsymbol{\mathcal{E}} \times \boldsymbol{\mathcal{H}} \qquad \text{W/m}^2 \tag{3.83}$$

which can be interpreted as the power flow density in the direction of $\boldsymbol{\mathcal{E}} \times \boldsymbol{\mathcal{H}}$.

We have considered a current source that provides the input power $p_{\text{in}}(t)$ by injecting a current density $\boldsymbol{\mathcal{J}}_{\text{a}}$ in the source region V'. A voltage source also can provide input power by injecting an electric field intensity $\boldsymbol{\mathcal{E}}_{\text{a}}$ in the source region. Explicit expression for the input power provided by the two types of source are

$$\text{current source:} \quad \boldsymbol{\mathcal{J}}_{\text{a}} \neq 0 \text{ in } V'$$
$$= \text{outside}$$

$$p_{\text{in}}(t) = -\int_V \boldsymbol{\mathcal{E}} \cdot \boldsymbol{\mathcal{J}}_{\text{a}}\, dv \tag{3.84}$$

$\boldsymbol{\mathcal{E}}$ is due to $\boldsymbol{\mathcal{J}}_{\text{a}}$.

voltage source: $\boldsymbol{\mathcal{E}}_{\text{a}} \neq 0$ in V'
$= $ outside

$$p_{\text{in}}(t) = -\int_V \boldsymbol{\mathcal{E}}_{\text{a}} \cdot \boldsymbol{\mathcal{J}}\, dv \tag{3.85}$$

$\boldsymbol{\mathcal{J}}$ due to $\boldsymbol{\mathcal{E}}_{\text{a}}$.

Application

A linear antenna carrying a current $I = I_0 \cos \omega t$ radiating in free space, $V = \infty,\ \sigma = 0$ lossless. What is the radiation resistance?

Solution

Assume the antenna element is made of prefect conductors ($\sigma_{\text{a}} \equiv \infty$). In the present case

$$\begin{aligned} p_{\text{L}}(t) &= \text{medium loss} + \text{antenna element loss} \\ &= \int_V \sigma |\boldsymbol{\mathcal{E}}|^2\, dv + \int \frac{\boldsymbol{\mathcal{J}}_{\text{a}} \cdot \boldsymbol{\mathcal{J}}_{\text{a}}}{\sigma_{\text{a}}^2}\, dv \\ &= 0 + 0 \end{aligned}$$

$$\overline{p_2(t)} = \text{time average} \equiv 0$$

$$\frac{\partial}{\partial t}\overline{(w_{\text{m}}(t) + w_{\text{e}}(t))} \equiv 0.$$

Thus, on the time average scale, (3.59a), (3.59b),

$$\overline{p_{\text{in}}(t)} = \overline{p_{\text{r}}(t)} \qquad \text{(i.e., time average radiated power } \neq 0).$$

Now assume

$$\overline{p_{\text{in}}(t)} = \frac{I_0^2}{2} R_{\text{r}}, \qquad I_0 - \text{peak current}.$$

Thus

$$\begin{aligned} R_{\text{r}} &= \frac{2\overline{p_{\text{r}}(t)}}{I_{\text{a}}^2} = \text{radiation resistance} \\ &= \frac{1}{I_{\text{a}}^2} \oint_S \boldsymbol{\mathcal{E}} \times \boldsymbol{\mathcal{H}} \cdot \hat{n} (= \hat{r})\, ds. \end{aligned} \tag{3.86}$$

Equation (3.86) is used to determine the radiation resistance of many antennas.

3.3.10 Uniqueness Theorem

Once an electromagnetic problem is solved by using Maxwell's equations with appropriate boundary conditions, then that solution is unique to the given situation. This is guaranteed by the uniqueness theorem that we state next. The electromagnetic field is uniquely determined within V bounded by S at all time $t > 0$ by the initial values of $\mathcal{E}$ and $\mathcal{H}$ throughout V and the value of either tangential $\mathcal{E}$ or $\mathcal{H}$ on S for $t > 0$. Proof and more detailed statement of the theorem are given in many places, for example, in [5, 8].

3.4 HARMONICALLY OSCILLATING FIELDS

3.4.1 Introduction

In the majority of cases we will deal with the steady state values of fields or signals harmonically oscillating in time at single frequency. Such signals (or sources producing them) are generally called time harmonic, monochromatic, or continuous wave (CW). It should be noted that because electromagnetic signals and sources are physical in nature, the variables representing them must also be real quantities.

For time harmonic problems it is convenient to express each real electromagnetic variable as the real part of a product of two appropriate complex variables such that one, called the phasor equivalent, is entirely space dependent and the other is entirely time dependent, having a complex exponential time dependency of the form $e^{j\omega t}$, ω being the radian frequency of the original real variable. We will see later that, with such representation of the variables, Maxwell's equations become independent of time. The space dependent equations thus obtained for the equivalent complex phasor quantities are called the time harmonic Maxwell's equations, and they are valid at any frequency ω. The solutions of the time harmonic equations are complex phasor quantities independent of time.

The desired time dependent real solution is obtained from the phasor solution by multiplying it with $e^{j\omega t}$, and then taking the real part of the product. As we will see later, this considerably simplifies the solution process. The present section discusses the phasor notation, the time harmonic Maxwell's equations, and a few other relevant relations useful for frequency domain analysis of electromagnetic problems.

3.4.2 Phasors

Let a time harmonic volume electric charge distribution be

$$\varrho(r, t) = \rho(r) \cos(\omega t + \phi), \tag{3.87}$$

where

$\rho(r), \phi$ are the space dependent amplitude and phase of $\varrho(r,t)$ both of which are real quantities.

ω is the radian frequency of the time variation of $\varrho(r,t)$.

We now define a complex quantity ρ such that

$$\rho = \rho(r)\, e^{j\phi} \tag{3.88}$$

with $\rho(r)$ and ϕ obtained form the given real time and space dependent physical quantity $\varrho(r,t)$. It can be seen now that

$$\varrho(r,t) = \mathrm{Re}[\rho e^{j\omega t}], \tag{3.89}$$

where Re[.] means the real part of the quantity within the brackets. The complex quantity ρ, defined by (3.88) is the phasor equivalent of the time harmonic $\varrho(r,t)$. Since $\varrho(r,t)$ is a scalar quantity, so is its phasor equivalent.

Similarly, for the time harmonic electric field $\boldsymbol{\mathcal{E}}(\mathbf{r},t)$, the equivalent vector phasor is a complex quantity $\boldsymbol{\mathcal{E}}(r)$ such that

$$\boldsymbol{\mathcal{E}}(\mathbf{r},t) = \mathrm{Re}[\mathbf{E}(\mathbf{r})\, e^{j\omega t}]. \tag{3.90}$$

Note that if we express the phasor $\mathbf{E}(\mathbf{r})$ as

$$\mathbf{E}(\mathbf{r}) = \mathbf{E}_{\mathrm{re}}(\mathbf{r}) + j\mathbf{E}_{\mathrm{im}}(\mathbf{r}), \tag{3.91a}$$

where $\mathbf{E}_{\mathrm{re}}$ and $\mathbf{E}_{\mathrm{im}}$ are the real and imaginary parts of $\mathbf{E}(\mathbf{r})$, then we obtain from (3.90) and (3.91a),

$$\mathbf{E}(\mathbf{r}) = \mathbf{E}_{\mathrm{re}}(\mathbf{r})\ \cos|\omega t - \mathbf{E}_{\mathrm{im}}(\mathbf{r})\ \sin|\omega t, \tag{3.91b}$$

which implies that the time harmonic field vector remains at all times parallel to the plane of $\mathbf{E}_{\mathrm{re}}$ and $\mathbf{E}_{\mathrm{im}}$. For future reference, the time harmonic physical electromagnetic variables and their phasor equivalents along with their equivalent notations are given in Table 3.1.

Note that in Table 3.1 the physical parameters are space and time dependent real quantities, whereas the phasor parameters are space dependent (i.e. time independent) complex quantities. It is important to note that Table 3.1 suggests the use of $e^{+j\omega t}$ with a phasor to convert it to its equivalent time harmonic physical variable with the help of (3.90). For example, if the x-component E_x of the phasor $\mathbf{E}(\mathbf{r}) \equiv \mathbf{E}$ is expressed in terms of its real and imaginary parts as

$$E_x = E_{\mathrm{r}x} + jE_{\mathrm{i}x}. \tag{3.92}$$

Table 3.1 Time Harmonic Physical Electromagnetic Variables And Their Equivalent Phasor Variables.

Physical Variable	Phasor Variable	Units
$\mathcal{E}(\mathbf{r},t) \equiv \mathcal{E}$	$\mathbf{E}(\mathbf{r}) \equiv \mathbf{E}$	V/m
$\mathcal{D}(\mathbf{r},t) \equiv \mathcal{D}$	$\mathbf{D}(\mathbf{r}) \equiv \mathbf{D}$	C/m^2
$\mathcal{H}(\mathbf{r},t) \equiv \mathcal{H}$	$\mathbf{H}(\mathbf{r}) \equiv \mathbf{H}$	A/m
$\mathcal{B}(\mathbf{r},t) \equiv \mathcal{B}$	$\mathbf{B}(\mathbf{r}) \equiv \mathbf{B}$	Wb/m^2 $= T$
$\mathcal{J}(\mathbf{r},t) = \mathcal{J}$	$\mathbf{J}(\mathbf{r}) = \mathbf{J}$	A/m^2
$\varrho(\mathbf{r},t) = \varrho$	$\rho(r) = \rho$	C/m^3
$\mathcal{J}(\mathbf{r},t) = \mathcal{J}_{\mathrm{m}}$	$J_{\mathrm{m}}(\mathbf{r}) = J_{\mathrm{m}}$	V/m^2
$\varrho_{\mathrm{m}}(\mathbf{r},t) = \varrho$	$\rho_{\mathrm{m}}(\mathbf{r}) = \rho_{\mathrm{m}}$	Wb/m^3
Note: Time variation: $\cos \omega t$	Assume time variation $e^{j\omega t}$.	

Then, by using (3.91a), (3.91b) in (3.90), we obtain the x-component of its time harmonic form as

$$\mathcal{E}_x(\mathbf{r},t) = \mathcal{E}_x = |E_x| \cos(\omega t + \phi_x), \tag{3.93}$$

where

$$|E_x| = (E_{\mathrm{r}x}^2 + E_{\mathrm{i}x}^2)^{1/2} \tag{3.94a}$$

$$\tan \phi_x = \frac{E_{\mathrm{i}x}}{E_{\mathrm{r}x}} . \tag{3.94b}$$

Note that (3.93) could have been obtained also by using $e^{-j\omega t}$ in (3.90). However, in Table 3.1 we suggest the use of the positive sign convention $e^{+j\omega t}$, as recommended by the IEEE. A time harmonic analysis using positive or negative sign convention yield correct results, although some details in each may appear different. Generally, engineers use the positive sign convention while physicists and some mathematicians use the negative sign convention. However, with $e^{-j\omega t}$ time dependence, one must accept that the inductive and capacitive reactances are negative and positive, respectively! Hence the IEEE recommendation for the convention $e^{+j\omega t}$.

3.4.3 Time Harmonic Relations

Use of complex electromagnetic variables or phasors associated with an assumed complex time dependence of $e^{j\omega t}$ renders Maxwell's and other relevant relations independent of time. As an illustration, we consider the time dependent Maxwell's equation for the electric field

$$\nabla \times \boldsymbol{\mathcal{E}}(\mathbf{r},t) = -\frac{\partial \boldsymbol{\mathcal{B}}(\mathbf{r},t)}{\partial t}\,. \tag{3.24a}$$

We now assume that $\boldsymbol{\mathcal{E}}(\mathbf{r},t) = \mathrm{Re}[\mathbf{E}(\mathbf{r})\; e^{j\omega t}]$ and $\boldsymbol{\mathcal{B}}(\mathbf{r},t) = \mathrm{Re}[\mathbf{B}(\mathbf{r})\; e^{j\omega t}]$, and after substituting them in (3.24a) we obtain

$$\mathrm{Re}[\nabla \times \mathbf{E}(r)\; e^{j\omega t}] = \mathrm{Re}[-j\omega\; \mathbf{B}(r)\; e^{j\omega t}], \tag{3.94}$$

which must be true for all time t. This implies that

$$\nabla \times \mathbf{E}(\mathbf{r}) = -j\omega\; \mathbf{B}(\mathbf{r}) \tag{3.95}$$

must be valid in general.

Similarly the integral form of (3.24a),

$$\oint_C \boldsymbol{\mathcal{E}}(\mathbf{r},t) \cdot d\boldsymbol{\ell} = -\int_S \frac{\partial \boldsymbol{\mathcal{B}}(\mathbf{r},t)}{\partial t} \cdot ds, \tag{3.24b}$$

is modified to

$$\oint_C \mathbf{E}(\mathbf{r}) \cdot d\boldsymbol{\ell} = -\int_S j\omega\; \mathbf{B}(\mathbf{r}) \cdot ds, \tag{3.96}$$

where the contour C encloses the open surface S. Equations (3.95) and (3.96) are the phasors or time harmonic Maxwell's equations equivalent to the time dependent Maxwell's equations (3.24a) and (3.24b), respectively. It is emphasized that the two time harmonic equations are independent of time and are valid for the frequency ω, meaning the equations are in the frequency domain. Note that the two phasor equations can be obtained by simply replacing $\boldsymbol{\mathcal{E}}(\mathbf{r},t)$ and $\boldsymbol{\mathcal{B}}(\mathbf{r},t)$ by $\mathbf{E}(\mathbf{r})$ and $\mathbf{B}(\mathbf{r})$, respectively, and by replacing $\partial/\partial t$ in the original differential and integral forms of the time dependent equations. Proceeding exactly in the same manner, all of the time dependent relations described in Section 3.3 can be modified into their corresponding time harmonic form. They are given as follows. It should be noted that for simplicity we have used the notation $\mathbf{E}(\mathbf{r}) \equiv \mathbf{E}$, $\mathbf{B}(\mathbf{r}) \equiv \mathbf{B}$, and so forth.

Time Harmonic Maxwell's Equations

Differential Form		**Integral Form**	
$\nabla \times \mathbf{E} = -j\omega\mathbf{B}$	(3.97a)	$\oint_C \mathbf{E} \cdot d\boldsymbol{\ell} = -\int_S j\omega\mathbf{B} \cdot d\mathbf{s}$	(3.97b)
$\nabla \times \mathbf{H} = \mathbf{J} + j\omega\mathbf{D}$	(3.98a)	$\oint_C \mathbf{H} \cdot d\boldsymbol{\ell} = \int_S (\mathbf{J} + j\omega\mathbf{D}) \cdot d\mathbf{s}$	(3.98b)
$\nabla \cdot \mathbf{D} = \rho$	(3.99a)	$\oint_S \mathbf{D} \cdot d\mathbf{s} = \int_V \rho \, dv$	(3.99b)
$\nabla \cdot \mathbf{B} = 0$	(3.100a)	$\oint_S \mathbf{B} \cdot d\mathbf{s} = 0$	(3.100b)

The current density $\mathbf{J}$ and the charge density ρ in the above equations refer to all currents and charges. In some cases it is found convenient to explicitly separate them into source current (which generates the field quantities) and induced (by the fields) current. For example, the current density term is written as $\mathbf{J}_\mathrm{a} + \mathbf{J}$, where $\mathbf{J}_\mathrm{a}$ is the source (applied) current that exists only in the source region and $\mathbf{J}$ is the induced current that may exist in the region under consideration. Similarly the charge distribution is also separated into source and induced quantities.

The Equation of Continuity

$$\nabla \cdot \mathbf{J} + j\omega\rho = 0 \tag{3.101a}$$

$$\oint_S \mathbf{J} \cdot d\mathbf{s} = -\int_V j\omega\rho \, dv. \tag{3.101b}$$

Note that in (3.97b) and (3.98b) the open surface S is bounded by the contour C and that the direction of the unit normal $\hat{n}$ in $d\mathbf{s}$ is related to that of the integration along C by the right-hand rule. In (3.98b), (3.100b), and (3.101b) the surface S encloses the volume V.

Constitutive Relations

For a linear and isotropic medium, the constitutive relations (3.35) are modified as

$$\mathbf{D} = \epsilon\mathbf{E} \tag{3.102a}$$

$$\mathbf{B} = \mu\mathbf{H} \tag{3.102b}$$

$$\mathbf{J} = \sigma\mathbf{E} \tag{3.102c}$$

where ϵ, μ, σ are the parameters characterizing the medium.

3.4.4 Complex Permittivity

With the source present, Maxwell's equation (3.98a) is written as

$$\nabla \times \mathbf{H} = \mathbf{J}_{\mathrm{a}} + \mathbf{J} + j\omega\mathbf{D}, \tag{3.103}$$

where

> $\mathbf{J}_{\mathrm{a}}$ is the source (applied) current density in the source region and it produces the fields,
>
> $\mathbf{J}$ is strictly the current density produced by the electric field, and it is given by Ohm's law

$$\mathbf{J} = \sigma\mathbf{E}. \tag{3.104}$$

Note that if $\sigma = 0$ (i.e., if the medium is lossless), then $\mathbf{J} = 0$. After using (3.102a), (3.104), and (3.103), we now have

$$\nabla \times \mathbf{H} = (\sigma + j\omega\epsilon)\mathbf{E} + \mathbf{J}_{\mathrm{a}} \tag{3.105a}$$

$$= j\omega\epsilon_{\mathrm{c}}\mathbf{E} + \mathbf{J}_{\mathrm{a}}, \tag{3.105b}$$

where ϵ_{c} is the complex permittivity of the medium defined by

$$\epsilon_{\mathrm{c}} = \epsilon - j\frac{\sigma}{\omega}. \tag{3.106}$$

Note that the permittivity is complex only when the loss in the medium is nonzero; in this case when $\sigma \neq 0$ in (3.106). Generally, a dielectric medium can have polarization damping loss in addition to that associated with its finite conductivity. Thus, even in (3.105a), the conduction current density is $\sigma\mathbf{E}$ and the displacement current density is $j\omega\epsilon\mathbf{E}$; the former is in phase with $\mathbf{E}$ and the latter leads $\mathbf{E}$ by $90°$. Frequently, the relative amplitude of the current densities are used to classify materials. For example: [4] gives the following arbitrary but useful classification:

Conductors: $\dfrac{\sigma}{\omega\epsilon} > 100$

Quasi-conductors: $\dfrac{1}{100} < \dfrac{\sigma}{\omega\epsilon} < 100$

Dielectric: $\dfrac{\sigma}{\omega\epsilon} < \dfrac{1}{100}$,

with

σ = conductivity of medium ℧/m
ϵ = permittivity of medium F/m
ω = radian frequency = $2\pi f$, where f is the frequency, Hz.

Observe that the ratio $\sigma/\omega\epsilon$ is dimensionless.

Thus, even if $\sigma = 0$ for a dielectric medium, it still can have complex permittivity. We can write the permittivity ϵ of the medium in (3.105a) as $\epsilon' - j\epsilon''$, where the real part ϵ' is the actual permittivity and the imaginary part ϵ'' accounts for the losses associated with the medium, and obtain

$$\nabla \times \mathbf{H} = (\sigma + \omega\epsilon'')\mathbf{E} + j\omega\epsilon'\mathbf{E} + \mathbf{J}_{\mathrm{a}}. \quad (3.107)$$

This indicates that the effective conductivity of the medium is now

$$\sigma_{\mathrm{eff}} = \sigma + \omega\epsilon'', \quad (3.108)$$

where σ is the conductivity of the medium and $\omega\epsilon''$ is the increase of its conductivity due to polarization damping. At $\omega = 0$, damping loss vanishes and $\sigma_{\mathrm{eff}} = \sigma$, which is usually small for a good dielectric. However, at high frequencies the damping loss dominates and

$$\sigma_{\mathrm{eff}} \simeq \omega\epsilon''. \quad (3.109)$$

We now generalize the definition of complex permittivity for a general dielectric medium by using (3.108) in conjunction with (3.105b) as

$$\begin{aligned} \epsilon_{\mathrm{c}} &= \epsilon' - \hat{j}\,\frac{\sigma_{\mathrm{eff}}}{\omega} \\ &= \epsilon' - \hat{j}\epsilon'' \end{aligned} \quad (3.110)$$

with σ_{eff}, given by (3.109). As mentioned earlier, it is frequently assumed that ϵ'' includes the effect of σ. In most cases (3.109) is a good approximation, and the second form of ϵ_{c} in (3.110) is generally accepted. We rewrite (3.110) as

$$\epsilon_{\mathrm{r}}^{\mathrm{c}} = \frac{\epsilon_{\mathrm{c}}}{\epsilon_0} = \frac{\epsilon' - j\epsilon''}{\epsilon_0} = \epsilon_{\mathrm{r}}' - j\epsilon_{\mathrm{r}}'', \quad (3.111)$$

where

ϵ_0 is the permittivity of free space

$\epsilon_{\mathrm{r}}^{\mathrm{c}}$ is the relative complex permittivity of the medium

ϵ_{c} is the complex permittivity of the medium

ϵ_r' is the relative permittivity or the *dielectric constant* of the medium

ϵ_r'' is the relative dielectric loss factor of the medium.

With ϵ_c given by (3.110), we obtain the induced conduction current density $\mathbf{J}$ and the displacement currant density $\mathbf{J}_D$ from (3.110) as

$$\mathbf{J} = \sigma_{\text{eff}}\mathbf{E} \tag{3.112a}$$

$$\mathbf{J}_D = j\omega\epsilon'\mathbf{E}. \tag{3.112b}$$

The losses associated with a dielectric medium are frequently expressed by the loss tangent of the medium defined by

$$\tan\delta = \frac{|\mathbf{J}|}{|\mathbf{J}_D|} = \frac{\sigma_{\text{eff}}}{\omega\epsilon'} = \frac{\sigma + \omega\epsilon''}{\omega\epsilon'} = \frac{\epsilon''}{\epsilon'} = \frac{\epsilon_r''}{\epsilon_r'}, \tag{3.113}$$

where the last two forms assume that the effect of nonzero σ are included in ϵ'' and ϵ_r''. Since ϵ' and σ_{eff} (or $\tan\delta$) are often determined by measurement, generally it is not necessary to distinguish between losses due to σ or $\omega\epsilon''$. The parameters ϵ_r', ϵ_r'', $\tan\delta$, and so forth, for dielectric materials are given in standard handbooks. For example, see tables in Appendix C.

The power loss associated with a dielectric medium illuminated by harmonically oscillating fields can be estimated by

$$P_{\text{av}} = \int_V \omega\epsilon'' \frac{|\mathbf{E}|^2}{2}\, dv \qquad \text{W} \tag{3.114}$$

where V is the volume containing $\mathbf{E}$. The expression above is the basis of dielectric heating including microwave heating. Incorporating $\mathbf{J}$, ρ into the medium characteristics, we rewrite Maxwell's equations (3.97a), (3.98a), (3.99a), and (3.100a) as

$$\begin{aligned} \nabla \times \mathbf{E} &= -j\omega\mathbf{B} \\ \nabla \times \mathbf{H} &= j\omega\epsilon_c\mathbf{E} + \mathbf{J}_a \\ \nabla \cdot \mathbf{D} &= \rho_a \\ \nabla \cdot \mathbf{B} &= 0 \end{aligned} \tag{3.115}$$

with $\mathbf{B} = \mu\mathbf{H}$, $\mathbf{D} = \epsilon_c\mathbf{E}$ and

$$\nabla \cdot \mathbf{J}_a + j\omega\rho_a = 0. \tag{3.116}$$

Note that the use of the complex permittivity has eliminated the induced current and charge densities from the set of Maxwell's equations. If desired, Maxwell's equations can also be modified for a magnetic medium by assuming complex permeability $\mu_0 = \mu' - j\mu''$.

3.4.5 Boundary Conditions Again

Because of their fundamental importance in many practical electromagnetic problems, we further discuss here certain aspects of boundary conditions and other related useful relations in the context of time harmonic fields.

General Conditions

The conditions for the time dependent field discussed in section 3.3.8 are applicable to the time harmonic fields provided the real field variables are replaced by their appropriate phasor field quantities, and any time derivative is replaced by $j\omega$, as mentioned earlier. For example, at the interface between two arbitrary media, the boundary conditions on the phasor field quantities are from (3.66a), (3.66d),

$$\hat{n}_1 \cdot (\mathbf{E}_1 - \mathbf{E}_2) = 0 \tag{3.117a}$$

$$\hat{n}_1 \times (\mathbf{H}_1 - \mathbf{H}_2) = \mathbf{J}_s \tag{3.117b}$$

$$\hat{n}_1 \cdot (\mathbf{D}_1 - \mathbf{D}_2) = \rho_s \tag{3.117c}$$

$$\hat{n}_1 \cdot (\mathbf{B}_1 - \mathbf{B}_2) = 0. \tag{3.117d}$$

The condition due to the equation of continuity is

$$\hat{n}_1 \cdot (\mathbf{J}_1 - \mathbf{J}_2) + \boldsymbol{\nabla}_s \cdot \mathbf{J}_s = -j\omega\rho_s, \tag{3.117e}$$

where the notations are as explained before.

It is important to realize that of the four boundary conditions (3.117a)–(3.117d), the first two are independent relations and the other two are not. Hence (3.117a) and (3.117b) form a necessary and sufficient set of conditions in this case. The uniqueness theorem mentioned earlier guarantees that among all possible solutions of Maxwell's equations, that solution is unique to a given configuration that also satisfies the boundary conditions. It is therefore important to determine the necessary and sufficient boundary conditions in order obtain a unique solution.

For example, if the two media are a perfect dielectric, $\mathbf{J}_s = 0$, $\rho_s = 0$ and from (3.117a)–(3.117e) we conclude that the necessary and sufficient boundary conditions at the interface between two perfect dielectric media are

$$\begin{aligned} \hat{n}_1 \times (\mathbf{E}_1 - \mathbf{E}_2) &= 0 \\ \hat{n}_1 \times (\mathbf{H}_1 - \mathbf{H}_2) &= 0, \end{aligned} \tag{3.118}$$

meaning the tangential components of the electric and magnetic fields are continuous.

Another important case is the interface between a perfect dielectric and a perfect conductor, when $\sigma_1 = 0$ and $\sigma_2 = \infty$. In this case all fields vanish in medium 2. So we can ignore the subscript 1 and obtain the conditions as

$$\hat{n} \times \mathbf{E} = 0 \tag{3.119a}$$

$$\begin{aligned} \hat{n} \times \mathbf{H} &= \mathbf{J}_s \\ \hat{n} \cdot \mathbf{D} &= \rho_s \\ \hat{n} \cdot \mathbf{B} &= 0, \end{aligned} \tag{3.119b}$$

where all field quantities are in the dielectric medium, $\mathbf{J}_s$ and ρ_s, are on the conducting surface and $\hat{n}$ points from the conducting to the dielectric media. The necessary and sufficient boundary condition in the present case is (3.119a), meaning the tangential component of $\mathbf{E}$ at the interface is zero. The conditions (3.119b) follow from (3.119a). For example, ρ_s, $\mathbf{J}_s$ are known only when $\mathbf{E}$, $\mathbf{H}$ are determined in the normal manner from the condition (3.119a).

Impedance Boundary Condition

The interface between a perfect dielectric ($\sigma_1 = 0$) and an imperfect conductor ($\sigma_2 = \sigma$ is large but not infinity) occurs in many practical problems. As we will discuss later, the fields in such cases cannot penetrate much into medium 2 and are in fact confined to within a skin depth δ (to be discussed later) from the interface given by

$$\delta = \left(\frac{2}{\omega \mu_0 \sigma} \right)^{1/2}. \tag{3.120}$$

If the radius of curvature of the interface surface S is much greater than the skin depth, then the following approximate condition, called the Leontovich impedance boundary condition holds [2,4,10]:

$$\hat{n} \times \mathbf{E} = \mathbf{E}_t = Z_s \mathbf{J}_s = Z_s(\hat{n} \times \mathbf{H}), \tag{3.121}$$

where $\mathbf{E}$, $\mathbf{H}$ are the fields in medium 1, $\mathbf{J}_s$ is the surface current density on the interface, $\hat{n}$ is the unit positive normal to the interface from medium 2 to medium 1, and Z_s is the surface impedance defined by

$$\begin{aligned} Z_s &= \frac{\text{tangential } \mathbf{E}}{\text{tangential } \mathbf{H}} \quad \text{at the interface} \\ &= \frac{1+j}{\sigma \delta}, \end{aligned} \tag{3.122}$$

where δ is defined by (3.120) and σ is the conductivity of medium 2. The impedance boundary condition eliminates the need to consider the fields in the second medium, and this leads to considerable mathematical simplification.

Other Conditions

If the problem under consideration contains a sharp edge, the fields can be infinitc at the edges but the energy stored around the edge must be finite. In such cases the fields are required to satisfy Meixner's edge condition to account for the finiteness of the stored energy. In radiation problems (e.g., antennas) where the region of interest extends to infinity, the fields are required to satisfy Sommerfield's radiation condition, which is discussed in advanced electromagnetic textbooks [13]. The radiation condition essentially requires that the wave must be outgoing toward infinity and the Poynting vector is directed outward and decreases as $1/r^2$ from the source, where r is the distance from the source. Thus a complete mathematical description (or solution) of a general electromagnetics problem entails the solution of Maxwell's equations where the solution must satisfy the boundary conditions, the edge conditions, and the radiation condition in general.

3.4.6 Notes on the Solution

Time Average Power Flow

It is clear now that the time harmonic frequency domain solution of Maxwell1s equations subject to boundary and other conditions appropriate for a given problem is generally a complex phasor quantity. For example, a typical solution for the electric field can be written as

$$\mathbf{E} = |\mathbf{E}|e^{j\phi}, \tag{3.123}$$

where $|\mathbf{E}|$, ϕ are the amplitude and phase of the phasor electric field, and by definition, they are independent of time but may both be functions of frequency. The time dependent electric field corresponding to (3.123) is

$$\begin{aligned} \mathcal{E}(\mathbf{r}, t) &= \text{Re}[|\mathbf{E}|e^{j\omega t}] \\ &= |\mathbf{E}| \cos(\omega t + \phi). \end{aligned} \tag{3.124}$$

The quantities $|\mathbf{E}|$, ϕ are also the amplitude and phase of the real field and hence they are measurable quantities. Therefore in most cases it is sufficient to know the phasor solution (3.123), unless complete time domain behavior is desired, in which case a solution in the form (3.124) is required. As long as the phasor solution deals with single phasors or addition and subtraction of phasors, the translation of results from the frequency domain to the time domain is straightforward, as outlined above. More care is needed in calculating time average power, where the product of two phasors is involved. We consider the time dependent Poynting vector given earlier,

$$\mathcal{S}(t) = \mathcal{E} \times \mathcal{H}, \tag{3.83}$$

which gives the instantaneous power flow in the direction of $\mathcal{E} \times \mathcal{H}$. The time average power flow for the harmonic time dependence case is

$$\begin{aligned}\langle \mathcal{S}(t) \rangle_{\text{av}} &= \frac{1}{T} \int_0^T \mathcal{S}(t)\, dt \\ &= \frac{1}{T} \int_0^T \mathcal{E} \times \mathcal{H}\, dt,\end{aligned} \tag{3.125}$$

where T is the time period of the harmonic fields $\mathcal{E}, \mathcal{H}$. If $\mathbf{E}, \mathbf{H}$ are complex, corresponding to $\mathcal{E}, \mathcal{H}$, respectively, then we obtain from (3.80),

$$\begin{aligned}\mathcal{S}(t) &= \text{Re}[\mathbf{E}e^{j\omega t}] \times \text{Re}[\mathbf{H}e^{j\omega t}] \\ &= (\mathbf{E}_\text{r} \times \mathbf{H}_\text{r}) \cos^2 \omega t + (\mathbf{E}_\text{i} \times \mathbf{H}_\text{i}) \sin^2 \omega t \\ &\quad - \frac{1}{2}[\mathbf{E}_\text{i} \times \mathbf{H}_\text{r} + \mathbf{E}_\text{r} \times \mathbf{H}_\text{i}] \sin 2\omega t,\end{aligned} \tag{3.126}$$

where we have used the real and imaginary parts of $\mathbf{E}$, $\mathbf{H}$ as

$$\begin{aligned}\mathbf{E} &= \mathbf{E}_\text{r} + j\mathbf{E}_\text{i} \\ \mathbf{H} &= \mathbf{H}_\text{r} + j\mathbf{H}_\text{i}.\end{aligned} \tag{3.127}$$

Introducing (3.125) into (3.80) and carrying out the integrals, we obtain

$$\mathbf{S}_\text{av} = \mathcal{S}(t)|_\text{av} = \frac{\mathbf{E}_\text{r} \times \mathbf{H}_\text{r}}{2} + \frac{\mathbf{E}_\text{i} \times \mathbf{H}_\text{i}}{2} \qquad \text{W/m}^2. \tag{3.128}$$

Let us define a quantity $\mathbf{S}^*$ such that

$$\mathbf{S}^* = \mathbf{E} \times \mathbf{H}^*. \tag{3.129}$$

It can be seen that

$$\begin{aligned}\text{Re}\left[\frac{\mathbf{S}^*}{2}\right] &= \text{Re}\frac{1}{2}(\mathbf{E} \times \mathbf{H}^*) \\ &= \frac{\mathbf{E}_\text{r} \times \mathbf{H}_\text{r}}{2} + \frac{\mathbf{E}_\text{i} \times \mathbf{H}_\text{i}}{2}.\end{aligned} \tag{3.130}$$

From (3.128) and (3.130) we find that

$$\mathcal{S}(t)_\text{av} = \langle \mathcal{E}(t) \times \mathcal{H}(t) \rangle_\text{av} = \frac{1}{2}\text{Re}[\mathbf{E} \times \mathbf{H}^*] = \frac{1}{2}\text{Re}[\mathbf{S}^*] = \mathbf{S}_\text{av}. \tag{3.131}$$

Equation (3.131) indicates that we can evaluate the time average power flow by using the complex phasor $\mathbf{S}^*$, thereby avoiding this integral in time which can often be quite complicated. We will see later the vector S^* is called the complex Poynting vector. It should be mentioned here that the real and complex Poynting vectors $\mathcal{S}(t)$ and S^* are not related to each other by the usual phasor relationship. That is,

$$\mathcal{S}(t)_{\text{av}} \neq \text{Re}[\mathbf{S}^* e^{j\omega t}]. \tag{3.132}$$

Solution for Arbitrary Time Dependence

In principle, if a time harmonic solution of a given problem is known at any frequency ω, then the solution can be obtained for arbitrary time dependence. This follows from Fourier (or Laplace) transform theory. If a function $f(t)$ is such that

$$\int_{-\infty}^{\infty} |f(t)|\, dt < \infty, \tag{3.133}$$

then its Fourier transform $F(\omega)$ is defined [12]

$$F(\omega) = \int_{-\infty}^{\infty} f(t)\, e^{-j\omega t}\, dt, \tag{3.134}$$

and the inverse Fourier transform is defined by

$$f(t) = \frac{1}{2\pi} \int_{-\infty}^{\infty} F(\omega)\, e^{j\omega t}\, d\omega. \tag{3.135}$$

If $f(t)$ is a real function, then it is known that

$$F(\omega) = F(-\omega)^*. \tag{3.136}$$

From (3.135) and (3.136) it follows that

$$f(t) = \frac{1}{\pi} \int_{0}^{\infty} F(\omega)\, e^{j\omega t}\, d\omega. \tag{3.137}$$

Note that in contrast with (3.135), (3.137) it involves only positive frequencies.

As an application of the relations above, we consider $\mathbf{E}_1(\omega)$ to be the phasor electric field for a time harmonic excitation function $e^{j\omega t}$. Now let the excitation function be $f(t)$ with a Fourier transform $F(\omega)$. We will assume here that $f(t)$ is a real function. Thus the frequency domain electric field at the frequency ω is

$$\mathbf{E}(\omega) = F(\omega)\, \mathbf{E}_1(\omega). \tag{3.138}$$

The time domain electric field corresponding to $\mathbf{E}(\omega)$ is

$$\begin{aligned} \mathcal{E}(\mathbf{r}, t) &= \frac{1}{\pi}\text{Re}\left[\int_{0}^{\infty} \mathbf{E}(\omega)\, e^{j\omega t}\, d\omega\right] \\ &= \frac{1}{\pi}\text{Re}\left[\int_{0}^{\infty} F(\omega)\, \mathbf{E}_1(\omega)\, e^{j\omega t}\, d\omega\right], \end{aligned} \tag{3.139}$$

where we have assumed $F(\omega) = \mathbf{F}(-\omega)^*$ and $\mathbf{E}_1(\omega) = \mathbf{E}_1(-\omega)^*$. Equation (3.139) finds application in many transient electromagnetic problems.

3.4.7 The Complex Poynting Theorem

The complex Poynting theorem for time harmonic fields cannot be obtained from its time dependent form, given by (3.78), by simply replacing the real field and source quantities by their corresponding phasor equivalents. The theorem must be obtained from the time harmonic Maxwell's equations. We begin with the geometry sketched in Figure 3.6 where we now assume that the field and source quantities are given by **E**, **D**, **B**, **H**, and $\mathbf{J}_\text{a}$ (or $\mathbf{E}_\text{a}$), respectively. The medium is characterized by μ, ϵ and σ. We now use the vector identity (Appendix A)

$$\nabla \cdot \mathbf{E} \times \mathbf{H}^* = \mathbf{H}^* \cdot \nabla \times \mathbf{E} - \mathbf{E} \cdot \nabla \times \mathbf{H}^*, \tag{3.140}$$

where $*$ above a quantity indicates its complex conjugate. Using Maxwell's equations (3.97a) and (3.98a) in (3.140), we obtain

$$\nabla \cdot \mathbf{E} \times \mathbf{H}^* = -\mathbf{E} \cdot \mathbf{J}_\text{a}^* - j\omega[\mathbf{H}^* \cdot \mathbf{B} - \mathbf{E} \cdot \mathbf{D}^*] - \mathbf{E} \cdot \mathbf{J}^*. \tag{3.141}$$

Referring to Figure 3.6, we integrate (3.141) over the volume V, apply divergence theorem to the term $\nabla \cdot \mathbf{E} \times \mathbf{H}^*$, and rearrange the terms to obtain the following:

$$\begin{aligned} -\int_V \left(\frac{\mathbf{E} \cdot \mathbf{J}_\text{a}^*}{2}\right) dv = & \oint_S \frac{\mathbf{E} \times \mathbf{H}^*}{2} \cdot d\mathbf{s} \\ & + 2j\omega \left[\frac{1}{2}\int_V \frac{\mathbf{H}^* \cdot \mathbf{B}}{2}\, dv - \frac{1}{2}\int_V \frac{\mathbf{E} \cdot \mathbf{D}^*}{2}\, dv\right] \\ & + \int_V \frac{\mathbf{E} \cdot \mathbf{J}^*}{2}\, dv, \end{aligned} \tag{3.142}$$

where we have divided each term by 2, the reason for which will be evident later. Equation (3.142) is the Poynting theorem in complex form. For the purpose of giving proper interpretations of (3.142), we define the following:

$$\text{Complex power input to volume } V = -\int_V \mathbf{E} \cdot \mathbf{J}_\text{a}^*\, dv \tag{3.143a}$$

$$\text{Time average complex power input to} V = P_\text{in}^\text{c} = -\int_V \frac{\mathbf{E} \cdot \mathbf{J}_\text{a}^*}{2}\, dv \tag{3.143b}$$

Complex power flow through the surface S

$$= \oint_S \mathbf{E} \times \mathbf{H}^* \cdot d\mathbf{s}$$

$$= \oint_S \mathbf{S}^* \cdot d\mathbf{s}, \tag{3.144a}$$

where $\mathbf{S}^*$ is the complex Poynting vector, defined as the complex power flow density

$$\mathbf{S}^* = \mathbf{E} \times \mathbf{H}^*. \tag{3.144b}$$

The time average complex power flow through the surface S is

$$P_{\mathrm{r}}^{\mathrm{C}} = \oint_S \frac{\mathbf{S}^*}{2} \cdot d\mathbf{s} = \oint_S \frac{\mathbf{E} \times \mathbf{H}^*}{2} \cdot d\mathbf{s}. \tag{3.145}$$

The time average energy stored in the magnetic field is

$$W_{\mathrm{m}} = \frac{1}{2} \int_V \frac{\mathbf{H}^* \cdot \mathbf{B}}{2}\, dv, \tag{3.146}$$

which is a real quantity. The time average energy stored in the electric field is

$$W_{\mathrm{e}} = \frac{1}{2} \int_V \frac{\mathbf{E} \cdot \mathbf{D}^*}{2}\, dv, \tag{3.147}$$

which is a real quantity. The time average energy dissipated in the medium is

$$P_{\mathrm{L}} = \frac{1}{2} \int_V \mathbf{E} \cdot \mathbf{J}^*\, dv = \int \frac{\sigma |\mathbf{E}|^2}{2}\, dv, \tag{3.148}$$

which is a real quantity. Introducing (3.143b), (3.146), (3.145), (3.147), and (3.148) in (3.142), we obtain the complex power balance equation as

$$P_{\mathrm{in}}^{\mathrm{C}} = P_{\mathrm{r}}^{\mathrm{C}} + j2\omega(W_{\mathrm{m}} - W_{\mathrm{e}}) + P_{\mathrm{L}}, \tag{3.149}$$

which is a succinct statement for the complex Poynting theorem. The real and imaginary parts of (3.149) are

$$\begin{aligned} \mathrm{Re}[P_{\mathrm{in}}^{\mathrm{C}}] &= \text{Real input power} \\ &= \mathrm{Re} \oint_S \frac{\mathbf{S}^*}{2} \cdot d\mathbf{s} + P_{\mathrm{L}} \end{aligned} \tag{3.150a}$$

$$\begin{aligned} \mathrm{Im}[P_{\mathrm{in}}^{\mathrm{C}}] &= \text{Reactive input power} \\ &= \text{stored reactive power.} \end{aligned} \tag{3.150b}$$

These three forms of the Poynting theorem find applications in variety of circuit, antenna, and radiation problems.

Example 3.1

Application of the complex Poynting theorem to an RLC circuit excited by a time harmonic voltage.

Solution

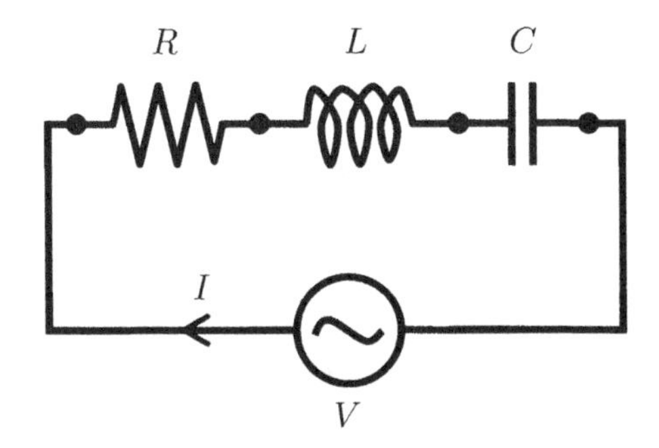

Figure 3.7 An RLC circuit excited by a harmonic voltage

Assume the RLC circuit shown in Figure 3.7 excited by a phasor voltage V exciting a phasor current I in the circuit as shown.

$$v(t) = \text{Re}(Ve^{j\omega t})$$
$$i(t) = \text{Re}(Ie^{j\omega t}).$$

We will assume that the amplitude of voltage and current are expressed in their peak values. From circuit theory we know that

$$V = ZI$$

where Z is the independence of the series circuit given by

$$Z = R + j\left(\omega L - \frac{1}{\omega c}\right).$$

The time average complex power input to the circuit is

$$P_{\text{in}}^{\text{C}} = \frac{1}{2}VI^* = \frac{1}{2}ZII^*.$$

The time average power lost in the circuit is

$$P_{\text{L}} = \frac{1}{2}II^*R$$

where we have assumed that L and C are lossless. The time average stored magnetic power in the inductance is

$$W_{\mathrm{m}} = \frac{1}{4} L I I^*$$

and the time-average stored electric power in the capacitance is

$$W_{\mathrm{e}} = \frac{1}{4} C V_{\mathrm{c}} V_{\mathrm{c}}^* = \frac{1}{4} C \frac{I I^*}{(j\omega C)(-j\omega C)},$$

where $V_{\mathrm{c}} = I/j\omega C$ is the voltage across C. It can be shown now that

$$\begin{aligned} P_{\mathrm{L}} + j2\omega(W_{\mathrm{m}} - W_{\mathrm{e}}) &= \frac{I I^*}{2} \left[R + j \left(\omega L - \frac{1}{\omega C} \right) \right] \\ &= \frac{1}{2} I I^* Z = P_{\mathrm{in}}^{\mathrm{C}}, \end{aligned}$$

which is the same as the complex Poynting theorem (3.149) with $P_{\mathrm{r}}^{\mathrm{C}} = 0$. Note that for the RLC circuit $P_{\mathrm{r}}^{\mathrm{C}} \equiv 0$.

Example 3.2

The total power P_{rad} radiated by a linear antenna in free space can be expressed as

$$P_{\mathrm{rad}} = \frac{1}{2} I_{\mathrm{in}}^2 R_{\mathrm{r}},$$

where I_{in} is the peak value of the input current and R_{r} is the radiation resistance of the antenna. Use the complex Poynting theorem to determine R_{r}.

Solution

We will use the Poynting theorem (3.150a), (3.150b) and assume that the antenna is located at the center of spherical surface S. With the medium being free space, we have $P_{\mathrm{L}} \equiv 0$, and we obtain from (3.150a), (3.150b),

$$\mathrm{Re}[P_{\mathrm{in}}^{\mathrm{c}}] = \mathrm{Re} \oint_S \frac{\mathbf{S}^*}{2} \cdot d\mathbf{s}.$$

We will see later that when S is at a sufficiently large distance from the antenna, the right-hand side of the above equation represents P_{rad}. Hence we obtain

$$P_{\mathrm{rad}} = \mathrm{Re} \oint_S \frac{\mathbf{S}^*}{2} \cdot d\mathbf{s} = \frac{1}{2} I_{\mathrm{in}}^2 R_{\mathrm{r}}.$$

Therefore

$$R_{\mathrm{r}} = \frac{2P_{\mathrm{rad}}}{I_{\mathrm{in}}^2} = \mathrm{Re} \oint_S \frac{\mathbf{S}^*}{2} \cdot d\mathbf{s}, \qquad (3.151)$$

which is the desired expression. In a later section we will use the expression above to determine R_{r} for a given antenna.

3.5 THE WAVE EQUATION

Using his original algebraic equations in free space, Maxwell showed that the two transverse components of time dependent electric and magnetic fields (e.g., $\mathcal{E}_x$ and $\mathcal{H}_y$) together form an electromagnetic wave propagating in the longitudinal (or z-) direction at the velocity of light in free space. In this section we describe how the wave equation arises from the modern vector form of Maxwell's equations given earlier.

3.5.1 Time Dependent Case

We assume that time dependent electric and magnetic fields ($\boldsymbol{\mathcal{E}}(\mathbf{r}, t)$, $\boldsymbol{\mathcal{D}}(\mathbf{r}, t)$ and $\boldsymbol{\mathcal{H}}(\mathbf{r}, t)$, $\boldsymbol{\mathcal{B}}(\mathbf{r}, t)$) exist in a source free simple medium characterized by μ, ϵ, and the medium being lossless, $\sigma = 0$.

In the present case the appropriate Maxwell's equations and constitutive relations are (see (3.24a)–(3.27a) and (3.37b)–(3.37b))

$$\nabla \times \boldsymbol{\mathcal{E}} = -\frac{\partial \boldsymbol{\mathcal{B}}}{\partial t} \qquad (3.152a)$$

$$\nabla \times \boldsymbol{\mathcal{H}} = -\frac{\partial \boldsymbol{\mathcal{D}}}{\partial t} \qquad (3.152b)$$

$$\nabla \cdot \boldsymbol{\mathcal{D}} = 0 \qquad (3.152c)$$

$$\nabla \cdot \boldsymbol{\mathcal{B}} = 0 \qquad (3.152d)$$

$$\boldsymbol{\mathcal{B}} = \mu \boldsymbol{\mathcal{H}} \qquad (3.152e)$$

$$\boldsymbol{\mathcal{D}} = \epsilon \boldsymbol{\mathcal{E}}, \qquad (3.152f)$$

where μ and ϵ are the permeability and permittivity of the infinite and homogeneous medium. Note that in view of μ, ϵ being constant, (3.152c), (3.152d) and (3.152e), (3.152f) imply that

$$\nabla \cdot \boldsymbol{\mathcal{E}} = \nabla \cdot \boldsymbol{\mathcal{H}} = 0. \qquad (3.153)$$

Now, after taking the curl of (3.152a) and making use of (3.152b), we obtain

$$\nabla \times \nabla \times \boldsymbol{\mathcal{E}} + \frac{1}{v^2} \frac{\partial^2 \boldsymbol{\mathcal{E}}}{\partial t^2} \qquad (3.154)$$

where the parameter v is determined by the medium as

$$v = \frac{1}{\sqrt{\mu\epsilon}} \quad \text{m/s} \tag{3.155}$$

Similarly, taking the curl of (3.152b) and making use of (3.152a), we obtain

$$\nabla \times \nabla \times \mathcal{H} + \frac{1}{v^2}\frac{\partial^2 \mathcal{H}}{\partial t^2} = 0. \tag{3.156}$$

Thus we find that $\mathcal{E}$ and $\mathcal{H}$ satisfy the same source free vector wave equation with the same velocity parameter. In the present case, since the medium is homogeneous, we have (3.153) and, using the vector identity for any vector $\mathbf{A}$,

$$\nabla \times \nabla \times \mathbf{A} = \nabla(\nabla \cdot \mathbf{A}) - \nabla^2 \mathbf{A}, \tag{3.157}$$

it can be shown that $\mathcal{E}, \mathcal{H}$ satisfy the alternate form of the vector wave equation

$$\nabla^2 \begin{pmatrix} \mathcal{E} \\ \mathcal{H} \end{pmatrix} - \frac{1}{v^2}\frac{\partial^2}{\partial t^2}\begin{pmatrix} \mathcal{E} \\ \mathcal{H} \end{pmatrix} = 0, \tag{3.158}$$

where the operator ∇^2 now operates only on the vector parameters $\mathcal{E}$ and $\mathcal{H}$. It should be noted that in the rectangular coordinate system (Appendix A)

$$\nabla^2 \mathcal{E} = \hat{x}\nabla^2 \mathcal{E}_x + \hat{y}\nabla^2 \mathcal{E}_y + \hat{z}\nabla^2 \mathcal{E}_z, \tag{3.159}$$

where $\mathcal{E}_x$, $\mathcal{E}_y$, $\mathcal{E}_z$ are the rectangular components of $\mathcal{E}$, and similarly for $\mathcal{H}$. Each rectangular component, for example, $\mathcal{E}_x$, satisfies the scalar wave equation

$$\nabla^2 \mathcal{E}_x - \frac{1}{v^2}\frac{\partial^2 \mathcal{E}_x}{\partial t^2} = 0 \tag{3.160}$$

and similarly of each rectangular component of $\mathcal{H}$.

It is appropriate now to discuss why (3.154), (3.158), and (3.160) are called the wave equations. For simplicity we will use the one-dimensional version of (3.160)), meaning $\mathcal{E}_x$ is a function of (z, t) instead of (x, y, z, t) and write

$$\begin{aligned} \mathcal{E}_x &= \mathcal{E}_x(z, t) \\ \frac{\partial \mathcal{E}_x}{\partial x} &= \frac{\partial \mathcal{E}_x}{\partial y} = 0. \end{aligned} \tag{3.161}$$

Using (3.161), we obtain the one-dimensional version of the scalar wave, equations as

$$\frac{\partial^2 \mathcal{E}_x}{\partial z^2} - \frac{1}{v^2}\frac{\partial^2 \mathcal{E}_x}{\partial t^2} = 0. \tag{3.162}$$

The general solution of (3.162) is

$$\mathcal{E}_x(z,t) = A_1 f_1(z - vt) + A_2 f_2(z + vt) \tag{3.163}$$

where A_1, A_2 are constants and f_1, f_2 are functions of the respective arguments shown. That $\mathcal{E}_x(z,t)$ given by (3.163) is a solution of (3.160) can be shown by simple substitution of (3.163) in the wave equation (3.160). The first term of (3.163) represents a wave propagating in the $x + z$ direction. To illustrate this, we assume two sets of parameters (z_1, t_1) and (z_2, t_2) with $z_2 > z_1$ and $t_2 > t_1$ such that $f_1(z_1 - vt_1) = f_1(z_2 - vt_2)$, which implies that $z_2 = z_1 + v(t_2 - t_1)$. This indicates that the waveform $f_1(z - vt)$ propagates undistorted in the $+z$ direction with velocity v. Similarly it can be shown that $f_2(z + vt)$ represents a wave propagating in the $-z$ direction with velocity v.

Some of the common forms of the wave functions are

$$\text{wave functions} = \begin{cases} \sin(\beta z \pm \omega t) \\ \cos(\beta z \pm \omega t) \\ e^{j(\omega t \pm \beta z)}. \end{cases} \tag{3.164}$$

3.5.2 Time Harmonic Case

We assume the time harmonic electric and magnetic field quantities **E**, **D**, **H**, and **B** exist in a source free simple medium characterized by μ, ϵ, and the medium being lossless, $\sigma = 0$.

In the present case, the appropriate Maxwell's equations and the constitutive relations are

$$\nabla \times \mathbf{E} = -j\omega\mu\mathbf{H} \tag{3.165a}$$

$$\nabla \times \mathbf{H} = j\omega\mu\mathbf{E} \tag{3.165b}$$

$$\nabla \cdot \mathbf{D} = 0 \tag{3.166a}$$

$$\nabla \cdot \mathbf{B} = 0 \tag{3.166b}$$

$$\mathbf{D} = \epsilon\mathbf{E} \tag{3.167a}$$

$$\mathbf{B} = \mu\mathbf{H}. \tag{3.167b}$$

Note that in view of μ, ϵ being constant, (3.167a) and (3.167b) imply that

$$\nabla \cdot \mathbf{E} = \nabla \cdot \mathbf{H} = 0. \tag{3.168}$$

In a manner similar to that used to obtain the vector wave equations in Section 3.5.1, it can be shown from (3.165a)–(3.167b) that **E**, **H** satisfy the following:

$$\nabla \times \nabla \times \mathbf{E} - \beta^2\mathbf{E} = 0 \tag{3.169a}$$

$$\nabla \times \nabla \times \mathbf{H} - \beta^2\mathbf{H} = 0 \tag{3.169b}$$

with

$$\beta = \omega\sqrt{\mu\epsilon} = \frac{\omega}{v} \tag{3.170}$$

$$v = \frac{1}{\sqrt{\mu\epsilon}} .$$

As we will see later, β is called the phase constant. Again, using the vector identity (3.157) and (3.169a)–(3.169b) it can be shown that **E**, **H** satisfy

$$\nabla^2\mathbf{E} + \beta^2\mathbf{E} = 0 \tag{3.171a}$$

$$\nabla^2\mathbf{H} + \beta^2\mathbf{H} = 0. \tag{3.171b}$$

Equations (3.169a), (3.169b) and (3.171a), (3.171b) are the time harmonic (or time independent) source free vector wave equations that are commonly called vector Helmholtz equations. It is instructive to compare the Helmholtz equations with the corresponding vector wave equations discussed in the previous section. In the next section we will describe how wave solutions for **E** and **H** are obtained from the Helmholtz equations.

3.6 UNIFORM PLANE WAVES

3.6.1 General Considerations

We assume a lossless ($\sigma \equiv 0$), homogeneous and isotropic unbounded medium characterized by μ, ϵ. We also assume that the region under consideration is source free and that there exists an electric field in the region such that it has only one rectangular component that varies with respect to one of the other two rectangular coordinates:

$$\mathbf{E} = \hat{x}\ E_x(z) \text{ and } E_y = E_z \equiv 0, \tag{3.172}$$

We now determine a magnetic field configuration **H** compatible with the assumed **E** such that **E**, **H** both satisfy the Helmholtz equation (3.171a), (3.171b). We then describe the detailed nature of the solution so obtained.

Under the assumptions (3.171a), (3.171b), and (3.171a) give the following equation for E_x:

$$\frac{\partial^2 E_x}{\partial z^2} + \beta^2 E_x = 0. \tag{3.173}$$

The solution of (3.173) is

$$\begin{aligned} E_x &= E_0^+ e^{-j\beta z} + E_0^- e^{+j\beta z} \\ &= E_x^+ + E_x^- , \end{aligned} \tag{3.174}$$

where E_x^+, E_x^- are two constants, β is as defined before. The time dependent form of E_x is:

$$\begin{aligned}\mathcal{E}(z,t) &= \text{Re}[\hat{x}E_x e^{j\omega t}] \\ &= \hat{x}E_0^+ \cos(\omega t - \beta z) \\ &\quad + \hat{x}E_0^- \cos(\omega t + \beta z).\end{aligned} \tag{3.175}$$

It is now clear that the first and second terms of (3.175) represent waves propagating in the $+z$ and $-z$ directions, respectively. To obtain the velocity of the wave propagating in the positive z direction, the total phase of the wave is assumed to be constant at combinations of z, t, that is

$$\omega t - \beta z = \text{constant}. \tag{3.176}$$

Differentiating (3.176) with respect to time and defining the phase velocity of the wave as

$$v_\text{p} = \frac{dz}{dt}, \tag{3.177a}$$

we obtain the following for v_p:

$$v_\text{p} = \frac{\omega}{\beta} = \frac{1}{\sqrt{\mu\epsilon}} = \text{velocity of light in the medium}. \tag{3.177b}$$

It should be noted that the phase velocity v_p, is associated with the propagation of the constant phase front of a strictly time harmonic field. Note that $z =$ constant, meaning the $x - y$ plane surfaces represent the constant phase surfaces of the wave at any time. These are called the wavefronts, and they are perpendicular to the directions of propagation. In the present case the wavefronts are plane surfaces.

Thus the complete **E** field for the time harmonic form of the forward (moving in the $+z$ direction) wave is

$$\mathbf{E} = \hat{x}E_x^+ = \hat{x}E_0^+ e^{-j\beta z}, \tag{3.178}$$

where E_0^+ is the amplitude of the field and β determines the phase of the wave; hence it is all called the phase (propagation) constant for the wave.

What is the **H** field compatible with (3.178)? This can be obtained from the **E** field by using Maxwell's equations (3.165b) as shown:

$$\begin{aligned}\mathbf{H} &= -\frac{1}{j\omega\mu}[\nabla \times \mathbf{E}] \\ &= -\frac{1}{j\omega\mu}\begin{vmatrix} \hat{x} & \hat{y} & \hat{z} \\ \frac{\partial}{\partial x} & \frac{\partial}{\partial y} & \frac{\partial}{\partial z} \\ E_x^+ & 0 & 0 \end{vmatrix} \\ &= \frac{E_0^+}{\omega\mu/\beta}\, e^{-j\beta z}\hat{y} \\ &= \hat{y}\frac{E_0^+}{\eta}\, e^{-j\beta z},\end{aligned} \tag{3.179}$$

where

$$\eta = \frac{\omega\mu}{\beta} = \sqrt{\frac{\mu}{\epsilon}} = \text{the intrinsic impedance of the medium.} \tag{3.180}$$

The complete expression for the forward traveling wave can now be written as

$$\mathbf{E} = \hat{x}E_x = \hat{x}E_0^+ e^{-j\beta z} \tag{3.181a}$$

$$\mathbf{H} = \hat{y}H_y = \hat{y}\,\frac{E_0^+}{\eta}\, e^{-j\beta z} = \hat{y}H_0^+ j^{-j\beta z}, \tag{3.181b}$$

where we have included various alternate notations that are in use in the literature. Note that that the **E** and **H** fields of the wave described by (3.181a), (3.181b) are in phase with each other and are mutually orthogonal, and that both are perpendicular (i.e., transverse) to the direction of propagation. The wave fronts in (3.181a), (3.181b) being plane surfaces, the wave that it represents is called a plane wave.

The ratio of the transverse electric field and the transverse magnetic field, namely E_x/H_y in the present case, is generally called the wave independence. In our case (3.181a), (3.181b) indicate the wave independence

$$\begin{aligned}&= \frac{E_x}{H_y} = \eta = \sqrt{\frac{\mu}{\epsilon}} \\ &= \text{intrinsic impedance of the medium.}\end{aligned} \tag{3.182}$$

The wave impedance and phase velocity of the plane waves in a simple medium with μ, ϵ are generally related to those in free space and to the medium's material constants.

For example, if the medium has $\mu = \mu_0,\ \epsilon = \epsilon_0\epsilon_r$, and $\sigma = 0$,

$$\eta = \sqrt{\frac{\mu}{\epsilon}} = \frac{\eta_0}{\sqrt{\epsilon_r}}, \tag{3.183a}$$

where

$$\begin{aligned} \eta_0 = \sqrt{\frac{\mu_0}{\epsilon_0}} &= 120\pi \simeq 377\ \Omega \\ &= \text{intrinsic impedance of free space} \end{aligned} \tag{3.183b}$$

and

$$v_p = \frac{1}{\sqrt{\mu\epsilon}} = \frac{c}{\sqrt{\epsilon_r}} \tag{3.184a}$$

with

$$\begin{aligned} c = \frac{1}{\sqrt{\mu_0\epsilon_0}} &= 3 \times 10^8 \qquad \text{m/s} \\ &= \text{velocity of light in free space.} \end{aligned} \tag{3.184b}$$

The real time dependent form for the forward traveling ($+z$ direction) wave can now be written from (3.181a), (3.181b) as

$$\mathcal{E}(z,t) = \hat{x}E_0^+ \cos(\omega t - \beta z) \tag{3.185a}$$

$$\mathcal{H}(z,t) = \hat{y}\,\frac{E_0^+}{\eta} \cos(\omega t - \beta z), \tag{3.185b}$$

where it should be noted that $\mathcal{E}, \mathcal{H}$ are in phase at all time, and that they are uniform in the transverse (xy) plane. Equations (3.185a), (3.185b) indicate that the fields are periodic in time as well as in space in the z direction. The period in time in (3.185a), (3.185b) is designated as T, such that

$$\begin{aligned} \omega T &= 2\pi \\ \text{or} \quad T &= \frac{1}{f}, \end{aligned} \tag{3.186}$$

where f is the frequency of the wave. The period in space in (3.185a), (3.185b) is called the wavelength, λ, such that

$$\begin{aligned} \beta\lambda &= 2\pi \\ \text{or} \quad \lambda &= \frac{2\pi}{\beta} = \text{wavelength in the medium.} \end{aligned} \tag{3.187}$$

By definition, we then have

$$\beta = \frac{2\pi}{\lambda} = \frac{\omega}{v_p}, \tag{3.188a}$$

yielding

$$v_p = f\lambda. \tag{3.188b}$$

Also note that

$$\lambda = \frac{\lambda_0}{\sqrt{\epsilon_r}}, \tag{3.189}$$

where λ is the wavelength in a medium having $\epsilon = \epsilon_0\epsilon_r$, and

$$\lambda_0 = \frac{c}{f} = \text{wavelength in free space.}$$

Similarly, for the backward wave, which is the wave traveling in the negative direction, the appropriate phasor fields are

$$\begin{aligned} \mathbf{E} &= \hat{x} E_0^- e^{+j\beta z} \\ \mathbf{H} &= -\hat{y}\, \frac{E_0^-}{\eta}\, e^{+j\beta z}, \end{aligned} \tag{3.190}$$

and the corresponding time dependent fields are

$$\begin{aligned} \boldsymbol{\mathcal{E}}(z,t) &= \hat{x} E_0^- \cos(\omega t + \beta z) \\ \boldsymbol{\mathcal{H}}(z,t) &= -\hat{y}\, \frac{E_0^-}{\eta} \cos(\omega t + \beta z). \end{aligned} \tag{3.191}$$

General behavior of the wave represented by (3.190) or (3.191) is similar to that of the forward wave described earlier, except for the fact the former travels in the opposite direction.

3.6.2 Energy Considerations

Once the electric and magnetic fields of a wave are known, the time average energy (or power) carried by the wave can be determined by two alternate methods, as indicated by (3.131). The first method requires the knowledge of the time dependent fields and yields the power average as the time average of the instantaneous Poynting vector $\boldsymbol{\mathcal{S}}(t) = \boldsymbol{\mathcal{E}}(t) \times \boldsymbol{\mathcal{H}}(t)$. The second method requires the knowledge of phasor (complex) fields and yields the power average as one half of the real part of the complex Poynting vector $\mathbf{S}^* = \mathbf{E} \times \mathbf{H}^*$.

We will use the second method here to obtain power. With the phasor fields given by (3.181) we find the average energy carried by the forward wave is

$$\begin{aligned} \mathbf{S}_{\text{av}} &= \text{Re}\left[\frac{\mathbf{E}\times\mathbf{H}^*}{2}\right] \\ &= \hat{z}\,\frac{|E_0^+|^2}{2\eta} \quad \text{W/m}^2. \end{aligned} \tag{3.192}$$

Similarly, for the backward traveling wave we use the fields (3.190) and obtain the power carried by the wave,

$$\mathbf{S}_{\text{av}} = -\hat{z}\,\frac{|E_0^+|^2}{2\eta} \quad \text{W/m}^2. \tag{3.193}$$

We now describe a few other properties of the forward wave only; the backward wave will have similar properties with slight modification (because of its opposite direction of propagation). The time average stored electric and magnetic energy densities for the forward wave are

$$\begin{aligned} W_{\text{E}}(t)|_{\text{av}} &= \frac{1}{2}\,\text{Re}\left[\frac{\mathbf{E}\cdot\mathbf{D}^*}{2}\right] \\ &= \frac{1}{2}\,\frac{\epsilon|E_0^*|^2}{2} \quad \text{J/m}^3 \end{aligned} \tag{3.194a}$$

$$\begin{aligned} W_{\text{H}}(t)|_{\text{av}} &= \frac{1}{2}\,\text{Re}\left[\frac{\mathbf{H}\cdot\mathbf{B}^*}{2}\right] \\ &= \frac{1}{2}\,\mu\frac{|H_0^*|^2}{2} = \frac{1}{2}\,\frac{\epsilon|E_0^*|^2}{2} \quad \text{J/m}^3 \end{aligned} \tag{3.194b}$$

Thus, we find that the average stored energy in the electric field equals the average stored energy in the magnetic field. The total energy W_{av}^{T} stored in the forward wave is

$$\begin{aligned} W_{\text{av}}^{\text{T}} &= W_{\text{E}}(t)|_{\text{av}} + W_{\text{H}}(t)|_{\text{av}} \\ &= \frac{1}{2}\,\epsilon|E_0^+ \quad \text{W/m}^3. \end{aligned} \tag{3.195}$$

If we now define a velocity $\mathbf{v}_{\text{E}} = \hat{z}v_{\text{e}}$, then it can be seen that the dimension of the quantity $[W_{\text{av}}^{\text{T}}\mathbf{v}_{\text{E}}]$ is

$$\left[\frac{\text{J}}{\text{m}^3}\,\frac{\text{m}}{\text{s}}\right] = \text{W/m}^2,$$

that is, it represents some power flow in the direction of $\mathbf{v}_{\mathrm{E}}$. Now equating $W_{\mathrm{av}}^{\mathrm{T}}\mathbf{v}_{\mathrm{E}}$ to the average energy carried by the wave (i.e., to $\mathbf{S}_{\mathrm{av}}$ given by (3.192)), we obtain

$$W_{\mathrm{av}}^{\mathrm{T}}\mathbf{v}_{\mathrm{E}} = \mathbf{S}_{\mathrm{av}}, \tag{3.196}$$

provided that

$$v_{\mathrm{E}} = \frac{1}{\sqrt{\mu\epsilon}} = v, \tag{3.197}$$

where v is the velocity of light in the medium. Here v_{E} is called the energy flow velocity [1,3,6]. In the present case, for a forward wave in the medium, we can make the following general statement: phase velocity of the wave (v_{P}) = energy flow velocity, (v_{E}) = velocity of light (v) in the medium.

3.6.3 Group Velocity

In the propagation of electromagnetic signals an important question arises concerning the velocity of a wave packet containing a band of frequency components: for example, why the information carried by a modulated signal is represented by the shape or the envelope of the wave. The measure of the speed of propagation of a group of frequencies that constitute a wave packet is the group velocity, defined as [1–4];

$$v_{\mathrm{g}} = \frac{1}{d\beta/d\omega}, \tag{3.198}$$

where the phase velocity v_{P} is defined by

$$v_{\mathrm{p}} = \frac{\omega}{\beta} \tag{3.199}$$

with β being the phase constant for the wave defined earlier. The velocity at which information (envelope of a modulated signal) travels is the group velocity. If the phase velocity v_{p} is a function of frequency, it leads to distortion in the signal, by dispersing its frequency components. This is known as dispersion. For a linear, passive, and isotropic dielectric medium we have

$$v_{\mathrm{p}} = \frac{\omega}{\beta}, \beta = \omega\sqrt{\mu\epsilon}.$$

Therefore

$$v_{\mathrm{g}} = \frac{1}{d\beta/d\omega} = \frac{1}{\sqrt{\mu\epsilon}} = v_{\mathrm{p}}, \tag{3.200}$$

which indicates that the phase velocity and group velocity are both equal to the velocity of light in the medium. In a dispersive medium, v_{p} and v_{g} are generally different. Detailed discussion of dispersion is given in [1–4]. We only summarize below the main definitions and observations.

A dispersive medium is one in which the phase velocity is a function of frequency (or, of the free space wavelength λ).

1. Normally dispersive medium

$$\text{if} \quad \frac{dv_p}{d\lambda} > 0 \quad \text{and} \quad v_g < v_p$$

2. Anomalously dispersive medium

$$\text{if} \frac{dv_p}{d\lambda} < 0, v_g < v_p.$$

The observations above follow from the following equivalent relations with a small bandwidth signal:

$$v_g = v_p + \beta\frac{dv_p}{d\beta} \tag{3.201a}$$

or

$$v_g = v_p - \lambda\frac{dv_p}{d\lambda} . \tag{3.201b}$$

The last expression indicates that for a normally dispersive medium, $v_g < 0$ and v_g is identified as the signal velocity, details in [1–4].

3.6.4 Summary

For future reference, the characteristics of uniform plane electromagnetic or transverse electromagnetic (TEM) waves in an unbounded simple medium are summarized a follows:

1. Equiphase surfaces, also called wavefronts or wave planes, are plane surfaces and hence the name plane waves. Wave fronts are perpendicular or transverse to the directions of propagation.

2. **E** and **H** fields of the wave are orthogonal to each other and are in the wave plane, meaning they are transverse to the direction of propagation and hence the name TEM.

3. **E** and **H** are in phase at all times.

4. **E** and **H** do not vary in the wave plane, meaning they are uniform and hence the name uniform.

5. In the case we considered earlier, $\mathbf{E}$ was entirely in the x direction, that is $\mathbf{E} = \hat{x}E_x$ with $\mathbf{H} = \hat{y}H_x$. The direction of propagation is thus z, i.e., in the direction of $\mathbf{E} \times \mathbf{H}$. We will see later, the polarization of a wave is identified with the orientation of the vector $\mathbf{E}$ in space; in the present case it is linearly polarized along $\hat{x}$.

6. In an unbounded space, the wavefronts being of infinitive extent, the wave carries infinite power, and hence they are impractical to generate.

3.6.5 General Representation of TEM Waves

In many situations we need to consider a uniform plane or TEM wave propagating in an arbitrary direction in a homogeneous, lossless, and isotropic medium.

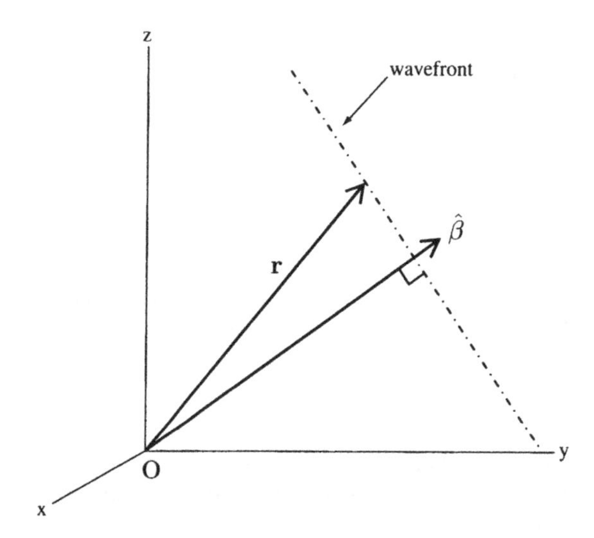

Figure 3.8 A plane wavefront propagating in the $\hat{\beta}$ direction.

For example, for the arbitrary wavefront shown in Figure 3.8 assume a propagation direction $\hat{\beta} = \hat{\beta}(\theta_0, \phi_0)$, meaning θ_0, ϕ_0 are the spherical coordinates with respect to the rectangular coordinate system, with origin at 0. Thus we define the propagation vector as

$$\boldsymbol{\beta} = \beta\hat{\beta} \tag{3.202}$$

with

$$\beta = \frac{2\pi}{\lambda} = \frac{\omega}{v}, v = \frac{1}{\sqrt{\mu\epsilon}}$$

as was used before.

If $\mathbf{r}$ is the position vector of any point on the wave front, then the representation of the wave front for any $\mathbf{r}$ is

$$\mathbf{r} \cdot \boldsymbol{\beta} = \text{constant for any } \mathbf{r}, \tag{3.203a}$$

which follows from the condition that wavefront be perpendicular to $\hat{\beta}$, and

$$\mathbf{r} = x\hat{x} + y\hat{y} + z\hat{z}, \tag{3.203b}$$

where x, y, z are the coordinates of any point on the phasefront. Now we choose the direction of the electric field $\hat{e}$ such that $\hat{e}$ is perpendicular to $\hat{\beta}$. For a wave moving away from O, we have the electric field

$$\mathbf{E} = \hat{e} E_0 e^{-j\boldsymbol{\beta} \cdot \mathbf{r}}, \tag{3.204a}$$

and

$$\mathbf{H} = \frac{\hat{\beta} \times \mathbf{E}}{\eta} \tag{3.204b}$$

with

$$\eta = \sqrt{\frac{\mu}{\epsilon}} = \text{intrinsic impedance of the medium.}$$

We can now obtain the average power flow for the wave by using the complex Poynting vector and it can be represented by

$$\begin{aligned} \mathbf{S}_{\text{av}} &= \frac{1}{2} \operatorname{Re}[\mathbf{E} \times \mathbf{H}^*] \\ &= \frac{1}{2} \frac{|E_0|^2}{\eta} (\hat{e} \times \hat{\beta} \times \hat{e}) \\ &= \frac{1}{2} \frac{|E_0|^2}{\eta} \hat{\beta}. \end{aligned} \tag{3.205}$$

Equations (3.204a), (3.204b), and (3.205) give the complete representation for a uniform plane wave propagating in an arbitrary direction, and note that the wave in (3.204a), (3.204b) is polarized in the $\hat{e}$ direction.

Example 3.3

Determine the fields and power flow for a $\hat{z}$ polarized TEM wave propagating in the y-direction.

Solution

Assume $\hat{\beta} = \hat{y}$ and $\hat{e} = \hat{z}$, that is a $\hat{z}$-polarized wave propagating in the $\hat{y}$ direction.

$$\begin{aligned}\boldsymbol{\beta} \cdot \mathbf{r} &= \beta\hat{y} \cdot (x\hat{x} + y\hat{y} + z\hat{z}) \\ &= \beta y.\end{aligned}$$

Thus

$$\begin{aligned}\mathbf{E} &= \hat{z}E_0 e^{-j\beta y} \\ \mathbf{H} &= \frac{\hat{y} \times \hat{z}}{\eta} E_0 e^{-j\beta y} = \hat{x}\, \frac{E_0}{\eta}\, e^{-j\beta y} \\ \mathbf{S}_{\text{av}} &= \text{Re}\left[\frac{1}{2}\, \mathbf{E} \times \mathbf{H}^*\right] \\ &= \frac{1}{2}\, \frac{|E_0|^2}{\eta} \quad \text{W/m}^2.\end{aligned}$$

Example 3.4

The electric field of a plane electromagnetic wave propagating in a lossless simple and unbounded medium with μ, ϵ and $\sigma_{\text{eq}}0$ is given by

$$\boldsymbol{\mathcal{E}}(z,t) = \hat{x}\, 100 \cos\left(2\pi \times 10^6\, t - \beta\, z\right) \quad \text{V/m}.$$

where the notations are as explained in the text. Determine the following for the wave:

Case 1. The wavefront, the direction of wave propagation, the frequency of the wave, and the phase velocity of the wave.

Case 2. The phasor electric field $\mathbf{E}$. Use the appropriate time harmonic Maxwell's equation to obtain the corresponding phasor magnetic field $\mathbf{H}$. Then determine the time dependent quantity $\boldsymbol{\mathcal{H}}(z,t)$.

Case 3. The direction of propagation β. Use the appropriate time harmonic Maxwell's equation to obtain $\mathbf{E}$ from $\mathbf{H}$ as determined in (ii), and equate it to $\mathbf{E}$ as in (ii). Now determine β.

Case 4. The explicit phasor expressions for $\mathbf{E}$ and $\mathbf{H}$.

Case 5. The wave impedance $\mathbf{E}/\mathbf{H}$. Show that it is equal to the intrinsic impedance of the medium.

Case 6. The average power flow $\mathbf{S}_{\text{av}}$ carried out by the wave. Determine it by using the phasor fields and the time dependent real field quantities.

Solution

Case 1. Let the direction of propagation be $\hat{\beta}$; define the propagation vector $\boldsymbol{\beta} = \beta\hat{\beta}$, where β is the phase (the propagation constant in the present case). The total instantaneous phase for a harmonic wave with propagation vector $\boldsymbol{\beta}$, with $\omega\, t_{\text{eq}}\boldsymbol{\beta}\dot{\mathbf{r}}$, ω being the radian frequency and $\mathbf{r}$ the position vector of a point on the wavefront.

It is given that for the wave under consideration, $\boldsymbol{\beta} \cdot \mathbf{r} = \beta\, z$, which implies that $\hat{\beta} = \hat{z}$, meaning the wave propagation is in the $+\hat{z}$ direction.

The wavefront at any time t is given by $\beta\, z =$ constant surface, which implies that the wavefront is parallel to the $z =$ constant or xy-plane surfaces.

The frequency is $f = \omega/2\pi = 1$ MHz. Since in the present case $\omega\, t - \beta\, z =$ constant, we obtain the phase velocity v_{p} as

$$\frac{d}{dt}\left(\omega t - \beta z\right) = 0$$

$$\text{or } \omega - \beta\frac{dz}{dt} = 0$$

$$\text{or } V_{\text{p}} = \frac{\omega}{\beta}.$$

Thus

$$\beta = \frac{\omega}{V_{\text{p}}} = \frac{2\,\pi, f}{V_{\text{p}}} = \frac{2\,\pi}{\lambda}$$

where λ is the wavelength, and $V_{\text{p}} = \lambda\, f$.

Case 2.

$$\mathbf{E} = \hat{x}\, 100\, e^{-j\,\beta\, z} \quad \text{V/m}$$

$$\mathbf{H} = \frac{1}{-j\,\omega\,\mu}\,\Delta \times \mathbf{E} = \frac{1}{-j\,\omega\,\mu}\begin{vmatrix} \hat{x} & \hat{y} & \hat{z} \\ \frac{\partial}{\partial x} & \frac{\partial}{\partial y} & \frac{\partial}{\partial z} \\ E_x & 0 & 0 \end{vmatrix}$$

$$= \frac{\hat{y}}{-j\,\omega\,\mu}\,\frac{\partial\, E_x}{\partial x}$$

$$= \hat{y}\,\frac{100}{\frac{\omega\,\mu}{\beta}}\,e^{-j\,\beta\, z} \quad \text{A/m}$$

$$\mathcal{H}(z,t) = \hat{y}\,\frac{100}{\frac{\omega\,\mu}{\beta}}\,\cos\left(\omega t - \beta z\right) \quad \text{A/m.}$$

Case 3. We have $\mathbf{E} = \hat{x}\,100\,e^{-j\beta z}$ with

$$\mathbf{H} = \hat{y}\,\frac{100}{\omega\mu/\beta}\,e^{-j\beta z}$$

as in Case 2. We obtain

$$\mathbf{E} = \frac{1}{j\,\omega\,\epsilon}\,\nabla\times\mathbf{H} = \frac{\hat{x}}{j\,\omega\,\epsilon}\left(-\frac{\partial H_y}{\partial z}\right) \tag{A}$$

$$= \hat{x}\,\frac{100\,\beta^2}{\omega^2\,\mu\,\epsilon}\,e^{-j\beta z}. \tag{B}$$

After equating the E fields given by (A) and (B), we obtain

$$\beta = \omega\,\sqrt{\mu\,\epsilon}.$$

Note that the phase velocity V_{p} now is

$$V_{\mathrm{p}} = \frac{\beta}{\omega} = \frac{1}{\sqrt{\mu\,\epsilon}}$$

$$= \frac{v}{\sqrt{\mu_{\mathrm{r}}\,\epsilon_{\mathrm{r}}}} = \frac{v}{\sqrt{\epsilon_{\mathrm{r}}}}$$

with $\mu_{\mathrm{r}} = 1$, and $v = 1/\sqrt{\mu_0\,\epsilon_0}$ being the velocity of light in free space.

Case 4.

$$\mathbf{E} = \hat{x}\,100\,e^{-j\beta z} \quad \text{V/m}$$

$$\mathbf{H} = \hat{y}\,\frac{100}{\eta}\,e^{-j\beta z} \quad \text{A/m;}$$

where

$$\eta = \sqrt{\frac{\mu}{\epsilon}} = \frac{\eta_0}{\sqrt{\epsilon_r}} \text{for } \mu = \mu_0,\ \epsilon = \epsilon_0\epsilon_{\mathrm{r}}$$

with

$$\eta_0 = \sqrt{\frac{\mu_0}{\epsilon_0}} = 120\,\pi = 377\,\Omega$$

being the intrinsic impedance of free space.

Case 5.

$$\frac{|\mathbf{E}|}{|\mathbf{H}|} = \eta = \text{intrinsic impedance of the medium.}$$

Case 6. Now in phasor notation

$$\begin{aligned}
\mathbf{S}_{\text{av}} &= \text{Re.}\left[\frac{1}{2}\,(\mathbf{E}\times\mathbf{H}^*)\right] \\
&= \text{Re.}\left[\frac{1}{2}\left(\hat{x}\,100\,e^{-j\,\beta\,z}\right)\times\left(\hat{y}\,\frac{100}{\eta}\,e^{-j\,\beta\,z}\right)^*\right] \\
&= \frac{10^4}{2\,\eta}\,\hat{z}\quad \text{W/m}^2.
\end{aligned}$$

$\mathbf{S}_{\text{av}}$ with time dependent fields

$$\begin{aligned}
\mathcal{E}(z,t) &= \hat{x}\,100\,\cos\left(\omega\,t-\beta\,z\right)\quad \text{V/m} \\
\mathcal{H}(z,t) &= \hat{y}\,\frac{100}{\eta}\,\cos\left(\omega\,t-\beta\,z\right)\quad \text{A/m} \\
\mathcal{S}_{\text{av}} &= \frac{1}{T}\int_0^T \mathcal{E}(z,t)\times\mathcal{H}(z,t)\,dt\ \text{ with } T=\frac{2\pi}{\omega}.
\end{aligned}$$

It can be shown that

$$\begin{aligned}
\mathcal{S}_{\text{av}} &= \hat{z}\,\frac{10^4}{2\,\eta}\int_0^T \left[1+\cos\,2\,\left(\omega\,t-\beta\,z\right)\right]\,dt \\
&= \frac{10^4}{2\,\eta}\,\hat{z}\quad \text{W/m}^2.
\end{aligned}$$

3.6.6 Plane Waves in Lossy Media

We consider plane waves in an unbounded, lossy, isotropic, and homogeneous medium characterized by μ, ϵ and σ. The appropriate equations to obtain the wave equation for the source free case are (from (3.97a)–(3.100a))

$$\nabla\times\mathbf{E} = -j\omega\mu H \tag{3.206a}$$

$$\nabla\times\mathbf{H} = (\sigma + j\omega\epsilon\mathbf{E}) \tag{3.206b}$$

$$\nabla\cdot\mathbf{B} = 0, \qquad \nabla\cdot\mathbf{D} = 0, \tag{3.206c}$$

supplemented by the constitutive relations

$$\mathbf{B} = \mu\mathbf{H}, \qquad \mathbf{D} = \epsilon\mathbf{E}. \tag{3.207}$$

Following the procedures outlined in Sections 3.5.1 and 3.5.2, it can be shown that **E**, **H** satisfy the vector wave equation

$$\nabla \times \nabla \times \begin{pmatrix} \mathbf{E} \\ \mathbf{H} \end{pmatrix} + \gamma^2 \begin{pmatrix} \mathbf{E} \\ \mathbf{H} \end{pmatrix} + \gamma^2 = 0, \quad (3.208a)$$

which can also be written as

$$\nabla^2 \begin{pmatrix} \mathbf{E} \\ \mathbf{H} \end{pmatrix} - \gamma^2 \begin{pmatrix} \mathbf{E} \\ \mathbf{H} \end{pmatrix} + \gamma^2 = 0, \quad (3.208b)$$

where the complex propagation constant γ is defined as

$$\gamma^2 = (j\omega\mu)(\sigma + j\omega\epsilon). \quad (3.209a)$$

It is important to note that for the time convention $e^{j\omega t}$ (which we are using) γ is defined to be the positive root of γ^2 given by (3.209a):

$$\begin{aligned} \gamma &= +\sqrt{(j\omega\mu)(\sigma + j\omega\epsilon)} \\ &= \alpha + j\beta, \alpha, \beta > 0. \end{aligned} \quad (3.209b)$$

Equating the real and imaginary parts of the appropriate sides of (3.209b), the quantities α and β can be expressed in terms of the medium parameters and the frequency ω. They are

$$\begin{aligned} \alpha &= \omega\sqrt{\mu\epsilon} \left[\frac{1}{2} \left(\sqrt{1 + \left(\frac{\sigma}{\omega\epsilon} \right)^2} - 1 \right) \right]^{1/2} \quad \text{Nepers/m} \\ \beta &= \omega\sqrt{\mu\epsilon} \left[\frac{1}{2} \left(\sqrt{1 + \left(\frac{\sigma}{\omega\epsilon} \right)^2} + 1 \right) \right]^{1/2} \quad \text{radians/m.} \end{aligned} \quad (3.209c)$$

In cases where the medium is characterized by its complex permittivity $\epsilon_c = \epsilon' - j\epsilon''$ or by its permittivity ϵ' and $\tan\delta = \epsilon''/\epsilon'$. Then the appropriate α, β can be obtained by making the substitutions: $\epsilon = \epsilon'$ and $\sigma/\omega = \epsilon''$ in (3.209c).

As was done in Section 3.6.1, we assume $\mathbf{E} = \hat{x}\, E_x(z)$, $E_y = E_z \equiv 0$, and $\partial/\partial x = \partial/\partial y \equiv 0$ for the anticipated wave. So it can be shown that the E field satisfies the equation

$$\frac{\partial^2 E_x}{\partial z^2} - \gamma^2 E_z = 0. \quad (3.210)$$

The general solution of (3.210) is

$$E_x = E_0^+ e^{-\gamma z} + E_0^- e^{+\gamma z}. \quad (3.211a)$$

Now after using (3.211a) and (3.208a), we obtain the corresponding H field as

$$H_y = \frac{E_0^+}{j\omega\mu/\gamma}\, e^{-\gamma z} - \frac{E_0^-}{j\omega\mu/\gamma}\, e^{+\gamma z}. \tag{3.211b}$$

Equations (3.211a) and (3.211b) give the complete solution, namely the total **E** and **H** fields. We now consider only the forward moving wave given by

$$\mathbf{E}^+ = \hat{x} E_0^+ e^{-\gamma z} \tag{3.212a}$$

$$\mathbf{H}^+ = \hat{y} \frac{E_0^+}{\eta}\, e^{-\gamma z}, \tag{3.212b}$$

where

$$\eta = \sqrt{\frac{j\omega\mu}{\sigma + j\omega\epsilon}} \tag{3.213}$$

is the intrinsic impedance of the medium and note that it is now a complex quantity. Writing the explicit expression for γ in (3.212a), (3.212b), we obtain

$$\mathbf{E}^+ = \hat{x} E_0^+ e^{-\alpha z} e^{-j\beta z} \tag{3.214a}$$

$$\mathbf{H}^+ = \hat{y} \frac{E_0^+}{\eta}\, e^{-\alpha z} e^{-j\beta z}, \tag{3.214b}$$

where α, β are given by (3.209c). Equation (3.209a), (3.209b) should be compared with the fields for the lossless case given by (3.182). In the present case the amplitudes of both **E**, **H** decay as the wave propagates in the z direction. The parameter α is commonly called the attenuation constant providing the amplitude decay, and β is called the phase constant, which provides the phase propagation for the wave. Note that in the present lossy case, the wave impedance looking in the direction of propagation is complex and is given by

$$Z^+ = \frac{E_x}{H_y} = \eta = \sqrt{\frac{j\omega\mu}{\sigma + j\omega\epsilon}}\,. \tag{3.215}$$

Equation (3.215) indicates that the two fields **E** and **H** are no longer in phase. In fact it can be shown that for a good conducting medium $\sigma \gg \omega\epsilon$, the H field will lag the E field by $\pi/4$.

By the fields (3.214a), (3.214b) and the complex Poynting vector, it can be shown that the average power carrier by the positive wave is

$$\mathbf{S}_{\text{av}} = \frac{1}{2}\, |E_0^+|^2 e^{-2\alpha z}\, \text{Re}\left(\frac{1}{\eta^*}\right) \text{ W/m}^2, \tag{3.216}$$

where we again see that the power flow decreases exponentially in the direction of propagation. We have seen that the amplitude of the fields as well as the power flow decay exponentially in the direction of wave propagation in a lossy medium. The decay is determined by the attenuation constant α as in (3.214a), (3.214b). For example, the amplitude ratio for the electric fields at distances z and $z + d$ is

$$R = \left| \frac{E^+(z)}{E^+(z+d)} \right| = e^{\alpha d}, \tag{3.217}$$

which is the attenuation suffered by the field after traveling through a distance d. We express the attenuation as

$$\alpha\, d = \ln \left| \frac{E^+(z)}{E^+(z+d)} \right| = \ln R \quad \text{Nepers} \tag{3.218}$$

If d is expressed in meters then the unit of α is Nepers/meter = Np/m. In engineering practice it is customary to express attenuation in dB units, which in the present case is

$$\begin{aligned} &= 20 \log_{10} R = (\alpha d)\, 20 \log_{10} e \\ &= (\alpha d)\, 8.6589 \text{ dB}. \end{aligned} \tag{3.219a}$$

Thus we give below the important relationship

$$\alpha \text{ in Neper/meter} = 8.6589 \text{ dB/m}. \tag{3.219b}$$

In radio wave propagation through lossy media, the term penetration depth is often used to specify a distance over which the field amplitude is decreased by a factor $1/e$. Thus, from (3.217), the penetration depth d_p occurs when $R = e$. That is, from (3.218) we have

$$d_\text{p} = \frac{1}{\alpha} \quad \text{m}, \tag{3.220}$$

where α must be expressed in Np/m, and is given by (3.209c).

Some Approximate Relations

Lossy and conducting media occur in many practical situations. Knowledge of the propagation constant $\gamma = \alpha + j\beta$. Is needed for the purpose of investigating the attenuation and the propagation characteristics of waves in such media. The explicit rigorous expressions for α, β, given by (3.209c), can be used to determine accurate numerical values under various situations. However, it is often found convenient and sufficient to use some form of approximate expression for certain investigations. We describe here a few useful approximations.

If $\sigma \gg \omega\epsilon$, which is when the conduction current in the medium dominates the displacement current, we have

$$\gamma = \sqrt{(j\omega\mu)(\sigma + j\omega\epsilon)} \simeq \sqrt{j\omega\mu\sigma}$$
$$= \sqrt{\frac{\omega\mu\sigma}{2}}\,(1+j) = a + j\beta. \tag{3.221}$$

This indicates that in such cases

$$\alpha = \beta \simeq \sqrt{\frac{\omega\mu\sigma}{2}} = \frac{1}{\delta} \tag{3.222}$$

where

$$\delta = \sqrt{\frac{2}{\omega\mu\sigma}} = \sqrt{\frac{1}{\pi f\mu\sigma}} \tag{3.223}$$

is the skin depth in the medium (to be discussed later). The skin depth only represents a distance over which the wave amplitude is reduced by by $1/e$:, namely by 1 Np $=$ 8.5589 dB. Another expression of importance is the intrinsic independence of the medium, which is defined by

$$\eta = \sqrt{\frac{j\omega\mu}{\sigma + j\omega\epsilon}} = \frac{\sqrt{\mu/\epsilon}}{\left[1 + \left(\frac{\sigma}{\omega\epsilon}\right)^2\right]^{1/4}}\, e^{j/z\tan^{-1}(\sigma/\omega\epsilon)}. \tag{3.224}$$

Again, when $\sigma \gg \omega\epsilon$, η may be approximated by several alternate expressions. That is, for $\sigma \gg \omega\epsilon$,

$$\eta \simeq \sqrt{\frac{j\omega\mu}{\sigma}} = \sqrt{\frac{\omega\mu}{\sigma}}\, e^{j\pi/4}$$
$$= \sqrt{\frac{\omega\mu}{2\sigma}}\,(1+j)$$
$$= \frac{1}{\sigma\delta}\,(1+j), \tag{3.225}$$

where δ is the skin depth defined before. We have occasion to use the above expressions later.

3.6.7 Skin Effect

Plane waves attenuate as they propagate through a lossy medium. From (3.220) it is found that the depth of penetration of fields in a lossy medium is inversely proportional to the square-root of the conductivity σ of the medium.

In the extreme case of $\sigma = \infty$, this depth vanishes and in fact time-varying fields and induced currents cannot exist within the medium. In other words, all fields and induced currents are confined near the skin region of the medium. This is the skin effect phenomenon, and it manifests in variety of electromagnetic problems. A good example of skin effect is the increased high-frequency resistance of a piece of round metallic wire, which occurs as a result of current confinement through a smaller cross section due to the skin effect.

We will describe the phenomenon with the help of a simple example illustrated in Figure 3.9 that defines a plane interface between free space ($x < 0$) and a conducting medium ($x \geq 0$). We assume that an electric field $\mathbf{E} = \hat{z}E_0$ is maintained at surface $x = 0$. This could be done, for example, with the help of a uniform plane electromagnetic wave propagating in the positive x direction from the free space side, as shown in Figure 3.9, such that the incident fields are

$$\begin{aligned} \mathbf{E}^{\mathrm{i}} &= \hat{z}E_z = \hat{z}E_0 e^{-j\beta x} \\ \mathbf{H}^{\mathrm{i}} &= -\hat{y}H_z = -\hat{y}\,\frac{E_0}{\eta_0}\,e^{-j\beta x} \end{aligned} \tag{3.226}$$

with $\beta = \omega\sqrt{\mu_0\epsilon_0}$ and $\eta = \sqrt{\mu_0/\epsilon_0}$.

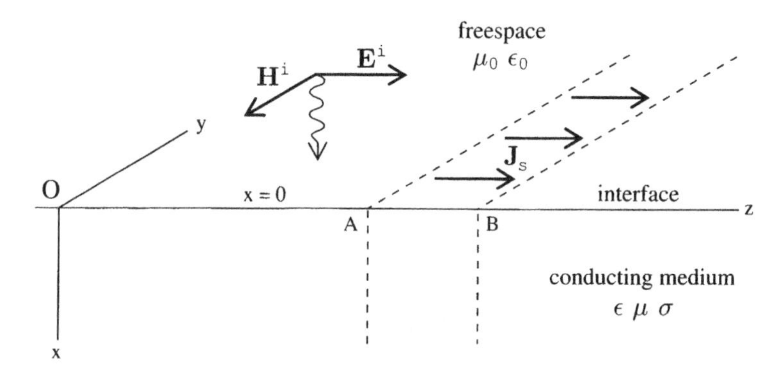

Figure 3.9 Plane wave normally incident on an interface between free space and a conducting medium.

It is clear that the wave will propagate into the conducting medium, with the electric field $E_z = E_0$ at $x = 0$. The geometry also indicates that $\partial/\partial y = d/dz \equiv 0$ in the conducting medium. It is clear that the wave will propagate into the conducting medium, with the electric field E_z. Thus, the E field within the conducting medium satisfies the wave equation

$$\frac{d^2E_z}{dx^2} - \gamma^2 E_z = 0 \tag{3.227}$$

with

$$\gamma = \alpha + j\beta = \frac{1+j}{\delta}$$

given by (3.221).

We transform (3.227) into a corresponding equation to the induced current density **J** in the conducting medium as

$$\mathbf{J} = \sigma \mathbf{E} \text{ or } J_z = \sigma E_z \tag{3.228}$$

in the present case. Assuming σ to be constant, we obtain from (3.227) and (3.228),

$$\frac{d^2 J_z}{\partial x^2} - \gamma^2 J_z = 0. \tag{3.229}$$

The general solution of (3.229) is

$$J_z = A_1 e^{\alpha x} e^{j\beta x} + A_2 e^{-\alpha x} e^{-j\beta x}. \tag{3.230}$$

Discarding the wave traveling in the $-x$ direction, we obtain from (3.230),

$$J_z = J_0 e^{-\alpha x} e^{-j\beta x}, \tag{3.231a}$$

where

$$J_0 = J_x|_{\text{at } x=0} = \sigma E_0. \tag{3.231b}$$

Using the values of α and β from (3.221), we obtain from (3.231a)

$$\begin{aligned} J_z &= J_0 e^{-x/\delta} e^{-jx/\delta} \\ &= \sigma E_0 e^{-x/\delta} e^{-jx/\delta}, \end{aligned} \tag{3.232}$$

which indicates that the current density in $x \geq 0$ induced by the progressing wave decays experientially in the $x \geq 0$ direction. It should be noted that the amplitudes of the E and H fields associated with the progressing wave in the conducting medium both decay similarly. Observe that $J_z = J_0 = \sigma E_0$ is maximum at $x = 0$, that is, on the skin of the conducting medium, (3.232) indicates that at $x = \delta$, $J_z = J_0/e$ so the current density is reduced by $1/e$, or by about 37% or 8.69 dB, at distance $x = \delta$ from the surface. This distance δ is defined as the skin depth for the conducting medium is

$$\delta = \sqrt{\frac{2}{\omega\mu\sigma}} = \frac{1}{\sqrt{\pi f \mu \sigma}} = \frac{1}{\sqrt{\pi f \mu_0 \sigma}}, \tag{3.233}$$

where we have assumed that $\mu = \mu_0$ for the medium. Similar effects take place during propagation of a plane wave through a lossy dielectric where the

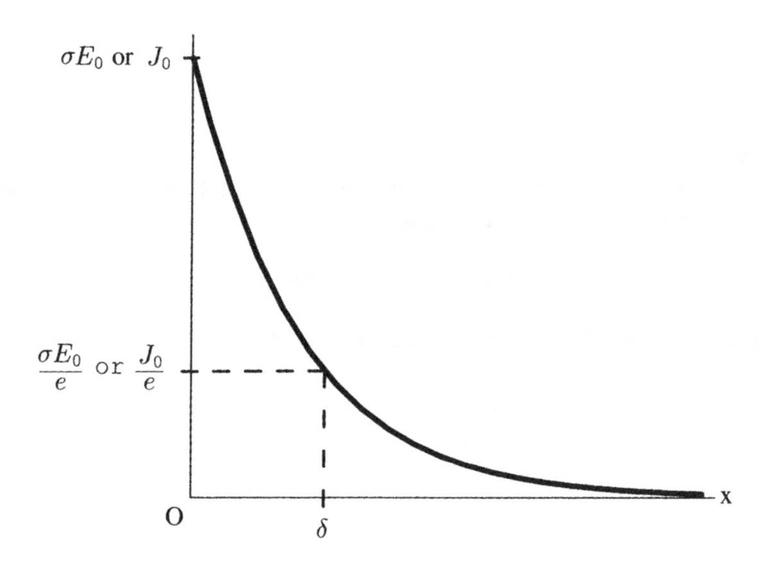

Figure 3.10 $J_x(\sigma E_x)$ versus x inside the conducting medium.

field amplitudes decay in a similar manner, and the depth of penetration (see (3.220)) given there should be compared with δ.

Figure 3.10 gives a plot of J_z (or $\sigma\ E_x$) in relation to x indicates. As the plot indicates, all field quantities tend to be confined near the interface when δ is small and particularly when σ is very large. Similar effects can occur for a lossy dielectric medium at sufficiently high frequencies. We now determine the total current is the slab section AB in Figure 3.10, per unit width in the y direction and extending from $x = 0$ to $x = \infty$, given by

$$\begin{aligned} I_z &= \int\int_S \mathbf{J} \cdot d\mathbf{s} \\ &= \int_0^1 \int_0^\infty J_0 e^{-x/\delta} e^{-jx/\delta} \cdot \hat{z} \cdot \hat{z}\ dx\ dy \\ &= \frac{\sigma \delta E_0}{1+j}, \end{aligned} \tag{3.234}$$

where we have used $J_0 = \sigma E_0$, E_0 being the tangential electric field at the surface $x = 0$. We define the slab independence per unit length in the z direction and per unit width in the y direction (see Figure 3.9) as:

$$\text{slab impedance} = \frac{\text{tangential } E\text{-field (at the surface)}}{\text{tangential current (i.e., } z\text{-direction) through the slab section}}.$$

Alternatively

$$\text{slab impedance} = \frac{E_0}{I_z} = \frac{1+j}{\sigma\delta} . \tag{3.235}$$

We note in (3.235) that E_0 is the electric field parallel to, and at the surface of, the conductor (i.e. tangential E field at $x = 0$). Since the current is mostly confined near the surface (Figure 3.10), we may represent the total current I_z per unit width by an equivalent surface current J_s per unit width (Figure 3.10) Now the surface impedance Z_s of an imperfectly conducting surface is defined by

$$Z_s = \frac{E_t}{J_s}, \tag{3.236}$$

where E_t is the electric field strength parallel to and at the surface of the conductor and J_s is the surface current density caused by E_t.
Thus from (3.235) and (3.236) we obtain

$$Z_s = R_s + jX_s = \frac{1+j}{\sigma\delta} \qquad \Omega \text{ per square}, \tag{3.237}$$

where R_s, X_s refer to the surface resistance and reactance, respectively, for the surface, and Ω per square means Ω per unit length in the direction of current flow and unit width means across the direction of current flow on the surface.

The skin effect impedance (surface impedance) has the dimensions of ohms per square. Its value does not depend on the units used to measure length and width as long as they are the same [2].

The vector form of (3.236) can be written as

$$\mathbf{E}_t = Z_s \mathbf{J}_s, \tag{3.238}$$

which is the same as the impedance boundary condition (3.121)–(3.122). The surface impendence Z_s can be used to approximate the impedance of various configurations made with lossy materials, including any good conductors, so long as the radius of curvature of the surface is several skin depths. For example, for a rectangular plate carrying a current density J, the impedance Z between the terminals (Figure 3.11) is

$$Z = R + jX_L = \frac{Z_s \ell}{w} \tag{3.239}$$

$$R_s = \frac{1}{\sigma\delta}\frac{\ell}{w} \tag{3.240a}$$

$$X_s = \frac{1}{\sigma\delta}\frac{\ell}{w} . \tag{3.240b}$$

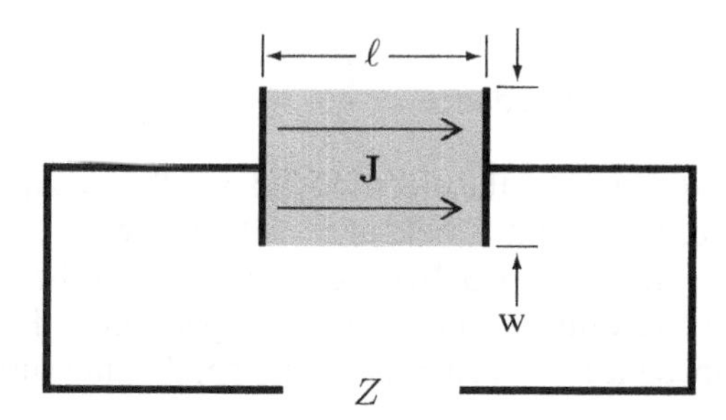

Figure 3.11 Current through a conducting rectangular plate.

3.6.8 Polarization of Plane Waves

Polarization of a plane electromagnetic wave indicates the time behavior of its electric vector in space when viewed in the direction of the wave propagation. Let the electric field of a plane wave propagating in the z direction be represented at any time t in a simple medium by

$$\begin{aligned}\boldsymbol{\mathcal{E}}(z,t) &= \hat{x}\mathcal{E}_x + \hat{y}\mathcal{E}_y \\ &= \hat{x}E_1\cos(\omega t - \beta z) + \hat{y}E_2\cos(\omega t - \beta z + \alpha), \end{aligned} \tag{3.241a}$$

where it is assumed that the phase of the component $\mathcal{E}_y$ leads that of the component $\mathcal{E}_x$ by an amount amount α, and E_1, E_2 are the amplitudes of the x and y, respectively. The other parameters in (3.241a) are as defined earlier. The phasor equivalent of $\boldsymbol{\mathcal{E}}(z,t)$ is

$$\mathbf{E} = \hat{x}E_1e^{-j\beta z} + \hat{y}E_2e^{-j\beta z + j\alpha}. \tag{3.241b}$$

At any wave plane, $\boldsymbol{\mathcal{E}}(z,t) \equiv \boldsymbol{\mathcal{E}}$ can be represented by the vector diagram shown in Figure 3.12.

For simplicity we assume $z = 0$. It can be shown that in the wave plane (in this case xy plane)

$$\frac{\mathcal{E}_y^2}{E_2^2} - \frac{2\mathcal{E}_y\mathcal{E}_x}{E_1E_2}\cos\alpha + \frac{\mathcal{E}_x^2}{E_1^2} = \sin^2\alpha \tag{3.242}$$

$$\tan\theta = \frac{\mathcal{E}_y}{\mathcal{E}_x} = \frac{E_2\cos(\omega t + \alpha)}{E_1\cos\omega t}. \tag{3.243}$$

Equation (3.242) is an equation of an ellipse in the $\mathcal{E}_x, \mathcal{E}_y$ plane that indicates that any harmonic wave is elliptically polarized in general. The plane of the ellipse and its shape and orientation in that plane are independent of time. Also note that the direction of $\boldsymbol{\mathcal{E}}$ changes with time (see (3.243)) as its tip moves around the ellipse.

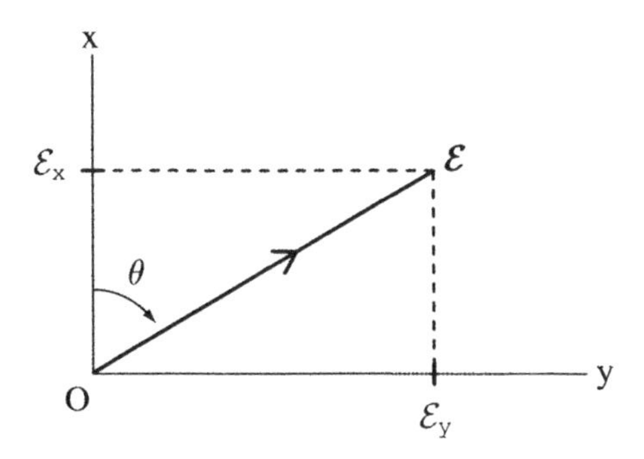

Figure 3.12 Vector Diagram showing the components of E at any (z, t).

We now consider a few special cases:

Case 1. $\alpha = 0$, meaning $\mathcal{E}_x$, $\mathcal{E}_y$ are in phase and

$$\tan\theta = \frac{\mathcal{E}_y}{\mathcal{E}_x} = \frac{E_2}{E_1}$$

is independent of time. The above yields linear polarization, which occurs at any time the E vector moves along a fixed line. For example, $E_2 = 0$ gives $\theta = 0$, which is x polarization; $E_1 = 0$ gives $\theta = \pi/2$, which is y polarization.

Case 2. $\alpha = \pm\frac{\pi}{2}$, meaning $\mathcal{E}_x$ and $\mathcal{E}_y$ are orthogonal in time and we also assume that they are of equal amplitude, which means $E_1 = E_2 = E_0$.

Under the conditions above, we have from (3.241a), (3.241b), and (3.242)

$$\frac{\mathcal{E}_y^2}{E_0^2} + \frac{\mathcal{E}_x^2}{E_0^2} = 1 \tag{3.244}$$

and

$$\begin{aligned} \theta &= -\omega t \text{ for } \alpha = \frac{\pi}{2} \\ &= +\omega t \text{ for } \alpha = -\frac{\pi}{2}\,. \end{aligned} \tag{3.245}$$

Equations (3.244) and (3.245) show that for both $\alpha = \pi/2$ and $-\pi/2$ the tip of the electric vector moves along a circle, meaning the wave is circularly polarized in both cases. The two cases are represented diagrammatically in Figure 3.13.

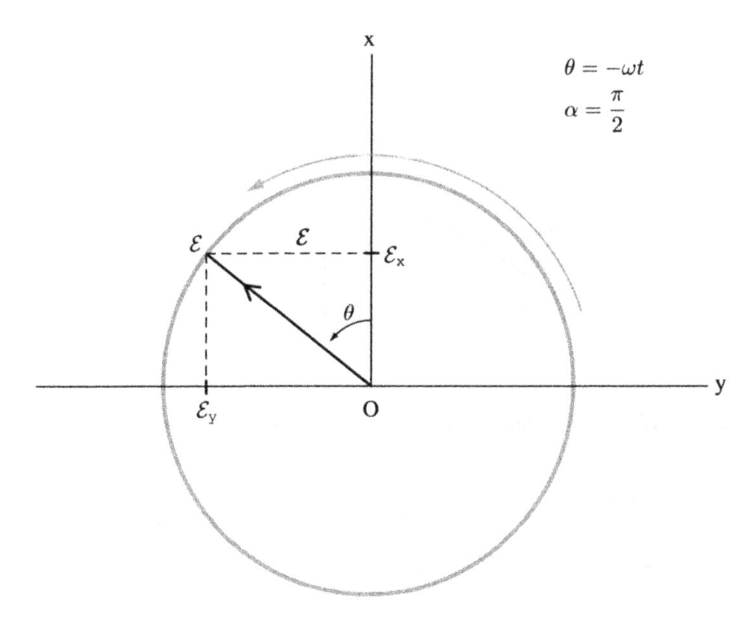

(a) lefthanded or CCW polarization

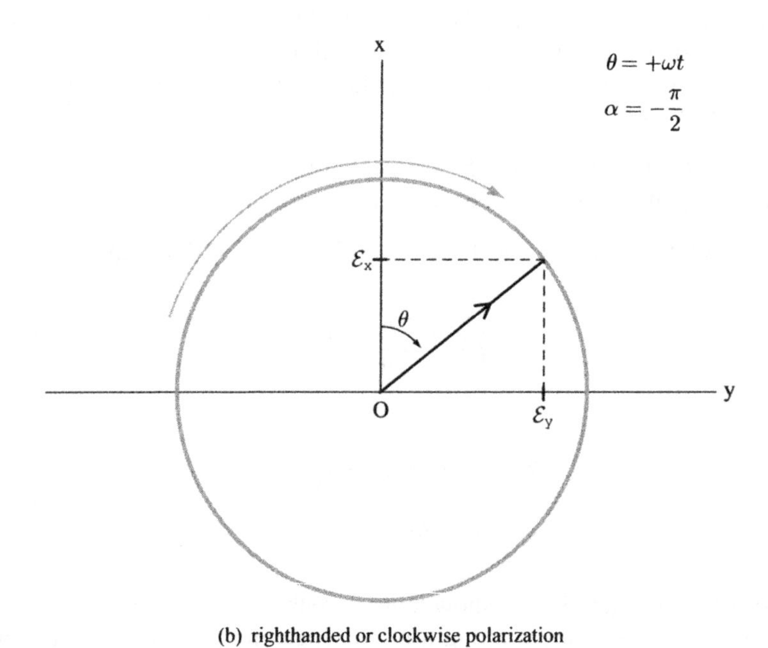

(b) righthanded or clockwise polarization

Figure 3.13 Polarization of a plane wave propagating in the z-direction.

Viewed in the direction of the wave propagation, which is through the positive z direction, we find that the tip of the $\mathcal{E}$-vector moves anticlockwise along the circle for $\alpha = \pi/2$ and clockwise for $\alpha = -\pi/2$. Thus, according to IEEE standards, Figure 3.13a represents anticlockwise or a left-handed polarization and case Figure 3.13b represents clockwise or a right-handed polarization. This helps us to understand how to generate circular polarization. We need to combine two linearly polarized plane waves orthogonal in space (i.e., E vectors perpendicular to each other) and in time (E vectors differing in phase by $\pi/2$). In general, for $E_1 \neq E_2$, $\alpha \neq 0$ elliptic polarization is obtained; negative and positive values of α provide left-handed and right-handed polarizations, respectively. Polarizations of waves are discussed in more detail in [4, 8]

We have described the right-hand and left-hand circular polarizations as recommended by the IEEE. However, in physics, right-hand polarization refers to the wave where the E vector rotates in the clockwise directions when observed by the receiver, or as the wave approaching. The engineering definitions of right hand is based on the rotation of the E vector as viewed by the transmitter, meaning when the wave is receding. Therefore the right polarization in engineering is left polarization in physics, and vice versa.

3.7 REFLECTION AND REFRACTION (TRANSMISSION) OF PLANE WAVES

In this section we describe the basic aspects of reflection and refraction or transmission on effects suffered by plane electromagnetic waves at a planar interface between two different semi-infinite, homogeneous, linear, and isotropic media. We describe some selected topics having application to radio wave propagation, operation of wave guiding structures and of antennas above the earth. Further discussion in this area may be found in [2–4, 7].

3.7.1 Normal Incidence on a Plane Interface

Figure 3.14 shows a plane interface between two semi-infinite media characterized by μ_1, ϵ_1, σ_1 and μ_2, ϵ_2, σ_2, respectively. A plane wave in medium 1 propagating in the $\hat{z}$ direction in incident normally at the interface as shown. The incident wave suffers some reflection and transmission at the interface. Symmetry indicates that the reflected and transmitted waves will also be plane waves with polarization similar to that of the incident wave. The reflected wave in medium 1 and the transmitted wave in medium 2 will travel in the negative and positive z direction, as shown in Figure 3.14.

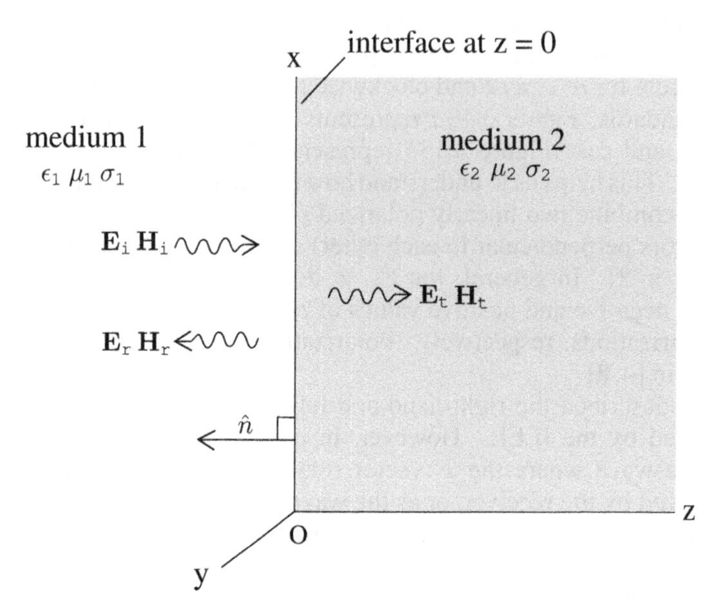

Figure 3.14 Geometry of the incident, reflected, and transmitted waves at an interface between two media.

The incident, reflected, and transmitted waves are represented as follows

$$\text{incident wave} = \begin{cases} \mathbf{E}_i = \hat{x} E_i e^{-d_1 z} \\ \mathbf{H}_i = \hat{y}\, \frac{E_i}{\eta_1}\, e^{-d_1 z} \end{cases} \tag{3.246a}$$

$$\text{reflected wave} = \begin{cases} \mathbf{E}_r = \hat{x} E_r e^{d_1 z} \\ \mathbf{H}_r = -\hat{y}\, \frac{E_r}{\eta_1}\, e^{d_1 z} \end{cases} \tag{3.246b}$$

$$\text{transmitted wave} = \begin{cases} \mathbf{E}_t = \hat{x} E_t e^{-d_2 z} \\ \mathbf{H}_t = \hat{y}\, \frac{E_t}{\eta_2}\, e^{-d_2 z} \end{cases} \tag{3.246c}$$

where the reversal of $\mathbf{H}_r$ in the $-\hat{y}$ direction is needed to ensure propagation in the negative z direction. The propagations constants γ_1, γ_2 and the intrinsic impedances η_1, η_2 for the two media are given by

$$\gamma_1 = \alpha_1 + j\beta_1 = \sqrt{j\omega\mu_1(\sigma_1 + j\omega\epsilon_1)} \tag{3.247a}$$

$$\eta_1 = \sqrt{\frac{j\omega\mu_1}{\sigma_1 + j\omega\epsilon_1}}, \qquad \eta_1 = |\eta_1| e^{j\theta_{\eta_1}}, \tag{3.247b}$$

$$\gamma_2 = \alpha_2 + j\beta_2 = \sqrt{j\omega\mu_2(\sigma_2 + j\omega\epsilon_2)} \tag{3.248a}$$

$$\eta_2 = \sqrt{\frac{j\omega\mu_2}{\sigma_2 + j\omega\epsilon_2}}, \eta_2 = |\eta_2|e^{j\theta_{\eta 2}}. \tag{3.248b}$$

Note that the parameters above were introduced earlier (see Section 3.6.6), so they will not be discussed further. It is assumed that all the parameters given by (3.247a), (3.247b) and (3.248a), (3.248b) characterizing the two media are known.

In the set of equations (3.246a)–(3.246c), E_{i} and H_{i} are known. We are interested in knowing E_{r}, E_{t} because this will enable us to determine $\mathbf{E}_{\mathrm{r}}$, $\mathbf{H}_{\mathrm{r}}$ and $\mathbf{E}_{\mathrm{t}}, \mathbf{H}_{\mathrm{t}}$. We apply the boundary conditions, that the tangential components of $\mathbf{E}$ and $\mathbf{H}$ are continuous at the interface $z = 0$. Then, using (3.246a) and (3.246b), we obtain

$$E_{\mathrm{i}} + E_{\mathrm{r}} = E_{\mathrm{t}} \tag{3.249}$$

$$\frac{E_{\mathrm{i}}}{\eta_1} - \frac{E_{\mathrm{r}}}{\eta_1} = \frac{E_{\mathrm{t}}}{\eta_2}. \tag{3.250}$$

The reflection and transmission of the incident waves are generally characterized by the voltage reflection and transmission coefficients, defined by $\Gamma = E_{\mathrm{r}}/E_{\mathrm{i}}$ and $T = E_{\mathrm{t}}/E_{\mathrm{i}}$, respectively. It is straightforward to determine Γ and T from (3.246a)–(3.246c), and they are given by

$$\Gamma = \frac{E_{\mathrm{r}}}{E_{\mathrm{i}}} = \frac{\eta_2 - \eta_1}{\eta_2 + \eta_1} \tag{3.251a}$$

$$T = \frac{E_{\mathrm{t}}}{E_{\mathrm{i}}} = \frac{2\eta_2}{\eta_2 + \eta_1}. \tag{3.251b}$$

Note that both Γ and T can be complex quantities in general, and it follows from (3.251a), (3.251b) that

$$1 + \Gamma = T. \tag{3.251c}$$

We now use the complex Poynting vector

$$\mathbf{S}_{\mathrm{av}} = \mathrm{Re}\left[\frac{\mathbf{E} \times \mathbf{H}^*}{2}\right]$$

to determine the average power carried by the three waves. The average power $\mathbf{S}_{\mathrm{av,i}}$ carried by the incident wave (3.246a) is

$$\begin{aligned}\mathbf{S}_{\mathrm{avi}} &= \mathrm{Re}\left[\frac{\mathbf{E}_{\mathrm{i}} \times \mathbf{H}_{\mathrm{i}}^*}{2}\right] \\ &= \frac{1}{2}|E_{\mathrm{i}}|^2 e^{-2\alpha_1 z}\,\mathrm{Re}\left(\frac{1}{\eta_1^*}\right) \\ &= \frac{1}{2}\frac{|E_{\mathrm{i}}|^2}{|\eta_1|}\cos\theta_{\eta_1}\, e^{-2\alpha_1 z}, \mathrm{W/m^2}, \end{aligned} \tag{3.252a}$$

where it should be noted that

$$\mathrm{Re}\left(\frac{1}{\eta_1^*}\right) \neq \frac{1}{\mathrm{Re}(\eta_1^*)}.$$

Proceeding similarly, it can be shown that the average powers carried by the reflected and transmitted waves are

$$\mathbf{S}_{\mathrm{avr}} = -\frac{|\Gamma|^2|E_{\mathrm{i}}|^2}{2|\eta_1|}\cos\theta_{\eta_1}\, e^{+2\alpha_1 z}\hat{z} \quad \mathrm{W/m^2} \tag{3.252b}$$

$$\mathbf{S}_{\mathrm{avt}} = \frac{1}{2}\frac{|T|^2|E_{\mathrm{i}}|^2}{2|\eta_2|}\cos\theta_{\eta_2}\, e^{-2\alpha_2 z}\hat{z} \quad \mathrm{W/m^2}, \tag{3.252c}$$

where we have used Γ and T to eliminate E_{r} and E_{t} from (3.252a) and (3.252c), respectively. The above are the most general results. We now apply them to some special cases.

Special Cases

Case 1. **Interface between Two Perfect Dielectric Media**

We have $\sigma_1 = \sigma_2 \equiv 0$ and $\theta_{\eta_1} = \theta_{\eta_2} = 0$. That is, the η_1 and η_2 are real quantities and are

$$\eta_1 = \sqrt{\frac{\mu_1}{\epsilon_1}} \qquad \text{and} \qquad \eta_2 = \sqrt{\frac{\mu_2}{\epsilon_2}}.$$

Using the relations (3.252a)–(3.252c) and (3.251a)–(3.251c), we show that

$$\begin{aligned}|\mathbf{S}_{\mathrm{avr}}| + |\mathbf{S}_{\mathrm{avt}}| &= |\mathbf{S}_{\mathrm{avi}}|\left[\Gamma^2 + \frac{\eta_1}{\eta_2}T^2\right] \\ &= |\mathbf{S}_{\mathrm{avi}}|, \end{aligned} \tag{3.253}$$

which indicates power balance, meaning the power in the reflected and transmitted waves is equal to that in the incident wave.

If the dielectric constants of the two media are ϵ_{r_1} and ϵ_{r_2}, then the reflection and transmission coefficients from (3.251a)–(3.251c) are

$$\Gamma = \frac{\sqrt{\epsilon_{r_1}} - \sqrt{\epsilon_{r_2}}}{\sqrt{\epsilon_{r_1}} + \sqrt{\epsilon_{r_2}}} \tag{3.254a}$$

$$T = \frac{2\sqrt{\epsilon_{r_1}}}{\sqrt{\epsilon_{r_1}} + \sqrt{\epsilon_{r_2}}} . \tag{3.254b}$$

If we assume medium 1 to be air, $\epsilon_{r_1} = 1$, and for the conducting medium 2, $\epsilon_{r_2} = \epsilon_r$, then we obtain the useful relations

$$\Gamma = \frac{1 - \sqrt{\epsilon_r}}{1 + \sqrt{\epsilon_r}} \tag{3.255a}$$

$$T = \frac{2}{1 + \sqrt{\epsilon_r}} . \tag{3.255b}$$

Case 2. **Interface between a Dielectric and a Conductor**

If medium 1 is free space and medium 2 is a perfect conductor ($\sigma_2 \equiv$ inf), then $\eta_1 = \eta_0$ and $\eta_2 \equiv 0$. It follows from (3.251a)–(3.251c) that

$$\Gamma = -1, T = 0, \tag{3.256}$$

which means that the $E_r = -E_i$ and $H_r = H_i$ fields are not transmitted. Thus we find that with the field (3.246a) incident on the perfectly conducting medium, the tangential components of the electric and magnetic fields in the interface are

$$E_{\text{TAN}} = 0 \tag{3.257}$$

$$H_{\text{TAN}} = 2H_i, \tag{3.258}$$

where in the present case, H_i, is the magnetic field of the incident wave tangent to the surface. When medium 2 is imperfectly conducting and medium 1 is free space, we have $\sigma_2 = \infty$, and $\eta_1 = \eta_0 \gg \eta_2$. It follows from (3.251a)–(3.251c) that

$$\Gamma \simeq -1, T \simeq \frac{2\eta_2}{\eta_1} . \tag{3.259}$$

Using the value above for T, we obtain

$$E_t = TE_i \simeq \frac{2\eta_2}{\eta_1} E_i \tag{3.260a}$$

or

$$\eta_2 H_t \simeq \frac{2\eta_2}{\eta_1} \eta_1 H_i, \qquad (3.260b)$$

which yields

$$H_{TAN} = H_t \simeq 2H_i. \qquad (3.261)$$

From the results above we observe that the tangential component of **H** at the interface is approximately twice the tangential component of the incident magnetic field at the surface. For the perfectly conducting case, all the incident power is reflected back to medium 1, while for the imperfectly conducting case, a certain amount of power will be lost as Joule heating in the conductor. We can estimate the lost power by using the complex Poynting vectors (3.132) along with the impendence boundary conditions (see (3.121)). In the present case the power lost per unit area is given by

$$P_{av} = \mathbf{S}_{av} \cdot \hat{z}, \mathbf{S}_{av} = \frac{1}{2} \operatorname{Re}(\mathbf{E} \times \mathbf{H}^*) \cdot \hat{z} \quad \text{W/m}^2, \qquad (3.262)$$

where **E**, **H** are the fields at the interface.

Defining a unit normal $\hat{n} = -\hat{z}$, directed from the interface toward the free space side (see Figure 3.14), we have the following expression for the power loss per unit area:

$$S_{av} = -1/2 \operatorname{Re}[\mathbf{E} \times \mathbf{H}^*] \cdot \hat{n}. \qquad (3.263)$$

After using the vector identity $\mathbf{a} \times \mathbf{b} \cdot \mathbf{c} = \mathbf{c} \times \mathbf{a} \cdot \mathbf{b} = \mathbf{b} \times \mathbf{c} \cdot \mathbf{a}$ and the impendence boundary relation (3.121), it can be shown that average power lost per unit area by the imperfectly conducting surface is

$$S_{av} = 1/2\, |\mathbf{J}_s|^2 \cdot \operatorname{Re}(Z_s) \text{or} 1/2\, |\mathbf{H}_{TAN}|^2 \operatorname{Re}(Z_s) \quad \text{W/m}^2 \qquad (3.264)$$

where, as defined earlier,

$$\mathbf{J}_s = \hat{n} \times \mathbf{H} \quad \text{A/m}$$

is the surface current induced an the interface by the normally incident wave and Z_s is the surface impendence of the interface, also defined earlier as

$$Z_s = \eta_2 \simeq \sqrt{\frac{j\omega\mu_0}{\sigma}} = \sqrt{\frac{\omega\mu}{2\sigma}}\,(1+j)$$
$$= \frac{1+j}{\sigma\delta} = R_s + jx_s \quad \Omega/\text{square}\,.$$

Here R_s, X_s are the surface resistance and surface reactance in Ω/square for the imperfectly conducting surface, and $\mathbf{H}_{TAN}$ is the component of the incident magnetic field, tangential to the interface (3.261). Both expressions for S_{av} can be used to estimate the power loss in a conducting surface illuminated by a plane wave.

Example 3.5

Derive equation (3.264).

Solution

We start from (3.263) and write

$$S_{av} = -\frac{1}{2}\,\text{Re}[\mathbf{E} \times \mathbf{H}^*] \cdot \hat{n},$$

Then we apply the identity $\mathbf{a} \times \mathbf{b} \cdot \mathbf{c} = \mathbf{c} \times \mathbf{a} \cdot \mathbf{b} = \mathbf{b} \cdot \mathbf{c} \times \mathbf{a}$ and obtain

$$\begin{aligned}
S_{av} &= -1/2\ \text{Re}[\hat{n} \times \mathbf{E} \cdot \mathbf{H}^*] \\
&= -1/2\ \text{Re}[Z_s \times \mathbf{J}_s \cdot \mathbf{H}^*] \text{by impedance boundary condition (3.121)} \\
&= 1/2\ \text{Re}[Z_s \mathbf{J}_s \cdot \hat{n} \times \mathbf{H}^*] \\
&= 1/2\ \text{Re}[Z_s \mathbf{J}_s \cdot \mathbf{J}_s^*] \text{or} 1/2\ \text{Re}[Z_s(\hat{n} \times \mathbf{H}) \cdot (\hat{n} \times \mathbf{H}^*)] \\
&\qquad \text{use of } \mathbf{J}_s = \hat{n} \times \mathbf{H} \\
&= 1/2\ \text{Re}[Z_s] |\mathbf{J}_s|^2 \text{or} 1/2\ \text{Re}(Z_s)\ |\mathbf{H}_{TAN}|^2 \\
&\qquad \text{use of } \mathbf{H}_{TAN} = \hat{n} \times \mathbf{H}.
\end{aligned}$$

3.7.2 Oblique Incidence

We consider the reflection and refraction of linearly polarized uniform plane waves obliquely incident an a plane interface between two semi-infinite, linear, homogeneous, and isotropic lossless dielectric media. The geometry and the coordinate systems used are shown in Figure 3.15. As the figure shows, the incident ray along a plane wave propagating in the $\hat{\beta}_i$ direction is incident on the plane interface ($z = 0$) at an angle θ_i and is reflected and refracted at angles θ_r and θ_t, respectively. All angles are measured from the $\hat{n}$, which is also the unit positive normal $\hat{n}$, $\hat{\beta}_i$, $\hat{\beta}_r$, and $\hat{\beta}_t$ to the interface.

From optics we define the *plane of incidence* as the plane containing the incident ray and the normal to the interface surface $\hat{n}$; in this case yz is the plane of incidence. It is assumed that β_i, β_r, and β_t are in the plane of incidence (i.e., the incident), since reflected and refracted (transmitted) rays

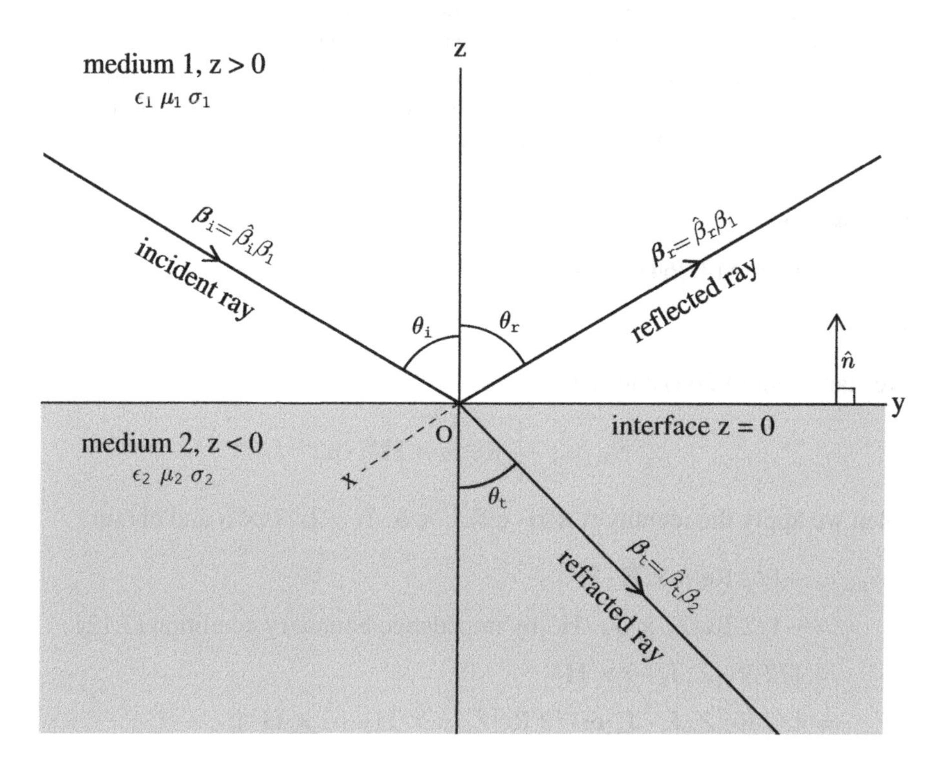

Figure 3.15 Geometry of the refection and refraction of waves at the plane interface between two semi-infinite media.

are all in the plane of incidence. Note that β_{i}, β_{r}, and β_{t} are the phase (in this case propagation) constants for the incident, reflected, and refract wave, respectively.

We will consider the following two cases involving the polarization of the incident wave:

Case 1. The E-field vector is perpendicular to the plane of incidence, which is where the electric vector is parallel to the boundary or the interface surface. In the radio wave propagations area where the flat earth surface acts as the interface, the perpendicular polarization is also called the horizontal polarization.

Case 2. The E-field vector is parallel to the plane of incidence (or the H-field vector is perpendicular to the plane of incidence), and in the radio propagations area this case is referred to as vertical polarizations. It should be noted that although the E-field vector is in the plane of incidence, it is not necessarily in the vertical direction with respect to the horizontally oriented interface. We will consider the two cases separately.

The reflection and transmission coefficients for plane waves discussed here are commonly known as Fresnel coefficients in the physics community, in honor of Fresnel who originally derived them for optical waves.

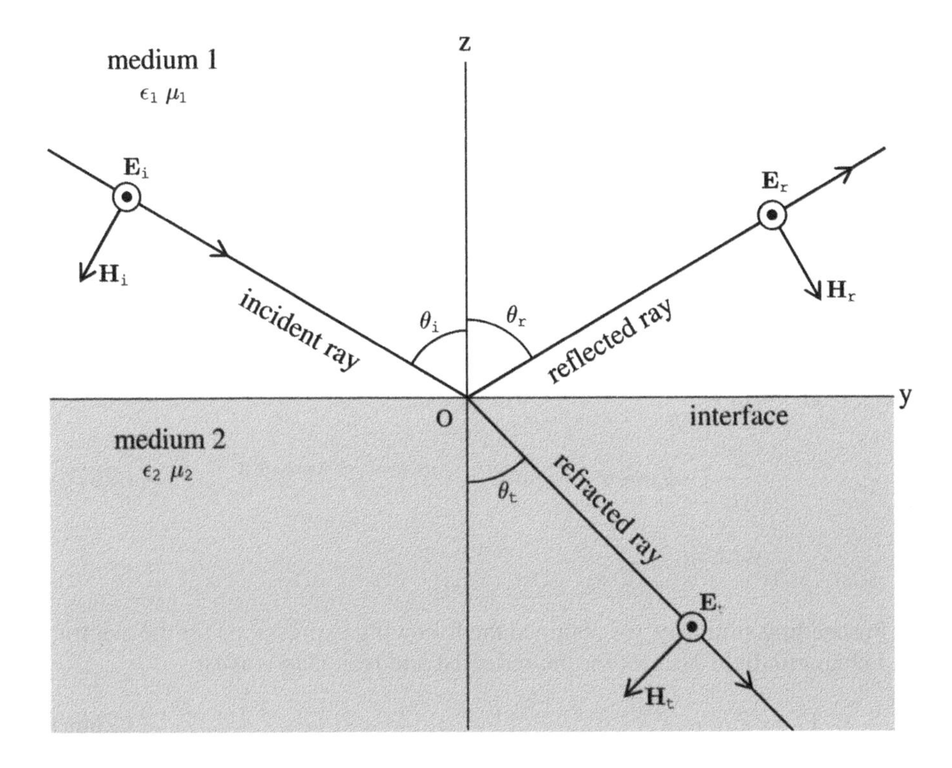

Figure 3.16 Incident reflected and refracted rays and the orientation of the **E** and **H** fields for perpendicular polarization.

Perpendicular or Horizontal Polarization

The orientation of the electric and magnetic field vectors along the incident, reflected and refracted rays are sketched in Figure 3.16. There it can be seen that all the electric vectors are perpendicular to the plane of the paper (which is the plane of incidence) and are chosen to be directed toward the reader. The magnetic vectors in the plane of incidence are then as shown so that the energy flow for each ray is in its respective direction.

Explicit expressions for the $\mathbf{E}$ and $\mathbf{H}$ vectors along the various rays can be obtained by using (3.204a), (3.204b) in Section 3.6.5. With the given polarization of the electric field, the incident electric field can be assumed as

$$\mathbf{E}^{\mathrm{i}} = \hat{x}E^{\mathrm{i}}e^{-j\boldsymbol{\beta}_{\mathrm{i}}\cdot\mathbf{r}}, \tag{3.265}$$

where

$$\boldsymbol{\beta}_{\mathrm{i}} = \hat{\beta}_{\mathrm{i}}\beta_1$$

with

$$\begin{aligned}\hat{\beta}_{\mathrm{i}} &= \hat{y}\sin\theta_{\mathrm{i}} - \hat{z}\cos\theta_{\mathrm{i}} \\ \beta_{\mathrm{i}} &= \omega\sqrt{\mu_1\epsilon_1} = \beta_1 \\ \boldsymbol{\beta}_{\mathrm{i}}\cdot\mathbf{r} &= \beta_1 y\sin\theta_{\mathrm{i}} - \beta_1 z\cos\theta_{\mathrm{i}}.\end{aligned} \tag{3.266}$$

Thus we obtain the incident wave,

$$\mathbf{E}^{\mathrm{i}} = \hat{x}E^{\mathrm{i}}e^{-j\beta_1(y\sin\theta_{\mathrm{i}} - z\cos\theta_{\mathrm{i}})} \tag{3.267a}$$

$$\begin{aligned}\mathbf{H}^{\mathrm{i}} &= \hat{\beta}_{\mathrm{i}} \times \frac{\mathbf{E}^{\mathrm{i}}}{\eta_1} \\ &= \frac{E^{\mathrm{i}}}{\eta_1}(-\hat{y}\cos\theta_{\mathrm{i}} - \hat{z}\sin\theta_{\mathrm{i}})e^{-j\beta_1(y\sin\theta_{\mathrm{i}} - z\cos\theta_{\mathrm{i}})} \\ &\text{with } \eta_1 = \sqrt{\frac{\mu_1}{\epsilon_1}}\,.\end{aligned} \tag{3.267b}$$

Proceeding similarly, we obtained the following expressions for the electric and magnetic field vectors for the reflected and refracted waves:

$$\mathbf{E}^{\mathrm{r}} = \hat{x}E^{\mathrm{r}}e^{-j\beta_1(y\sin\theta_{\mathrm{r}} + z\cos\theta_{\mathrm{r}})} \tag{3.268a}$$

$$\mathbf{H}^{\mathrm{r}} = \frac{E^{\mathrm{r}}}{\eta_1}(\hat{y}\cos\theta_{\mathrm{r}} - \hat{z}\sin\theta_{\mathrm{r}})e^{-j\beta_1(y\sin\theta_{\mathrm{r}} + z\cos\theta_{\mathrm{r}})}, \tag{3.268b}$$

$$\mathbf{E}^{\mathrm{t}} = \hat{x}E^{\mathrm{t}}e^{-j\beta_2(y\sin\theta_{\mathrm{t}} - z\cos\theta_{\mathrm{t}})} \tag{3.269a}$$

$$\begin{aligned}\mathbf{H}^{\mathrm{t}} &= \frac{E^{\mathrm{t}}}{\eta_2}(-\hat{y}\cos\theta_{\mathrm{t}} - \hat{z}\sin\theta_{\mathrm{t}})e^{-j\beta_2(y\sin\theta_{\mathrm{t}} + z\cos\theta_{\mathrm{t}})} \\ &\text{with } \beta_2\omega\sqrt{\mu_2\epsilon_2} \text{ and } \eta_2 = \sqrt{\frac{\mu_2}{\epsilon_2}}\,.\end{aligned} \tag{3.269b}$$

To determine the unknown quantities E^{r}, E^{t} in (3.268a), (3.268b) and (3.269a), (3.269b) in terms of E^{i}, we apply the boundary conditions that the tangential components of **E** and **H** are continuous at the interface ($z = 0$), which yields

$$E^{\mathrm{i}} e^{-j\beta_1 y \sin\theta_{\mathrm{i}}} + E^{\mathrm{r}} e^{-j\beta_1 y \sin\theta_{\mathrm{r}}} = E^{\mathrm{t}} e^{-j\beta_2 y \sin\theta_{\mathrm{t}}} \tag{3.270a}$$

$$-\frac{E^{\mathrm{i}}}{\eta_1}\cos\theta_{\mathrm{i}}\, e^{-j\beta_1 y \sin\theta_{\mathrm{i}}} + \frac{E^{\mathrm{r}}}{\eta_1}\cos\theta_{\mathrm{r}}\, e^{-j\beta_1 y \sin\theta_{\mathrm{r}}} = -\frac{E^{\mathrm{t}}}{\eta_2}\cos\theta_{\mathrm{t}}\, e^{-j\beta_2 y \sin\theta_{\mathrm{t}}}. \tag{3.270b}$$

The two relations in (3.270a), (3.270b) must be true for all values of y, which requires

$$\beta_1 y \sin\theta_{\mathrm{i}} = \beta_1 y \sin\theta_{\mathrm{r}} = \beta_2 y \sin\theta_{\mathrm{t}}. \tag{3.271}$$

It now follows from (3.271) that

$$\theta_{\mathrm{i}} = \theta_{\mathrm{r}} \tag{3.272}$$

$$\beta_1 \sin\theta_{\mathrm{i}} = \beta_2 \sin\theta_{\mathrm{t}}. \tag{3.273}$$

After using (3.272) and (3.273), we obtain from (3.270a), (3.270b),

$$E^{\mathrm{i}} + E^{\mathrm{r}} = E^{\mathrm{t}} \tag{3.274a}$$

$$-\frac{E^{\mathrm{i}}}{\eta_1}\cos\theta_{\mathrm{i}} + \frac{E^{\mathrm{r}}}{\eta_1}\cos\theta_{\mathrm{i}} = -\frac{E^{\mathrm{t}}}{\eta_2}\cos\theta_{\mathrm{t}}. \tag{3.274b}$$

Using the relations (3.274a), (3.274b), we define the reflection and transmission coefficients $\Gamma_\perp$ and $\tau_\perp$, respectively, for the case of perpendicular polarization as

$$\Gamma_\perp = |\Gamma_\perp| e^{j\phi_\perp} = \frac{E^{\mathrm{r}}}{E^{\mathrm{i}}} = \frac{\eta_2\cos\theta_{\mathrm{i}} - \eta_1\cos\theta_{\mathrm{t}}}{\eta_2\cos\theta_{\mathrm{i}} + \eta_1\cos\theta_{\mathrm{t}}} \tag{3.275a}$$

$$\tau_\perp = \frac{E^{\mathrm{t}}}{E^{\mathrm{i}}} = \frac{2\eta_2\cos\theta_{\mathrm{i}}}{\eta_2\cos\theta_{\mathrm{i}} + \eta_1\cos\theta_{\mathrm{t}}}, \tag{3.275b}$$

and note that we have

$$1 + \Gamma_\perp = \tau_\perp. \tag{3.276}$$

Sometimes it is found convenient to express (3.275a), (3.275b) in terms of the incident angles only. For lossless, nonmagnetic media the following expressions are commonly used instead of (3.275a), (3.275b):

$$\Gamma_\perp = \frac{\cos\theta_\mathrm{i} - \sqrt{\dfrac{\epsilon_2}{\epsilon_1} - \sin^2\theta_\mathrm{i}}}{\cos\theta_\mathrm{i} + \sqrt{\dfrac{\epsilon_2}{\epsilon_1} - \sin^2\theta_\mathrm{i}}} \tag{3.277a}$$

$$\tau_\perp = \frac{\cos\theta_\mathrm{i}}{\cos\theta_\mathrm{i} + \sqrt{\dfrac{\epsilon_2}{\epsilon_1} - \sin^2\theta_\mathrm{i}}}, \tag{3.277b}$$

and (3.276) also applies to (3.277a), (3.277b).

Laws of Reflection and Refraction

Equation (3.272) is the law of reflection. The law of reflection states that the angle of reflection equals the angle of incidence. Equation (3.273) is Snell's law of refraction, which for the case $\mu_1 = \mu_2$ implies

$$\frac{\sin\theta_\mathrm{i}}{\sin\theta_\mathrm{t}} = \sqrt{\frac{\epsilon_2}{\epsilon_1}} \tag{3.278}$$

and thereby relates the angle of refractions with the angle of incidence.

In optics the index of refraction for a medium characterized by μ, ϵ is defined as

$$\begin{aligned} n &= \frac{\text{velocity of light in free space}}{\text{velocity of light in the medium}} \\ &= \frac{\dfrac{1}{\sqrt{\mu_0\epsilon_0}}}{\dfrac{1}{\sqrt{\mu\epsilon}}} = \sqrt{\frac{\epsilon}{\epsilon_0}} \text{ for } \mu = \mu_0 \\ &= \sqrt{\epsilon_\mathrm{r}}\,, \end{aligned} \tag{3.279}$$

where ϵ_r is the dielectric constant of the medium. Thus we obtain from (3.278),

$$\frac{\sin\theta_\mathrm{i}}{\sin\theta_\mathrm{t}} = \frac{\sqrt{\epsilon_2}}{\sqrt{\epsilon_1}} = \frac{n_2}{n_1}\,, \tag{3.280}$$

where n_1, n_2 are the refractive indexes of medium 1 and medium 2, respectively. Equation (3.279) indicates that for incidence from the rare medium to a dense medium, the angle of refraction is less than the angle of incidence, and vice versa. It should be noted that (3.279) was originally introduced by Maxwell [11], and it is often referred to as Maxwell's relation.

Total Reflection and Critical Angle

Angle, critical We consider the case when the wave is incident from a dense medium to a rare medium, meaning when $\epsilon_1 > \epsilon_2$ (or $n_1 > n_2$) in Figure 3.16. We see from (3.280) $\theta_i > \theta_t$ in this case and also that as θ_i increases, θ_t also increases. It is clear from Figure 3.16 that the maximum value for the angle γ refraction can be $\pi/2$. The angle of incidence $\theta_i = \theta_c$ for which the maximum angle of refraction occurs is called the critical angle. The critical angle is obtained from (3.280) as

$$\sin\theta_c = \sqrt{\frac{\epsilon_2}{\epsilon_1}} = \frac{n_2}{n_1} \tag{3.281}$$

when $n_1 > n_2$, so the incidence is from the more dense medium side.

It can be seen from (3.277a) and (3.280) that for $\theta_i = \theta_c$, $\Gamma_\perp = 1$, that is the wave is totally reflected. Thus for $\theta_i \geq \theta_c$ there is total reflection and there is no power transmitted in the z direction through the interface into the less dense medium. However, this does not mean there is no field in medium 2. In fact it can be shown that the transmitted E field (3.269a) in medium 2 can be given by

$$\begin{aligned} \mathbf{E}^t &= \hat{x}E^t e^{+\beta_2 z\sqrt{(\epsilon_1/\epsilon_2)\sin^2\theta_i - 1}} \\ &\quad \times e^{-j\beta_2 y \sin\theta_t} \\ \text{with } \sin\theta_t &= \sqrt{\frac{\epsilon_1}{\epsilon_2}}\sin\theta_i. \end{aligned} \tag{3.282}$$

Equation (3.282) indicates a wave decaying in amplitude in the z direction and propagating in the y direction. In other words, it is a surface wave in the less dense medium and propagating along the interface.

Example 3.6

Total reflection explains the operating principle for optical transmission through fibers. The optical signal enters the fiber A–A at an angle θ_i, suffers total reflections as it propagates along the fiber, and finally exists at an angle θ_i through the end B–B (see Figure 3.17).

What should ϵ_r be so that the fiber can handle all angles of incidence?

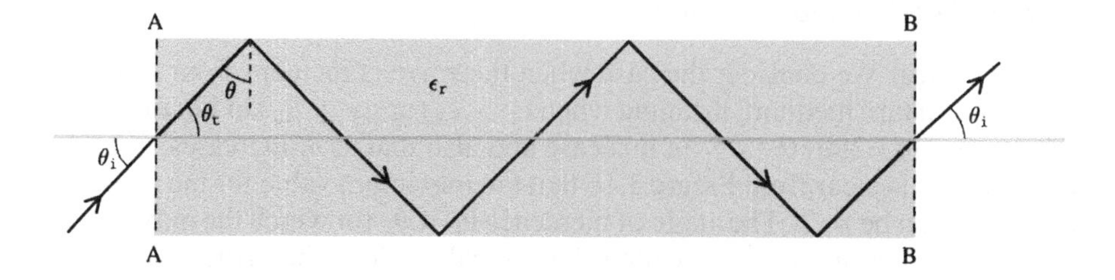

Figure 3.17 Optical transmission through a fiber using total reflection.

Solution

For total reflection we need $\theta \geq \theta_c$ or $\sin\theta \geq \frac{1}{\sqrt{\epsilon_r}}$, meaning $\sin^2\theta \geq 1/\sqrt{\epsilon_r}$. Snell's law requires

$$\begin{aligned}\sin\theta_i &= \sqrt{\epsilon_r}\,\sin\theta_t \\ &= \sqrt{\epsilon_r}\,\cos\theta, \qquad \theta_t = \frac{\pi}{2} - \theta.\end{aligned}$$

So for Snell's law we need

$$\epsilon_r \geq 1 + \sin^2\theta_i.$$

The maximum value of the incident angle is $\pi/2$. Thus $\epsilon_r \geq 2$ can accommodate all angles incidence.

Commonly used materials for fibers are quartz, for which $\epsilon_r \simeq 3.78$, and glass, for which $\epsilon_r \simeq 6.78$.

Total Transmission and Brewster Angle

The Brewster angle is the angle of incidence for which the reflection coefficient vanishes, meaning a total transmission occurs. Equation (3.274a) indicates that $\Gamma_\perp = 0$ if $\eta_2\cos\theta_i = \eta_1\cos\theta_t$, which in conjunction with Snell's law (3.168) yields the Brewster angle $\theta_{B_\perp}$ for the perpendicular polarization case:

$$\theta_{B_\perp} = \sin^{-1}\left[\frac{1 - \dfrac{\mu_1\epsilon_2}{\mu_2\epsilon_1}}{1 - \dfrac{\mu_1}{\mu_2}^2}\right]^{1/2}. \tag{3.283}$$

Now, if $\mu = \mu_2 = \mu_0$, where the media are nonferromagnetic, (3.283) indicates that no Brewster angle exists.

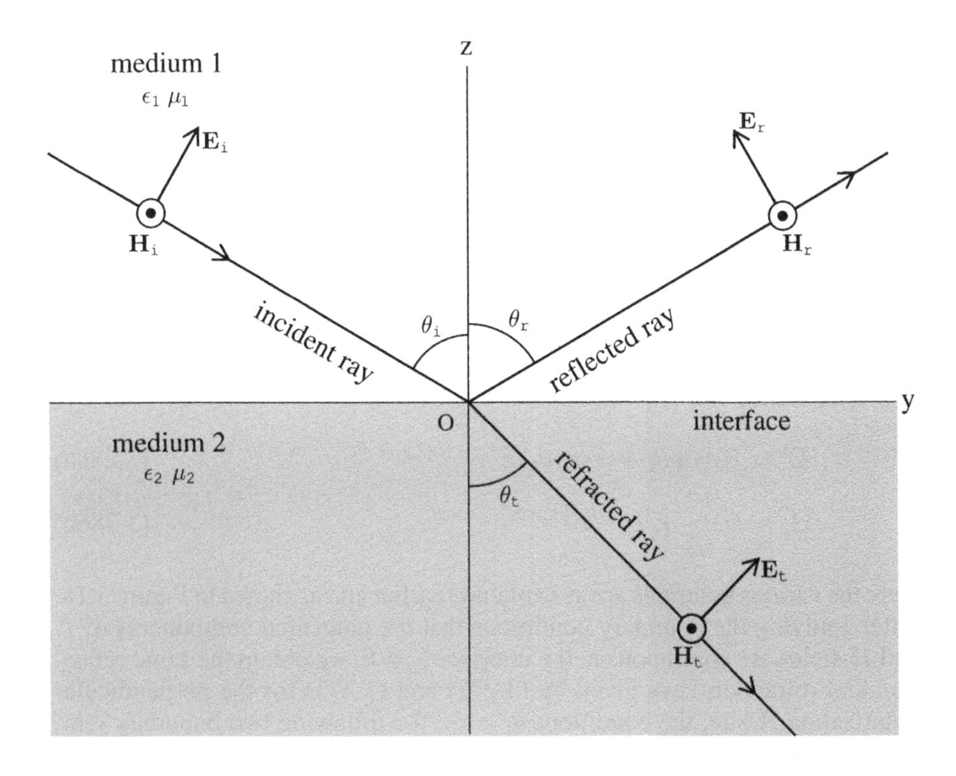

Figure 3.18 Incident, reflected, and refracted rays and the orientation of E and H field vectors for parallel polarization.

Parallel or Vertical Polarization

The orientation of the electric and magnetic field vectors along the incident, reflected and refracted rays are sketched in Figure 3.18. In the figure all three H vectors are chosen in the positive x direction, so they are perpendicular to the plane of incidence and are horizontal. The E vectors are in the plane of incidence and are oriented such that the three waves propagate in their respective directions. It should be noted that although the E vectors are in the plane of incidence (or parallel to it), they are not necessarily vertically oriented with respect to the interface. However, the convention is to call this a vertical polarization case.

With the H vectors oriented along the positive x direction, it can be shown that the complete field expressions for the incident, reflected, and refracted waves are as follows:

Incident wave,

$$\mathbf{E}^{\mathrm{i}} = (\hat{y}\cos\theta_{\mathrm{i}} + \hat{z}\sin\theta_{\mathrm{i}})E^{\mathrm{i}}e^{-j\beta_1(y\sin\theta_{\mathrm{i}} - z\cos\theta_{\mathrm{i}})} \tag{3.284a}$$

$$\mathbf{H}^{\mathrm{i}} = \hat{x}\,\frac{E^{\mathrm{i}}}{\eta_1}\,E^{\mathrm{i}}e^{-j\beta_1(y\sin\theta_{\mathrm{i}} - z\cos\theta_{\mathrm{i}})}. \tag{3.284b}$$

Reflected wave,

$$\mathbf{E}^{\mathrm{r}} = (-\hat{y}\cos\theta_{\mathrm{r}} + \hat{z}\sin\theta_{\mathrm{i}})E^{\mathrm{r}}e^{-j\beta_1(y\sin\theta_{\mathrm{i}} + z\cos\theta_{\mathrm{r}})} \tag{3.285a}$$

$$\mathbf{H}^{\mathrm{r}} = \hat{x}\,\frac{E^{\mathrm{r}}}{\eta_1}\,E^{\mathrm{i}}e^{-j\beta_1(y\sin\theta_{\mathrm{i}} + z\cos\theta_{\mathrm{r}})}. \tag{3.285b}$$

Refracted wave,

$$\mathbf{E}^{\mathrm{t}} = (\hat{y}\cos\theta_{\mathrm{t}} + \hat{z}\sin\theta_{\mathrm{t}})E^{\mathrm{t}}e^{-j\beta_2(y\sin\theta_{\mathrm{t}} - z\cos\theta_{\mathrm{t}})} \tag{3.286a}$$

$$\mathbf{H}^{\mathrm{t}} = \hat{x}\,\frac{E^{\mathrm{t}}}{\eta_2}\,E^{\mathrm{i}}e^{-j\beta_2(y\sin\theta_{\mathrm{t}} - z\cos\theta_{\mathrm{t}})}. \tag{3.286b}$$

Here the various notations are as explained earlier and as shown in Figure 3.18. After applying the boundary conditions that the tangential components of E and H-fields are continuous at the interface $z = 0$, we obtain the same reflection and refraction laws given by (3.272) and (3.273) for the perpendicular polarization. Using the conditions $\theta_{\mathrm{i}} = \theta_{\mathrm{r}}$, the following two boundary relations are then obtained.

$$E^{\mathrm{i}}\cos\theta_{\mathrm{i}} - E^{\mathrm{r}}\cos\theta_{\mathrm{i}} = E^{\mathrm{t}}\cos\theta_{\mathrm{t}} \tag{3.287}$$

$$\frac{E^{\mathrm{i}}}{\eta_1} + \frac{E^{\mathrm{r}}}{\eta_1} = \frac{E^{\mathrm{t}}}{\eta_2}. \tag{3.288}$$

From (3.185a), (3.185b) and (3.286a), (3.286b) we obtain the following for the reflection and transmission coefficients for the parallel polarization:

$$\Gamma_{\|} = |\Gamma_{\|}|e^{j\phi_{\|}} = \frac{E^{\mathrm{r}}}{E^{\mathrm{i}}} = \frac{\eta_1\cos\theta_{\mathrm{i}} - \eta_2\cos\theta_{\mathrm{t}}}{\eta_1\cos\theta_{\mathrm{i}} + \eta_2\cos\theta_{\mathrm{t}}} \tag{3.289a}$$

$$\tau_{\|} = \frac{E^{\mathrm{t}}}{E^{\mathrm{i}}} = \frac{2\eta_2\cos\theta_{\mathrm{i}}}{\eta_1\cos\theta_{\mathrm{i}} + \eta_2\cos\theta_{\mathrm{t}}}. \tag{3.289b}$$

Note for the present case that

$$1 - \Gamma_{\|} = \left(\frac{\cos\theta_{\mathrm{t}}}{\cos\theta_{\mathrm{i}}}\right)\tau_{\|}. \tag{3.290}$$

The alternate expressions for $\Gamma_{\|}$ and $\tau_{\|}$ in terms of the incident angles are

$$\Gamma_{\|} = \frac{\dfrac{\epsilon_2}{\epsilon_1}\cos\theta_{\rm i} - \sqrt{\dfrac{\epsilon_2}{\epsilon_1} - \sin^2\theta_{\rm i}}}{\dfrac{\epsilon_2}{\epsilon_1}\cos\theta_{\rm i} + \sqrt{\dfrac{\epsilon_2}{\epsilon_1} - \sin^2\theta_{\rm i}}} \tag{3.291a}$$

$$\tau_{\|} = \frac{2\sqrt{\dfrac{\epsilon_2}{\epsilon_{\rm r}}}\cos\theta_{\rm i}}{\dfrac{\epsilon_2}{\epsilon_1}\cos\theta_{\rm i} + \sqrt{\dfrac{\epsilon_2}{\epsilon_1} - \sin^2\theta_{\rm i}}}\,. \tag{3.291b}$$

Expressions (3.289a)–(3.291b) for $\Gamma_{\|}$ and $\tau_{\|}$ should be compared with the corresponding expressions (3.275a)–(3.277b) for the perpendicular polarization.

Total Reflection and Critical Angle

The total reflection phenomenon also occurs in the parallel polarization case when the wave propagation takes place from from a dense medium to a less dense medium, which is when $\mu_2\epsilon_2 < \mu_1\epsilon_1$ or $\epsilon_2 < \epsilon_1$ when $\mu_1 = \mu_2$. The critical angle is given by expression (3.281) for the perpendicular polarizations case; thus it is not a function of polarization.

Total Transmission–Brewster Angle

$\Gamma_{\|}$ expression (3.186) indicates that $\Gamma_{\|} = 0$ and $\tau_{\|} = 1$ if $\eta_1\cos\theta_{\rm i} = \eta_2\cos\theta_{\rm t}$, which yields

$$\theta_{B_{\|}} = \sin^{-1}\left[\frac{1 - \dfrac{\mu_1\epsilon_2}{\mu_2\epsilon_1}}{1 - \dfrac{\mu_1}{\mu_2}^2}\right]^{1/2}. \tag{3.292}$$

If $\mu_1 = \mu_2$, we have

$$\theta_{B_{\|}} = \sin^{-1}\left[\frac{1 - \dfrac{\epsilon_1}{\epsilon_2}}{1 - \dfrac{\epsilon_1}{\epsilon_2}^2}\right]^{1/2}$$
$$= \sin^{-1}\left(\frac{\epsilon_2}{\epsilon_1 + \epsilon_2}\right)^{1/2} = \tan^{-1}\sqrt{\frac{\epsilon_2}{\epsilon_1}}\,. \tag{3.293}$$

In contrast to perpendicular polarization, the Brewster angle is possible in nonferromagnetic media. This has significant implications in radio wave propagation.

Plane Earth Reflection

The Fresnel reflection coefficients are extensively used in radio wave propagation above a plane earth and also in a variety of electromagnetic problems involving antennas operating above the plane earth. In view of such applications, we recast the reflection coefficient expressions described earlier into suitable form so that they can be directly applied to wave propagation and antenna problems.

We assume a plane interface between the earth and free space. Thus, with reference to Figure 3.15 through 3.18, we have for medium 1, $\epsilon_1 = \epsilon_0,\ \mu_1 = \mu_0$, and $\sigma_1 = 0$, and for medium 2, $\epsilon_2 = \epsilon_0\epsilon_r,\ \mu_2 = \mu_0$, and $\sigma_2 = \sigma$. As mentioned earlier, it is found convenient to assume complex permittivity for the lossy earth, and we have in the present case for medium 1:

$$\frac{\epsilon_2}{\epsilon_1} = \frac{\epsilon_2^c}{\epsilon_0} = \epsilon_r' - j\epsilon_r'' = \epsilon_r - j\,\frac{\sigma}{\omega\epsilon_0}, \tag{3.294a}$$

where

ϵ_r is the dielectric constant of earth

σ is the conductivity of earth in S/m.

For computational purposes, we express the loss term as

$$\chi = \frac{\sigma}{\omega\epsilon_0} = \frac{18 \times 10^3}{f_{\text{MHz}}}\,\sigma = 60\lambda\sigma, \tag{3.294b}$$

where λ is in meters, and σ is in S/m. Then rewriting (3.294a) obtains

$$\frac{\epsilon_2^c}{\epsilon_0} = (\epsilon_r - j\chi). \tag{3.295}$$

After replacing ϵ_2/ϵ_1 in (3.277a) and (3.291a) by $(\epsilon_r - j\chi)$, we have the following modified expressions for $\Gamma_\perp$ and $\Gamma_\parallel$:

$$\Gamma_\perp = |\Gamma_\perp| e^{j\phi_\perp} = \frac{\sin\psi - \sqrt{(\epsilon_r - j\chi) - \cos^2\psi}}{\sin\psi + \sqrt{(\epsilon_r - j\chi) - \cos^2\psi}} \tag{3.296a}$$

$$\Gamma_\parallel = |\Gamma_\parallel| e^{j\phi_\parallel} = \frac{(\epsilon_r - j\chi)\sin\psi - \sqrt{(\epsilon_r - j\chi) - \cos^2\psi}}{(\epsilon_r - j\chi)\sin\psi + \sqrt{(\epsilon_r - j\chi) - \cos^2\psi}}, \tag{3.296b}$$

where we express the incident angle θ_i in terms of the grazing angle ψ measured from the horizon, $\theta_i = \frac{\pi}{2} - \psi$. The amplitude and phase of the reflection coefficients for both polarizations have been computed for various value of ϵ_r, σ and f. They are described in various places, for example in [2, 10].

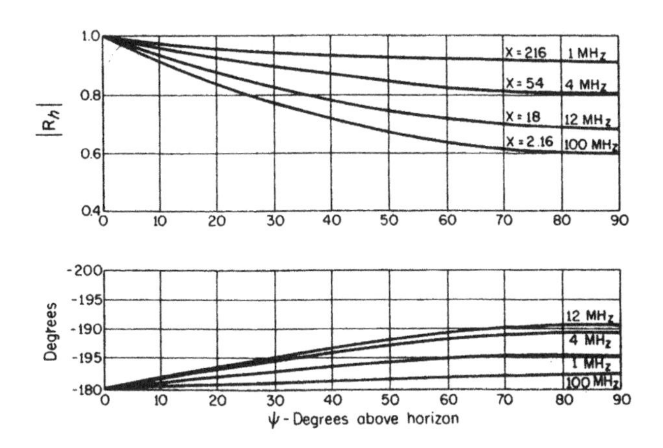

Figure 3.19 Magnitude and phase of the plane wave reflection coefficient for horizontal (i.e. perpendicular) polarization. The cures are for relatively good earth ($\sigma = 12 \times 10^{-}3, \epsilon_r = 15$) but can be used to give approximate results for other earth conductivities and other frequencies by calculating the appropriate values for $x = 18 \times 10^3\sigma/f_{\text{MHz}}$, Source: From [2], p. 632.

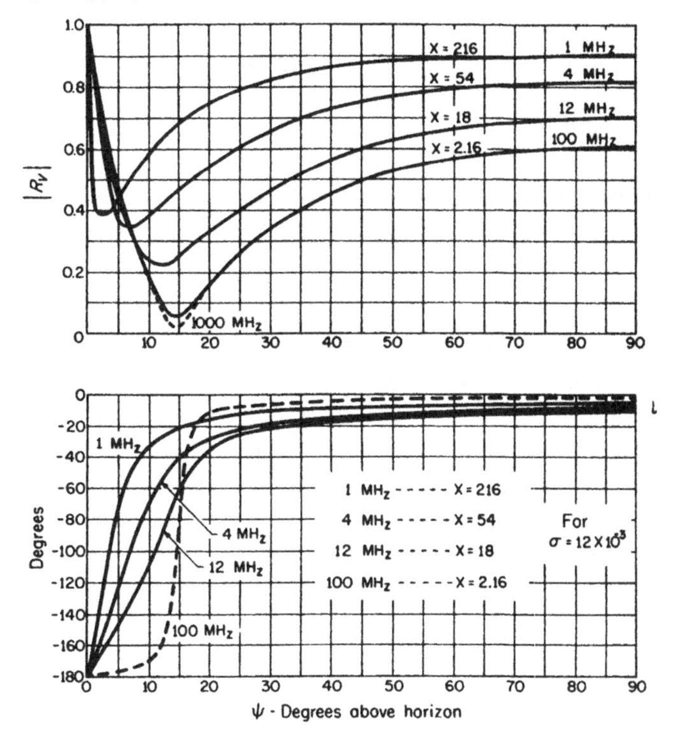

Figure 3.20 Magnitude (a) and phase (b) of the plane wave reflection coefficient for vertical (or parallel) polarization (see legend of Figure 3.19). Source: From [2], p. 633.

Figures 3.19 and 3.20 show $|\Gamma_\perp|$, $\phi_\perp$ as opposed to ψ and $|\Gamma_\parallel|$ $\phi_\parallel$ as opposed to ψ, respectively, for relatively good earth and free space plane interface. Note that at the low elevation angle ($\psi \sim 0$) both $\Gamma_\perp$ and $\Gamma_\parallel$are close to -1 for all frequencies and conductivities. Also for the vertical polarization (Figure 3.20) the Brewster angle effects cause $|\Gamma_\parallel|$ to go through a minimum value for $\psi \leq 15^\circ$ over which $\phi_\parallel$ undergoes rapid change from -180° to 0°. Further discussions are given in [2, 10].

REFERENCES

1. S. Ramo, T.R. Whinnery, and T. Van Duzer, *Electromagnetic Fields and Waves in Communication Electronics*, 3rd ed., Wiley, New York, 1994.
2. E. C. Jordan and K. A. Balmain, *Electromagnetics Waves and Radiating Systems*, 2nd ed., Prentice-Hall, Englewood Cliffs, NJ, 1968.
3. M. N. O. Sadiku, *Elements of Electromagnetics*, 2nd ed., Saunders/Harcourt Brace, Fort Worth, TX, 1994.
4. J. D. Kraus and D. A. Fleisch, *Electromagnetics with Applications*, 5th ed., WCB/McGraw-Hill, Boston, 1999.
5. D. H. Staelin, A. W. Morgenthaler, and J. A. Kong, *Electromagnetic Waves*, Prentice-Hall, Englewood Cliffs, NJ, 1994.
6. R. Plonsey and R. E. Collin, *Principles and Applications of Electromagnetic Fields*, McGraw-Hill, New York, 1961.
7. U. S. Inan and A. S. Inan, *Engineering Electromagnetic*, AddisonWesley, Boston, 1998.
8. U. S. Inan and A. S. Inan, *Electromagnetic waves*, Prentice-Hall, Upper Saddle River, NJ, 2000.
9. C. A. Balanis, *Advanced Engineering Electromagnetic*, Wiley, New York, 1989.
10. R. E. Collin, *Antennas and Propagation*, McGraw-Hill, New York, 1985.
11. D. L. Sengupta and T. K. Sarkar, "Maxwell, Hertz, Maxwellians and the Early History of Electromagnetic waves," IEEE Ant. and Propag. Mag., 45 (April 2003): 13–19.
12. J. C. Maxwell, A Dynamical Theory of Electromagnetic Field, *Phil. Trans. Roy. Soc.*, 166 (1865): 459–512, reprinted in *The Scientific Papers of James Clerk Maxwell*, Vol. 1, Dover, New York, (1952): 528–597.
13. A. A. Smith, *Radiofrequency Principles and Applications*, IEEE Press/Chapman and Hill, New York, 1998.

PROBLEMS

3.1 **(a)** Express the following time dependent quantities into their phasor equivalent form:

$$\mathcal{E} = 5\sin(\omega t - 2z)\hat{x}$$
$$\mathcal{H} = 10^{-2z}\cos(\omega t - 3z)\hat{y}$$
$$\mathcal{E} = 2\cos\omega t\,\hat{y} - 3\sin(\omega t - x)\hat{x}$$
$$\mathcal{H} = 5\cos\left(\omega t - \frac{\pi}{6}\right)\hat{x} + \sin\left(\omega t + \frac{\pi}{4}\right)\hat{y}.$$

(b) Express the following time dependent quantities into their harmonic time dependent form:

$$\mathbf{A} = 3je^{-j\pi/4}\hat{z} - (3 + j4)x\hat{y}$$
$$\mathbf{B} = 10e^{-j\beta z}\hat{z} + j5e^{-j(\beta z + \pi/4)}\hat{y}$$
$$\mathbf{C} = \frac{3}{j}\,e^{-j2}\sin x + e^{3x - j4x}\hat{y}$$
$$\mathbf{D} = \hat{x}\,\frac{E_0}{j}\,e^{j(\delta - \pi/2)}.$$

Answer:

(a) $\mathbf{E} = 5e^{-j(2z+\pi/2)}\hat{x}$; $\mathbf{H} = 10e^{-2z}e^{-j3z}\hat{y}$; $\mathbf{E} = 2\hat{y} - 3e^{-j(x+\pi/2)}\hat{x}$; $\mathbf{H} = 5e^{-j\pi/6}\hat{x} + e^{-j\pi/4}\hat{y}$.

(b) $\mathbf{A}(t) = 3\cos(\omega t + \frac{\pi}{4})\hat{z} - 5x\cos(\omega t - \tan^{-1}53.1°)\hat{y}$; $\mathbf{B}(t) = 10\cos(\omega t - \beta z)\hat{z} + 5\cos(\omega t - \beta z + \frac{\pi}{4})\hat{z}$; $\mathbf{C}(t) = 3\sin x\cos(\omega t - \frac{\pi}{2} - 2)\hat{x} + e^{3x}\cos(\omega t - 4x)$; $\mathbf{D}(t) = E_0\cos(\omega t + \delta - \pi)\hat{x}$.

3.2 A small 50-turn coil having a diameter of 15 cm is illuminated by an incident plane electromagnetic wave of frequency 1 MHz. The incident electric field has 20 mV/m peak value. The loop is oriented such that the induced voltage at a small gap in the loop is maximized. Assume the medium to be free space. Apply Faraday's law (justify) to determine the induced voltage at the gap. What would be the voltage if the coil has a core with $\mu_r = 60$?

Answer: with $\mu = \mu_0$: $V = 2.9 \times 10^{-5}$ V, with $\mu_r = 60$, $V = 174 \times 10^{-5}$ V.

3.3 Assume that seawater has the following characteristics: $\mu_r = 1$, $\epsilon_r = 81$, and $\sigma = 4$ S/m, and that these characteristics are independent of frequency. Determine **(a)** the frequency at which the displacement current equals the conductor current, **(b)** the range of frequencies at which the conduction current exceeds the displacement current by at least 10%, and **(c)** the range of frequencies at which the the displacement current exceeds the conduction current by at least 10%.

Answer: **(a)** 0.889 GHz; **(b)** $\leq$ 0.823 GHz; **(c)** $\geq$ 1.00 GHz.

3.4 A uniform plane electromagnetic wave with a peak E field $\hat{x}$ 100 V/m is traveling in the positive z direction. It is given that the wavelength in the medium is 10 cm and the velocity of the wave is 2.5×10^8 m/s. Assume that $\mu = \mu_0$. Determine the following: **(a)** the frequency of the wave, **(b)** dielectric constant of the medium, **(c)** the magnetic field intensity **H**, and **(d)** the time dependent expressions for **E** and **H**.

Answer: **(a)** 2.5 GHz; **(b)** 1.44 GHz; **(c)** $\mathbf{H} = \hat{y}\,0.266$ A/m; **(d)** $\boldsymbol{\mathcal{E}} = \hat{x}\,100\cos(15.7\times 10^9\,t - 0.628\,z)$ and $\boldsymbol{\mathcal{H}} = \hat{y}\,0.266\cos\left(15.7 \times 10^9\,t - 0.628\,z\right)$

3.5 Give the wavelength in meters for the various signals whose frequencies (or frequency bands) are as follows:

Power system: 60 Hz, 50 MHz; AM Radio: 535–1605 kHz; FM Radio: 88–108 MHz; VHF TV Channel 2: 54–60 MHz, VHF TV Channel 13: 210–216 MHz; VHF TV Channel 14: 470–476 MHz; VHF TH Channel 83: 884–890 MHz; US Cordless Telephone: 46–49 MHz; European Cordless Telephone: 1.880–1.990 GHz; Mobile Telephone: 824–894 MHz; PCS: 1850–1990 MHz; IEEE Bands: HF 3–30 MHz VHF 30–300 MHz; VHF 300–1000 MHz; L 1–2 GHz; C 4–8 GHz; X 8–12 GHz; Visible Light: 1.75×10^{14} Hz – 7.5×10^{14} Hz; X-Ray: $\sim 10^{18}$ Hz; Y-Ray: $\sim 10^{21}$ Hz.

Answer: Use the relationship $\lambda = f/c = 3 \times 10^8/f$, where c is the velocity of light in free space, λ is the wavelength in meters, f is the frequency in Hz.

3.6 Give in Hz the frequencies corresponding to the following wavelengths: 1 km, 1 m, 1 mm, 1 μm, 1 nm, and 1 Angstrom $= 10^{-10}$ m.

Answer: Use the relationship given in the answer of Problem 3.4.

3.7 A plane electromagnetic wave is propagating through an unbounded lossy medium with $\epsilon_{\rm r} = 4, \mu_{\rm r} = 1$. The E field of the wave is $\mathbf{E} = \hat{x}\,1.0 \times 10^{-z/3}\cos\left(10^8\,t - \beta\,z\right)$. Determine **(a)** β, **(b)** loss tangent $\tan\,\delta$, **(c)** wave impedance, and **(d)** wave velocity.

Answer: **(a)** 1. 4238 rad/m; **(b)** $\tan\,\delta = 0.8011$; **(c)** $83.27\angle 19.3^\circ$ Ω; **(d)** 7.023×10^7 m/s.

3.8 A plane electromagnetic wave of frequency $f = 10^4$ Hz is propagating through seawater having $\sigma = 4$ S/m and $\epsilon_r = 81$, $\mu_r = 1$. Determine the following: **(a)** the complex propagation constant $\gamma = \alpha + j\beta$, **(b)** phase velocity, **(c)** the wavelength, **(d)** α in Nepers/m, and **(e)** the depth of penetration.

Answer: **(a)** $83.88 + j\,83.88$; **(b)** 7.5×10^3 m/s; **(c)** 0.075 m; **(d)** 83.88 Np/m = 728.7 dB/m; **(e)** 0.0119 m.

3.9 A uniform plane electromagnetic wave of frequency 1 GHz is propagating through a conducting medium like Cu with $\sigma = 5.76 \times 10^7$ S/m, $\epsilon = \epsilon_0$, $\mu = \mu_0$. Determine the following: **(a)** intrinsic impedance of the medium, **(b)** the skin depth, **(c)** the complex propagation constant, **(d)** wavelength λ_{Cu}, and **(e)** the phase velocity.

Answer: **(a)** $\eta = \frac{1+j}{\sigma\,\delta} = (1+j)\,(0.549)\ \Omega$; **(b)** $\delta = 3.16 \times 10^{-6}$ m; **(c)** $\gamma = \alpha + j\,\beta = \frac{1+j}{\delta} = (1+j) \times 0.316 \times 10^6$; **(d)** $\lambda_{\mathrm{Cu}} = 2\,\pi\delta = 19.85 \times 10^{-6}$ m; **(e)** $v_{\mathrm{p}} = 1.985 \times 10^4$ m/s.

3.10 Find the polarization of the following uniform plane waves whose E-fields are **(a)** $\mathbf{E} = E_0(j\hat{x} + \hat{y})e^{-j\beta z}$ and **(b)** $\mathbf{E} = E_0(\hat{x} - j\hat{y})e^{+j\beta z}$.

Answer: **(a)** Right circular; propagation in the z direction. **(b)** Left circular; propagation in the $-z$ direction.

CHAPTER 4

SIGNAL WAVEFORM AND SPECTRAL ANALYSIS

4.1 INTRODUCTION

The waveform (the time domain behavior) and the spectrum (the frequency content) of a signal determine many of its practical applications, namely its information-carrying capacity and its transmission characteristics through a variety of guiding structures and through open media like free space.

A time-varying signal is mathematically described by a function of time $f(t)$. It may represent an electric field, a magnetic field, voltage, or current. Generally, two types of signals are encountered: energy signals and power signals (to be defined later), each of which can be classified as deterministic and random [1–3].

Deterministic signals can be represented by instantaneous values which can be specified at each instant of time. Random signals, also called noise, cannot be represented by a predictable function of time. They can only be described in terms of statistical parameters.

Applied Electromagnetics and Electromagnetic Compatibility. By D. L. Sengupta, V. V. Liepa
ISBN 0-471-16549-2

In this chapter we will use Fourier transform and Fourier series techniques [4–7] to determine and describe the spectra of a selected number of energy and power signals having known waveforms.

4.2 CLASSIFICATION OF SIGNALS

Let $f(t)$ be an arbitrary signal. The total energy W and the average power P associated with the signal are [1,4,5]

$$W = \lim_{T \to \infty} \alpha \int_{-T/2}^{T/2} f^2(t)\, dt \qquad \text{W-s or J} \tag{4.1}$$

$$P = \lim_{T \to \infty} \frac{\alpha}{T} \int_{-T/2}^{T/2} f^2(t)\, dt \qquad \text{W or J/s} \tag{4.2}$$

where α is a constant determined by the particular system. Expressions (4.1) and (4.2) are used to classify signals in two broad categories.

It is clear that by definition the energy signals are time limited; that is they exist over a finite interval of time and they are nonperiodic. An energy signal's total energy W is finite, and hence the average power P is zero by (4.2). Examples of energy signals are single pulses, impulses, a band of pulses, and sinusoidal radar pulses. We will see later that deterministic energy signals possess continuous frequency spectra and that they can be described by Fourier transform techniques.

Power signals — the other class of signals — exist over infinite time (i.e., for all time) or, in practice, a time period that is long compared with the measurement time. For such signals, W is infinite by (4.1), but the average power (over one period) is finite by (4.2). Examples of deterministic and periodic power signals are sine waves, pulse trains, and so forth. Such signals have line (or discrete) spectra of harmonically related frequencies with the fundamental being that of the original waveform. Power signals are described by the Fourier series.

Energy and power signals can be either analog or digital. Analog signals can be measured and specified at any instant of time. They are described by a continuous function. There are cases where the signals are known only at specific intervals of time or at multiples of basic sampling time interval. These are known as sampled or discrete signals. The magnitude of each sample of discrete signals can be represented by a binary number [3]. A common representation is to use 8 (eight) bits and thus a pulse train of eight pulses of magnitude 1 or 0 to represent each sample. Such a signal is described as a digital signal.

The signal described above can be analyzed by Fourier techniques, which are discussed in many textbooks, including [6, 7]. We will apply these techniques to determine the spectra of a selected number of signals having known waveform.

4.3 ENERGY SIGNALS

4.3.1 Definitions

The Fourier transform $F(\omega)$ of an energy signal $f(t)$ is defined as

$$F(\omega) = \int_{-\infty}^{\infty} f(t)\, e^{-j\omega t}\, dt \tag{4.3}$$

and its inverse transform is

$$f(t) = \frac{1}{2\pi} \int_{-\infty}^{\infty} f(\omega)\, e^{+j\omega t}\, d\omega, \tag{4.4}$$

where ω is the radian frequency $= 2\pi f$, f being the frequency expressed in cycles per second or Hz. Observe that $F(\omega)$ is in general a complex quantity and (4.4) indicates the use of negative frequency.

Generally, we write

$$F(\omega) = |F(\omega)| e^{j\phi(\omega)} \tag{4.5}$$

where $|F(\omega)|$ is called the *Amplitude, spectrum* (or the *two-sided amplitude spectrum*) of the signal $f(t)$, and $\phi(\omega)$ is the *phase spectrum* of the signal. Note that both amplitude and phase spectra for energy signals are continuous functions of frequency (which is a continuous variable).

All physical functions representing practical signals are real. It is known from Fourier transform theory [4] that

$$F(-\omega) = F^*(\omega). \tag{4.6}$$

for real $f(t)$.

From (4.6) it can be shown that

$$\begin{aligned} f(t) &= \frac{1}{2\pi} \mathrm{Re} \int_{-\infty}^{\infty} 2F(\omega)\, e^{j\omega t}\, d\omega \\ &= \frac{1}{2\pi} \mathrm{Re} \int_{0}^{\infty} 2|F(\omega)| e^{j(\omega t + \phi)}\, d\omega, \end{aligned} \tag{4.7}$$

For later use, we define the single-sided transform $S(\omega)$ (i.e., for $\omega \geq 0$) that is related to the amplitude spectrum as

$$S(\omega) = 2|F(\omega)|, \qquad \omega \geq 0, \tag{4.8}$$

where $S(\omega)$ is called the *spectral intensity* defined for all positive frequencies. Since negative frequencies are fictitious, it is more satisfying to use (4.8) for response calculations using spectrum and network analyzers [9].

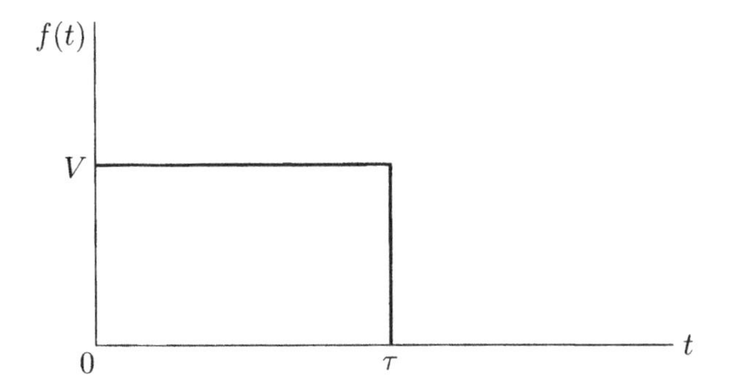

Figure 4.1 A rectangular voltage pulse signal.

4.3.2 A Rectangular Pulse

Consider a signal consisting of a voltage pulse defined by

$$f(t) = \begin{cases} V, & 0 \leq t \leq \tau \\ 0, & t > 0 \end{cases} \tag{4.9}$$

where τ is the pulse width as shown in Figure 4.1

Using (4.9) in (4.3), we obtain the Fourier transform (spectrum) of the voltage pulse as

$$\begin{aligned} F(t) &= \int_0^\tau V e^{-j2\pi ft}\, dt \\ &= V\tau \frac{\sin(\pi ft)}{\pi f\tau} e^{-j\pi f\tau} \\ &= |F(t)| e^{j\phi(f)}, \end{aligned} \tag{4.10}$$

where $\phi(f) = \mp(\pi f\tau)$ and the negative sign is selected when $f\tau < 1$.

The amplitude spectrum, the phase spectrum, and the spectral intensity appropriate for the pulse are shown in Figure 4.2.

The $\sin x/x$ *Function*

The Fourier transform of the rectangular pulse contains the function $f(x) = \frac{\sin x}{x}$, which occurs frequently in signal analysis, antenna theory, and other areas. Although its properties are discussed in [8] and in other places, we will briefly discuss the function.

Figure 4.3 shows a plot of $\frac{\sin x}{x}$ versus x which indicates that $\frac{\sin x}{x}$ is an even function of x. Starting from the origin ($x = 0$) for increasing and decreasing

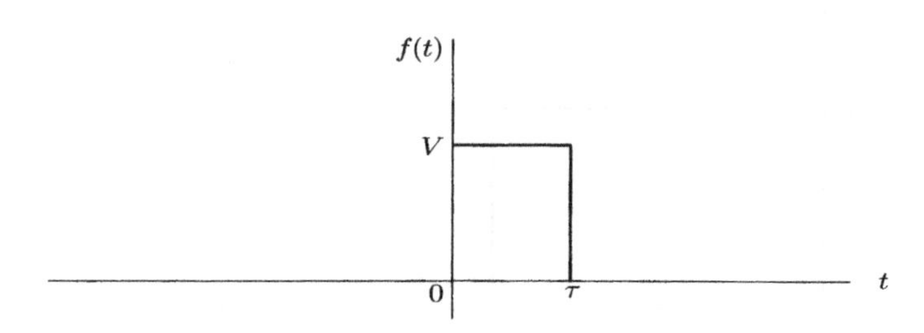

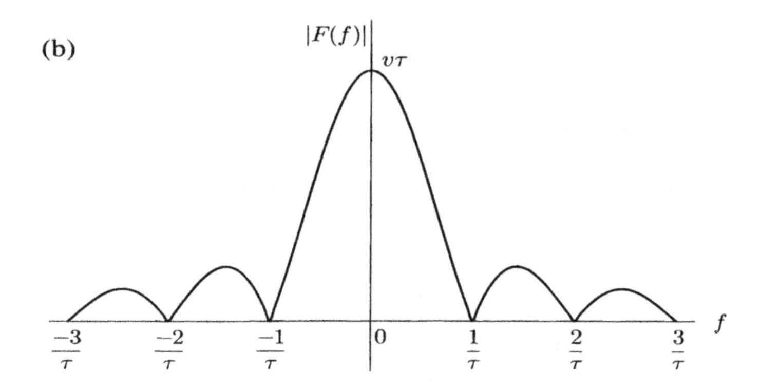

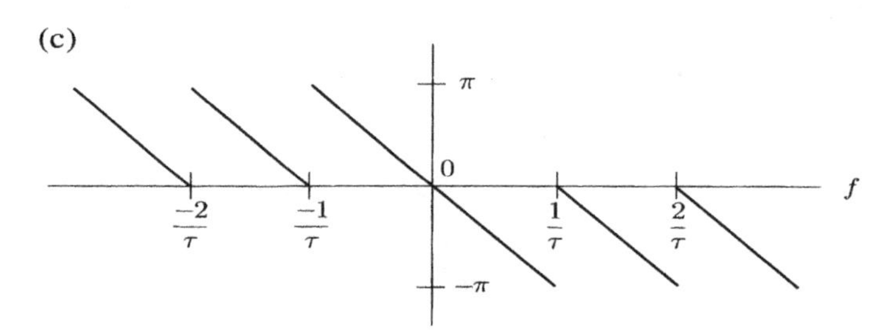

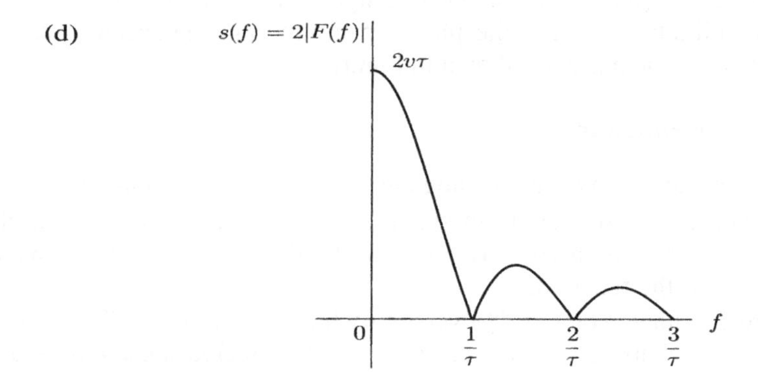

Figure 4.2 Fourier transform of a rectangular pulse: (*a*) pulse; (*b*) amplitude spectrum; (*c*) phase spectrum; (*d*) spectral intensity.

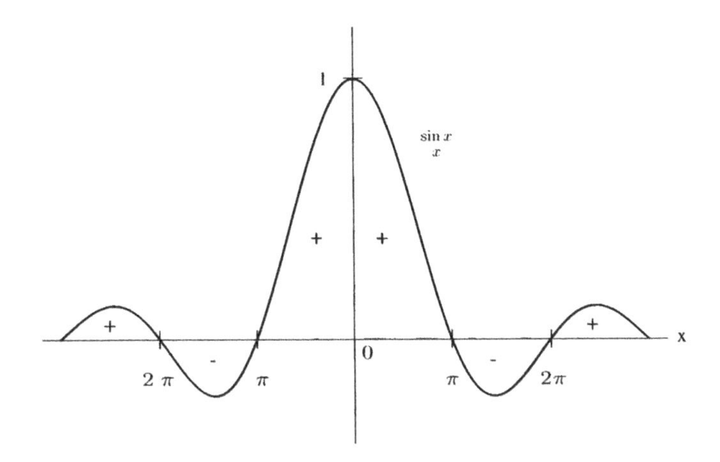

Figure 4.3 Plot for $\frac{\sin x}{x}$ versus x.

values of x, the function assumes alternatively positive and negative values over consecutive intervals of $\Delta x = \pi$, as shown in Figure 4.3. The other properties of $\frac{\sin x}{x}$, denoted by $f(x)$ for discussion, are as follows

1. at $x = \pm m\pi$ for $m = 1, 2, 3, \ldots$

$$f(x) = 0, \tag{4.11}$$

2. the principal maximum of $f(x)$ at $x = 0$

$$f(x) = 1. \tag{4.12}$$

The secondary maxima (maxima and minima) of $f(x)$,

$$f(x) = 2\pi \frac{(-1)^n}{(2n+1)}, \tag{4.13}$$

occur at $x = \pm(2n+1)\pi/2$ for $n = 1, 2, 3, \ldots$

These properties are identified in Figure 4.3. Figure 4.4 shows a plot of $\left|\frac{\sin x}{x}\right|$ versus x, where the principal and secondary maxima and their location and values are identified.

The variation of the envelope of the $\left|\frac{\sin x}{x}\right|$ versus x shown in Figure 4.5 with respect to the principal maximum is of considerable practical importance. It is used to estimate the amplitude of various frequency components of a certain signal.

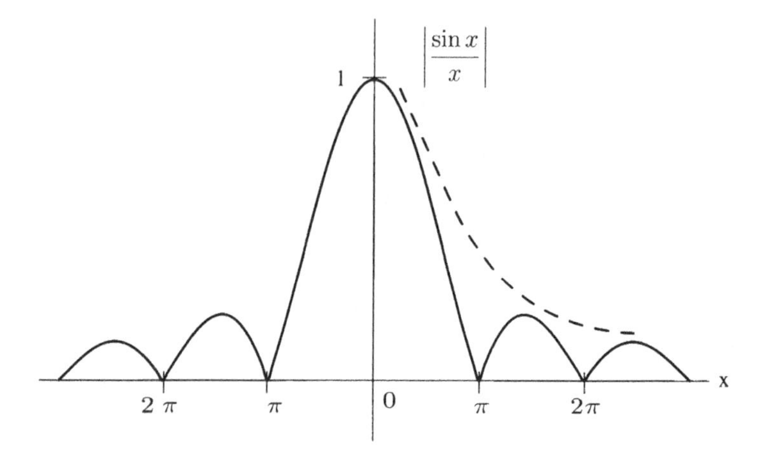

Figure 4.4 Plot for $\left|\frac{\sin x}{x}\right|$ versus x.

For spectrum analysis, the envelope mentioned above is represented with the help of two asymptotes approximating the logarithm of the amplitudes of the principal maximum and secondary maxima. This representation is called the Bode plot and is used extensively in spectral and circuit analysis [6–8]. For this purpose we use the small and large argument asymptotic behavior of $\frac{\sin x}{x}$ as

$$\left|\frac{\sin x}{x}\right| \simeq \begin{cases} 1 & \text{for } |x| \ll 1 \\ \frac{1}{|x|} & \text{for } |x| \gg 1. \end{cases} \tag{4.14}$$

Assuming that the two asymptotes of $\left|\frac{\sin x}{x}\right|$ in (4.14) meet at $x = 1$, we obtain the Bode plot for $\left|\frac{\sin x}{x}\right|$ shown in Figure 4.5.

The first asymptote (representing the principal maximum) has a slope of 0 dB/decade of $|x|$ (or $-20 \log |x|$ with $x = 1$) and the second asymptote representing the secondary maxima has a slope of -20 dB/decade of x [$-20 \log |x|$, i.e., -20 dB/decade of $|x|$].

The spectral intensity of the rectangular pulse shown in Figure 4.1 can be represented as

$$s(f) = 2v\tau \left|\frac{\sin \pi f \tau}{\pi f \tau}\right|, \tag{4.15}$$

which was shown graphically in Figure 4.2(d). The Bode plot for $s(f)$ can now be represented as in Figure 4.6.

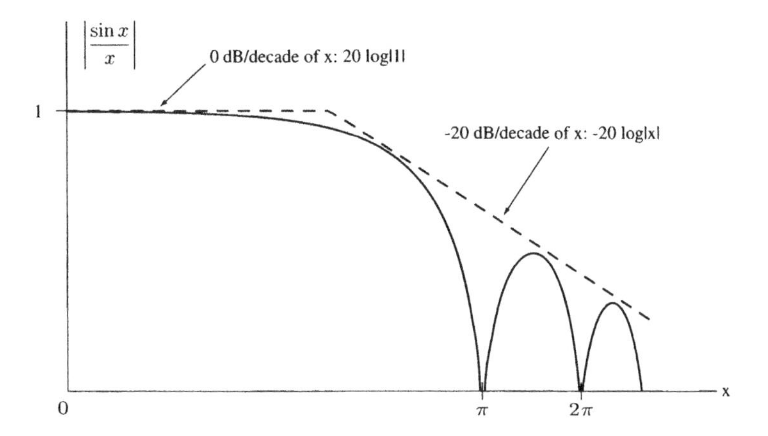

Figure 4.5 Bode plot for $\left|\frac{\sin x}{x}\right|$ function.

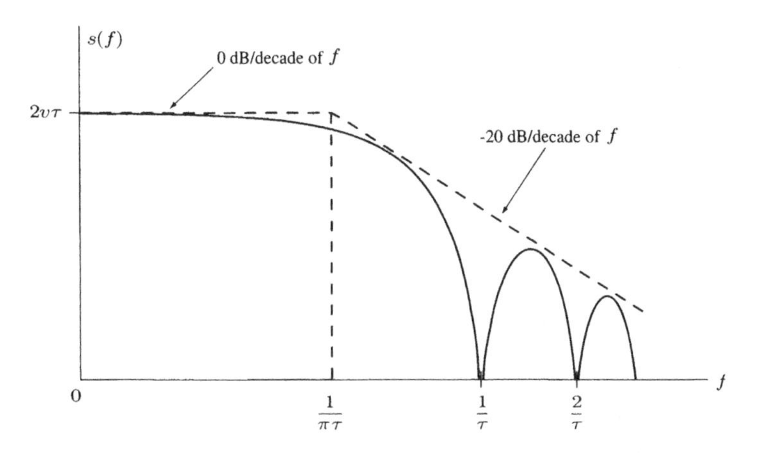

Figure 4.6 Bode plot for the spectral intensity of the rectangular pulse.

4.4 POWER SIGNALS

We consider here the class of periodic signals used for information transmission in a variety of practical cases. Specifically, we consider waveforms representative of signals used in digital electronic systems. Our discussion of the frequency spectrum of such signals will be brief. A detailed discussion of the spectral analysis of power signals can be found in the references cited; for example, in [7].

4.4.1 Periodic Signals

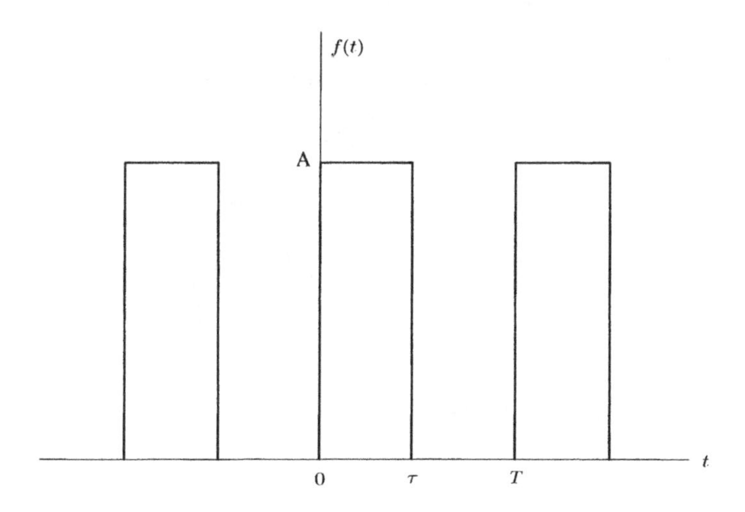

Figure 4.7 Periodic rectangular pulse train.

Assume a signal $f(t)$ consisting of a periodic rectangular pulse train where each pulse has time-width of τ and the pulses repeat with time period T. The signal is sketched in Figure 4.7 and can be represented by

$$f(t) = \begin{cases} A, & 0 \leq t \leq \tau \\ 0, & \tau < t \leq \tau. \end{cases} \tag{4.16}$$

The periodicity of the pulses requires that

$$f(t + n\tau) = f(t) \tag{4.17}$$

with $n = 0, \pm 1, \pm 2, \pm 3, \ldots$ The time period τ defines the fundamental frequency of the signal as

$$f_0 = \frac{1}{T} \qquad \text{Hz} \tag{4.18}$$

with $\omega_0 = 2\pi f_0$ being the fundamental frequency expressed in radians per second.

The signal $f(t)$ can be represented [7] by a complex Fourier series given below:

$$f(t) = \sum_{n=-\infty}^{\infty} a_n e^{jn\omega_0 t}, \tag{4.19}$$

where

$$a_n = \frac{1}{T} \int_{t_1}^{t_1+T} f(t) \, e^{-jn\omega_0 t} \, dt. \tag{4.20}$$

In (4.20), where t_1 is arbitrary, it is important that the integration be carried out over a time interval T. We assume $t_1 = 0$, which then simplifies the analysis.

We make the following observations from (4.19) and (4.20):

1. $f(t)$ consists of a dc component (i.e., of zero frequency) and an infinite number of discrete positive and negative frequencies, given by $\pm n\omega_0$, that are harmonically related to the fundamental frequency ω_0.
2. Since $f(t)$ is real, from (4.20), with $t_1 = 0$ for negative frequency $-n\omega_0$ we obtain

$$a_{-n} = \frac{1}{T}\int_0^T f(t)\, e^{+jn\omega_0 t}\, dt = a_n^*, \tag{4.21}$$

which is similar to that obtained for energy signals (Section 4.3).

We wish to express $f(t)$ in terms of positive frequencies only. This is done by assuming $a_n = |a_n|e^{j\alpha_n}$ and using (4.21) in (4.19) to obtain

$$f(t) = a_0 + \sum_{n=1}^{\infty} 2|a_n| \cos(n\omega_0 t + \alpha_n). \tag{4.22}$$

From (4.20) we obtain the Fourier coefficients

$$a_n = \frac{A\tau}{T}\, e^{-jn\omega_0\tau/2}\, \frac{\sin(n\omega_0\tau/2)}{n\omega_0\tau/2}. \tag{4.23}$$

Taking the results, we obtain

$$|a_n| = \frac{A\tau}{T}\left|\frac{\sin(\pi n f_0 \tau)}{\pi n f_0 \tau}\right| \tag{4.24}$$

$$a_n = -\pi n f_0 \tau + \arg\sin\left(\frac{\pi n f_0 \tau}{\pi n f_0 \tau}\right) \tag{4.25}$$

$$n = 0, \pm 1, \pm 2, \ldots,$$

where we have expressed the coefficients in terms of the fundamental frequency f_0. Equations (4.24) and (4.25) give the two-sided amplitude and phase spectrum for the rectangular pulse signal. The one-sided spectral intensity of the same signal is given by

$$S = \begin{cases} a_0 & \text{for } n = 0 \\ 2|a_n| & \text{for } n \neq 0, \end{cases} \tag{4.26a}$$

that is,

$$S = \begin{cases} \frac{A\tau}{T} & \text{for } n = 0 \\ = \frac{2a\tau}{T}\left|\frac{\sin(\pi n f_0 \tau)}{(\pi n f_0 \tau)}\right| & \text{for } n \neq 0. \end{cases} \tag{4.26b}$$

Various spectra for the rectangular pulse signal are shown in Figure 4.8, where we have indicated the use of a continuous frequency variable $f = nf_0$. Note that the envelopes of the amplitude spectrum and spectral intensity can be expressed in terms of the $\sin x/x$ function. For example,

$$|a_n(f)| = \left|\frac{\sin(\pi f \tau)}{\pi f \tau}\right|, \tag{4.27}$$

with f as a continuous variable. Equation (4.27) indicates that the zero a_n's occur at

$$f = \frac{m}{\tau}, \qquad m = \pm 1, \pm 2, \ldots. \tag{4.28}$$

and for the spectral intensity shown in Figure 4.8*c*.

The following observations are made from an examination of (4.27), (4.29) and Figure 4.8:

1. All component frequencies or spectral lines are located at integer multiples of the fundamental frequency, i.e., they are given by $nf_0 = \frac{n}{T}$.
2. As the fundamental frequency increases (or T decreases), the density of spectral lines decreases and the amplitude of the lines increases.
3. The shapes of the envelopes of the amplitude spectrum and the spectral intensity are primarily of the form $\frac{\sin x}{x}$.

A Special Case

We consider a special case having application to clock signals used in digital systems. Assume that the duty cycle $\frac{\tau}{T} = \frac{1}{2}$, meaning 50%. Using (4.23) and (4.24), we find that

$$a_0 = \frac{A}{2}, \qquad |a_n| = f2A|n|\pi\left|\sin\frac{n\pi}{2}\right|, \qquad n = 1, 2, \ldots \tag{4.29}$$

$$|a_1| = \frac{2A}{\pi}, \qquad |a_2| = 0, \qquad |a_3| = \frac{2A}{3\pi}, \tag{4.30}$$

and

$$\alpha_1 = -\frac{\pi}{2}, \alpha_3 = -\frac{\pi}{2}, \qquad \text{etc.}$$

Thus, (4.22) reduces to

$$f(t) = \frac{A}{2} + \frac{2A}{\pi}\sin(\omega_0 t) + \frac{2A}{3\pi}\sin(3\omega_0 t) + \cdots \tag{4.31}$$

We conclude from here that a periodic rectangular pulse having a 50% duty cycle will have only odd harmonics, although there is a constant term present.

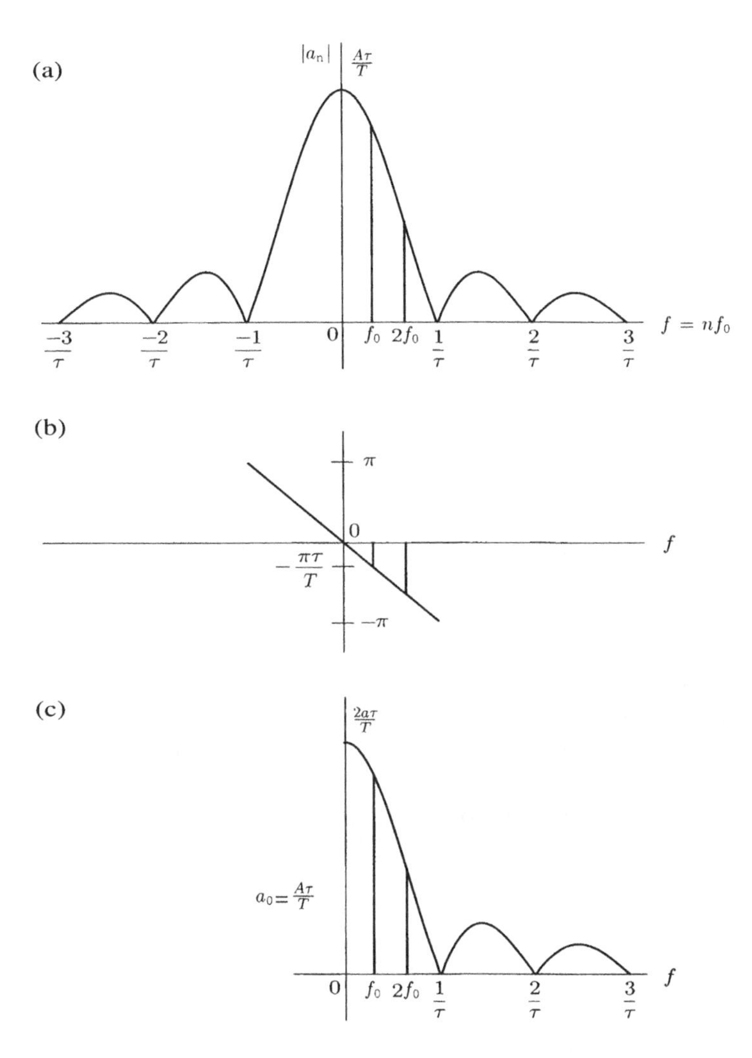

Figure 4.8 (a) amplitude spectrum, (b) phase spectrum, and (c) spectral intensity of a rectangular pulse signal. Pulse width τ and time period $T = 1/f_0$.

4.4.2 Trapezoidal Waveform

Generally, clock signals in digital circuits are periodic pulses of trapezoidal waveforms sketched in Figure 4.9 where the key parameters of the waveform are also indicated.

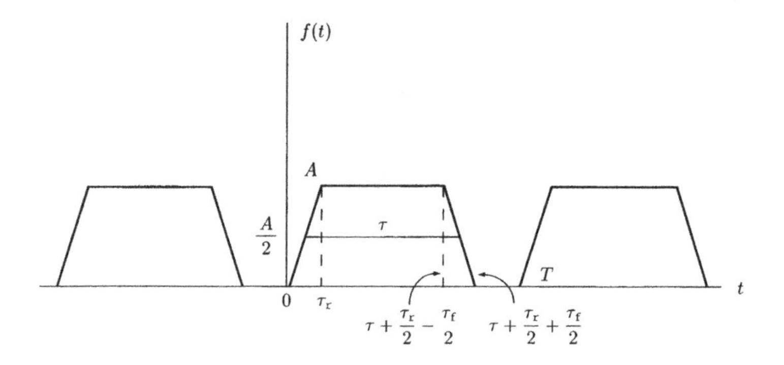

Figure 4.9 Periodic trapezoidal pulse train representing a clock signal.

The pulse width τ is defined as the time interval between the 50% points of the waveform amplitude A. τ_r, τ_f are the pulse rise time and fall time from level 0 to level A, and vice versa. T is the time period of the repeating signal.

In [7] the required Fourier coefficients for the signal $f(t)$ are derived by decomposing the given function into linear combinations of simpler functions. This simplifies the calculation.

We will represent $f(t)$ as a Fourier series in a standard manner described earlier. Since the integrations required for the determination of the Fourier coefficients a_n are rather lengthy, we will present here the key steps involved.

From Figure 4.9 the complete representation of $f(t)$ in the time interval $0 \leq t \leq T$ is

$$\begin{aligned} f(t) = f_1(t) &= \frac{At}{\tau_r}, 0 \leq t \leq \tau_r \\ &= f_2(t) = A, \tau_r \leq t \leq \tau + \frac{\tau_r}{2} - \frac{\tau_f}{2} \\ &= f_3(t) = -\frac{At}{\tau_f} + \frac{A}{\tau_f}\left(\tau + \frac{\tau_r}{2} + \frac{\tau_f}{2}\right), \\ &\qquad \left(\tau + \frac{\tau_r}{2} - \frac{\tau_f}{2}\right) \leq f \leq \tau + \frac{\tau_r}{2} + \frac{\tau_f}{2} \\ &= 0, \tau + \frac{\tau_r}{2} + \frac{\tau_f}{2} \leq t \leq T. \end{aligned} \tag{4.32}$$

From the preceding representation of $f(t)$, we can now express the required a_n as

$$a_n = I_1 + I_2 + I_3, \tag{4.33}$$

where

$$I_1 = \frac{1}{T}\int_0^{\tau_r} f_1(t)\, e^{-jn\omega_0 t} \tag{4.34a}$$

$$I_2 = \frac{1}{T}\int_{\tau_r}^{\tau+\tau_r/2-\tau_f/2} f_2(t)\, e^{-jn\omega_0 t} \tag{4.34b}$$

$$I_3 = \frac{1}{T}\int_{\tau+\tau_r/2-\tau_f/2}^{\tau+\tau_r/2+\tau_f/2} f_3(t)\, e^{-jn\omega_0 t} \tag{4.34c}$$

and $\omega_0 = 2\pi/T$, $T = 1/f_0$, f_0 being the fundamental frequency.

Introducing the expressions for $f_1(t)$, $f_2(t)$, and $f_3(t)$ from (4.32) into the appropriate expressions of (4.34a)–(4.34c) we can perform the integrations. After rearrangement of certain terms we obtain the following

$$\begin{aligned}
I_1 &= \frac{A}{-jn\omega_0}\,\frac{1}{T}\,e^{-jn\omega_0(\tau+\tau_r/2)} \\
&\quad \cdot \left[e^{jn\omega_0\tau} - e^{-jn\omega_0\tau}\frac{\sin(n\omega_0\tau_r/2)}{n\omega_0\tau_r/2}\right], \\
I_2 &= \frac{A}{-jn\omega_0}\,\frac{1}{T}\,e^{-jn\omega_0(\tau+\tau_r/2)}\left[e^{jn(\omega_0/2)\tau_f} - e^{jn\omega_0\tau}\right], \\
I_3 &= \frac{A}{-jn\omega_0}\,\frac{1}{T}\,e^{-jn\omega_0(\tau+\tau_r/2)} \\
&\quad \cdot \left[e^{jn(\omega_0\tau_f/2)} + \frac{\sin(n\omega_0\tau_f/2)}{n\omega_0\tau_f/2}\right].
\end{aligned}$$

Inserting the values of I_1, I_2 and I_3 in (4.33), we obtain

$$a_n = \frac{A}{-jn\omega_0 T}\,e^{-jn\omega_0(\tau+\tau_r/2)}\left[\frac{\sin(n\omega_0\tau_f/2)}{n\omega_0\tau_f 2} - e^{-jn\omega_0\tau}\frac{\sin(n\omega_0\tau_r/2)}{n\omega_0\tau_r/2}\right]. \tag{4.35}$$

With $\tau_f = \tau_r$, a_n is given by

$$a_n = \frac{A\tau}{T}\,e^{-jn\omega_0(\tau+\tau_r/2)}\left[\frac{\sin(n\omega_0\tau/2)}{n\omega_0\tau/2}\cdot\frac{\sin(n\omega_0\tau_r/2)}{n\omega_0\tau_r/2}\right]. \tag{4.36}$$

Observe that for $\tau_r = \tau_f = 0$, the signal is a train of rectangular pulses and a_n given by (4.36) reduces to that given earlier by (4.23), as it should be.

The amplitude spectrum, the phase spectrum and the spectral intensity for the trapezoidal waveform signal with equal rise and fall time can now be obtained from (4.36) in a manner similar to that for the rectangular pulse signal (see (4.27)–(4.29)). For example, the spectral intensity for the trapezoidal clock signal (with $\tau_r = \tau_f$) is

$$S(f) = \frac{2A\tau}{T}\left|\frac{\sin(\pi\tau f)}{(\pi\tau f)}\right|\left|\frac{\sin(\pi\tau_0 f)}{(\pi\tau_0 f)}\right|, \tag{4.37}$$

where we have assumed $f = nf_0$ and note that $S(0) = A\tau/T$, as explained earlier.

We are now in a position to sketch the envelope of the spectral intensity or its Bode plot for the assumed trapezoidal pulse signal. It is sketched in Figure 4.10. A brief description of bode plots is given in Section 7.8.1.

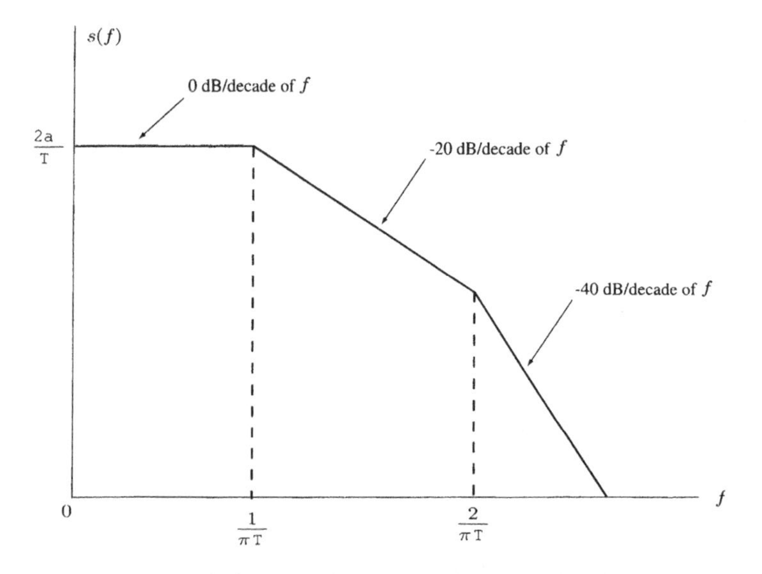

Figure 4.10 Bode plot of the envelope of the spectral intensity function of a trapezoidal pulsed signal.

The clock signals play an important role in the EMC of any digital electronic system. Therefore for future reference we make the following observations from an examination of the spectral intensity envelope shown in Figure 4.10:

1. The three regions of the envelope – the CD region with a slope of 0 dB/decade of frequency f, the DE region with a slope of -20 dB/decade of frequency f, and the EF region with a slope of -40 dB/decade of frequency f – can be explained by the logarithmic behavior of $S(f)$ versus f, with relation to Figure 4.6.

2. The high-frequency content of the trapezoidal pulse (mainly for $f \geq 1/\pi\tau_r$) is mainly determined by the rise/fall time of the pulse.

3. The extent of the high-frequency spectrum can be reduced by increasing the rise/fall time, which in turn would reduce the high-frequency emission from the system using the clock signal waveform parameters (τ, τ_r, τ_p). This is important from the viewpoint of the high-frequency interference signal generated by the system.

A good discussion of numerically and experimentally obtained bounds on the spectral intensity envelope of clock signals having a variety of duty cycles and rise/ fall times is given in [7].

4.5 EXAMPLES OF SOME SIGNALS

As an illustration, we discuss some classes of signals which may be typical of signals used for information transmission and some of which may act as undesirable or interfering signals. Figure 4.11 illustrates the corresponding waveforms (time domain behavior) and the amplitude spectrum (frequency domain Fourier spectrum) for a selected class of signals [10].

Signal 1 is a continuous sinusoidal signal of amplitude A and frequency f_0 existing for all time t. The frequency domain spectrum of this wave consists of a line of infinite amplitude at the frequency f_0, which is a delta function $\delta(f_0)$.

Signal 2 is similar to signal 1 but it is limited in time; that is it exists from time t_0 to t_1. The spectrum of the signal is as shown, which indicates a peak at f_0 in addition to other frequency components.

Signal 3 is a single pulse and transient in nature starting at time t_0 and slowly decaying with time. The frequency spectra indicate that it is a wideband signal having significant energy at low frequencies.

Signal 4 is a periodic train of pulses, each pulse being similar to signal 3, with period $T\,(= 1/f_0)$. The spectrum has many peaks occurring at frequencies $f_n = nf_0,\ n = 1, 2, \ldots$.

Signal 5 is an impulse (δ-function) in time having a flat frequency spectrum, that is it has all frequency components with equal amplitude. This is an ideal signal but is useful for analysis.

Signal 6 is a random transient typical of white noise superimposed on a dc signal whose spectrum amplitude is also random.

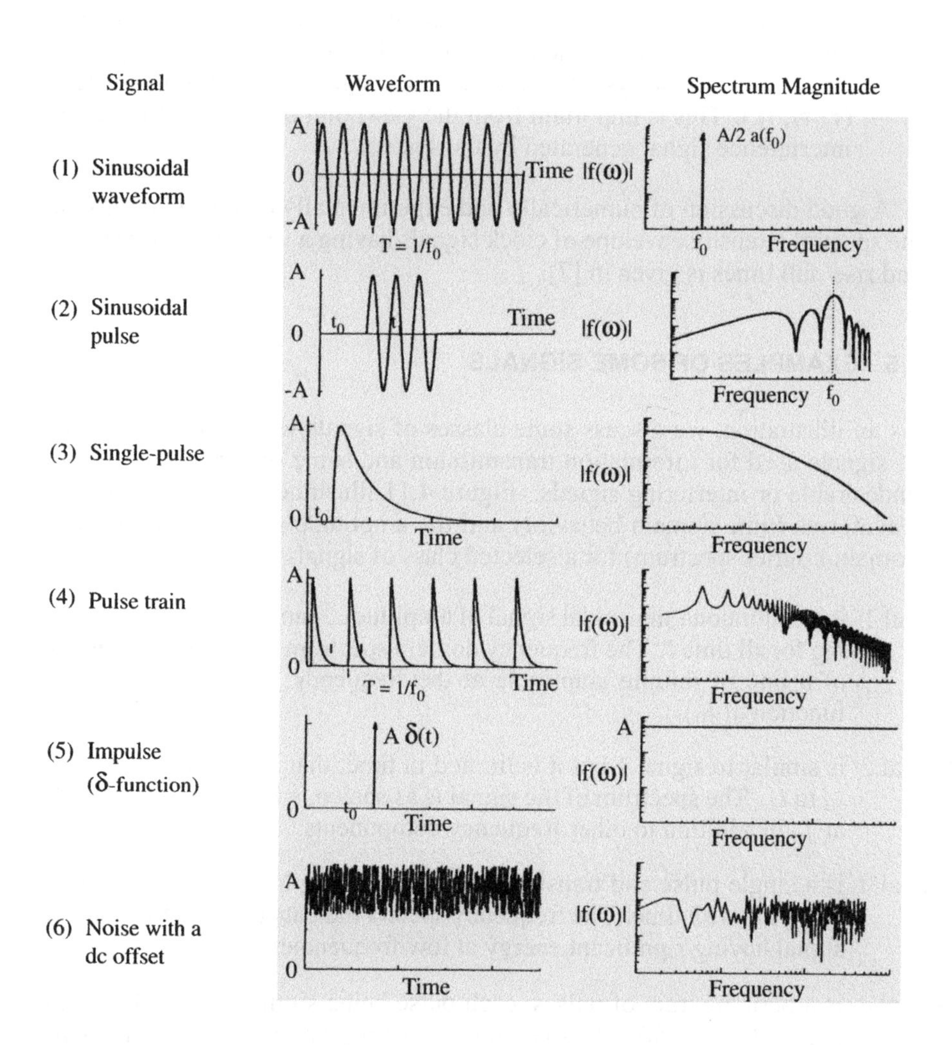

Figure 4.11 Waveforms and spectra of selected class of signals of importance in EMC. (Source: From [10], p.17.)

Note that the function $f(t)$ representing each of the signals need not be continuous but must be a single-valued function of time.

REFERENCES

1. D. K. Frederick and A. B. Carlson, *Linear Systems in Communications and Control*, Wiley, New York, 1971.
2. D. F. Mix, *Random Signal Analysis*, Addison-Wesley, Reading, MA, 1969.
3. M. Schwartz, *Information Transmission, Modulation and Noise*, 4th Edition, McGraw-Hill, New York, 1990.
4. A. Papoulis, *The Fourier Integral and Its Application*, McGraw-Hill, New York, 1962.
5. A. Papoulis, *Signal Analysis*, McGraw-Hill, New York, 1997.
6. C. R. Paul, *Analysis of Linear Circuits*, McGraw-Hill, New York, 1989.
7. C. R. Paul, *Introduction to Electromagnetic Compatibility*, Wiley, New York, 1992.
8. S. J. Mason and H. J. Zimmerman, *Electronic Circuits, Signals and Systems*, Wiley, New York, 1960.
9. A. A. Smith Jr., *Radio Frequency Principles and Applications*, IEEE Press/ Chapman and Hall, New York, 1998.
10. F. M. Tesche, M. V. Ianoz and T. Karlsson, *EMC Analysis Methods and Computational Models*, Wiley, New York, 1997.

PROBLEMS

4.1 Derive (4.22) from the complex Fourier series representation (4.19) for a real periodic function $f(t)$.

4.2 Given a sawtooth waveform sketched in Figure 4.12, determine the following for $f(t)$:

(a) Average value for $f(av)$

(b) The rms value $f(rms)$

(c) Complex Fourier series coefficients $a(n)$

(d) Explicit series representation.

Answer:
(a) $f_{\text{av}} = \frac{A}{2}$; **(b)** $f_{\text{rms}} = \frac{A}{\sqrt{3}}$; **(c)** $a_0 = \frac{A}{2} = f_{\text{av}}$; $a_n = \frac{jA}{2n\pi}$, $n = \pm 1, \pm 2, \ldots$;
(d) $f(t) = 2\, f_{\text{av}} \left\{ \frac{1}{2} - \sum_1^{\infty} \frac{1}{n\pi} \sin \frac{n2\pi t}{T} \right\}$.

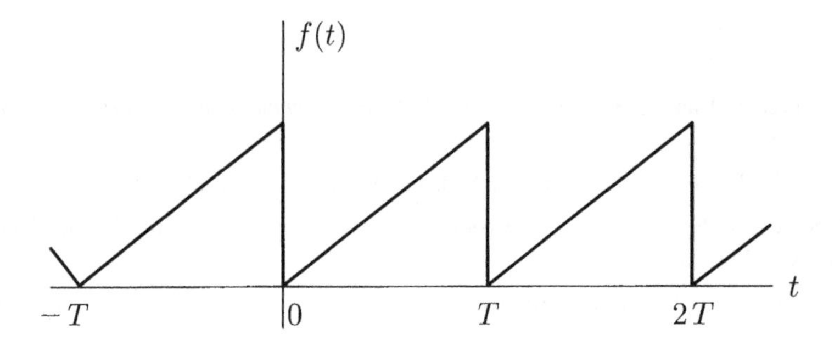

Figure 4.12 Sawtooth Waveform.

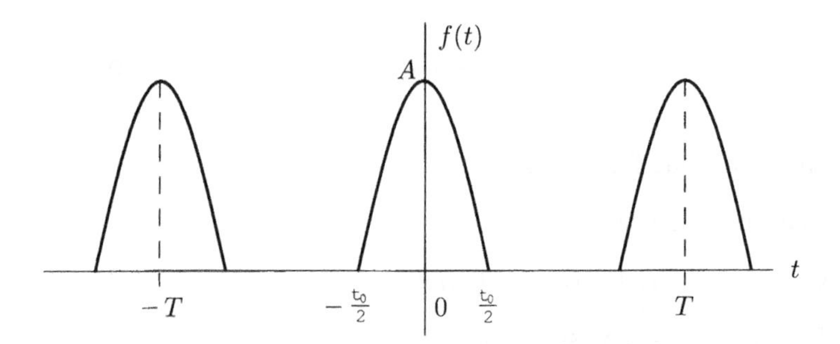

Figure 4.13 Train of cosine pulses.

4.3 Given a waveform $f(t)$ consisting of a train of cosine pulses sketched in Figure 4.13, determine the following for $f(t)$

(a) Average value $f(av)$

(b) The rms value $f(rms)$

(c) Complex Fourier series coefficients $a(n)$

(d) Explicit series representation.

Answer:
(a) $f_{\rm av} = \frac{2A}{\pi}\frac{t_0}{T}$; **(b)** $f_{\rm rms} = A\left(\frac{t_0}{2T}\right)^{\frac{1}{2}}$; **(c)** $a_0 = \frac{2A}{\pi}\frac{t_0}{T} = f_{\rm av}$, $a_n = \frac{2A}{\pi}\frac{t_0}{T}\frac{\cos\left(\frac{n\pi t_0}{T}\right)}{1-\left(\frac{2nt_0}{T}\right)^2}$, note $a_n = a_{-n}$; **(d)** $f(t) = s\, f_{\rm av}\left\{\frac{1}{2} + \sum_{n=1}^{\infty}\frac{\cos\left(\frac{n\pi t_0}{T}\right)}{1-\left(\frac{2nt_0}{T}\right)^2}\cos\frac{2n\pi t}{T}\right\}$

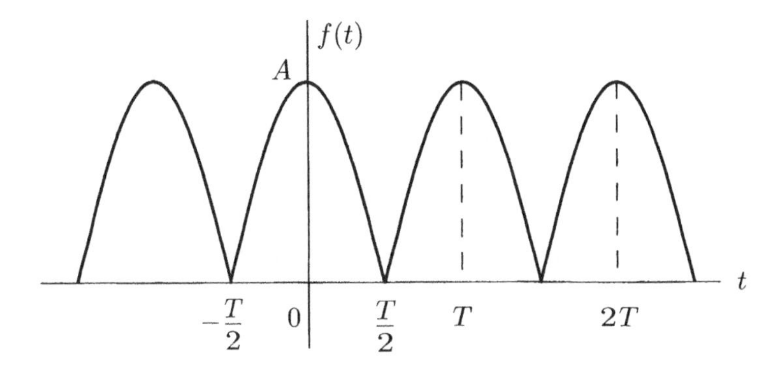

Figure 4.14 Full-wave rectified sine wave.

4.4 Given a full-wave rectified sine sketched in Figure 4.14, determine the following for $f(t)$:

(a) Average value $f(av)$

(b) The rms value $f(rms)$

(c) Complex Fourier series coefficients $a(n)$

(d) Explicit series representation.

(*Hint*: Use $t(0)$=T in (4.3))

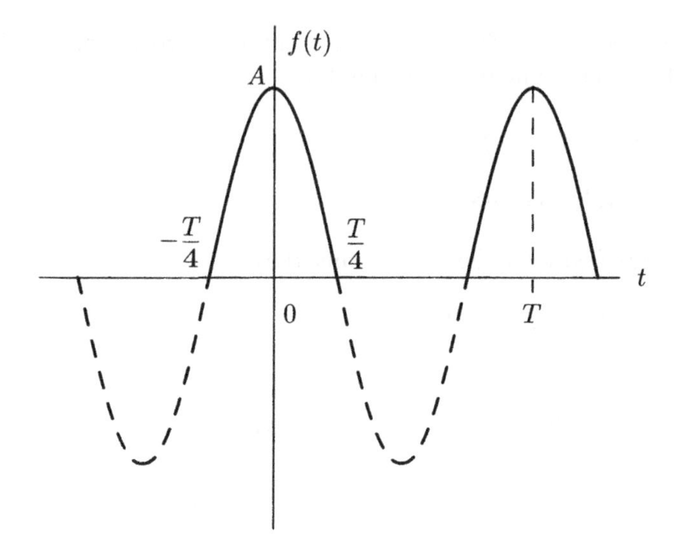

Figure 4.15 Half-wave rectified sine waveform.

Answer:
(a) $f_{\text{av}} = \frac{A}{\pi}$; $f_{\text{rms}} = \frac{A}{\sqrt{2}}$; **(b)** $a_n = -\frac{2A}{\pi}\frac{(-1)^n}{4n^2-a}$, **(c)** $a_0 = \frac{2A}{\pi} = f_{\text{av}}$;
(d) $f(t) = 2f_{\text{av}}\left\{\frac{1}{2} - \sum \frac{(-1)^n}{4n^2-a}\cos\left(\frac{n2\pi t}{T}\right)\right\}$.

4.5 Given a half-wave rectified sine waveform sketched in Figure 4.15, determine the following for $f(t)$:

(a) Average value $f(av)$

(b) The rms value $f(rms)$

(c) Complex Fourier series coefficients $a(n)$

(d) Explicit series representation.

(*Hint*: Use $t(0)$=T/2 in (4.3))

Answer:
(a) $f_{\text{av}} = \frac{A}{\pi}$; **(b)** $f_{\text{rms}} = \frac{A}{2}$; **(c)** $a_0 = \frac{A}{\pi} = f_{\text{av}}$, $a_n = \frac{A}{\pi}\frac{\cos\left(\frac{n\pi}{2}\right)}{1-n^2}$, $a_{2n+1} = 0$ except for $a_1 = f_{\text{av}}\frac{\pi}{4}$. for even $n = 2m$: $a_{2m} = -\frac{A}{\pi}\frac{(-1)^m}{4m^2-1}$; **(d)** $f(t) = 2\,f_{\text{av}}\left\{\frac{1}{2} + \frac{\pi}{4}\cos\frac{2\pi t}{T} + \frac{1}{3}\cos\frac{2\pi t}{T} - \frac{1}{15}\cos 4\frac{2\pi t}{T} \ldots - \frac{(-1)^m}{4m^2-1}\cos 2m\frac{2\pi t}{T}\right\}$.

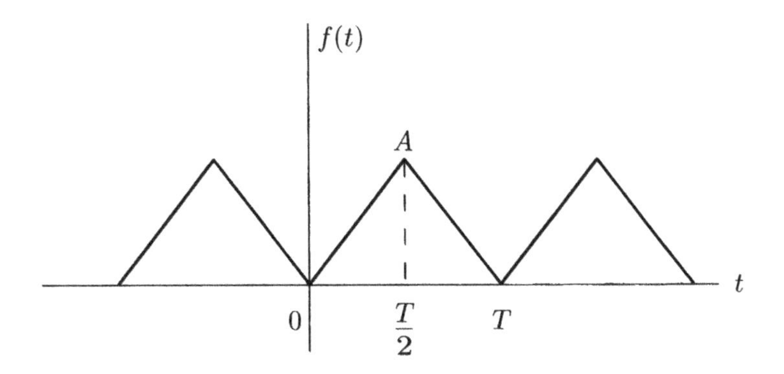

Figure 4.16 Triangular waveform.

4.6 Given a triangular waveform sketched as shown in Figure 4.16, determine the following for $f(t)$:

(a) Average value $f(av)$

(b) The rms value for $f(rms)$

(c) Complex Fourier coefficients $a(n)$

(d) Explicit series representation.

Answer:
(a) $f_{\text{av}} = \frac{A}{2}$; **(b)** $f_{\text{rms}} = \frac{A}{\sqrt{3}}$; **(c)** $a_0 = \frac{A}{2} = f_{\text{av}}$; $a_n = -\frac{A}{\pi^2 n^2}$, $n \neq 0$;
(d) $f(t) = 2\, f_{\text{av}} \left\{ \frac{1}{2} - \sum_{n=1}^{\infty} \frac{2}{\pi^2 n^2} \cos\left(\frac{n 2\pi f}{T}\right) \right\}$, for n as odd numbers.

CHAPTER 5

TRANSMISSION LINES

5.1 INTRODUCTION

Generally, the transfer of electromagnetic energy or signals from source to destination is accomplished in two ways. When dealing with shorter distances, some kind of continuous structure capable of guiding electromagnetic waves through the desired distance is used. Examples of such structures include transmissions lines, waveguides, printed circuit board (PCB) traces, and fibers. When the distances involved are larger, the propagation of waves through free space is utilized for transmission purposes. Reception of commercial broadcast signals through air provides a good example of wave propagation through free space. Both methods are in use singly or in combination in many communications systems.

In this chapter we will discuss the commonly used transmission lines that form only one specific class of guiding structures used for transmission of electromagnetic signals. Such transmission lines include two-wire lines, coaxial lines, and telephone lines that may carry audio and video information to the

Applied Electromagnetics and Electromagnetic Compatibility. By D. L. Sengupta, V. V. Liepa
ISBN 0-471-16549-2

TV receivers or digital data to computer monitors. Although two-wire lines are widely used by the utility industry for transmission of power, our discussion will not include such low frequency application. Transmission lines are discussed in many books, for example, [1–7]. Our brief discussion will touch upon the basic aspects of transmission lines and their application in the context of EMC. Further details can be found in the references cited.

5.2 BASIC DISCUSSION

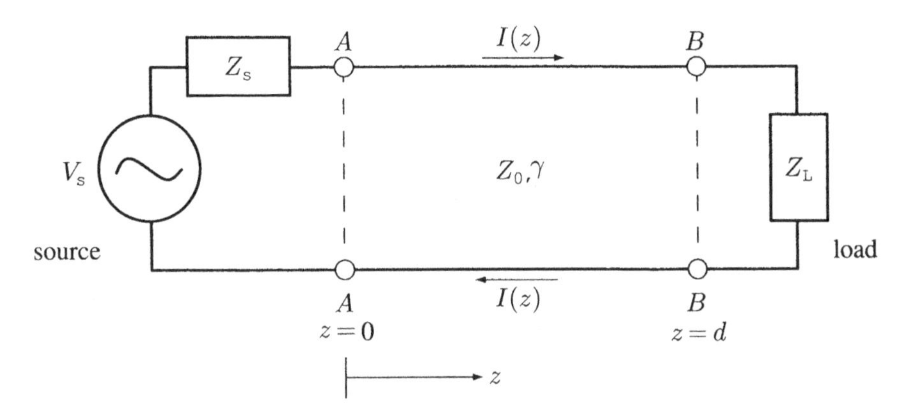

Figure 5.1 Typical transmissions line connecting a source to a load.

Figure 5.1 shows diagrammatically a typical arrangement for a time harmonic source of frequency $\omega = 2\pi f$ connected to a load Z_{L} at the receiving end through a transmission line of length d.

The voltage across the lines, the current through each line, and the relevant field quantities **E**, **H** on the line travel along the longitudinal z direction as $e^{-j\beta z}$. $\beta = \omega/v_{\mathrm{p}}$, v_{p} is the phase velocity of the wave propagating in the line. It is clear that a signal of frequency ω will suffer a phase delay after travelling from the input side A–A' to the output side B–B', which is given by

$$\phi = \beta d = \frac{\omega d}{v_{\mathrm{p}}} = \frac{2\pi d}{\lambda}, \tag{5.1}$$

where λ is the wavelength on the line.

At low frequencies, $d/\lambda \ll 1$, and (5.1) indicates that the phase delay is negligible. More important, it implies that the signal reaches instantaneously from the source to the load at all frequencies. This is generally the case at low-frequency circuits, where the finite velocity of signal propagation is insignificant due to the small values of the maximum linear dimensions of the circuits compared with the wavelengths involved. Under such conditions ordinary lumped circuit analysis involving lumped parameters R, L, C, etc., are sufficient for analysis.

However, when $d/\lambda \gg 1$, the phase or time delay effects mentioned become significantly large and cannot be ignored. Moreover the variation of the phase velocity with frequency during transmission may be such that distortions will be introduced through dispersion of the signal if it contains multiple frequencies. For such conditions an analysis of transmissions lines is carried out by applying field theory concepts. We will consider the important class of transverse electromagnetic (TEM) lines that are widely used in a variety of applications. In such lines the propagating waves contain electric and magnetic fields that are predominantly transverse to the direction of propagation (z direction in Figure 5.1).

5.3 TRANSVERSE ELECTROMAGNETIC (TEM) TRANSMISSION LINES

Longitudinal and transverse plane views of selected TEM lines are shown in Figure 5.2. It is assumed that the lines are excited at the left hand side so that energy is propagated along the z direction.

As shown, the z direction is defined as the longitudinal direction along which the energy propagates; also observe that each line has a constant (or uniform) transverse cross section (in the xy plane) along the longitudinal direction. This is why the lines shown here are also called uniform transmission lines.

For the purpose of illustration, consider the coaxial line in Figure 5.2b is fed at the left-hand side. It can be shown that the field configuration within the coaxial region is entirely transverse, meaning the fields consist of $\mathbf{E} = \hat{\rho}E_\rho$, $\mathbf{H} = \hat{\phi}H_\phi$, and $E_z = H_z \equiv 0$ in the present case. The average power flow in the line is $\mathbf{S}_{av} = Re[\frac{1}{2}\mathbf{E} \times \mathbf{H}] = E_\rho H_\phi \hat{z}$, meaning in the longitudinal z direction, and the fields are TEM type.

Similar considerations indicate that the two-wire line of Figure 5.2a ideally sustains a predominantly TEM type of field configuration. It can be shown that the fields here consist of $\mathbf{E} = \hat{x}E_x$, $\mathbf{H} = \hat{y}H_y$, and $E_z = H_z = 0$, thereby propagating energy in the z direction.

Neglecting the edge effects, it is simple to visualize that the fields within the parallel plate region of the line in Figure 5.2c will consist of $\mathbf{E} = \hat{x}E_x$, $\mathbf{H} = \hat{y}H_y$, $E_z = H_z = 0$, and hence the energy propagation is in the z direction.

The microstrip line in Figure 5.2d does not support strictly a TEM mode of fields; it does contain a small E_z component. Generally, E_z is neglected and

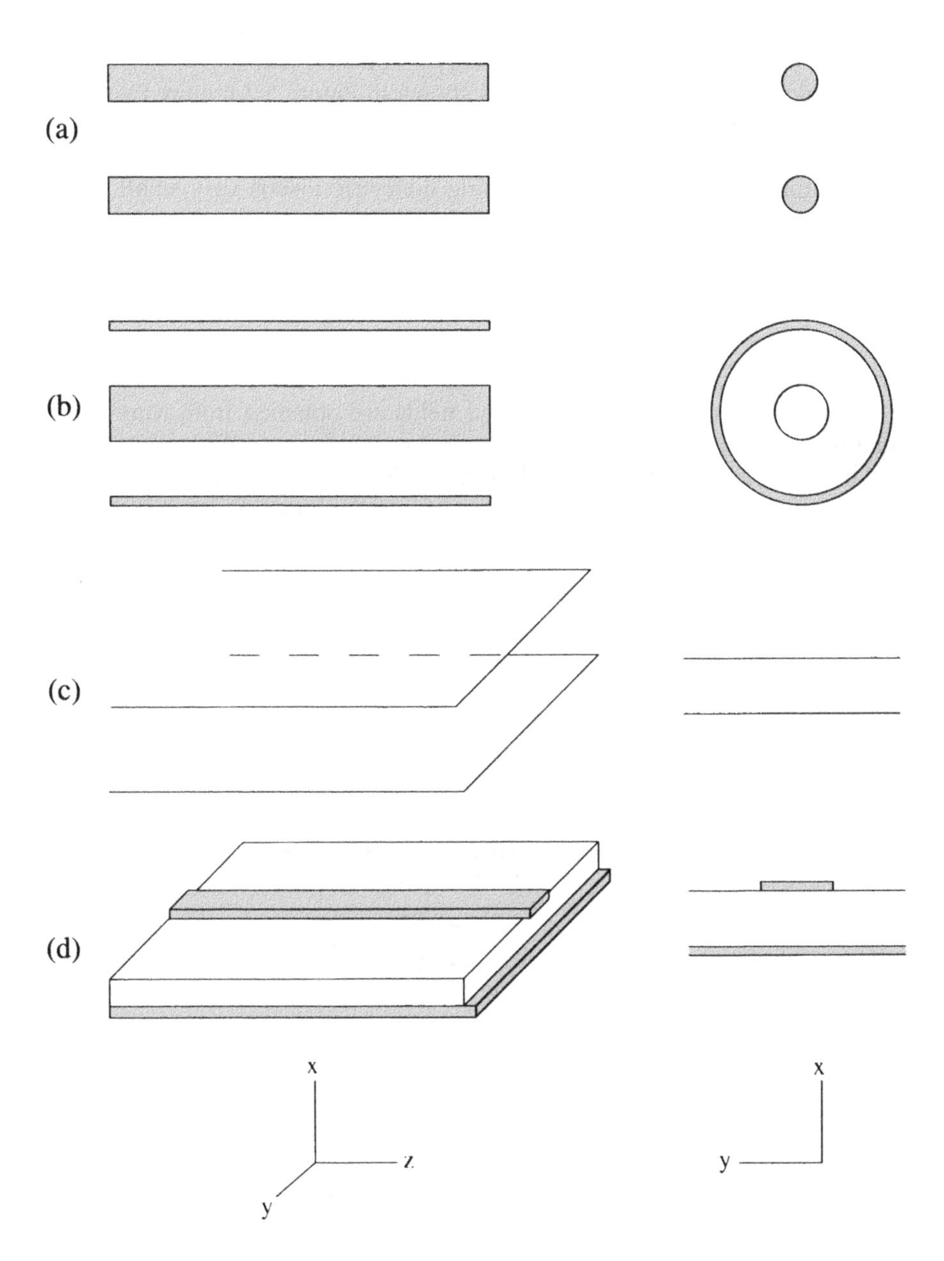

Figure 5.2 Longitudinal and transverse views of selected TEM transmission lines: (*a*) open two-wire; (*b*) coaxial; (*c*) parallel plate; (*d*) microstrip.

it is assumed that the fields within the microstrip region are only $\mathbf{E} = \hat{x}E_x$ and $\mathbf{H} = \hat{y}H_y$ thereby transmitting energy in the z direction.

It should be noted that all four lines shown in Figure 5.2 require two conductors, as is true for all TEM transmission lines. We will assume that the medium within the line is homogeneous and is characterized by μ_0, $\epsilon_0\epsilon_r$, σ; σ is assumed to be very small, meaning dielectric loss is very small. The conductivity of the conductors used is σ_c.

5.4 TELEGRAPHER'S EQUATIONS: QUASI-LUMPED CIRCUIT MODEL

A rigorous way to analyze the lines in Figure 5.2 is to consider them as boundary value problems where the required fields are obtained from Maxwell's equations and then apply the proper boundary conditions. We will not proceed in this complicated way but follow a simpler method called *quasi-lumped circuit theory*. This method involves modifying the lumped circuit model so that the finite time of propagation through the line is taken into account [1,3,4].

Intuitively, it is argued that the resistance, conductance, inductance, and capacitance associated with the line are distributed continuously along the length of the line. We shall represent them by lumped parameters R, G, L, C, respectively on a per unit length basis. Although these parameters are different for different lines shown in Figure 5.2, it is then possible to use a single quasi-lumped circuit model for each.

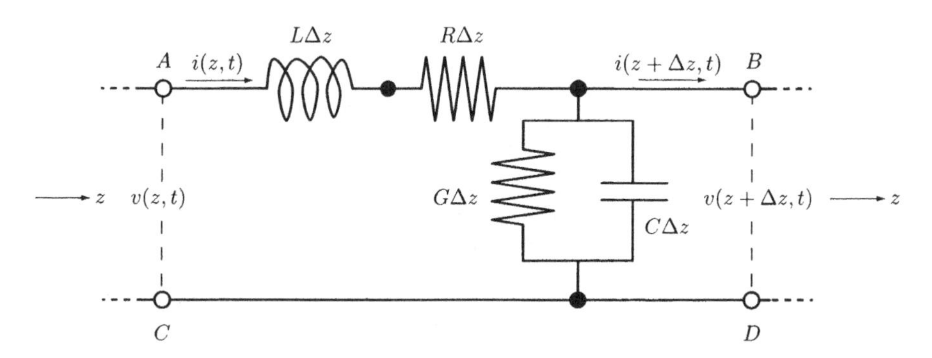

Figure 5.3 Quasi-lumped circuit model for the line between z and $z + dz$.

Thus the model for a section of the line between z and $z + dz$ (or between AD and BC) is shown in Figure 5.3 at time t, where the instantaneous values of the voltage and currents or $v(z,t)$ and $i(z,t)$ are as shown, and

R is the resistance in ohms per unit length of the line,

G is the conductance in mhos per unit length of the line,

L is the inductance in henries per unit length of the line, and

C is the capacitance in farads per unit length of the line.

Now, applying Kirchoff's voltage law (i.e., KVL) along the loop $ABCD$, we obtain

$$-v(z,t) + L\Delta z \frac{\partial i(z,t)}{\partial t} + R\Delta z\, i(z,t) + v(z+\Delta z, t) = 0. \tag{5.2a}$$

Applying Kirchoff's Current Law (KCL) at the node A, we obtain

$$\begin{aligned} i(z,t) &= i(z+\Delta z, t) + di \\ &= i(z+\Delta z, t) + C\Delta z \frac{\partial v(z,t)}{\partial t} + G\Delta z\, v(z,t). \end{aligned} \tag{5.2b}$$

Using the equations (5.2a) and (5.2b) and taking the limit $\Delta z \to 0$, we see that they reduce to

$$\frac{\partial v(z,t)}{\partial z} = -L\,\frac{\partial i(z,t)}{\partial t} - R\, i(z,t) \tag{5.3}$$

$$\frac{\partial i(z,t)}{\partial z} = -C\,\frac{\partial v(z,t)}{\partial t} - G_2\, v(z,t). \tag{5.4}$$

Equations (5.3) and (5.4) are called the *time dependent telegrapher's equations* (or *transmission line equations*).

To obtain the corresponding phasor equations from the time harmonic case we assume

$$\begin{aligned} v(z,t) &= \mathrm{Re}[V(z)\, e^{j\omega t}] \\ i(z,t) &= \mathrm{Re}[I(z)\, e^{j\omega t}]. \end{aligned} \tag{5.5}$$

We obtain from (5.3) and (5.4) the following:

$$\frac{dV}{dz} = -(R + j\omega L)I = -ZI \tag{5.6}$$

$$\frac{dI}{dz} = -(G + j\omega C)V = -YV, \tag{5.7}$$

where

$$Z = R + j\omega L = \text{series impedance per unit length for the line} \tag{5.8a}$$

$$Y = G + j\omega C = \text{series admittance per unit length (mhos = siemens).} \tag{5.8b}$$

Equations (5.6) and (5.7) are the frequency domain telegrapher's equations. The time and frequency domain equations given above are appropriate for solutions in the respective domains, that is for solutions giving time and frequency behavior of the solutions, respectively. In both cases we assume for lossless lines $R = G \equiv 0$.

5.5 WAVE EQUATIONS

For a lossless line, (5.3) and (5.4) reduce to

$$\frac{\partial v(z,t)}{\partial z} = -L\,\frac{\partial i(z,t)}{\partial t} \tag{5.9}$$

$$\frac{\partial i(z,t)}{\partial z} = -C\,\frac{\partial v(z,t)}{\partial t}. \tag{5.10}$$

Differentiating (5.9) with respect to z and using (5.10), we obtain an equation for $v(z,t)$ as

$$\begin{aligned}\frac{\partial^2 v(z,t)}{\partial z^2} &= -L\,\frac{\partial^2 i(z,t)}{\partial z\,\partial t} = -L\,\frac{\partial}{\partial t}\left(\frac{\partial i}{\partial z}\right)\\ &= LC\,\frac{\partial^2 v}{\partial t^2}\,.\end{aligned} \tag{5.11}$$

Similarly eliminating $v(z,t)$, we obtain

$$\frac{\partial^2 i(z,t)}{\partial z^2} = LC\,\frac{\partial^2 i(z,t)}{\partial t^2}. \tag{5.12}$$

We define a velocity parameter $v = 1/\sqrt{LC}$ m/s, and rewrite (5.11) and (5.12) as

$$\frac{\partial^2 v(z,t)}{\partial z^2} - \frac{1}{v^2}\,\frac{\partial^2 v(z,t)}{dt^2} = 0 \tag{5.13}$$

$$\frac{\partial^2 i(z,t)}{\partial z^2} - \frac{1}{v^2}\,\frac{\partial^2 (i,z,t)}{dt^2} = 0, \tag{5.14}$$

which are the standard one-dimensional wave equations and which indicate that the voltage and current along the line are solutions of the wave equation. It can be shown that solution of (5.13) is

$$v(z,t) = A\,V_+\left(t - \frac{z}{v}\right) + B\,V_-\left(t + \frac{z}{v}\right), \tag{5.15}$$

where A, B are constant. The first and second terms of (5.14) represent waves propagating in the $+z$ and $-z$ directions, respectively. Similar results from the current wave can be obtained from (5.14).

Thus quasi-lumped circuit model (theory) indicates that the voltage and current along the line must be interpreted as waves propagating along the forward ($+z$ direction) and backward ($-z$ direction) when the line is excited from the left side in Figure 5.3.

From the frequency domain equations (5.6) and (5.7), it can be shown that voltage and current along the lines satisfy the same equation

$$\frac{d^2V}{dz^2} + \gamma^2 V = 0 \tag{5.16}$$

$$\frac{d^2I}{dz^2} + \gamma^2 I = 0 \tag{5.17}$$

with

$$\gamma = \sqrt{ZY}. \tag{5.18}$$

Equations (5.16) and (5.17) are the time independent or frequency domain wave equations equivalent to (5.13) and (5.14), respectively. Equations (5.16) and (5.17) are the basis for frequency domain analysis of all transmission lines.

5.6 FREQUENCY DOMAIN ANALYSIS

In this section we discuss the case of a transmission line excited by a time harmonic (i.e., sinusoidal) signal of angular frequency $\omega = 2\pi f$, (f being the frequency) which is of most practical interest. Typical references for this section are [1–4, 7].

5.6.1 General Solution

We can write the general solution of (5.16) as

$$V(z) = V^+e^{-dz} + V^-e^{+dz}, \tag{5.19}$$

where we write explicitly γ (from (5.18)) as

$$\gamma = \sqrt{ZY} = \sqrt{(R + j\omega L)(G + j\omega C)} = \alpha + j\beta, \tag{5.20}$$

where $\alpha, \beta > 0$. The parameter γ is called the complex propagation constant for the line. After using (5.6)–(5.7) and (5.19), we obtain the current as

$$I(z) = I^+e^{-\gamma z} - I^-e^{+\gamma z} \tag{5.21}$$

with

$$I^+ = \frac{V^+}{\sqrt{Z/Y}}, \tag{5.22a}$$

$$I^- = \frac{V^-}{\sqrt{Z/Y}}. \tag{5.22b}$$

From (5.19) it can be interpreted that at any point across the line, the voltage consists of a forward moving wave (first term) in the $+z$ direction and a backward moving wave (second term) in the $-z$ direction, if it is assumed that the line is excited on the left-hand side. Similar interpretation applies to equation (5.21).

The time dependent instantaneous voltage along the line is obtained from (5.19) as

$$\begin{aligned} v(z,t) &= \text{Re}[V(z)\, e^{j\omega t}] \\ &= V^+ e^{-\alpha z} \cos(\omega t - \beta z) + V^- e^{+\alpha z} \cos(\omega t + \beta z). \end{aligned} \tag{5.23}$$

Similarly, using (5.21), we obtain the instantaneous current on the line as

$$\begin{aligned} i(z,t) &= \text{Re}[I(z)\, e^{j\omega t}] \\ &= I^+ e^{-\alpha z} \cos(\omega t - \beta z) + I^- e^{+\alpha z} \cos(\omega t + \beta z), \end{aligned} \tag{5.24}$$

with I^+, I^- as given earlier by (5.22a) and (5.22b) respectively.

Observe that first terms of (5.23) and (5.24) represent a voltage and a current wave, respectively, and which are in phase (at all times), and they propagate together in the forward (i.e., $+z$) direction. Similarly the second terms (V^-, I^- terms) can be interpreted as forming a backward wave propagating in the $-z$ direction. Equations (5.23) and (5.24) indicated that in the direction of propagation the amplitude of the forward and backward waves varies as $e^{-\alpha z}$ and $e^{+\alpha z}$, respectively, and the phase as $-\beta z$ and βz, respectively. We define α, β, as the attenuation and phase constants, respectively; they will be discussed further later.

We define the characteristic impedance Z_0 of the line as the ratio of the voltage and current for the forward or backward wave at any point in the line. Using (5.19) and (5.21), or (5.23) and (5.24), we obtain in alternate forms

$$Z_0 = \frac{V^+}{I^+} = -\frac{V^-}{I^-} = \frac{Z}{\gamma} = \frac{\gamma}{Y}. \tag{5.25}$$

We write (5.25) in the following form:

$$Z_0 = \sqrt{\frac{Z}{Y}} = \sqrt{\frac{R + j\omega L}{G + j\omega C}}. \tag{5.26}$$

5.6.2 Further Discussion of Propagation Constant and Characteristic Impedance

It is of practical interest to know the propagation constant parameters and the characteristic impedance of a line in terms of its basic parameters R, G, L, and C.

Taking the square of (5.20) and equating the real and imaginary parts of both sides, we obtain

$$\alpha^2 - \beta^2 = RG - \omega^2 LC \tag{5.27}$$

$$2\alpha\beta = \omega LG + \omega CR. \tag{5.28}$$

From (5.27) and (5.28) we obtain

$$(\alpha^2 + \beta^2) = \sqrt{(R^2 + \omega^2 L^2)(G^2 + \omega^2 C^2)}. \tag{5.29}$$

From (5.27) and (5.29) we obtain the following explicit expressions for α and β:

$$\alpha = +\sqrt{\frac{1}{2}\left[\sqrt{(R^2 - \omega^2 L^2)(G^2 + \omega^2 C^2)} + (RG - \omega^2 LC)\right]} \quad \text{Nepers/m} \tag{5.30}$$

$$\beta = +\sqrt{\frac{1}{2}\left[\sqrt{(R^2 - \omega^2 L^2)(G^2 + \omega^2 C^2)} - (RG - \omega^2 LC)\right]} \quad \text{rad/m.} \tag{5.31}$$

For a lossless line, $R = G = 0$, we can now list the following important parameters:

$$\begin{aligned} \alpha &= 0 \\ \beta &= \omega\sqrt{LC} \quad \text{rad/m} \\ Z_0 &= \sqrt{\frac{L}{C}} \quad \Omega \\ v_\text{p} &= \frac{1}{\sqrt{LC}} \quad \text{m/s.} \end{aligned} \tag{5.32}$$

It should be noted that the phase velocity in a lossless line is independent of frequency as a consequence of the phase constant (β) varying linearly with ω. Because of this characteristic TEM lines can ideally transmit all frequencies without distortion.

5.6.3 Voltage, Current, and Impedance Relations

Voltage and Current Relations [4, 5, 7]

Figure 5.4 shows a typical transmission line fed by a time harmonic source and terminated by an impedance Z_L. The source V_s with internal impedance Z_s is connected to the input (sending) end of the line at $z = 0$ and the load Z_L is connected at the output (receiving) end at $Z = d$. The characteristic impedance and complex propagation constant of the line are Z_0, γ as shown.

The general expressions for the voltage and current along the line at any point z are given by (5.19) and (5.21), and we repeat them below:

$$V(z) = V^+ e^{-\gamma z} + V^- e^{\gamma z}$$
$$I(z) = I^+ e^{-\gamma z} - I^- e^{\gamma z},$$

where V^+ and V^- are unknown constants to be determined by applying boundary conditions. The appropriate boundary conditions at $z = 0$ are

$$V(0) = V_s \tag{5.33}$$
$$I(0) = I_s, \tag{5.34}$$

which are referred to as the sending end voltage and current, respectively. Applying the boundary conditions at $z = 0$, we determine V^+ and V^- as

$$V^+ = \frac{V_s + I_s Z_0}{2}, \quad V^- = \frac{V_s - I_s Z_0}{2}. \tag{5.35}$$

Using (5.35), we recast (5.19) and (5.21) as follows [4]:

$$V(Z) = V_s \cosh \gamma z - I_s G_0 \sinh \gamma z \tag{5.36}$$
$$I(Z) = I_s \cosh \gamma z - \frac{V_s}{Z_0} \sinh \gamma z. \tag{5.37}$$

Equations (5.36) and (5.37) are used to analyze transmission line performance in terms of the sending end voltage and current, and where distance z is measured from $z = 0$ and increasing toward the load, as in Figure 5.1.

Frequently it is found convenient to refer voltage and current in terms of the values at the load (receiving) end, and where the distance is measured from the receiving end and is referred to as ℓ as shown in Figure 5.4, i.e., $\ell = -\ell$. By this convention, (5.19) and (5.21) can be written as

$$\begin{aligned} V(\ell) &= V^+ e^{\gamma \ell} + V^- e^{-\gamma \ell} \\ I(\ell) &= \frac{V^+}{Z_0} e^{\gamma \ell} - \frac{V^-}{Z_0} e^{-\gamma \ell}. \end{aligned} \tag{5.38}$$

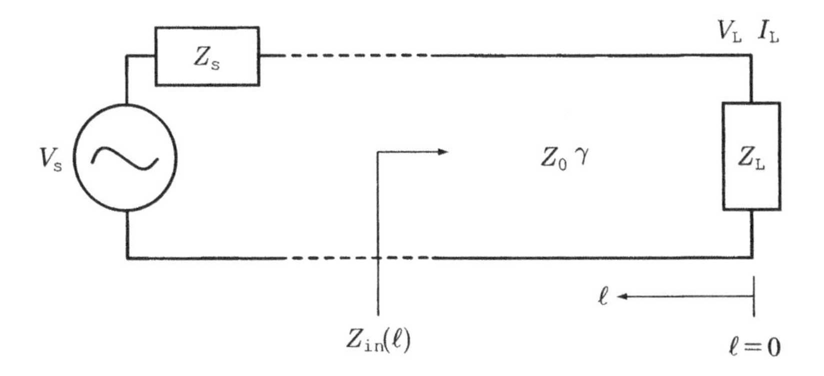

Figure 5.4 Terminated transmission line excited by a source. Distance ℓ is measured from the termination.

Note that since the source position is not changed, the forward and backward waves in (5.38) are represented by the terms containing $e^{\gamma\ell}$ and $e^{-\gamma\ell}$ respectively.

We now apply the boundary conditions

$$\begin{aligned} V(0) &= V_L = V^+ + V^- \\ I(0) &= I_L = \frac{V^+}{Z_0} - \frac{V^-}{Z_0} \end{aligned} \tag{5.39}$$

to (5.38) and obtain

$$V(\ell) = V_L \cosh \gamma\ell + I_L Z_0 \sinh \gamma\ell \tag{5.40}$$

$$I(\ell) = I_L \cosh \gamma\ell + \frac{V_L}{Z_0} \sinh \gamma\ell. \tag{5.41}$$

Note that if we write $z = -\ell$, $V_S = V_L$, $I_S = I_L$ in (5.36) and (5.37), we obtain (5.40) and (5.41).

Impedance Relations

The input impedance of a line of length ℓ and terminated by an impedance Z_L in Figure 5.4 is defined using (5.40) and (5.41) as

$$\begin{aligned} Z_{in}(\ell) &= \frac{V(\ell)}{I(\ell)} \\ &= \frac{V_L \cosh \gamma\ell + I_L Z_0 \sinh \gamma\ell}{I_L \cosh \gamma\ell + (V_L/Z_0) \sinh \gamma\ell}. \end{aligned} \tag{5.42}$$

Equation (5.42) is frequently written in the following popular form:

$$Z_{\text{in}}(\ell) = Z_0 \frac{Z_L + Z_0 \tanh \gamma\ell}{Z_0 + Z_L \tanh \gamma\ell}. \tag{5.43a}$$

The corresponding input admittance is

$$\begin{aligned} Y_{\text{in}}(\ell) &= \frac{I(\ell)}{V(\ell)} \\ &= Y_0 \frac{Y_L + Y_0 \tanh \gamma\ell}{Y_0 + Y_L \tanh \gamma\ell}, \end{aligned} \tag{5.43b}$$

where $Y_0 = 1/Z_0$ = characteristic admittance of the line and $Y_L = 1/Z_L$.

The impedance or admittance form of the input relations (5.43a) and (5.43b) are used depending on the circuit configuration under consideration. Using (5.43a), we obtain the following:

$$\begin{aligned} Z_{sc}\ell &= Z_0 \tanh \gamma\ell \\ Z_{oc}\ell &= \frac{Z_0}{\coth \gamma\ell} \end{aligned} \tag{5.44}$$

which implies $Z_{sc}Z_{oc} = Z_0^2$, or

$$Z_0 = \sqrt{Z_{sc}Z_{oc}}. \tag{5.45}$$

Equation (5.45) is an important relationship used for the determination of characteristic impedance of a section of line from its measured short circuit and open circuit impedances.

Again, from (5.44) we obtain

$$\tanh \gamma\ell = \sqrt{\frac{Z_{sc}}{Z_{oc}}}, \tag{5.46}$$

and thus

$$\alpha\ell + j\beta\ell = \tan^{-1} \sqrt{\frac{Z_{sc}}{Z_{oc}}}, \tag{5.47}$$

which can be used to determine α, β for a line from its measured open- and short-circuit impedances [4].

Inspection of (5.43a) reveals the following observations having practical significance:

1. The input impedance of a line terminated in its characteristic impedance (i.e., for $Z_L = Z_0$) is equal to the characteristic impedance of the line irrespective of its length, that is

$$\text{for } Z_L = Z_0, \qquad Z_{\text{in}} = Z_0 \text{ for all } \ell.$$

2. The input impedance of a line of infinite length is equal to its characteristic impedance, that is

$$\text{for } \ell \to \infty, \qquad Z_{\text{in}}(\ell) = Z_0.$$

In fact, this is sometimes used as the definition of characteristic impedance of a line.

3. For a lossless line, $R = G = 0$, and hence $\alpha = 0$, $\gamma = j\beta = jw\sqrt{LC}$ as shown in (5.32). For future reference we give the relations

$$Z_0 = \sqrt{\frac{L}{C}} = \text{real quantity} \tag{5.48}$$

$$Z_{\text{in}}(\ell) = Z_0 \frac{Z_{\text{L}} + jZ_0 \tan \beta\ell}{Z_0 + jZ_{\text{L}} \tan \beta\ell}. \tag{5.49}$$

Input Impedance of Short- and Open-Circuited Lines

Lengths of lossless lines with short and open terminations are used to provide desired reactive input impedances in a variety of radio-frequency circuits.

Under short-circuit conditions (i.e., with $Z_{\text{L}} = 0$) we obtain from (5.49) the following expression for the input impedance $Z_{\text{in}}^{\text{S}}(\ell)$ for a length ℓ of the line

$$Z_{\text{in}}^{\text{S}}(\ell) = jZ_0 \tan \beta\ell = jX_{\text{in}}^{\text{S}}(\ell), \tag{5.50}$$

where $X_{\text{in}}^{\text{S}}(\ell)$ is the input reactance offered by a short-circuited line of length ℓ. Similarly for an open-circuited line of length ℓ we obtain the following for the input impedance $Z_{\text{in}}^{\text{o}}(\ell)$:

$$Z_{\text{in}}^{\text{o}}(\ell) = -jZ_0 \cot \beta\ell = +jX_{\text{in}}^{\text{o}}(\ell), \tag{5.51}$$

where $+X_{\text{in}}(\ell)$ is the input reactance of the open-circuited line of length ℓ. Input reactance vs. length in wavelengths for the short and open-circuited sections of lines are shown in Figure 5.5(a) and 5.5(b).

Note that it follows from (5.50) and (5.51) that

$$\begin{aligned} Z_{\text{in}}^{\text{s}}(\ell) &= jZ_0 \tan \beta\ell \\ Z_{\text{in}}^{\text{o}}(\ell) &= -jZ_0 \cot \beta\ell \\ \therefore \quad Z_{\text{in}}^{\text{s}}(\ell)\, Z_{\text{in}}^{\text{o}} &= Z_0^2 \\ Z_0 &= \sqrt{Z_{\text{in}}^{\text{s}} Z_{\text{in}}^{\text{o}}}. \end{aligned} \tag{5.52}$$

This indicates that the characteristic impedance of a line is the geometric mean of the open- and short-circuit impedance of the line of any length. This provides a basis for the measurement of the impedance of lossless lines.

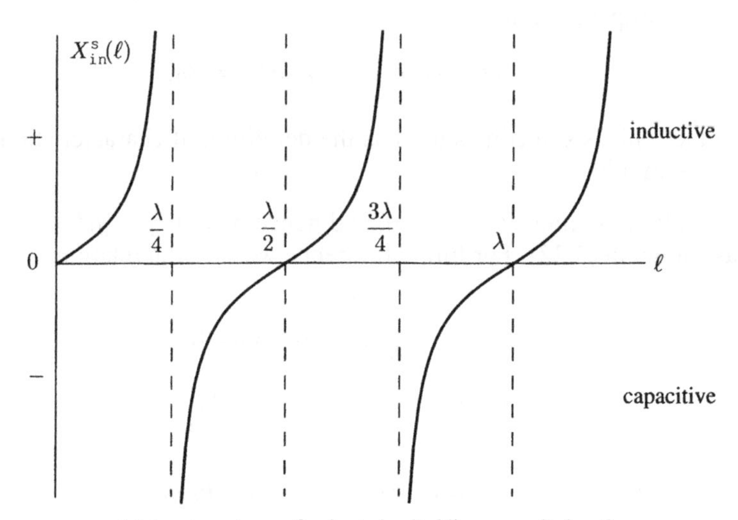

(a) Input reactance of a short-circuited line versus its length

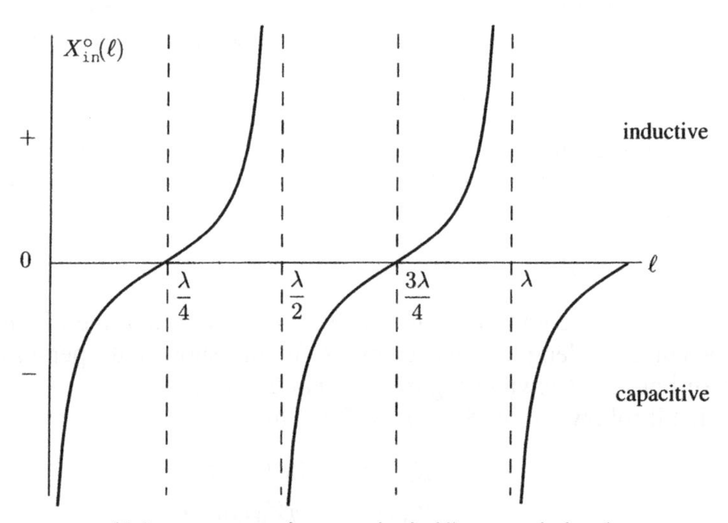

(b) Input reactance of an open-circuited line versus its length

Figure 5.5 Input reactance of short- and open-circuited lines.

The results of Figure 5.5 indicate that both short- and open-circuited sections of lines are capable of providing inductive and capacitive reactance of variable amount. For example,

$$
\begin{aligned}
X_{\text{in}}^{\text{s}}(\ell) \text{ is inductive for } &\le \ell \le \frac{\lambda}{4}, \quad \frac{\lambda}{2} \le \ell \le \frac{3\lambda}{4}, \ldots \\
\text{capacitive for } &\frac{\lambda}{4} \le \ell \le \frac{\lambda}{2}, \quad \frac{3\lambda}{4} \le \ell \le \lambda, \ldots \\
X_{\text{in}}^{\text{o}}(\ell) \text{ is capacitive for } &\le \ell \le \frac{\lambda}{4}, \quad \frac{\lambda}{2} \le \ell \le \frac{3\lambda}{4}, \ldots \\
\text{inductive for } &\frac{\lambda}{4} \le \ell \le \frac{\lambda}{2}, \quad \frac{3\lambda}{4} \le \ell \le \lambda.
\end{aligned}
$$

For practical reasons it is found convenient to use the minimum length for section of transmission line to achieve a desired amount of reactance. From the above, we find that with a section of line $\ell \le \lambda/4$, any amount of inductive and capacitive reactance can be obtained from the following:

$$
\left.
\begin{aligned}
X_n^{\text{s}}(\ell) &= Z_0 \tan \beta\ell \\
X_n^{\text{o}}(\ell) &= -Z_0 \cot \beta\ell
\end{aligned}
\right\}, \qquad 0 \le \ell \le \frac{\lambda}{4}. \tag{5.53}
$$

Quarter Wave ($\lambda/4$) Impedance Transformer

Lossless lines of length $\lambda/4$ are used for matching two resistive impedances provided its characteristic impedance is appropriately designed. From (5.49) it follows that when $\ell = \lambda/4$,

$$
Z_{\text{in}}\left(\frac{\lambda}{4}\right) = \frac{Z_0^2}{Z_{\text{L}}} \tag{5.54a}
$$

from which

$$
Z_0 = \sqrt{Z_{\text{in}}\left(\frac{\lambda}{4}\right) Z_{\text{L}}}\,. \tag{5.54b}
$$

Equations (5.54a) and (5.54b) indicate that two resistances $Z_{\text{in}}(\lambda/4)$ and Z_{L} are matched by the $\lambda/4$-long section of a transmission line if its characteristic impedance is the geometric mean of the two resistances.

Lossy Lines

We saw earlier (Section 5.5) that with a lossy line (i.e., $R \neq 0,\ G = 0$) the current and voltage waves on the line suffer attenuation as they progress toward the load; moreover the characteristic impedance becomes complex. Thus the parameters α, β, and Z_0 vary with frequency in such a way that they contribute to the distortion suffered by a multifrequency signal arriving at that load. We consider a line with small losses such that

$$\frac{R}{\omega L} \ll 1, \qquad \frac{G}{\omega C} \ll 1. \tag{5.55}$$

Under these assumptions we can approximate Z_0 and γ given by (5.26) and (5.20) as follows:

$$Z_0 \simeq \sqrt{\frac{L}{C}}\left[1 - j\left(\frac{R}{\omega L} - \frac{G}{2\omega C}\right)\right] \tag{5.56}$$

and

$$\gamma = \alpha + j\beta \simeq \omega\sqrt{LC}\,\frac{RC + LG}{2\omega LC} + j\omega\sqrt{LC}. \tag{5.57}$$

From these equations we obtain

$$\alpha \simeq \frac{1}{2}\left[R\sqrt{\frac{C}{L}} + G\sqrt{\frac{L}{C}}\right] \tag{5.58}$$

$$\beta \simeq \omega\sqrt{LC}\,. \tag{5.59}$$

If we now assume that

$$\frac{R}{L} = \frac{G}{C}, \tag{5.60}$$

then we obtain

$$\begin{aligned} Z_0 &= \sqrt{\frac{L}{C}} \\ \alpha &= \sqrt{RG} \\ \beta &= \omega\sqrt{LC} \\ v_{\mathrm{p}} &= \frac{1}{\sqrt{LC}}\,. \end{aligned} \tag{5.61}$$

Under the condition of (5.60) it is found from (5.61) that although the signal suffers some attenuation ($\alpha \simeq \sqrt{RG}$), it will not suffer any distortion due to dispersion effects because v_{p}'s independent of frequency. The Heaviside condition (5.60) was used to design a distortionless line by loading it artificially.

Reflection, Standing Waves

Transmission line measurements utilize the reflection and interference effects of waves on the line under consideration. We consider a section of a lossless line terminated by an impedance Z_L as shown in Figure 5.4. Since the line is lossless we have $\gamma = j\beta$, and we have the current and voltage at a distance ℓ from the load by (5.30) as

$$\begin{aligned} V(\ell) &= V^+ e^{j\beta\ell} + V^- e^{-j\beta\ell} \\ I(\ell) &= \frac{V^+}{Z_0}\, e^{j\beta\ell} - \frac{V^-}{Z_0}\, e^{-j\beta\ell}. \end{aligned} \tag{5.62}$$

Define the voltage reflection coefficient at ℓ as

$$\begin{aligned} \Gamma(\ell) &= \frac{\text{reflected voltage at } \ell}{\text{incident voltage at } \ell} \\ &= \frac{V^- e^{-j\beta\ell}}{V^+ e^{+j\beta\ell}} \\ &= \frac{V^-}{V^+}\, e^{-j2\beta\ell}. \end{aligned} \tag{5.63}$$

Note that Γ_L is a complex quantity $= |\Gamma_L| e^{j\delta}$, etc. Let $\Gamma(0) = V^-/V^+ =$ reflection coefficient at the load, at $\ell = 0$.
Rewrite (5.63) as

$$\Gamma(\ell) = \Gamma(0)\, e^{-j2\beta\ell} = \Gamma_L e^{-j2\beta\ell}, \tag{5.64}$$

where $\Gamma(0) = \Gamma_L$ is defined to be the reflection coefficient at the load. By the definitions above, we obtain from (5.62),

$$V(\ell) = V^+ e^{j\beta\ell}[1 + \Gamma_L e^{-j2\beta\ell}] \tag{5.65a}$$

$$I(\ell) = \frac{V^+ e^{j\beta\ell}}{Z_0}[1 - \Gamma_L e^{-j2\beta\ell}]. \tag{5.65b}$$

We now obtain the important relationship between Z_L and Γ_L using the relationship $Z_L = V(0)/I(0)$ and (5.64)

$$Z_L = Z_0 \frac{1 + \Gamma_L}{1 - \Gamma_L}. \tag{5.66}$$

From (5.66) we obtain

$$\Gamma_L = |\Gamma_L| e^{j\delta} = \frac{Z_L - Z_0}{Z_L + Z_0}, \tag{5.67}$$

and note that $|\Gamma_L| \le 1$, as it should be. Equation (5.67) is valid for all loads, real and complex.

Standing Waves

It is clear from (5.65a) and (5.65b) that when $\Gamma_L \neq 0$, the incident and reflected waves on the line will interfere with each other. Depending on the nature of the load, the interference effects will produce certain maxima and minima (of voltage and currents) at certain locations on the lines depending on the load. We rewrite (5.65a) and (5.65b) as

$$V(\ell) = V^+ e^{j\beta\ell}[1 + |\Gamma_L| e^{-j(2\beta\ell - \delta)}] \tag{5.68a}$$

$$I(\ell) = \frac{V^+}{Z_0} e^{j\beta\ell}[1 - |\Gamma_L| e^{-j(2\beta\ell - \delta)}] \tag{5.68b}$$

and obtain the amplitudes of the voltage and current on the line as

$$|V(\ell)| = |V^+|[1 + |\Gamma_L|^2 + 2|\Gamma_L| \cos(2\beta\ell - \delta)]^{1/2} \tag{5.69a}$$

$$|I(\ell)| = \frac{|V^+|}{Z_0}[1 + |\Gamma_L|^2 - 2|\Gamma_L| \cos(2\beta\ell - \delta)]^{1/2}. \tag{5.69a}$$

Equations (5.69a) and (5.69a) give the standing wave distributions of voltage and current on the line; they are called standing waves because there is no phase propagation and at all points on the line the amplitudes do not vary with time, meaning they are standing in time.

It is of interest to know the locations of the maxima and minima of voltage and current on the line. These locations are determined by the load Z_L through the parameters δ which is the phase of the reflection coefficient Γ_L (see (5.67)). Equation (5.69a) predicts that the voltage maxima and minima occur at

$$2\beta\ell_n^{\max} - \delta = 2n\pi \tag{5.70}$$

$$2\beta\ell_n^{\min} - \delta = (2n + 1)\pi \tag{5.71}$$

with $n = 1, 2, 3, \ldots$, where we have omitted open- and short-circuit terminations (to be discussed later), for $Z_L \neq 0, \infty$. From (5.70) and (5.71) the following are found:

1. Consecutive maxima (two adjacent maxima) and minima are separated by $\lambda/4$. A maximum and its nearest minimum are separated by $\lambda/4$. The observations above are often used to measure unknown wavelength or frequency.

2. Equation (5.71) gives the location of the $V_{\min}$ nearest the load as

$$2\beta\ell_1^{\min} - \delta = \pi \tag{5.72}$$

$$\delta = 2\beta\ell_1^{\min} - \pi = \frac{2\pi\ell_1^{\min}}{\lambda} - \pi. \tag{5.73}$$

The observations above are used during various VSWR measurements to be discussed later.

Short-Circuited Line

We have $Z_L = 0$, $\Gamma_L = -1$ from (5.67), which also means that $|\Gamma_L| = 1$, $\delta = \pi$. Thus (5.68a) and (5.68b) give the voltage and current distributions as:

$$V(\ell) = 2|V^+|\,|\sin(\beta\ell)| \tag{5.74a}$$

$$= 2|V^+|\left|\sin\left(\frac{2\pi\ell}{\lambda}\right)\right|$$

$$I(\ell) = 2\left|\frac{V^+}{Z_0}\right|\left|\cos\left(\frac{2\pi\ell}{\lambda}\right)\right|. \tag{5.74b}$$

The voltage and current distributions on the line are shown in Figure 5.6. Note that at the load (being a short) there is a voltage minimum and current maximum. The distance between two adjacent maxima is $\lambda/2$ and that between two adjacent minima is also $\lambda/2$. The separation between a maximum and an adjacent minimum is $\lambda/4$. These observations have significant applications during measurements.

Open-Circuited Line

We have $Z_L = \infty$, $\Gamma_L = 1$, $|\Gamma_L| = 1$ from (5.67), which means that $\delta = 0$. In this case we have

$$V(\ell) = 2V^+ \cos\left(\frac{2\pi\ell}{\lambda}\right) \tag{5.75a}$$

$$I(\ell) = \frac{2V^+}{Z_0}\sin\left(\frac{2\pi\ell}{\lambda}\right). \tag{5.75b}$$

The voltage and current distributions are shown in Figure 5.7. Note here that the voltage is maximum and current is minimum at the load end $\ell = 0$ (note $Z_L = \infty$, open), as they should. Although the locations of the maxima and minima are different from the short-circuit ones, their separation distance behaves in a similar fashion, as indicated.

Voltage Standing Wave Ratio (VSWR)

The voltage standing wave ratio (VSWR) is an important parameter and is frequently used for various RF and microwave measurements as well as for impedance matching and other purposes.

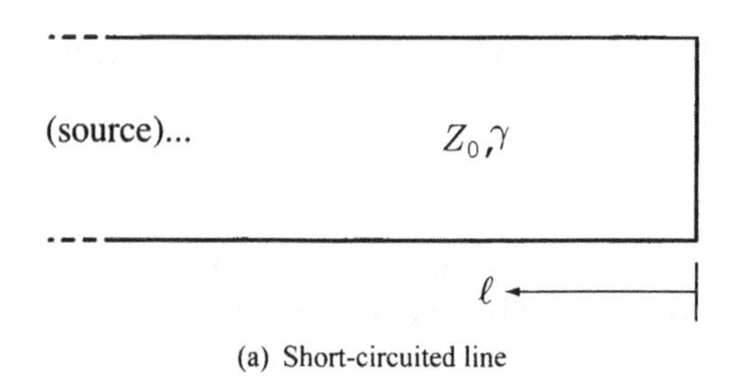

(a) Short-circuited line

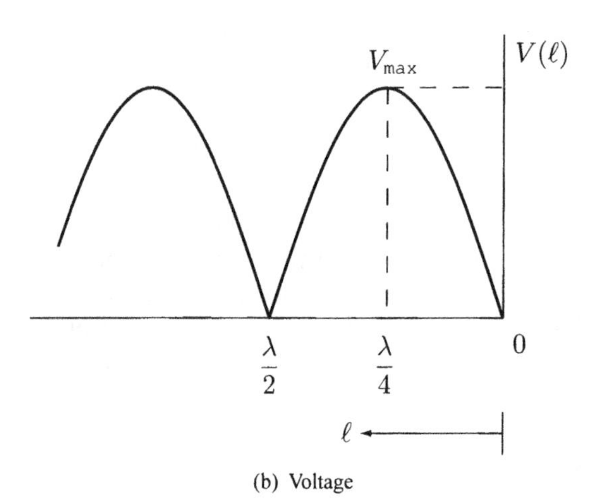

(b) Voltage

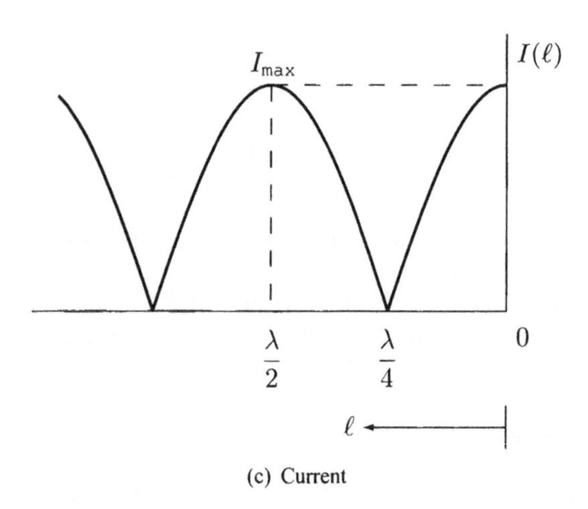

(c) Current

Figure 5.6 Voltage and current distribution on an short-circuited line.

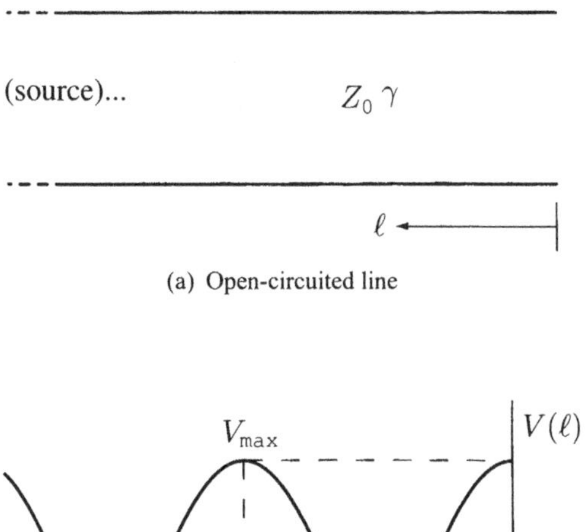

(a) Open-circuited line

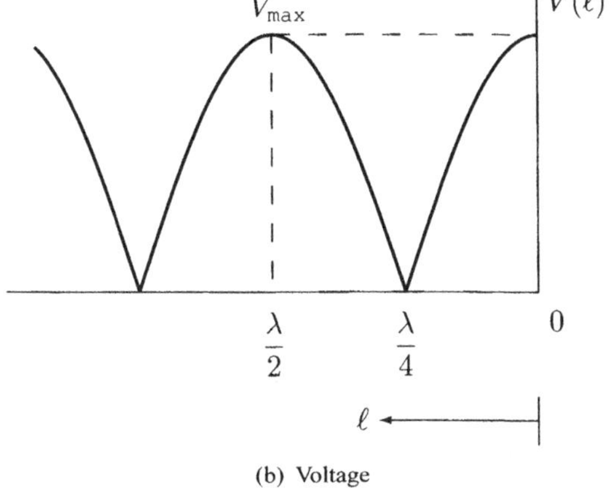

(b) Voltage

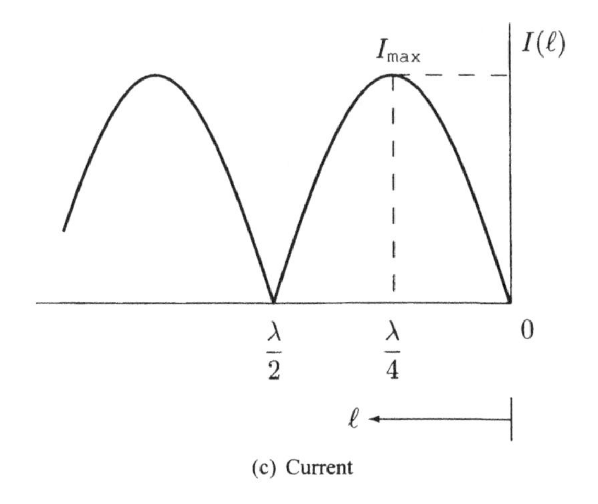

(c) Current

Figure 5.7 Voltage and current distributions on an open-circuited line.

From the voltage and current standing wave distributions on a line terminated by an arbitrary level, given by (5.68a) and (5.68b), we obtain the following expressions for the voltage and current maxima and minima on the line:

$$\begin{aligned} |V_{\max}| &= |V^+|[1+|\Gamma_L|] \\ |V_{\min}| &= |V^+|[1-|\Gamma_L|], \end{aligned} \tag{5.76}$$

$$\begin{aligned} |I_{\max}| &= \frac{|V^+|}{Z_0}[1+|\Gamma_L|] \\ |I_{\min}| &= \frac{V^+}{Z_0}[1-|\Gamma_L|]. \end{aligned} \tag{5.77}$$

The voltage standing wave ratio S is defined as

$$S = \frac{|V_{\max}|}{|V_{\min}|} = \frac{1+|\Gamma_L|}{1-|\Gamma_L|}\,. \tag{5.78}$$

Note that by definition $1 \leq S \leq \infty$, since $0 \leq |\Gamma_L| \leq 1$. The maximum resistive impedance occurs at the location of voltage maximum and current minimum, that is

$$Z_{\max} = \frac{V_{\max}}{I_{\min}} = Z_0\,\frac{1+|\Gamma_L|}{1-|\Gamma_L|} = Z_0 S \tag{5.79}$$

Similarly the minimum resistance occurs at the voltage minimum and current maximum, that is

$$Z_{\min} = \frac{V_{\min}}{I_{\max}} = Z_0\,\frac{1-|\Gamma_L|}{1+|\Gamma_L|} = \frac{Z_0}{S}. \tag{5.80}$$

Also, for Z_L real, the appropriate realtime for $|\Gamma_L|$ and S in terms of Z_L and Z_0 are

$$|\Gamma_L| = \frac{Z_L - Z_0}{Z_L + Z_0}\,, \qquad \text{for } Z_L \geq Z_0 \tag{5.81a}$$

$$S = \frac{Z_L}{Z_0} \tag{5.81b}$$

and

$$|\Gamma_L| = \frac{Z_0 - Z_L}{Z_0 + Z_L}\,, \qquad \text{for } Z_L < Z_0 \tag{5.82a}$$

$$S = \frac{Z_0}{Z_L}\,. \tag{5.82b}$$

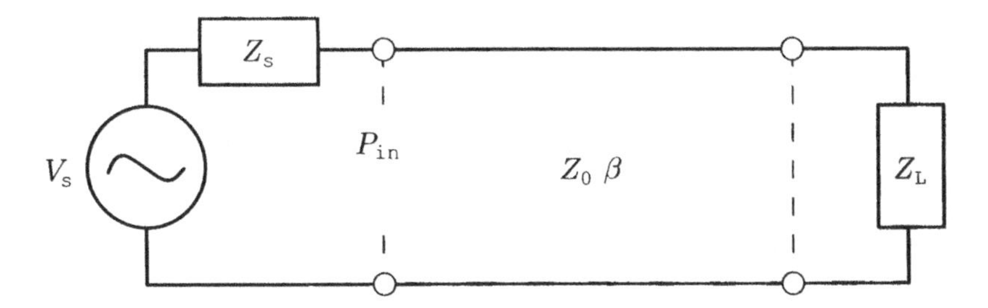

Figure 5.8 Lossless line terminated by a load. P_{in} signifies the average power supplied by the source of the line.

Power Flow

Consider a source V_s connected to a section of a lossless transmission line having Z_0, β and terminated by a load Z_L (Figure 5.8). We have seen before the voltage and current on the line at a distance ℓ from the load are given by (5.65), and we rewrite them here:

$$\begin{aligned} V(\ell) &= V^+ e^{j\beta\ell}[1 + \Gamma_L e^{-j2\beta\ell}] \\ I(\ell) &= \frac{V^+}{Z_0} e^{j\beta\ell}[1 - \Gamma_L e^{-j2\beta\ell}]. \end{aligned} \tag{5.83}$$

Average power supplied by the source is

$$\begin{aligned} P_{av} &= \frac{1}{2}\text{Re}[V(L)\ I(L)^*] \\ &= \frac{1}{2}\frac{|V_+|^2}{Z_0} - |\Gamma_L|^2 \frac{|V_+|^2}{2Z_0}. \end{aligned} \tag{5.84}$$

Note that in Figure 5.8,

$$P_{in} = \frac{1}{2}\frac{|V_+|^2}{Z_0}$$

is the incident power launched by the source into the line, and the power flows along the lossless line and that reflected at the load is

$$P_{ref} = |\Gamma_L|^2 \frac{|V_+|^2}{2Z_0} = |\Gamma_L|^2 P_{in}.$$

Thus we can write (5.84) as

$$\begin{aligned} P_{av} &= P_{in} - P_{ref} = P_{in}(1 - |\Gamma_L|^2) \\ &= P_{in}\left(1 - |\Gamma_L|^2\right) = P_L, \end{aligned} \tag{5.85}$$

where P_L is the power consumed by the load. Equation (5.85) indicates that for maximum power transfer to the load, Γ_L should be as small as possible, or in other words, the VSWR S should be as close to unity as possible, which means that the load should be matched to the line (for an ideal match, $\Gamma_L = 0$).

5.7 LINE PARAMETERS

So far in our discussion we have used a transmission line model characterized by a continuous distribution of lumped circuit parameters R, L, C, and G per unit length of the line. This model applies to all TEM lines shown in Figure 5.2; however, the circuit parameters applicable to each line are determined by its geometry, material constituents, and environment. The relevant parameters can be determined analytically by treating the line as an electromagnetic boundary value problem. Such treatment can be found in most textbooks on electromagnetics and transmission lines [1, 2, 4, 5]. We will follow here simplified and approximate methods [8] to obtain the required values that are sufficiently accurate for practical use. Detailed derivation and analysis may be found in the references cited.

5.7.1 Coaxial Line

Longitudinal and transverse views of a typical coaxial line are shown in Figure 5.2*b*. The infinitely long lines are made of conductors having conductivity σ_c, and the medium within the line is characterized by μ_0, $\epsilon = \epsilon_0\epsilon_r$ and σ (σ is assumed to be very small). The outer wall of the outer conductors is assumed to be grounded.

We place a line charge q per unit length on the inner conductor. Since the outer surface of the outer conductor is grounded, the induced negative charge on the outer conductor will be uniformly distributed as $-q$ per unit length over the inner surface. Using Gauss's law we obtain the transverse component of the electric field within the coaxial region as

$$\mathbf{E} = \hat{\rho}\,\mathbf{E}_\rho = \hat{\rho}\,\frac{\mathbf{D}}{\epsilon_0\epsilon_r} = \hat{\rho}\,\frac{\mathbf{q}}{2\pi\epsilon_0\epsilon_r\rho}, \qquad \mathbf{a} \leq \rho \leq \mathbf{b}, \tag{5.86}$$

where q is the charge per unit length on the inner conductor, ρ is the radial distance from the axis (z) of the inner conductor, and the other parameters are as defined elsewhere. The potential difference (or voltage) between the inner and outer conductors is determined as

$$V_{ab} = -\int_b^a \mathbf{E}\cdot d\boldsymbol{\ell} = \frac{q}{2\pi\epsilon}\ln\left(\frac{b}{a}\right). \tag{5.87}$$

The capacitance C per unit length for the line is now obtained as

$$C = \frac{q}{V_{ab}} = \frac{2\pi\epsilon}{\ln(b/a)} \quad \text{F/m}. \tag{5.88}$$

To obtain the inductance, we assume a current I flowing along the $\hat{z}$ direction in the inner conductor. By induction a current I will be flowing along the outer conductor.

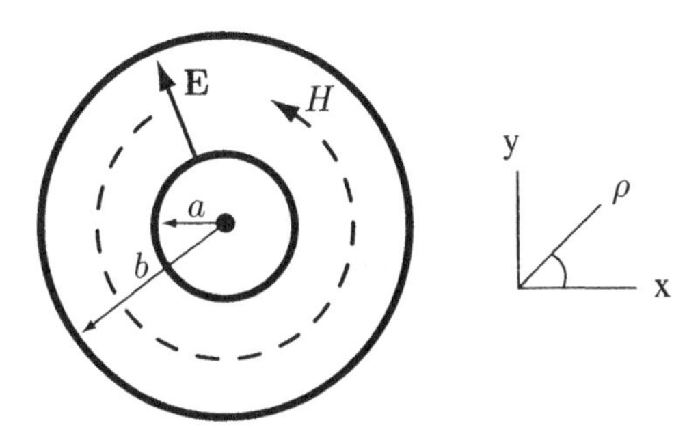

Figure 5.9 Transverse field distribution in a coaxial line. The field components are E_ρ and H_ϕ.

We now apply Ampère's law $\oint_C \mathbf{H} \cdot d\boldsymbol{\ell} = I$, using a contour of radius ρ within the line (Figure 5.9) which gives

$$H_\phi = \frac{I}{2\pi\rho} \, . \tag{5.89}$$

The flux linkage per unit length of the line is obtained as:

$$\begin{aligned} \psi = \int_S \mathbf{B} \cdot \mathbf{s} &= \int_0^1 \int_a^b \hat{\phi} \, \frac{\mathbf{I}}{2\pi\rho} \cdot \hat{\phi} \, \mathbf{d}\rho \, \mathbf{dz} \\ &= \frac{\mu I}{2\pi} \ln \frac{b}{a} \, . \end{aligned} \tag{5.90}$$

The inductance L per unit length of the line is

$$\begin{aligned} L &= \frac{\text{flux linkage per unit length}}{I} \\ &= \frac{\mu}{2\pi} \ln\left(\frac{b}{a}\right) \qquad \text{H/m}. \end{aligned} \tag{5.91}$$

We obtain the resistance R per unit length of the line by using the surface impedance concept

$$R = \frac{1}{\sigma_c 2\pi a\delta} + \frac{1}{\sigma_c 2\pi b\delta} \, , \tag{5.92}$$

where δ is the skindepth in the conductor. Introducing the value of $\delta = \sqrt{2/\omega\mu\sigma_c}$, we obtain

$$R = \frac{1}{2}\sqrt{\frac{\delta\mu}{\sigma_c}}\left(\frac{1}{a} + \frac{1}{b}\right) \quad \Omega/\text{m}. \tag{5.93}$$

The parameter G is associated with the loss in the dielectric medium within the line. Assuming a conductivity σ for the medium, it can be shown that G_c can be obtained by using the relation

$$\frac{C}{G_c} = \frac{\epsilon}{\sigma}\,. \tag{5.94}$$

Using (5.94) and (5.88), we now obtain

$$G = \frac{2\pi\sigma}{\ln b/a} \quad \text{S/m}, \tag{5.95}$$

which indicates that G could be obtained by replacing ϵ with σ in (5.88). For a lossless coaxial line the characteristic impedance is given by

$$Z_0 = \sqrt{\frac{L}{C}} = 60\sqrt{\frac{\mu_r}{\epsilon_r}}\ln\left(\frac{b}{a}\right), \tag{5.96a}$$

which under the assumption $\mu_r = 1$ is

$$Z_0 = \frac{60}{\sqrt{\epsilon_r}}\ln\left(\frac{b}{a}\right) = \frac{138}{\sqrt{\epsilon_r}}\log_{10}\left(\frac{b}{a}\right) \quad \Omega, \tag{5.96b}$$

where both (5.96a) and (5.96b) are in use.

5.7.2 Parallel Wire Line

The cross-sectional or transverse view of a balanced parallel wire transmission line is shown in Figure 5.10. For generality we have assumed that the line consists of two conducting wires of radii a and b as shown. Rigorous derivation of the line constants for this configuration is difficult and is described in the references. We follow here approximate methods.

For the purpose of determining C, we place two line charges carrying $+\rho_\ell$ per unit length and $-\rho_\ell$ per unit length on the lines #1 and #2, respectively. With the coordinate system as defined we obtain the total electrical field at any point x along the x axis

$$\mathbf{E}(x) = \mathbf{E}_1(x) + \mathbf{E}_2(x),$$

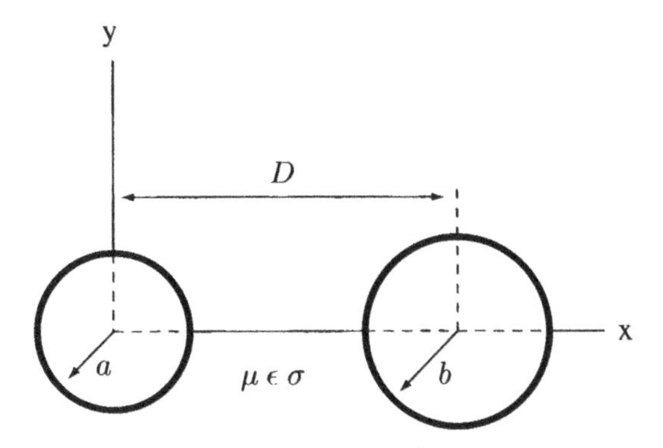

Figure 5.10 Transverse cross section of a two-wire line.

where $\mathbf{E}_1(x)$, $\mathbf{E}_2(x)$ are the electric fields at x produced by the charges #1 and #2, respectively. We can write $\mathbf{E}$ as follows:

$$\mathbf{E}(x) = \frac{\rho_\ell}{2\pi\epsilon}\left[\frac{1}{x} + \frac{1}{D-x}\right]\hat{x}. \tag{5.97}$$

The potential difference between the two conducting lines #1 and #2 is

$$\begin{aligned} V_{12} &= -\int_2^1 \mathbf{E}\cdot d\boldsymbol{\ell} = -\int_{D-b}^{a} \frac{\rho_\ell}{2\pi\epsilon}\left[\frac{1}{x} + \frac{1}{D-x}\right]\hat{x}\cdot\hat{x}\,dx \\ V_{12} &= \frac{\rho_\ell}{2\pi\epsilon}\ln\frac{(D-b)(D-a)}{ab} \\ &\simeq \frac{\rho_\ell}{2\pi\epsilon}\ln\frac{D^2}{ab} \qquad \text{for } D \gg b, a \\ &\simeq \frac{\rho_\ell}{\pi\epsilon}\ln\frac{D}{a} \qquad \text{for } a-b \ll D. \end{aligned} \tag{5.98}$$

Thus, for $a = b$, the capacitance C per unit length is

$$C = \frac{\rho_\ell}{V_{12}} = \frac{\pi\epsilon}{\ln(D/a)} \qquad \text{F/m}. \tag{5.99}$$

To obtain L for the line, we assume two line currents $+I$ and $-I$ placed along the conductors #1 and #2, respectively. The magnetic field along the x axis can be shown to be

$$\mathbf{H}(x) = \frac{I}{2\pi}\left[\frac{1}{x} + \frac{1}{D-x}\right]\hat{y}. \tag{5.100}$$

The flux linkage ψ per length ℓ of the line is then approximated by

$$\psi = \int_S \mathbf{B} \cdot d\boldsymbol{\ell} = \int_0^{\ell} \int_a^{D-b} \frac{\mu I}{2\pi} \left(\frac{1}{x} + \frac{1}{D-x} \right) \hat{y} \cdot \hat{y} \, dx \, dz$$
$$= \frac{\mu \ell I}{2\pi} \ln \frac{(D-b)(D-a)}{ab}$$
$$\simeq \frac{\mu \ell I}{2\pi} \ln \frac{D^2}{ab}, \quad D \gg b, a. \tag{5.101}$$

For $a = b$, we have

$$\psi = \frac{\mu l I}{\pi} \ln \frac{D}{a}. \tag{5.102}$$

Thus, for $a = b$,

$$L = \frac{\mu}{I} = \frac{\mu}{\pi} \ln \frac{D}{a} \qquad \text{H/m}. \tag{5.103}$$

It should be noted that the exact values of C and L are [1,4]

$$C = \frac{\pi \epsilon}{\cosh^{-1} \frac{D}{2a}} = \frac{\pi \epsilon}{\ln \left(\frac{D}{2a} + \sqrt{\frac{D^2}{4a^2} - 1} \right)}$$
$$\simeq \frac{\pi \epsilon}{\ln \frac{D}{a}}, \qquad \text{F/m} \tag{5.104}$$

for $D \gg a$

$$L = \frac{\mu}{\pi} \cosh^{-1} \frac{D}{2a} = \frac{\mu}{\pi} \ln \left(\frac{D}{2a} + \sqrt{\frac{D^2}{4a^2} - 1} \right)$$
$$\simeq \frac{\mu}{\pi} \ln \frac{D}{a}, \qquad \text{H/m} \tag{5.105}$$

for $D \gg a$

which are well approximated by the values given by (5.99) and (5.103), respectively.

Following the procedures similar to those used for the coaxial line, it can be shown that

$$G = \frac{\sigma}{\epsilon} C = \frac{\pi \sigma}{\ln \frac{D}{a}} \qquad \text{S/m}. \tag{5.106}$$

The resistance per unit length for the line is

$$R = \frac{1}{2\pi a \delta \sigma_c} + \frac{1}{2\pi a \delta \sigma_c}$$
$$= \sqrt{\frac{f\mu}{\pi \sigma_c}} \frac{2}{a} \qquad \Omega/\text{m}. \tag{5.107}$$

For a lossless line the characteristic impedance is

$$Z_0 = \sqrt{\frac{L}{C}} = \frac{120}{\sqrt{\epsilon_r}} \ln \frac{D}{a}$$
$$= \frac{276}{\sqrt{\epsilon_r}} \log_{10} \left(\frac{D}{a}\right) \quad \Omega. \tag{5.108}$$

Figure 5.11 gives the calculated characteristic impedances of open two-wire and coaxial transmission lines as functions of the geometrical parameters D/d of the lines [9]. Note that in Figure 5.11, D represents the separation distance between the two equal radii wires and $d = 2a$ for the two-wire line as in Figure 5.10; for the coaxial line, $D = 2b$ and $D = 2a$ as in Figure 5.9. The results shown in Figure 5.11 assume that the dielectric used is air and that the line conductors are perfect.

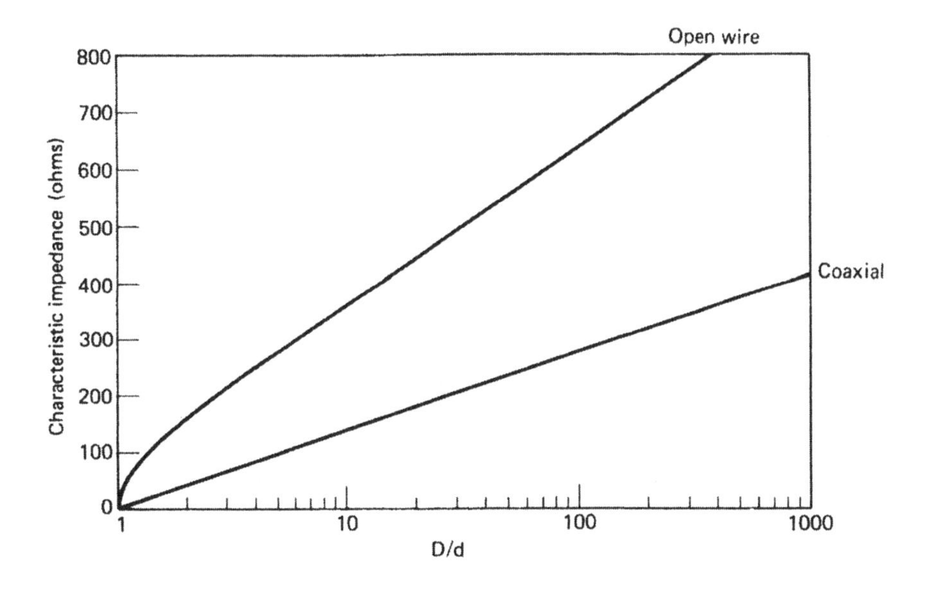

Figure 5.11 Characteristic impedance of open two-wire and coaxial lines. Both lines use air dielectric. (Source: From [9], p.5.)

5.7.3 Parallel Plate Line

The transverse cross section of the parallel plate line shown in Figure 5.2*c* is sketched in Figure 5.12 which shows that the parallel plates, each of width W, separated by a distance d, the longitudinal direction of the line is along the z direction as was mentioned earlier.

The parallel plate region is filled with a homogeneous medium characterized by η_0, $\epsilon_0\epsilon_r$, and σ, where the medium conductivity is assumed to be very small. The medium outside is free space. With appropriate excitation, and neglecting the fringe effects, it can be shown [2] that the fields of the TEM mode propagating in the line are $\vec{E} = \hat{x}E_x$ and $\vec{H} = \hat{y}H_y$, and they carry energy in the z direction. The circuit parameters for the line are

$$\begin{aligned} C &= \frac{\epsilon W}{d} \qquad \text{F/m} \\ L &= \frac{\eta_0 d}{W} \qquad \text{H/m} \\ G &= \frac{\sigma W}{d} \qquad \text{S/m} \\ R &= \frac{2}{\sigma_c W \sigma} = \frac{2}{W}\sqrt{\frac{\omega f \eta_0}{\sigma_c}} \qquad \Omega/\text{m}, \end{aligned} \tag{5.109}$$

$W \gg d$, and the skin depth $\delta \ll$ the thickness of the plate.

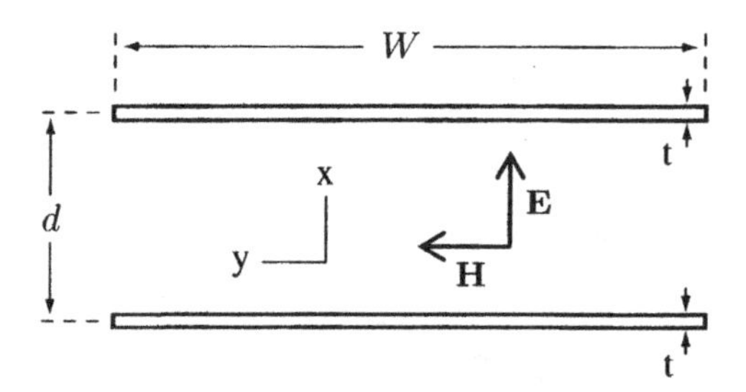

Figure 5.12 Transverse view of a parallel plate line. The field components are $\mathbf{E} = \hat{x}\, E_x$ and $\mathbf{H} = \hat{y}\, H_y$.

For the lossless case, $R = G = 0$, and the characteristic impedance is given by

$$Z_0 = \sqrt{\frac{L}{C}} = \frac{\eta_0}{\sqrt{\epsilon_r}} \frac{W}{d} \quad \Omega. \tag{5.110}$$

5.7.4 Circular Wire above a Ground Plane

A conducting wire above a ground plane occurs in circuit arrangements in a variety of electronic circuits. Such an arrangement is shown in Figure 5.13a, which by image theory is found to be equivalent to the upper half of a thin

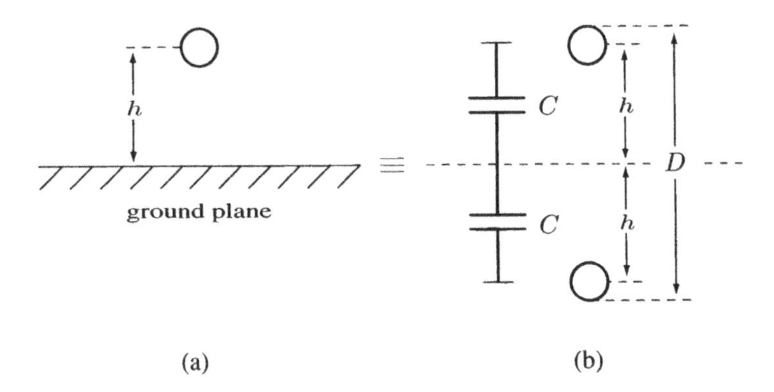

Figure 5.13 Wire above a ground and image theory: (*a*) a conducting wire above a ground (*b*) equivalent two-wire line by image.

wire transmission line as shown in Figure 5.13*b*. The latter can be utilized to determine the circuit parameters for the configuration. It is clear that the capacitance C per unit length from wire above ground can be obtained from (5.103) for the two-wire line as

$$\frac{C}{2} = \frac{\pi\epsilon}{\cosh^{-1} D/2a} \simeq \frac{\pi\epsilon}{\ln 2h/a} \qquad \text{if h} \gg \text{a},$$

which gives

$$C = \frac{2\pi\epsilon}{\ln 2h/a} \qquad \text{F/m} \tag{5.111}$$

It can now be shown that

$$L = \frac{\mu_0\varepsilon}{C} = \frac{\mu_0}{2\pi} \ln\left(\frac{2h}{a}\right) \qquad \text{H/m.} \tag{5.112}$$

In the lossless case, the characteristic impedance of the configuration shown in Figure 5.13 is

$$Z_0 = \frac{60}{\sqrt{\varepsilon}} \ln \frac{2h}{a} \qquad \Omega. \tag{5.113}$$

5.7.5 Microstrip Line

Microstrip lines (Figure 5.2*d*) are widely used as transmission lines for microwave integrated circuits and on printed circuit boards (PCBs). They are also used for circuit components such as filters, couplers, and resonators. Being a planar transmission line, they can be fabricated by pholithographic processes and are easily integrated with other passive and active microwave devices.

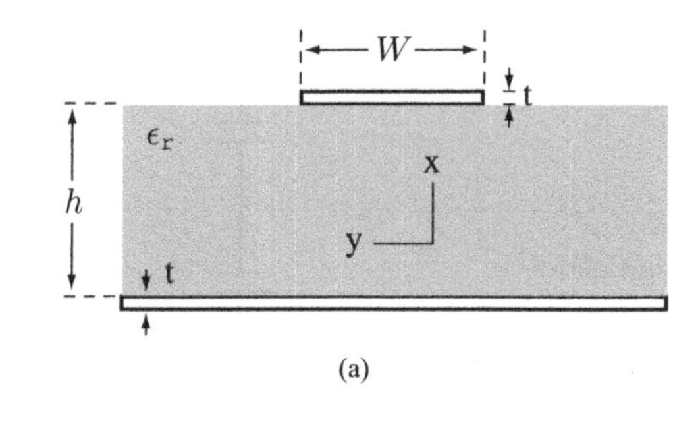

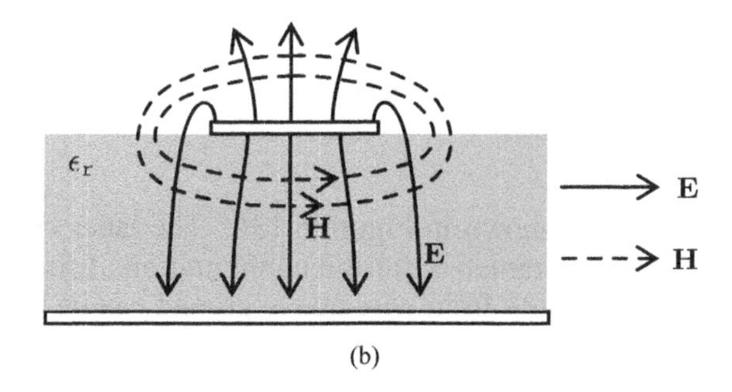

Figure 5.14 Transverse view of a microstrip line. (*a*) Geometry, (*b*) Electric and magnetic field lines.

The transverse view and geometry of a microstrip line is shown in Figure 5.14*a*. The line consists of a conducting strip of width W and thickness t mounted on the upper surface of a dielectric substrate with thickness h and dielectric constant ϵ_r. A sketch of the field lines in the transverse plane is given in Figure 5.14*b*.

In the absence of the dielectric ($\epsilon_r = 1$) or if the entire space were filled with a homogenous dielectric (ϵ_r), we could think of the microstrip as a parallel plate line (Section 5.7.3) consisting of two flat conducting strips of width W and separated by a distance $2h$ (the ground plane removed via image theory). In this case we would have a simple TEM transmission line as described in Section 5.7.3.

The presence of the dielectric, and particulary the fact that the dielectric does not fill the region above the substrate complicates the behavior and analysis of the microstrip line. As indicated in Figure 5.14*b*, most of the field lines are in the substrate region between the microstrip and the ground plane, and some fraction is in the air region above the substrate. For this reason, a microstrip line is incapable of supporting a pure TEM mode of wave propagation. Rigorous analysis shows that it supports hybrid TM-TE waves which may contain longitudinal components of $\mathbf{E}$ and $\mathbf{H}$ fields (that is components in the direction of propagation). In most cases the substrate thickness $h \ll \lambda$ and under this condition longitudinal field components become negligible compared to the transverse components. Thus, the fields maybe considered to be of quasi-TEM mode type; in other words, the field share essentially the same as those of the static case. Thus, good approximations for the characteristic impedance, phase velocity, propagation constant, and the line parameters L and C can be obtained from the quasi-static or static analysis [8].

Approximate expressions for various parameters are derived in [8] and given in [9, 10]. The expressions use the concept of the effective dielectric constant ϵ_r' which can be interpreted as the dielectric constant of a homogeneous medium that replaces the air and dielectric substrate regions of the microstrip line [9]. Since some of the field lines are in the dielectric region and some in the air (see Figure 5.14*b*), the effective dielectric constant satisfies the relation,

$$1 < \epsilon_r' < \epsilon_r$$

and it is dependant on the substrate thickness h, the conductor width W and the thickness t. The phase velocity v_p and the propagation constant β can then be expressed as

$$v_p = \frac{c}{\sqrt{\epsilon_r'}} \tag{5.114a}$$

$$\beta = \beta_0 \sqrt{\epsilon_r'} \tag{5.114b}$$

where β_0 is the propagation constant in free space and $c = 3 \times 10^8$ m/s is the velocity of light in free space.

Approximate expressions for the characteristic impedance of a microstrip line with strip thickness $t = 0$ is given by [8–10]

$$Z_0 = \frac{\eta_0}{2\pi\sqrt{\epsilon_r'}} \ln\left(\frac{8h}{W} + 0.25\frac{W}{h}\right) \tag{5.115a}$$

for $\frac{W}{h} \leq 1$ (a narrow strip) or

$$Z_0 = \frac{\eta_0}{\sqrt{\epsilon_r'}} \left\{\frac{W}{h} + 1.393 + 0.667 \ln\left(\frac{W}{h} + 1.444\right)\right\}^{-1} \tag{5.115b}$$

for $\frac{W}{h} \geq 1$ (a wide strip). Therein,

$$\eta_0 = 120\pi = 377 \quad \Omega,$$

$$\epsilon_r' = \frac{\epsilon_r + 1}{2} + \frac{\epsilon_r - 1}{2} F\left(\frac{W}{h}\right) \tag{5.116a}$$

$$F\left(\frac{W}{h}\right) = \left(1 + 12\frac{h}{W}\right)^{-1/2} + 0.04\left(1 - \frac{W}{h}\right)^2, \quad \text{if } \frac{W}{h} \leq 1$$
$$= \left(1 + 12\frac{h}{W}\right)^{-1/2}, \quad \text{if } \frac{W}{h} > 1. \tag{5.116b}$$

Approximations (5.116a)–(5.116b) and other better approximations are described in [8].

The design of a microstrip line having a characteristic impedance Zo and using a substrate of dielectric constant ϵ_r and a height h requires the determination of the width W of the strip (or trace). The required W/h ratio can be found from (5.116) and is given by [10]

$$\frac{W}{h} = \frac{8e^A}{e^{2A} - 2} \tag{5.117a}$$

for $W/h < 2$, where

$$A = \frac{Z_0}{60}\sqrt{\frac{\epsilon_r + 1}{2}} + \frac{\epsilon_r - 1}{\epsilon_r + 1}\left(0.23 + \frac{0.11}{\epsilon_r}\right),$$

$$B = \frac{377\,\pi}{2\,Z_0\sqrt{\epsilon_r}},$$

and

$$\frac{W}{h} = \frac{2}{\pi}\left\{B - 1 - \ln(2B - 1) + \frac{\epsilon_r - 1}{2\epsilon_r}\left[\ln(B - 1) + 0.39 - \frac{0.61}{\epsilon_r}\right]\right\} \tag{5.117b}$$

for $W/h \geq 2$.

With the knowledge of Z_0 and ϵ_r' and assuming only TEM modes in the line, the values for L and C per unit length and other parameters for the line can be determined in a manner described in Chapter 7.

For illustration, the characteristic impedance of a microstrip line with $t = 0$ and using substrate of typical glass epoxy ($\epsilon_r \simeq 4.8$), teflon epoxy ($\epsilon_r \simeq 2.55$) and alumina ($\epsilon_r = 10$) are shown in Figure 5.15. Note that for $W \gg h$, the results in Figure 5.15 can be approximated by the characteristic impedance of a transmission line consisting of a pair of conducting strips of width W and separated by a distance $2h$ and using a medium with a dielectric constant ϵ_r' (see (5.110)).

Detailed effects of the strip thickness on the parameters are discussed in [8]. An approximated expression for the characteristic impedance of a microstrip

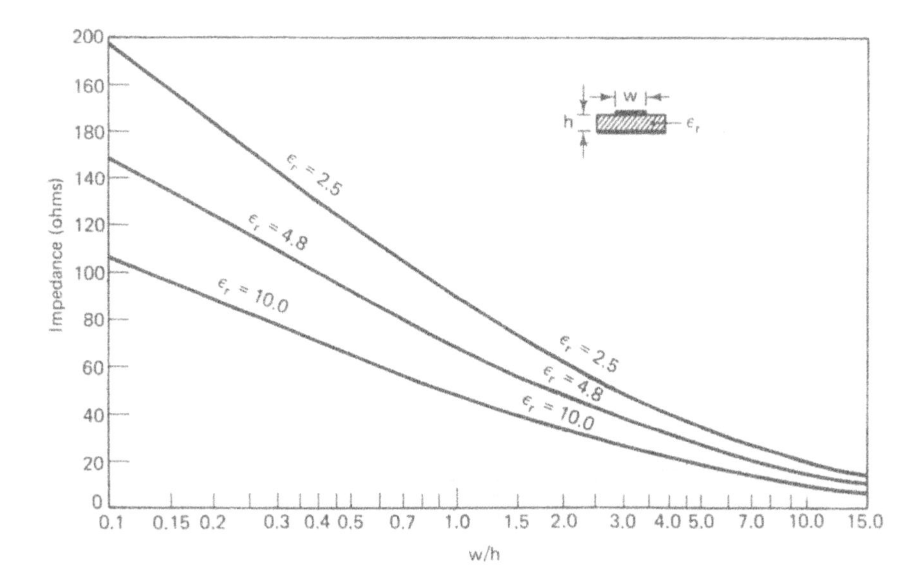

Figure 5.15 Characteristic impedance of a microstrip transmission line: glass epoxy ($\epsilon_r = 4.8$); Teflon epoxy ($\epsilon_r = 2.55$), and alumina ($\epsilon_r = 10$). (Source: from [11], p.16)

line with a strip of thickness t is [9, 12]

$$Z_0 = \frac{87}{\sqrt{\epsilon_r + 1.414}} \ln\left(\frac{5.98\,h}{0.8\,W + t}\right) \tag{5.118}$$

which is for strip thickness-to-width radios of $0.1 \leq t/W \leq 0.8$.

With knowledge of Z_0 and ϵ_r', and assuming only TEM mode of propagation in the line, the approximated values for L, C, and other parameters can be determined as described in Chapter 7.

5.7.6 Stripline

The transverse view and geometry of a single stripline using one signal plane is shown in Figure 5.16*a*. It consists of a central conducting strip (or trace) embedded in a dielectric material that is sandwiched between two conducting plates. It is a planar type of transmission line that lends itself very well to microwave integrated circuit and photolithographic fabrication. Single or stacked (multiple layers of) striplines are used for printed circuit board (PCB) arrangements [12]. Here we describe the simplest version shown in Figure 5.16*a*.

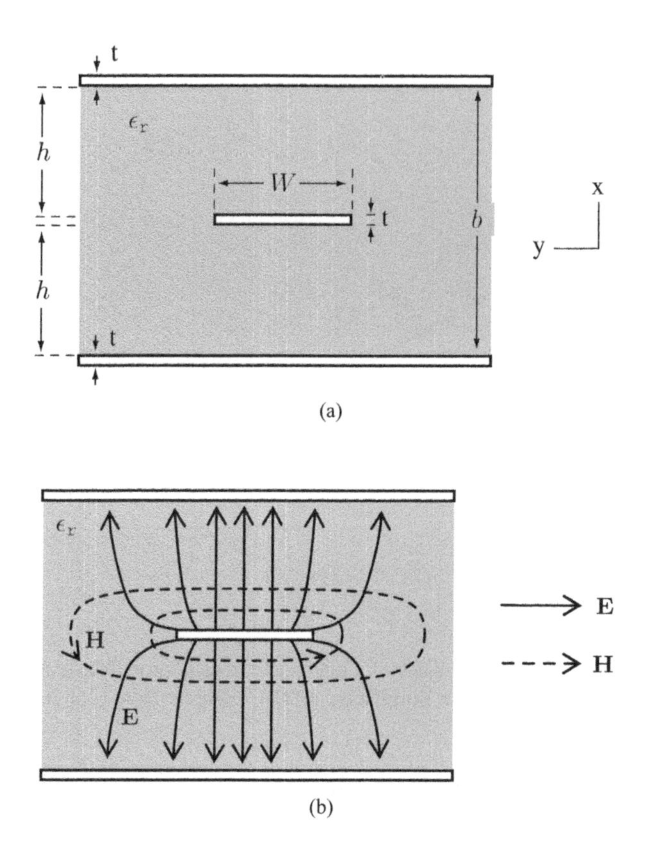

Figure 5.16 Transverse views of a stripline. (*a*) Geometry, (*b*) Electric and magnetic lines.

Intuitively, one can consider the stripline as a "flattened out" coaxial line; both have a central conductor and are uniformly filled with a dielectric medium. The main benefit of using a stripline is the complete shielding of RF energy generated from internal strips (traces) and consequent suppression of RF energy emissions outside. A sketch of the field lines from the stripline is shown in Figure 5.16*b*.

The stripline can support TEM mode of wave propagation and this is the usual mode of operation. It can also support higher order TE and TM modes. But these can be avoided in practice by restricting the ground plane spacing to $h < \lambda/4$ or by using shorting pins (vias) between the two ground planes.

The geometry of the stripline is such that it does not lend itself to simple analysis. However, we are primarily concerned with TEM or quasi-TEM modes behavior of the line where electrostatic analysis is sufficient to obtain the desired parameters. It is possible to determine the capacitance C per unit length for the line from an exact solution of Laplace's equation through a conformal mapping of the stripline configuration [13–15].

Assuming TEM mode of wave is propagating in the line, we obtain the phase velocity v_p and the propagation constant β for the wave as

$$v_p = \frac{c}{\sqrt{\epsilon_r}} \tag{5.119a}$$

$$\beta = \beta_0 \sqrt{\epsilon_r} \tag{5.119b}$$

where c is the velocity of light in free space, and the other symbols are as described earlier. The characteristic impedance of the line can now be formally written as

$$Z_0 = \sqrt{\frac{L}{C}} = \frac{1}{v_p C} \tag{5.120}$$

where L and C are the inductance and capacitance per unit length of the line.

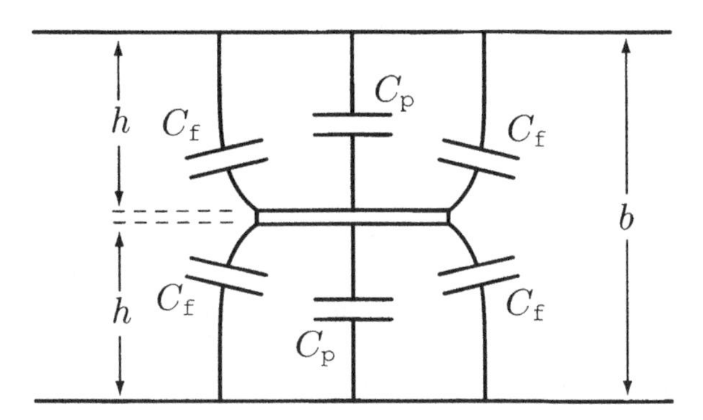

Figure 5.17 Capacitance model for stripline transmission lines.

The total capacitance C of a stripline can be modelled [15] as shown in Figure 5.17, and it is given by

$$C = 2C_p + 4C_f \tag{5.121}$$

where

$C =$ total capacitance per unit length
$C_p =$ parallel plate capacitance per unit length (in the absence of fringing fields)
$C_f =$ fringing capacitance per unit length.

Assuming that the medium between the plates in not ferromagnetic. Then by using (5.118) and (5.119), we obtain the following for the characteristic impedance

$$Z_0 = \frac{120\,\pi\,\epsilon}{\sqrt{\epsilon_r}\,C} \tag{5.122a}$$

or by using (5.120)

$$Z_0\,\sqrt{\epsilon_r} = \frac{30\,\pi}{\frac{W/h}{1-t/h} + \frac{C_f}{\epsilon}}. \tag{5.122b}$$

Approximate and exact values of C_f are discussed in [13, 14] for zero thickness center conductor ($t = 0$) the fringing capacitance $C_f \simeq 0.4413$. Thus, for $t = 0$ we obtain from (5.122b)

$$Z_0 = \frac{30\,\pi}{\sqrt{\epsilon_r}}\,\frac{1}{W/h + 0.4431} \tag{5.123}$$

For practical computations [10] quotes the following for characteristic impedance

$$Z_0 = \frac{30\,\pi}{\sqrt{\epsilon_r}}\,\frac{h}{W_e + 0.4431\,h} \tag{5.124}$$

where W_e is the effective width of the center conductor given by

$$W_e = \frac{W}{h} - \begin{cases} 0, & \text{for} \quad W/h > 0.35 \\ (0.35 - W/h)^2, & \text{for} \quad W/h \leq 0.35. \end{cases} \tag{5.125}$$

These expressions assume zero strip thickness, and are quoted as being accurate to about 1% of the exact results. It is seen from (5.122) that the characteristic impedance decreases as the strip width W increases.

Given the parameters Z_0, h, and ϵ_r, the required width W of the strip (or trace) for the stripline can be determined from the W/h ratio derived from (5.122) as

$$\frac{W}{h} = \begin{cases} x, & \text{for} \quad \epsilon_r\,Z_0 < 120 \\ \left(0.85 - \sqrt{0.6 - x}\right), & \text{for} \quad \epsilon_r\,Z_0 \geq 120. \end{cases} \tag{5.126}$$

Expressions for the attenuation constant of a stripline are described in [10].

5.7.7 Comments

It should be noted that for the transmission lines considered in this section, it has been assumed that they all support predominantly TEM mode of propagation. For such transmission lines, the line parameters generally satisfy the following relations:

$$\begin{aligned} LC &= \mu_0 \epsilon \\ LG &= \mu_0 \sigma, \end{aligned} \tag{5.127}$$

which can be utilized conveniently to determine some of the line parameters. For example, knowledge of C yields $L = \mu_0\epsilon/C$, as was done in (5.112) for the wire above a ground plane.

Example 5.1

The inner and outer conductors of the coaxial line RG-58U have the following dimensions $2a = 0.090$ cm and $2b = 0.295$ cm, and they are made of copper having $\sigma_c = 5.87 \times 10^7$ S/m. The medium between the conductors is polyethylene, having a complex dielectric constant $\epsilon_r = 2.25 - j\,7 \times 10^{-4}$ at 3×10^9 Hz. Determine the parameters R, L, G, C, Z_0, γ, α, β and v_p for the line.

Solution

Equation (5.93) indicates that the resistance R per unit length of the line is

$$R = \frac{1}{2\pi\delta\sigma_c}\left(\frac{1}{a} + \frac{1}{b}\right),$$

where the skin depth δ for copper is

$$\delta = \frac{1}{\sqrt{\pi f \mu_0 \sigma_c}}.$$

In the present case for copper can be expressed as

$$\delta = \frac{66.1}{\sqrt{f}} \quad \text{mm},$$

where f is expressed in Hz. Thus at $f = 3 \times 10^9$ Hz,

$$\delta = 12.06 \times 10^{-7} \quad \text{m}$$

$$\begin{aligned} R &= \frac{1}{(2\pi)\,(12.06 \times 10^{-7})\,(5.87 \times 10^7)}\left\{\frac{1}{0.45 \times 10^{-3}} + \frac{1}{1.475 \times 10^{-3}}\right\} \\ &= 6.67 \quad \Omega/\text{m} \end{aligned}$$

From (5.91)

$$L = \frac{\mu_0}{2\pi}\ln\frac{b}{a} = \frac{4\pi \times 10^{-7}}{2\pi}\ln\frac{1.475}{0.45} = 0.24 \quad \mu\text{H/m}.$$

From (5.88)

$$C = \frac{2\pi\epsilon_0\epsilon_r'}{\ln\frac{b}{a}} = \frac{(2\pi)\left(\frac{1}{36\pi}\times 10^{-9}\right)(2.25)}{\ln\frac{1.475}{0.45}} = 105\times 10^{-12} = 105 \quad \text{pF/m}.$$

The effective conductivity σ for the dielectric material at 3×10^9 Hz is

$$\sigma = \omega\epsilon_0\epsilon_r'' = (2\pi)\left(\frac{1}{36\pi}\times 10^{-9}\right)\left(7\times 10^{-4}\right) = 1.17\times 10^{-4} \quad \text{S/m}.$$

With Loss Considered

$$\begin{aligned}
Z_0' &= \sqrt{\frac{R+j\omega L}{G+j\omega C}} \\
R+j\omega L &= 4520\angle 89.92^\circ \\
G+j\omega C &= 0.62\times 10^{-3} + j1979\times 10^{-3} \\
&= 1979\times 10^{-3}\angle 89.98^\circ \\
\therefore Z_0' &= 47.79\angle -0.03^\circ \\
&= 47.47 - j0.03,
\end{aligned}$$

$$\begin{aligned}
\gamma = \alpha + j\beta &= \sqrt{(R+j\omega L)(G+j\omega C)} \\
&= \{(4520\angle 89.92^\circ)(1.979\angle 89.98^\circ)\}^{1/2} \\
&= 94.58\angle 89.95^\circ \\
&= 0.083 + j94.58 \\
\therefore \alpha &= 0.083 \text{ Np/m} = 0.083\times 8.686 = 0.74 \text{ dB/m} \\
\therefore \beta &= 94.58 \text{ rad/m}
\end{aligned}$$

$$\frac{\sigma}{\epsilon} = \frac{1.17\times 10^{-4}}{\left(\frac{1}{36\pi}\times 10^{-9}\right)2.25} = 58.8\times 10^5.$$

From (5.94) and (5.95),

$$\begin{aligned}
G = \frac{\sigma}{\epsilon}C &= \frac{2\pi\sigma}{\ln b/a} \\
&= \left(58.8\times 10^5\right)\left(105\times 10^{-12}\right) \\
&= 0.62 \quad \text{m℧/m} = \text{S/m}.
\end{aligned}$$

Thus at 3×10^9 Hz the basic line parameters are

$$R = 6.67\ \Omega/\text{m}$$
$$L = 0.24\ \mu\text{H/m}$$
$$C = 105\ \text{pF/m}$$
$$G = 0.62\ \text{S/m}.$$

Lossless Case

If the losses are neglected, $R = G = 0$, and we have the characteristic impedance Z_0 as

$$Z_0 = \sqrt{\frac{L}{C}} = \frac{60}{\sqrt{\epsilon_r'}} \ln \frac{b}{a} = 47.81\ \Omega$$
$$\gamma = j\beta,\ \alpha = 0$$
$$\beta = \omega\sqrt{LC} = 94.72\ \text{rad/s}$$
$$v_p = \frac{\omega}{\beta} = \frac{1}{\sqrt{LC}} = 1.99 \times 10^8\ \text{m/s}.$$

Example 5.2

A dipole antenna is fed by a two-wire transmission line having $Z_0 = 500\,\Omega$. The measured VSWR or the line is 4, and the location of the nearest (to the load) voltage minimum on the line is 2.8 m from the antenna feed point. The operating frequency is 112 MHz. What is the impedance of the antenna?

Solution

Experimental arrangement is sketched as below:
where d_{min} is the distance of the nearest voltage minimum at C–D from the antenna terminals A–B. Thus, assuming $Z_{\text{in}}\,(at d_{\text{min}} = \ell) = Z_0$ in (5.49) and solving for Z_L, we obtain

$$Z_L = Z_0 \frac{S - j\tan(\beta\, d_{\text{min}})}{1 - jS\tan(\beta\, d_{\text{min}})} \tag{A}$$

with the VSWR

$$S = \frac{|V_{\text{max}}|}{|V_{\text{min}}|}.$$

Similarly, if the location of the nearest voltage maximum d_{max} is used, then it can be shown that

$$Z_L = Z_0 \frac{1 - jS\tan(\beta\, d_{\text{max}})}{S - j\tan(\beta\, d_{\text{max}})}. \tag{B}$$

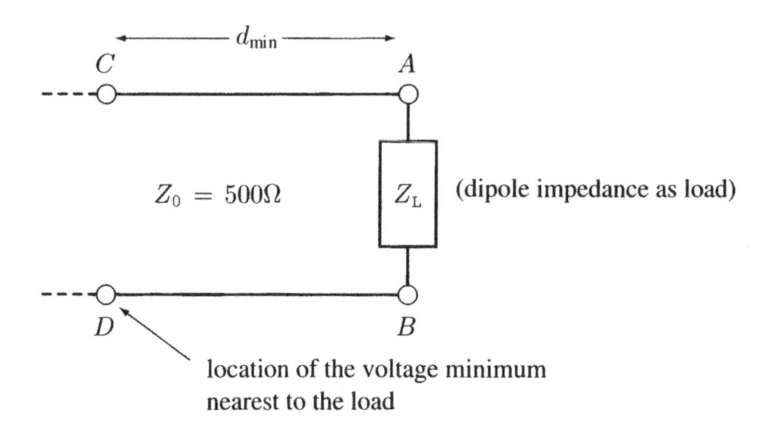

In the present case we use (A). We have $S = 4$, $d_{min} = 2.8$ m, $f = 112$ MHz, $\lambda = 2.679$ m, $Z_0 = 300\,\Omega$, and $\beta\, d_m = 6.6$ rad $= 376.2°$. Use of (A) gives

$$Z_L = 136 - j136\ \Omega.$$

5.8 TRANSIENTS ON TRANSMISSION LINES

When a source voltage or current is switched on to a transmission line for the purpose of transferring power to a load, it takes a finite time for the voltage and current on the line to reach their steady state values. This transitional period is called the transient. The results described earlier for time harmonic excitations apply only to the steady state condition.

Transient problems can be dealt with directly in the time domain as well as indirectly by way of frequency domain. In the latter case one requires the knowledge of the desired result(s) at all the appropriate frequencies contained in the input signal so that the desired time domain results can be obtained by utilizing the Fourier inversion technique (Chapter 3). In the present section we shall discuss the method of solving directly in the time domain the problem of a time dependent voltage or current transmission on a transmission line. More detailed discussions of transients on transmissions lines along with various examples are given in [1,4,5,7].

5.8.1 Initial and Final (Steady State) Values

We consider a voltage source $V_g(t)$ with internal impedance Z_g is connected through a switch to the input terminals A–A' of a lossless line of length d and terminated by a load impedance Z_L, at the output terminals B–B', as shown in Figure 5.18. The characteristic impedance and phase constant of the line are Z_0 and β, respectively.

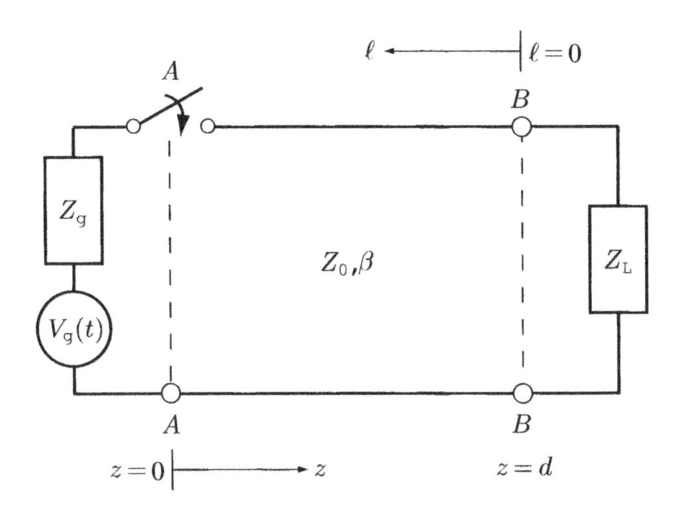

Figure 5.18 Transmission line terminated by load Z_L and excited by a time dependent voltage $V_g(t)$.

We assume that the input voltage $V_g(t)$ can be described as

$$V_g(t) = V_g u(t) \tag{5.128a}$$

$$\text{such that} \quad u(t) = \begin{cases} 1, & t \geq 0^+ \\ 0, & t < 0. \end{cases} \tag{5.128b}$$

Note that the equations above imply that the line is excited by a battery with a steady voltage v_g switched on a time $t = 0^+$. The switch is activated or closed at time $t = 0^+$, and we wish to follow the progress of the voltage and current waves on the line for $t \geq 0^+$. At time $t = 0^+$, the voltage $u(t)$ is on, and the source sees an infinite line at the terminal A–A', that is, it sees the characteristic impedance Z_0 connected across its terminals (i.e., the generator does not see the effect of the load Z_L). Thus the equivalent circuit for the system shown in Figure 5.18 can be represented as in Figure 5.19, where we have replaced the circuit beyond A–A' in Figure 5.18 by Z_0. Using Figure 5.19 we obtain the initial (i.e., at $t = 0^+$) values of voltage and current as

$$v(z,t) = v(0, 0^+) = V_0 = \frac{Z_0}{Z_g + Z_0} V_g \tag{5.129a}$$

$$i(z,t) = i(0, 0^+) = I_0 = \frac{V_g}{Z_g + Z_0}, \tag{5.129b}$$

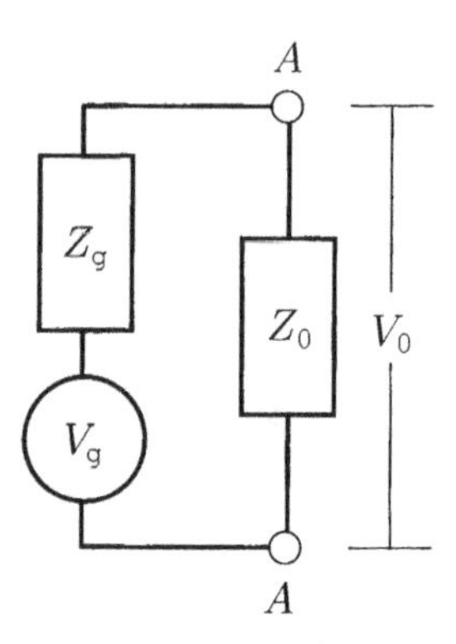

Figure 5.19 Equivalent circuit of Figure 5.18 at time $t = 0^+$.

and note that $v(z,t)/i(z,t) = V_0/I_0 = Z_0$, as it should be. Let us now assume that steady state or final state is obtained at time $t = t_0\ (= \infty)$. Under the steady state condition the equivalent circuit can be represented as shown in Figure 5.20.

Due to the nature of the voltage source, under steady state condition, the voltages at A–A' and B–B' are the same and using the circuit shown in Figure 5.20(b) we obtain them as

$$v(0,\infty) = v(d,\infty) = V_\infty = \frac{Z_\mathrm{L}}{Z_\mathrm{g} + Z_\mathrm{L}} V_\mathrm{g} \tag{5.130a}$$

$$i(0,\infty) = i(d,\infty) = I_\infty = \frac{Y_\gamma}{Z_\mathrm{g} + Z_\mathrm{L}} \ . \tag{5.130b}$$

5.8.2 Transient Values

We assume that at time $t = 0^+$, a voltage and a current wave of amplitude $V_0 = V^+$ and $I_0 = I^+$ are launched on the line, and they propagate toward the load Z_L along the z direction with velocity

$$u = \frac{1}{\sqrt{LC}} \, , \tag{5.131}$$

where L, C are the per unit length inductance and capacitance of the line. The time taken by both waves to reach the load is

$$t_1 = \frac{d}{u} \ . \tag{5.132}$$

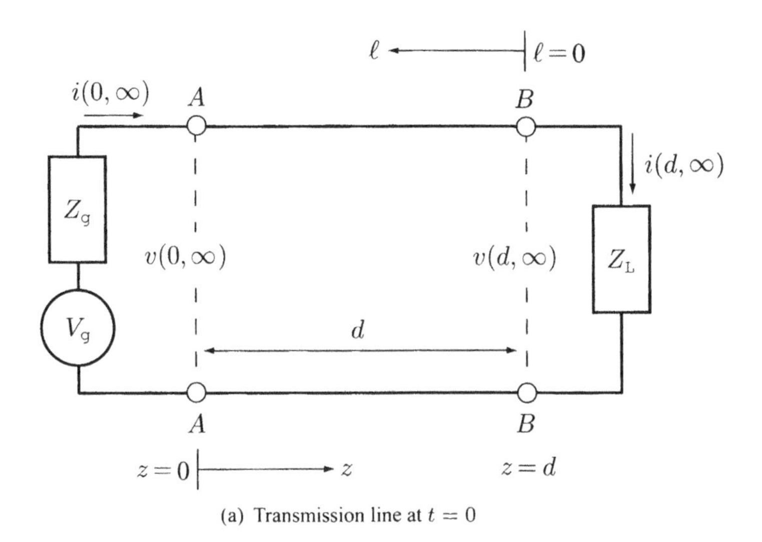

(a) Transmission line at $t = 0$

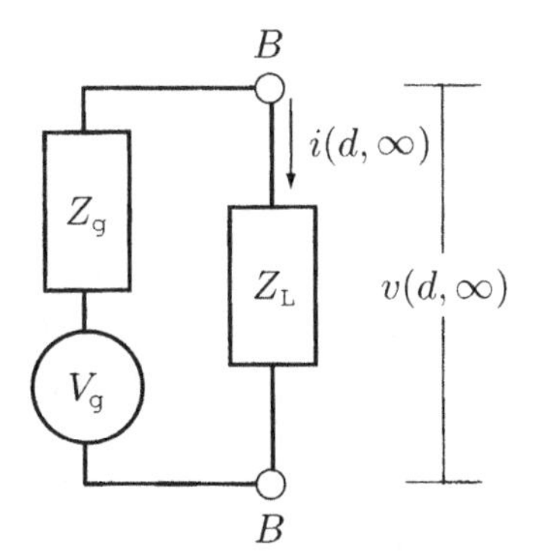

(b) Equivalent circuit and a final condition at time $t = \infty$

Figure 5.20 Transmission line under the steady state condition: (*a*) line circuit; (*b*) equivalent circuit.

On reaching the load, the waves suffer reflection at the load. Therefore at time $t = t_0 + 0^+$ the total voltage and current waves at the load are

$$\begin{aligned} v(d, t_1^+) &= V^+ + V^- = V_0(1 + \Gamma_\mathrm{L}) \\ i(d, t_1^+) &= I^+ + I^- = V_0(1 - \Gamma_\mathrm{L}), \end{aligned} \tag{5.133}$$

where voltage and current reflection coefficients at the load are Γ_L and $-\Gamma_\mathrm{L}$, respectively, and

$$\Gamma_L = \frac{Z_L - Z_O}{Z_L + Z_O}. \tag{5.134}$$

Note that the reflected voltage and current waves are

$$\begin{aligned} V^- &= \Gamma_\mathrm{L} V^+ = \Gamma_\mathrm{L} V_0 \\ I^- &= -\Gamma_\mathrm{L} V^+ = -\Gamma_\mathrm{L} I_0. \end{aligned} \tag{5.135}$$

The reflected voltage and current waves (5.135) will now travel back toward the generator and will suffer similar reflection at the terminals A–A' and travel toward the load again. The voltage and current reflection coefficients at the generator end are Γ_G and $-\Gamma_G$, respectively, where

$$\Gamma_G = \frac{Z_g - Z_o}{Z_g + Z_o}. \tag{5.136}$$

The voltage and current waves at A–A' at time $2t_1^+$ are

$$\begin{aligned} v(0, 2t_1^+) &= V^+ + V^-(1 + \Gamma_\mathrm{G}) \\ &= V_0 + \Gamma_\mathrm{L} V_0 + \Gamma_\mathrm{G}\Gamma_\mathrm{L} V_0 \end{aligned} \tag{5.137a}$$

$$\begin{aligned} i(0, 2t_1^+) &= I^+ + I^-(1 - \Gamma_\mathrm{G})I^- \\ &= I_0 - \Gamma_\mathrm{L} I_0 + \Gamma_\mathrm{G}\Gamma_\mathrm{L} I_0. \end{aligned} \tag{5.137b}$$

The reflected waves will proceed toward the load where they will be reflected and the process will continue back and forth until the steady state is obtained (which, ideally, will take infinite time in the present case).

Instead of tracing the voltage and current waves back and forth, it is easier to keep track of the process by using the bouncing (or lattice) diagrams shown in Figure 5.21 and 5.22, respectively. The diagrams are graphical representations of the voltage and current waves as they travel back and forth between the generator and the load. They can be used to estimate the transient duration before the steady state is reached.

For example, the steady state voltage at $(\ell, t \to \infty)$ or $z = (d - \ell,\ t \to \infty)$ is obtained by adding up the contributions at $A, B, C, D, E, \ldots$, and similarly for the current wave.

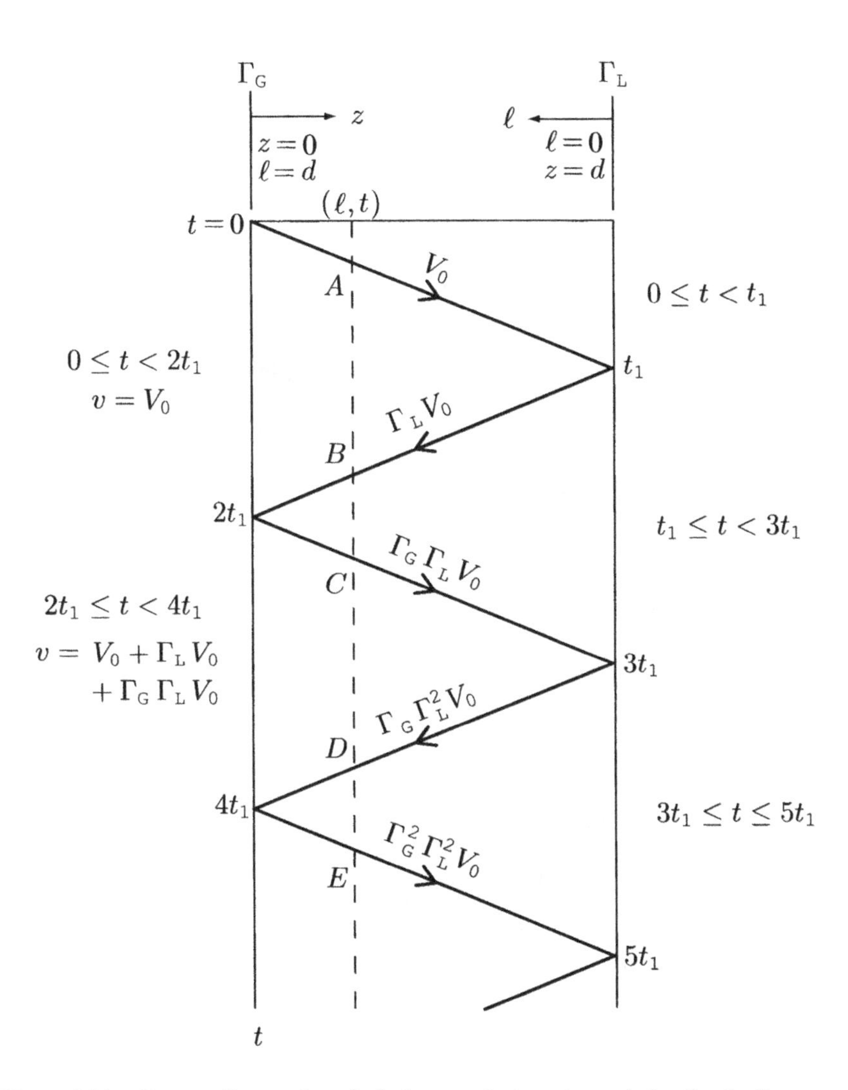

Figure 5.21 Bounce diagram for calculating transients on transmission line (voltage wave).

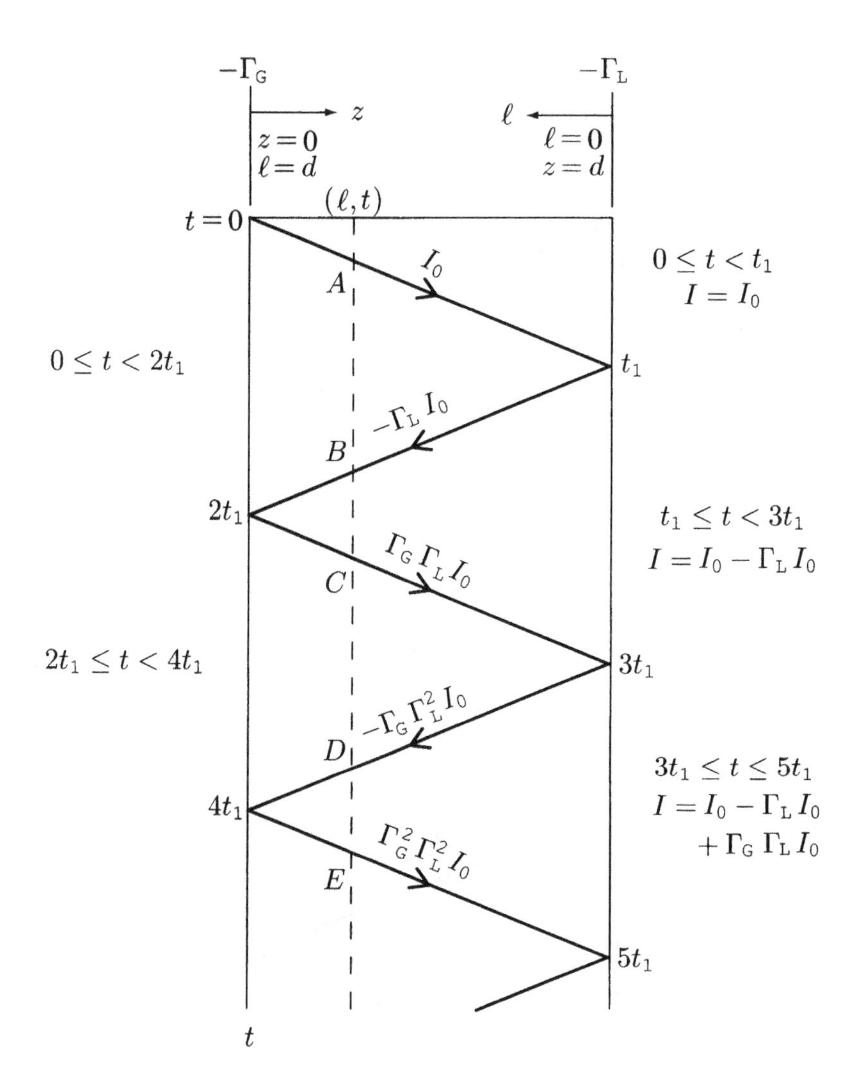

Figure 5.22 Bounce diagram for calculating transients on transmission line (current wave).

Figures. 5.21 and 5.22 apply to $v(\ell, t)$ or $v(z = d - \ell), t)$. Note that this incident wave starts at $z = 0$ ($\ell = d$), as shown, and travels a time t_1 to reach the load, as shown, where it gets reflected (to $\Gamma_L V_1^+ = \Gamma_L V_0$) at time $t_1 + 0^+$ and bounces back to the generator reaching A–A' at time $2t_1 + 0^-$. After reflection at $2t_1 + 0^+$ the incident wave becomes $\Gamma_L \Gamma_G V_0$, and the process goes on until steady state is reached.

Similar process takes place for the current wave, as shown in Figure 5.22. It should be noticed that the reflection coefficient at the load and generator ends for the current there are $-\Gamma_L$ and $-\Gamma_G$, respectively. The total voltage (or current) at any position ($z = d - \ell$, or ℓ) can be determined by drawing a vertical line through the point (z or ℓ) and then adding the voltages (or current) of all the zig-zag segments intersected by the line between time $t = 0$ and t. For example, the voltage and current buildup at the generator and load terminals are as indicated at the appropriate locations on Figure 5.21 and Figure 5.22, respectively. The voltage and current at the load and generator terminals can be obtained similarly from the bounce diagrams, and they are as given below:

At the load

$$v = \begin{cases} 0, & \text{for } 0 \le t < t_1 \\ V_0, & \text{for } t_1 \le t < 3t_1 \\ V_0 + \Gamma_L V_0 + \Gamma_G \Gamma_L V_0, & \text{for } 3t_1 \le t \le 5t_1. \end{cases}$$

At the generator

$$I = \begin{cases} I_0, & \text{for } 0 \le t < 2t_1 \\ I_0 - \Gamma_L I_0, & \text{for } 2t_1 \le t < 4t_1. \end{cases}$$

Steady State Values of Voltage and Current

From Figure 5.21 it is seen that the steady state value of the voltage at the load is

$$\begin{aligned} v(z = d,\ t \to \infty) &= V_0 + \Gamma_L V_0 + \Gamma_G \Gamma_L^2 V_0 + \Gamma_G^2 \Gamma_L^2 V_0 + \Gamma_G^2 \Gamma_L^3 V_0 + \cdots \\ &= V_0 (1 + \Gamma_L) \sum_{n=0}^{\infty} (\Gamma_G \Gamma_L)^n \\ &= V_0 \, \frac{1 + \Gamma_L}{1 - \Gamma_G \Gamma_L} \, . \end{aligned} \tag{5.138}$$

Introducing the value of Γ_L, Γ_G (from (5.134) and (5.136)) in (5.138) and using (5.129a) for V_0 we obtain:

$$v(d, \infty) = \frac{Z_L}{Z_g + Z_L} \, V_g$$

which is the same as V_∞ given by (5.129a). Hence it is compatible with the equivalent circuit in Figure 5.20*b*.

Similarly, from Figure 5.22 the steady state current at the load end ($z = d$, or $\ell = 0$) is

$$\begin{aligned} I(d,\, t = \infty) &= I_0(1 - \Gamma_\mathrm{L}) \sum_{n=0}^{\infty} (\Gamma_\mathrm{G}\Gamma_\mathrm{L})^n \\ &= I_0 \frac{1 - \Gamma_\mathrm{L}}{1 - \Gamma_\mathrm{G}\Gamma_\mathrm{L}} . \end{aligned} \tag{5.139}$$

Again by introducing the values of Γ_L, Γ_G (from (5.134) and (5.136)) and the value of I_0 from (5.129b) in (5.139), we obtain

$$I(d, \infty) = \frac{V_\mathrm{g}}{Z_\mathrm{g} + Z_\mathrm{L}} ,$$

which is the same as given in (5.130b) and is compatible with the equivalent circuit Figure 5.20(b).

Bounce diagrams can be used to determine the voltage and current at any point on the line for any time between $t = 0$ and steady state. Bounce diagrams and their applications are described in more details in [1,4,5,7].

Time Domain Reflectometer (TDR)

Time behavior of DC transients are utilized by a time domain reflectometer (TDR), which is used to locate faults (or damages) causing impedance discontinuities in underground and undersea cables. The TDR injects a step voltage at the input (or sending) end of the transmission line. The injected voltage wave propagating toward the receiving end will suffer reflection at the unknown impedance discontinuity; the reflected wave will reach the sending end after a certain interval of time depending on the distance of the fault from the sending end. Thus with the TDR it is possible to determine the location as well as the severity of the fault by observing the sending end voltage as a function of time [5].

5.9 MEASUREMENTS

General outlines of transmission line measurements are given in this section.

5.9.1 Slotted Line Measurements

The slotted line is a simple device that can be used to measure an unknown impedance (among other things) up to a few gigahertz frequency range. It consists of a section of air core (lossless) coaxial line with a longitudinal slot in its outer conductor. The line has probe oriented parallel to the transverse E field within and can be transported in the longitudinal direction along the line. With the line terminated by the unknown impedance standing waves are generated when it is excited at the other end. Sliding electric probe can be used to detect the location and amplitudes of voltage maxima and minima on the line as well as the VSWR. Slotted lines and their use for RF and microwave measurements are described in many places; for example, in [1,2,4].

Here we give the basic procedures to determine an unknown Z_L.

i) Terminate the line with Z_L.

ii) Use the probe to determine the amplitudes of $|V_{max}|$ and $|V_{min}|$ and obtain the VSWR $|V_{max}|/|V_{min}| = S$.

iii) Measure the location ℓ_1^{min} of the first voltage minimum, i.e., the minimum nearest the load.

It can be seen from (5.73) that the distance ℓ_1 of the nearest voltage minimum from the load is $2\beta\ell_1^{min} - \delta = \pi$, which gives

$$\delta = \frac{2\pi\ell_1^{min}}{\lambda} - \pi, \tag{5.140}$$

where δ is the phase of the reflection coefficient. Knowledge of S yields the magnitude of $\Gamma_L|$ as in (5.55). Thus, $\Gamma_L = |\Gamma_L|e^{j\delta}$, the reflection coefficient at the load is determined. Finally, the use of (5.42) determines the unknown impedance Z_L.

One does not need to confine the measurements at the nearest minimum. In fact, location of any minimum point will suffice. Also, ideally the location of a maximum point can also be used. However, from the measurement point of view, generally the maxima are broad and minima are sharp; hence, it is easier to identify the minimum points and hence they are preferred. More details can be found in the references cited.

5.9.2 Network Analyzer Measurement

A network analyzer, an instrument that was developed in 1980, does almost everything relating to transmission line and circuit parameter measurements. It is a self-contained unit, containing a signal source, tracking amplitude and phase detectors, and a signal processing facility. The analyzer measures reflection and transmission coefficients, and from these parameters such as $VSWR$, Γ and so-called S-parameters are obtained. It is an instrument of choice in microwave and circuit development work.

The basis of network analyzers and their application are described in [17]. Of course, the most appropriate reference for a network analyzer is it's own operator's manual and it's manufacturer's application/technical manual.

REFERENCES

1. M. N. O. Sadiku, *Electromagnetics*, 2nd ed., Harcourt Brace, New York, 1994, pp. 518–592.
2. S. Ramo, J. R. Whinnery and T. Van Duzer, *Fields and Waves in Communication Electronics*, 3rd ed., Wiley, New York, 1994, Chapter 5.
3. D. H. Staelin, A. W. Morgenthaler and J. A. Kong, *Electromagnetic Waves*, Prentice Hall, Englewood Cliffs, NJ, 1994, pp. 177–285.
4. W. C. Johnson, *Transmission Lines and Networks*, McGraw Hill, New York, 1950.
5. F. T. Ulaby, *Applied Electromagnetics*, Prentice Hall, Upper Saddle River, NJ, 1996, pp. 32–95.
6. J. D. Kraus, D. A. Fleisch and S. H. Russ, *Electromagnetics*, 5th ed., McGraw-Hill, New York, 1999, pp. 199–168.
7. U. S. Inan and A. S. Inan, *Engineering Electromagnetics*, Addison-Wesley, Menlo Park, CA, 1999, pp. 17–107, pp. 108–237 (extensive examples).
8. K. C. Gupta, R. Gang and I. J. Bahl, *Microstrip Lines and Slot Lines*, Artech, Debham, MA, 1974, Chapter 2.
9. C. R. Paul, *Introduction to Electromagnetic Compatibility*, Wiley, New York, 1992, Chapter 6.
10. D. M. Pozar, *Microwave Engineering*, Addison-Wesley, New York.
11. W. Sinnema, *Electronic Transmission Technology*, Prentice-Hall, Englewood Cliffs, NJ, 1979, Chapter 1.
12. M. I. Montrose, *Printed Circuit Board Design Techniques for EMC Compliance*, 2nd ed., IEEE Press, New York, 2000.
13. H. Howe Jr., *Stripline Circuit Design*, Artech, Debham, MA, 1974.
14. H. A. Wheeler, "Transmission-line Properties of Parallel Wide Strips by a Conformal Mapping Application", IEEE Trans. Microwave Theory Technique, Vol. MTT-12, No. 3, pp. 282-289, May 1964
15. H. A. Wheeler, "Transmission-line Properties of Parallel Strips Separated by Dielectric Sheet", IEEE Trans. Microwave Theory Technique, Vol. MTT-13, No. 2, pp. 172-185, May 1968
16. C. A. Balanis, *Advanced Engineering Electromagnetics*, Wiley, New York, 1989
17. R. A. Witte, *Spectrum and Network Measurements*, Prentice-Hall, Englewood Cliffs, NJ, 1991

PROBLEMS

5.1 Find the velocity at which a harmonic signal of 10^9 radians per second travels down a TEM transmission line for which $L = 0.4\,\mu$H/m, $C = 10$ pF/m and for which **(a)** $R = 0, G = 0$, **(b)** $R = 0.1\,\Omega$/m, $G = 10^{-5}$ S/m, and **(c)** $R = 300\,\Omega$/m, $G = 0$.

Answer: **(a)** 2.36×10^8 m/s; **(b)** 2.5×10^8m/s; **(c)** 2.4×10^8m/s.

5.2 The characteristics of the medium pertinent to a lossless TEM transmission line are $\mu_0 = 4\pi \times 10^{-7}\,\mu$H/m, $\epsilon_0 = (1/36\pi) \times 10^{-9}$ F/m. The line has $L = 1\,\mu$H/m. What is its capacitance C? Determine its characteristic impedance.

Answer: $C = 11.11$ pF/m, $Z_0 = 300\,\Omega$.

5.3 A coaxial line is operated at $f = 100$ MHz and is terminated in an impedance equal to its characteristic impedance of $90\,\Omega$. The power loss per meter is 0.1 %. Find the radii a and b.

Answer: $b/a \simeq 4.49$; $b = 2.56$ cm$\simeq 1.0''$; $a \simeq 0.57$ cm$\simeq 0.225''$.

5.4 The input impedance of a $75\,\Omega$ short-circuited transmission line is found to be $j150\,\Omega$ at 300 MHz. What is its length? What would be the impedance of the same length of line if its is open circuited.

Answer: $0.176\,\lambda$; $-j37.5\,\Omega$. Note: $(-j37.5 \times j156)^{1/2} = 75\,\Omega = Z_0$.

5.5 For a lossless transmission line section of length $\ell \ll \lambda/4$, show that the input reactance under short- and open-circuit conditions can be approximated by $X_{\text{in}}^{\text{short}}(\ell) = \omega\, L_{\text{in}} = \omega\, L\, \ell$, where L, C are the inductance and capacitance of the line per unit length and ω is the operating frequency in radians.

5.6 A lossless line of characteristic impedance $469\,\Omega$ is 30 cm long and is operated at $\omega = 10^9$ radians. **(a)** What inductance in microhenrys should be connected to the end of the line to make its input impedance infinite? **(b)** What capacitance in in pF should be connected at the end of the line to make the input impedance equal to zero?

Answer: **(a)** $L \simeq 0.305\,\mu$H; **(b)** $C \simeq 1.33$ pF.

5.7 What is the input impedance of a $\lambda/4$ long lossless transmission line whose $Z_0 = 200\,\Omega$, when the distant end is **(a)** (i) open, (ii) connected to a load of $65 + j100\,\Omega$ and (iii) short circuited; **(b)** when the line is $\lambda/2$ long?

Answer: **(a)** (i) $0\,\Omega$ (ii) $615.38\,\Omega$ (iii) $\infty\,\Omega$; **(b)** (i) $\infty\,\Omega$ (ii) $65\,\Omega$ (iii) $0\,\Omega$.

5.8 A lossless line with $Z_0 = 400\,\Omega$ is terminated by a load Z_L. Determine the VSWR S on the line if **(a)** $Z_L = 70\,\Omega$, **(b)** $Z_L = 800\,\Omega$, **(c)** $Z_L = 650 + j475\,\Omega$.

Answer: **(a)** 5.72; **(b)** 2; **(c)** 2.743.

5.9 A lossless line with $Z_0 = 70\,\Omega$ is terminated by a resistive load Z_L. What is the value of the load resistance when (a) the magnitude of the voltage on the line is same everywhere (b) the VSWR= 4 and $|V_{max}|$ occurs at the load and (c) VSWR = 4, and $|V_{max}|$ occurs $\lambda/4$ away from the load?

Answer: **(a)** $70\,\Omega$; **(b)** $280\,\Omega$; **(c)** $17.5\,\Omega$.

5.10 **(a)** What is the required length of a short circuited lossless coaxial line to provide an inductive reactance of $250\,\Omega$ if $\lambda = 80$ cm and $Z_0 = 70\,\Omega$. **(b)** What would be the length of the open circuit section of the same line to provide the same inductive reactance?

Answer: **(a)** 16.5 cm; **(b)** 36.5 cm.

5.11 A pair of parallel wires in free space each of diameter 1.6 mm and separated by 1 cm, acts as a transmission line. Determine the approximate Z_0, L, and C.

Answer: $Z_0 = 303\,\Omega$; $L = 1.01\,\mu$H/m; $C = 11$ pF/m.

5.12

(a) A lossless $75\,\Omega$ transmission line of certain length is terminated by a load $Z_L = 50 \pm j25\,\Omega$. For each case, determine the voltage reflection coefficient at the load and the VSWR on the line.

(b) The same line is terminated by $Z_L = 400\,\Omega$. Determine the impedance at a distance of $\lambda/8$ from the termination and looking toward the load.

Answer: **(a)** $Z_L = 50 + j25\,\Omega$: $\Gamma_L = 0.28\angle -56.31°$, $S = 1.77$; $Z_L = 50 - j25\,\Omega$: $\Gamma_L = 0.28\angle 56.31°$, $S = 1.77$; **(b)** $27.17 - j69.90\,\Omega$.

5.13 The input impedance $Z_{in}(\ell)$ looking toward the load Z_L for a terminated transmission line of characteristic impedance Z_0 is given by (5.49). Given that $f = 300$ MHz, $Z_0 = 200\,\Omega$, and $Z_L = 100\,\Omega$. Determine $Z_{in}(\ell)$ for: **(a)** $\ell = 15$ cm, **(b)** $\ell = 25$ cm, **(c)** $\ell = 50$ cm and **(d)** $\ell = 65$ cm. What conclusion is to be drawn from the similarity of some of the results obtained?

Answer: **(a)** $196 + j140\,\Omega$; **(b)** $400 + j0\,\Omega$; **(c)** $100 + j0\,\Omega$; **(d)** $196 + j40\,\Omega$. Input impedance for a length ℓ repeats every $\lambda/2$.

5.14 A transmission line terminated by a load Z_L is excited by a battery of steady voltage V_g, which is switched on at time $t = 0$, as in Figure 5.18. It is given that $d = 200$ m.

Develop the bounce diagram for the voltage bound-up at the terminals $A - -A'$ and $B - -B'$ in Figure 5.18. Use the diagram to determine $v(d, t)$ and $v(0, t)$. Estimate the transient duration and the stead state voltages $v(d, \infty)$ and $v(0, \infty)$.

Answer: Let $t_1 = d/v = 2\mu$s, transit time from the input to the output terminal.

$t = \infty$ $v(0, \infty) = v(d, \infty) = V_\infty = 6.66$ V.

$v(0, t)$ $0 \leq t < 4\,\mu$s (4 V), $4 \leq t < 8\,\mu$s (6.4 V), $8 \leq t < 12\,\mu$s (6.64 V), $12 \leq t < 16\,\mu$s (6.65 V), $16 \leq t < 20\,\mu$s (6.6531 V).

$v(d, t)$ $0 \leq t < 2\,\mu$s (0 V), $2 \leq t < 6\,\mu$s (6.0 V), $6 \leq t < 10\,\mu$s (6.6 V), $10 \leq t < 14\,\mu$s (6.65 V), $14 \leq t < 18\,\mu$s (6.653 V).

Transient duration $\sim 20\,\mu$s.

5.15 Repeat Problem 5.14 for the currents $i(0, t)$ and $i(d, t)$.

Answer:

$t = \infty$ $i(0, \infty) = i(d, \infty) = I_\infty = 44.44$ mA.

$i(0, t)$ $0 \leq t < 4\,\mu$s (80 mA), $4 \leq t < 8\,\mu$s (40 mA), $8 \leq t < 12\,\mu$s (44 mA), $12 \leq t < 16\,\mu$s (44.4 mA), $16 \leq t < 20\,\mu$s (44.44 mA).

$i(d, t)$ $0 \leq t < 2\,\mu$s (0 mA), $2 \leq t < 6\,\mu$s (40 mA), $6 \leq t < 10\,\mu$s (44 mA), $10 \leq t < 14\,\mu$s (44.4 mA), $14 \leq t < 18\,\mu$s (44.44 mA).

Transient duration $\sim 20\,\mu$s.

CHAPTER 6

ANTENNAS AND RADIATION

6.1 INTRODUCTION

Measurement and estimation of electromagnetic fields in the vicinity of an operating electronic system as well as the generation and radiation of known electromagnetic fields to estimate their impact on a system are of paramount importance in the area of EMC. An antenna, defined as a means for radiating and receiving electromagnetic waves, plays a key role in variety of applications.

We know from Maxwell's equations that a time-varying current or charge will radiate electromagnetic energy. In fact any such source configuration will radiate electromagnetic waves in space outside the region, and can be defined as a transmitting antenna. Any configuration capable of receiving (collecting) energy from electromagnetic waves incident on it is called a *receiving antenna*. In some systems like radar, frequently the same antenna is used to transmit and receive signals. However, the designs of antennas are generally governed by considerations related to the levels of power transmitted or received.

Applied Electromagnetics and Electromagnetic Compatibility. By D. L. Sengupta, V. V. Liepa
ISBN 0-471-16549-2

In this chapter we first describe the basic analytical techniques needed in antenna theory. Next, we determine the radiation from a short current element and a small loop of current, both of which are considered as basic antenna elements. Utilizing the results of the basic elements, we define certain parameters to characterize their radiation properties. These parameters can be used to describe the radiation characteristics of all antennas. Afterward, we describe radiation from selected linear antennas, simple antenna arrays, and a few other specific antennas that are of interest in the EMC area. We also discuss receiving antennas, and the equivalent circuits for receiving and transmitting antennas. Finally we briefly discuss the effect of ground on the radiation patters of elementary antennas.

We also briefly describe the effects of ground on the performance of elementary antennas. Receiving antennas and equivalent circuits for both transmitting and receiving antennas are also described.

The discipline of antennas and radiation is vast. Our discussion and choice of topics are limited and made in the context of EMC. More detailed information and description of variety of antennas may be obtained from the references [11, 12].

6.2 POTENTIAL FUNCTIONS

Direct solution of Maxwell's equations to obtain the fields produced by a given source distribution is not always possible. In cases where certain characteristics of the source current (or charges) are known or assumed, it is found convenient to determine the desired fields through some auxiliary functions, called *potential functions*, deduced from Maxwell's equations.

Consider an applied source distribution $\mathbf{J}(\mathbf{r}') = \mathbf{J}_a(\mathbf{r}')$, $\rho(\mathbf{r}') = \rho_a(\mathbf{r}')$ maintained in a source region V' located in a lossless, isotropic linear, and homogeneous medium shown in Figure 6.1, which also shows the coordinate system used. The medium is characterized by μ, ϵ and is assumed to be of infinite extent. We are interested in the fields produced at the field point P by the given source distribution.

The fields produced by the source are everywhere solutions of the time harmonic Maxwell's equations supplemented by the constitutive relations discussed in Section 3.4.3. We rewrite them in the context of the present problem as

$$\nabla \times \mathbf{E} = -j\omega \mathbf{B} \tag{6.1}$$

$$\nabla \times \mathbf{H} = \mathbf{J} + j\omega\epsilon \mathbf{E} \tag{6.2}$$

$$\nabla \cdot \mathbf{B} = 0 \tag{6.3}$$

$$\nabla \cdot \mathbf{D} = \rho \tag{6.4}$$

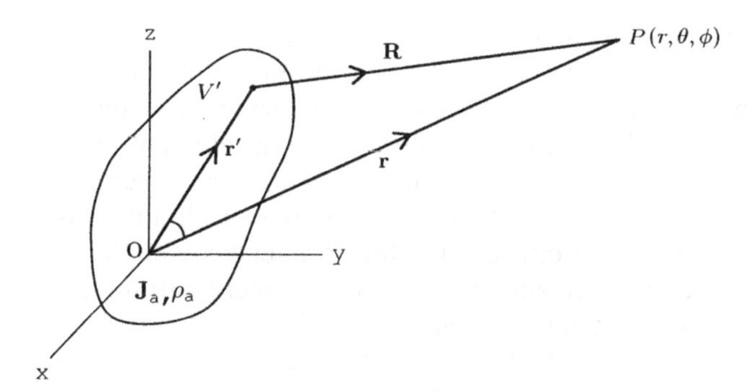

(a) for discrete distribution sources in V'.

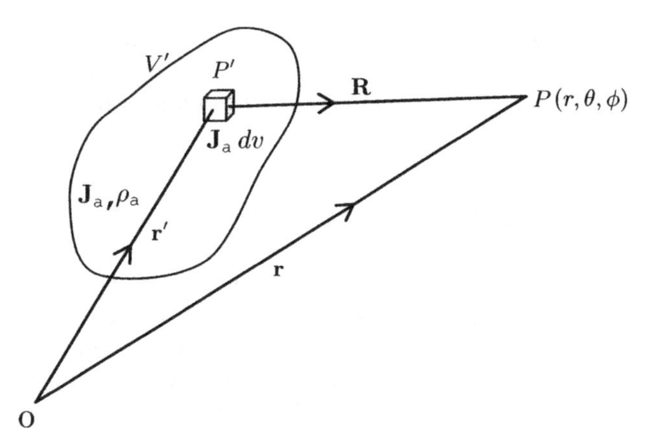

(b) for continuous distribution sources in V'.

Figure 6.1 Configuration showing the source region V' containing $\mathbf{J}(\mathbf{r}')$, $\rho(\mathbf{r}')$ in an infinite region. P' and P are the typical source and field points, respectively. Origin O of the coordinate system is arbitrary, and P is the field point located at (r, θ, ϕ). P' is the source point located at (r', θ', ϕ'). $\mathbf{r}$ is the position vector of the field point $\mathbf{r}'$ is the position vector of the source point. $\mathbf{R} = \mathbf{r} - \mathbf{r}'$ = vector from the source point to the field point.

and

$$\mathbf{B} = \mu \mathbf{H} \tag{6.5}$$

$$\mathbf{D} = \epsilon \mathbf{E}. \tag{6.6}$$

Note that the medium being lossless, $\mathbf{J}$, ρ in (6.2) and (6.4) refer only to the source currents and charges, respectively; the source current and charge distributions are maintained within the volume V' and are assumed zero outside V' (Figure 6.1). Also all field and source quantities are phasors, meaning they are complex quantities as described earlier. The source parameters $\mathbf{J}$ and ρ are not independent of each other; they must satisfy the equation of continuity discussed earlier in (Section 3.4.3).

We rewrite the equation (3.101a) here for convenience,

$$\nabla \cdot \mathbf{J} + j\omega\rho = 0. \tag{6.7}$$

Equation (6.3) indicates that $\mathbf{B}$ is a divergence less or solenoidal vector; hence we can assume (see Appendix A)

$$\mathbf{B} = \nabla \times \mathbf{A}, \tag{6.8}$$

where $\mathbf{A}$ is called the *magnetic vector potential*. It is important to note here that the vector $\mathbf{A}$ defined by (6.8) is not completely defined in the sense that its divergence is not specified; we will later choose a value for $\nabla \cdot \mathbf{A}$ to completely specify $\mathbf{A}$.

After using (6.1) and (6.8), we obtain

$$\nabla \times (\mathbf{E} + j\omega\mathbf{A}) = 0, \tag{6.9}$$

which indicates that $(\mathbf{E} + j\omega\mathbf{A})$ is an irrotational vector (see Appendix A), and hence we can assume

$$\mathbf{E} + j\omega\mathbf{A} = -\nabla\Phi, \tag{6.10}$$

where the negative sign on the right-hand side of (6.10) is chosen for convenience and ϕ is called the *electric scalar potential*. The electric field is commonly expressed by

$$\mathbf{E} = -j\omega\mathbf{A} - \nabla\Phi. \tag{6.11}$$

Equations (6.8) and (6.11) indicate that knowledge of $\mathbf{A}$ and Φ is required to determine the electric and magnetic fields produced by $\mathbf{J}$ and ρ.

We now obtain the relationship of the potentials with the sources. After taking the curl of (6.2) and using (6.1), (6.8) and (6.10) in conjunction with the two divergence and the two constitutive relations above along with the vector identity (3.159), it can be shown that

$$\nabla^2\mathbf{A} + j\omega^2\mu\epsilon\mathbf{A} - \mu\mathbf{J} + \nabla(\nabla \cdot \mathbf{A} + j\omega\mu\epsilon\Phi). \tag{6.12}$$

Using (6.4) and (6.10), it can be shown that

$$\nabla^2\Phi + j\omega\nabla \cdot \mathbf{A} = -\frac{\rho}{\epsilon}\,. \tag{6.13}$$

We now remove the arbitrariness of $\mathbf{A}$, mentioned earlier (i.e., its divergence still unspecified) by assuming the following in (6.12):

$$\nabla \cdot \mathbf{A} + j\omega\mu\epsilon\Phi = 0, \tag{6.14}$$

which is called the *Lorenz gauge condition.* It should be noted that the Lorenz condition is automatically satisfied if the charge and current satisfy the equation of continuity (6.7).

Under the Lorenz condition (6.12) and (6.13) reduce to

$$\nabla^2 \mathbf{A} + \beta^2 \mathbf{A} = -\mu \mathbf{J} \tag{6.15a}$$

$$\nabla^2 \Phi + \beta^2 \mathbf{A} = -\frac{\rho}{\epsilon}, \tag{6.15b}$$

with $\beta = \omega\sqrt{\mu\epsilon}$.

Equations (6.15a) and (6.15b) are the inhomogeneous Helmholtz vector and scalar equations, respectively. So they are the time harmonic forms of vector and scalar wave equations, respectively. It is now evident from (6.8), (6.11), and (6.13) that knowledge of $\mathbf{A}$ or ϕ is sufficient to determine $\mathbf{E}$ and $\mathbf{H}$. In many antenna problems it is customary to use the vector potential $\mathbf{A}$ and determine the desired fields by using the following expressions:

$$\begin{aligned} \mathbf{E} &= -j\omega\mathbf{A} - \nabla\phi \\ &= -j\omega\mathbf{A} + \frac{\nabla(\nabla \cdot \mathbf{A})}{j\omega\mu\epsilon}, \end{aligned} \tag{6.16}$$

where we have used (6.15a) to eliminate ϕ, and

$$\mathbf{H} = \frac{1}{\mu} \nabla \times \mathbf{A}, \tag{6.17}$$

which now indicates that the fields can be completely determined once $\mathbf{A}$ is obtained as a solution of (6.15a).

Equations (6.16) and (6.17) are fundamental equations used to determine $\mathbf{E}$ and $\mathbf{H}$ by a given current and charge distribution. They give the fields everywhere, including the source region V' (see Figure 6.1). In cases where one is interested only in the fields outside the source region, we can use the source-free Maxwell's equation (i.e., with $\mathbf{J} = 0$) to obtain $\mathbf{E}$ from $\mathbf{H}$. In such cases the following expressions are sufficient to determine the fields due to some current distributions:

$$\mathbf{H} = \frac{1}{\mu} \nabla \times \mathbf{A} \tag{6.18}$$

$$\mathbf{E} = \frac{1}{j\omega\epsilon} \nabla \times \mathbf{H}. \tag{6.19}$$

Outside the source region $\mathbf{A}$ is the solution of (6.15a) with $\mathbf{J} = 0$. Such a solution $\mathbf{A}$ of (6.15a) is discussed in many places [1–9], and we will assume here that the vector potential $\mathbf{A}$ at any field point P is given by

$$\mathbf{A} = \frac{\mu}{4\pi} \int_{v'} \mathbf{J}(\mathbf{r}') \frac{e^{-j\beta R}}{R}\, dv', \tag{6.20}$$

where the coordinate system representing the configuration is as shown in Figure 6.1.

For harmonic time dependence, the time dependent form of $\mathbf{A}$ can be obtained from (6.20) in a standard manner and is given by

$$\begin{aligned} \mathcal{A}(\mathbf{r}, t) &= \text{Re}\left[\mathbf{A} e^{j\omega t}\right] \\ &= \text{Re}\left[\frac{\mu}{4\pi} \int_{v'} \mathbf{J}(\mathbf{r}')\, e^{j(\omega t - R/v)}\, dv'\right], \end{aligned} \tag{6.21}$$

where $v = 1/\sqrt{\mu\epsilon}$ is the velocity of light in the medium. Equations (6.21) indicates that to calculate the vector potential at the field point P at time t due to the source current at P' we must take the value of the current at time $t - R/v$ (i.e., at a retarded time). This shows that the effect of current at the source point takes a finite time R/v to reach the field point P. (Note the excitation current is retarded by $e^{-j\omega R/v}$ in (6.20). For this reason the potential $\mathbf{A}$ is called the *retarded potential*, the time retardation, or the delay being exactly equal to the time taken by the wave to travel from the source to the field point. Similar interpretation is given to the electric scalar potential $\phi(\mathbf{r}, t)$ when used.

6.3 RADIATION FROM A SHORT CURRENT ELEMENT

A short conducting linear wire element carrying harmonically oscillating current is one of the fundamental sources of radiation in antenna theory. A Hertzian dipole consisting of two harmonically oscillating electric charges also behaves like an equivalent short current element. In antenna literature, the terms short current element, Hertzian dipole, and simply the electric dipole are often used interchangeably to represent a basic source of radiation.

6.3.1 Complete Fields

Consider a current element of moment $I\, d\boldsymbol{\ell}$ (where I is a phasor current) oriented along the z axis and placed symmetrically at the origin of a special coordinate system shown in Figure 6.2. The medium is assumed to be linear, homogeneous, isotropic, lossless, and of infinite extent. We wish to determine the fields produced by the current element at any field point $P(r, \theta, \phi)$. Note that with a similarly oriented Hertzian dipole of moment $q\ d\boldsymbol{\ell}$, q being the electric charge, the desired fields can be obtained from the current element fields by replacing I by $j\omega q$.

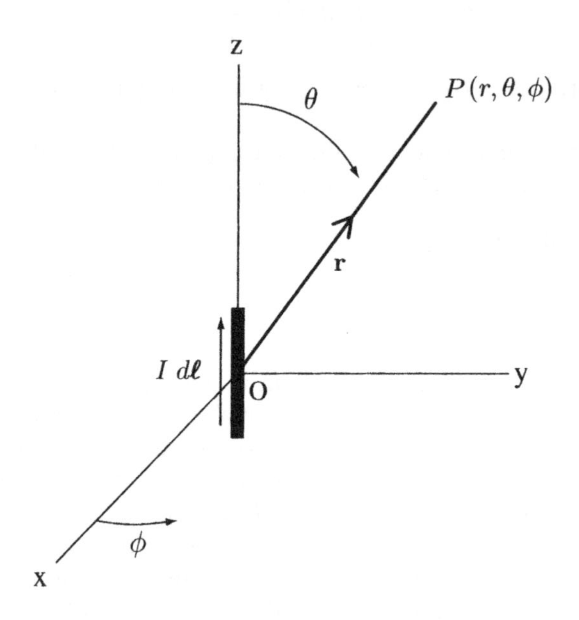

Figure 6.2 Current element $I\ d\boldsymbol{\ell}$ symmetrically located at the origin O and oriented along the z axis.

At first, we will use (6.20) to obtain $\mathbf{A}$ due to the current element. With reference to Figure 6.1 and equation (6.20), it can be seen that for the linear current element oriented along the z axis as in Figure 6.2, we have

$$\mathbf{J}\ dv' = I\ d\boldsymbol{\ell} = I(z')\ dz'\ \hat{z}, \tag{6.22}$$

where dependence of current on the source point location is explicitly indicated. In the present case we can write

$$R = |\mathbf{r} - \mathbf{r}'| \simeq r = r. \tag{6.23}$$

Thus for the present case the vector potential at the field point P is from (6.20) is

$$\mathbf{A} = \hat{z}\frac{\mu}{4\pi}\int_{-d\ell/2}^{d\ell/2} \hat{z}\ I(z')\ \frac{e^{-j\beta r}}{r}\ dz'. \tag{6.24}$$

We will assume $I(z') = I$, meaning the current amplitude is constant along the element and for sufficiently small dl. So we approximate (6.23) by

$$\mathbf{A} = A_z \hat{z} \simeq \hat{z}\, \frac{\mu I}{4\pi}\, d\ell\, \frac{e^{-j\beta r}}{r}\,. \tag{6.25}$$

Using the relationship of $\hat{z} = \hat{r}\cos\theta - \hat{\theta}\sin\theta$, we obtain the spherical components of $\mathbf{A}$ at P given by

$$\begin{aligned} A_r &= A_z \cos\theta \\ A_\theta &= -A_z \sin\theta \\ A_\phi &= 0, \end{aligned} \tag{6.26}$$

where A_z is obtained from (6.24) . It is interesting to note that the vector potential is entirely parallel to the current element in the present case.

From (6.18) we find that

$$\mathbf{H} = \frac{\hat{\phi}}{r}\left[\frac{\partial}{\partial r}(rA_\theta) - \frac{\partial A_r}{\partial \theta}\right],$$

with A_r, A_θ obtained by using (6.25) and (6.26). Thus we obtain

$$\begin{aligned} \mathbf{H} &= \hat{\phi}\,\frac{I\,d\ell}{4\pi}\,\beta^2 \sin\theta\left(\frac{j}{\beta r} + \frac{1}{\beta^2 r^2}\right) e^{-j\beta r} \\ \text{and}\quad H_r &= H_\theta \equiv 0, \quad \beta = \omega\sqrt{\mu\epsilon}\,. \end{aligned} \tag{6.27}$$

In the present case, since the field point P is outside the source, we can determine $\mathbf{E}$ from $\mathbf{H}$ by using (6.19) as follows:

$$\begin{aligned} \mathbf{E} &= \frac{1}{j\omega\epsilon}\,\nabla \times \mathbf{H} \\ &= \frac{1}{j\omega\epsilon}\left[\frac{\hat{r}}{r\sin\theta}\frac{\partial}{\partial\theta}(H_\phi \sin\theta) - \frac{\hat{\theta}}{r}\frac{\partial}{\partial r}(rH_\phi)\right], \end{aligned}$$

where H_ϕ is given by (6.27). The final form of $\mathbf{E}$ can now be written as

$$\begin{aligned} \mathbf{E} = \frac{I\,d\ell}{4\pi}\,\eta\beta^2 &\left[\hat{r}\left(\frac{1}{\beta^2 r^2} - \frac{j}{\beta^2 r^3}\right) 2\cos\theta \right. \\ &\left. + \hat{\theta}\left(\frac{j}{\beta r} + \frac{1}{\beta^2 r^2} - \frac{j}{\beta^3 r^3}\right)\sin\theta\right] e^{-j\beta r} \end{aligned} \tag{6.28}$$

and $E_\phi = 0$, where $\eta = \sqrt{\mu/\epsilon}$ = the intrinsic impedance of the medium. Observe that all fields given above are spherically symmetric; that is, they are independent of the coordinate ϕ, as it should be because the current is symmetric in ϕ. As mentioned earlier, the fields of a Hertzian dipole are also given by (6.27) and (6.28) provided that we make the replacement $I = j\omega q$.

Equations (6.27) and (6.28) give the complete field expressions due to a short current element or a Hertzian dipole obtained under the assumption of uniform (or constant) current over the element length. Although this type of current distribution is not physically realizable, the results are found quite useful, and they can be utilized to obtain the fields produced by longer and physically realizable current elements.

It is worth mentioning that sometimes a triangular current distribution is assumed for short current elements where the current is assumed to be maximum at the middle and falls linearly to zero at the end points. Such a dipole is called an *Abraham dipole*. The fields of an Abraham dipole are exactly half of those given by (6.27) and (6.28) [6].

The fields given by (6.27) and (6.28) are valid for all frequencies including $w = 0$ (or $\beta = 0$), meaning they are valid for the static case and for all distances outside the source. For a short current element or dipole of length $d\ell$ the fields are valid for $r \gg d\ell$, and for Hertzian dipole with $d\ell \to 0$ these fields are valid for all r.

In the static case we replace $I\, d\ell$ in (6.28) by $j\omega q$, where q is the charge involved for the Hertzian dipole. We obtain from (6.28),

$$\mathbf{E} = \frac{q\, d\ell}{4\pi\epsilon r^3}\left[2\cos\theta\, \hat{r} + \sin\theta\, \hat{\theta}\right], \tag{6.29}$$

which is the well-known result for the electric field produced by an electrostatic dipole [2]. Observe that to obtain (6.29), we have neglected the terms of the order of $1/\beta r$, $1/\beta^2 r^2$ compared with $1/\beta^3 r^3$ in (6.28). Similarly for $w \to 0$, (6.27) yields the following for the magnetic field:

$$\mathbf{H} = \hat{\phi}\, \frac{I\, d\ell \sin\theta}{4\pi r^2}, \tag{6.30}$$

which is the result for a dc current element given by Biot-Savart law [1,2].

Using the electric and magnetic fields of the current element given by (6.28) and (6.27), we can determine the complex Poynting vector for the element. It is given by

$$\begin{aligned} S^* &= \frac{\mathbf{E} \times \mathbf{H}^*}{2} \\ &= \hat{r}\left[\frac{\eta}{2}\left(\frac{I\, d\ell}{4\pi}\right)^2 \beta^4 \sin^2\theta \left(\frac{1}{\beta^2 r^2} - \frac{j}{\beta^5 r^5}\right)\right] \\ &\quad - \hat{\theta}\left[\frac{\eta}{2}\left(\frac{I\, d\ell}{4\pi}\right)^2 \beta^4 \sin 2\theta \left(\frac{-j}{\beta^3 r^3} - \frac{j}{\beta^5 r^5}\right)\right], \end{aligned} \tag{6.31}$$

which indicates that there is real power flow only in the $\hat{r}$ direction. The time average power flow from the radiating element is therefore

$$
\begin{aligned}
\mathbf{S}_{\text{av}} &= \text{Re}[\mathbf{S}^*] \\
&= \hat{r}\,\frac{\eta}{2}\left(\frac{\beta I\,d\ell}{4\pi r}\right)^2 \sin^2\theta \qquad \text{W/m}^2\,.
\end{aligned} \tag{6.32}
$$

6.3.2 Near Zone and Far Zone Considerations

The nature of the fields produced by the short current element and the Hertzian dipole depends significantly on the distance of the field point P from the source. The field amplitudes in (6.27) and (6.28) contain terms like $1/\beta r$, $1/(\beta r)^2$, and $1/(\beta r)^3$. A transition distance is defined at $\beta r = 1$ (i.e., $r = \lambda/2\pi$), where the three terms above are equal to unity. Thus, depending on whether βr is larger or smaller than unity, the entire space around the current element is divided into some district zones or regions where the field expressions are modified by appropriate approximations.

Near Field Region ($\beta r \ll 1$ or $r \ll \lambda/2\pi$)

This region is also referred to as the *induction field region* since the induction field dominates here. Since $\beta r \ll 1$ in this region, only the term of the order of $1/(\beta r)^2$ will provide sufficient approximation of the H field in (6.27). Similarly only the term of the order of $1/(\beta r)^3$ would be sufficient for the E field in (6.28). Thus we obtain the following field relations for the current element:

$$E_r \simeq -j\frac{I\,d\ell}{2\pi r^3}\cos\theta\; e^{-j\beta r} \tag{6.33a}$$

$$E_\theta \simeq -j\frac{I\,d\ell}{2\pi r^3}\sin\theta\; e^{-j\beta r} \tag{6.33b}$$

$$H_\phi \simeq -j\frac{I\,d\ell}{4\pi r^2}\sin\theta\; e^{-j\beta r}, \tag{6.33c}$$

and $E_\phi = H_r = H_\theta \equiv 0$.

Note that in the induction region E_θ and E_r are in time phase, but they are in time phase quadrature with H_ϕ. As given earlier, in the present case $\text{Re}[\mathbf{S}^*] \equiv 0$, implying that induction fields do not carry real power. Also, for $w \rightarrow 0$, the near zone fields reduce to the static fields, as we have seen earlier.

Intermediate Region ($\beta r > 1$ or $r > \lambda/2\pi$)

In this region, $1/\beta r$ terms tend to increase but do not completely dominate. The approximate fields are

$$E_r \simeq \frac{I\,d\ell}{4\pi}\,\eta\beta^2 \frac{2\cos\theta}{\beta^2 r^2}\,e^{-j\beta r} \tag{6.34a}$$

$$E_\theta \simeq j\frac{I\,d\ell}{4\pi}\,\eta\beta^2 \frac{\sin\theta}{\beta r}\,e^{-j\beta r} \tag{6.34b}$$

$$H_\phi \simeq j\frac{I\,d\ell}{4\pi}\,\beta^2 \frac{\sin\theta}{\beta r}\,e^{-j\beta r}, \tag{6.34c}$$

and $E_\phi = H_r = H_\theta = 0$.

In this region, the E-field phase components are in quadrature time phase, and E_θ and H_ϕ are in phase. Further discussion of these fields are given in [6].

Far Field or Radiation Region ($\beta r \gg 1$ or $r \gg \lambda/2\pi$)

For $\beta r \gg 1$ only the $1/\beta r$ terms are retained in the field expressions, and we obtain the radiation region fields as

$$E_\theta \simeq j\frac{\beta\eta I\,d\ell}{4\pi}\,\sin\theta\,\frac{e^{-j\beta r}}{r} \tag{6.35a}$$

$$H_\phi \simeq j\frac{\beta I\,d\ell}{4\pi}\,\sin\theta\,\frac{e^{-j\beta r}}{r}, \tag{6.35b}$$

and $E_r = E_\phi = H_r = H_\theta \equiv 0$.

Thus, the complete expressions for the far fields produced by a $\hat{z}$-directed short current element are

$$\mathbf{E} \simeq j\frac{\beta\eta I\,d\ell}{4\pi}\,\sin\theta\,\frac{e^{-j\beta r}}{r}\,\hat{\theta} \tag{6.36a}$$

$$\mathbf{H} \simeq j\frac{\beta I\,d\ell}{4\pi}\,\sin\theta\,\frac{e^{-j\beta r}}{r}\,\hat{\phi} \tag{6.36b}$$

with $\eta = \sqrt{\mu/\epsilon}$, and note that $\mathbf{H} = \mathbf{E}/\eta$. The time dependent far electric field is obtained in the usual manner, by $\boldsymbol{\mathcal{E}}(\mathbf{r},t) = \text{Re}[\mathbf{E}e^{j\omega t}]$, with $\mathbf{E}$ given by (6.35a). It can be shown that

$$\boldsymbol{\mathcal{E}}(\mathbf{r},t) = \frac{\beta\eta I\,d\ell}{4\pi r}\sin(\omega t - \beta r)(-\sin\theta\,\hat{\theta}) \tag{6.37a}$$

$$= \frac{\beta\eta I\,d\ell}{4\pi r}\sin(\omega t - \beta r)\,\hat{z}. \tag{6.37b}$$

It is interesting to observe from (6.37a) that the far electric field produced by a $\hat{z}$-directed current carrying a current $I \cos \omega t$, is maximum in the direction $\theta = \pi/2$ and that the E field in this direction is also directed in the $\hat{z}$ direction. This is because the far electric field is polarized parallel to the source current. This observation finds practical use in many applications.

Examination of (6.36a) and (6.36b) reveals the following about the far zone fields of a short current element: (1) **E** and **H** are in phase on $r =$ constant, surface, meaning on the spherical surfaces with origin at the location of the current element. (2) **E** is perpendicular to **H** on the constant phase surface. (3) The real power flow in the $\hat{r}$ direction is

$$\begin{aligned} \mathbf{S}_{\text{av}} &= \text{Re}\left[\frac{\mathbf{E} \times \mathbf{H}^*}{2}\right] \\ &= \frac{1}{2}\eta |H_\phi|^2 \hat{r} = \frac{1}{2}\frac{|E_\theta|^2}{\eta}\hat{r} = \frac{1}{2}\frac{\eta \beta^2 I^2\, d\ell^2}{16\pi^2 r^2}\sin^2\theta\, \hat{r}. \end{aligned} \tag{6.38}$$

(4) $|E_\theta|/|H_\phi| = \eta$ =the intrinsic impedance of the medium.

Thus the fields in the far zone constitute a spherical transverse electromagnetic wave propagating radically away from the current element, the phase center of the waves being located at the origin O or at the source. In contrast to the plane electromagnetic waves in free space, the power carried by the spherical waves varies inversely as $1/r^2$, which implies that the total power carried by the wave is finite, and hence spherical waves are amenable to generation in unbounded space. Power considerations will be taken up in a later section.

6.3.3 Near Zone and Far Zone Fields

Examination of the total field amplitudes versus distance from the source in a given direction can be used to characterize the behavior of the near zone and far zone fields. For this purpose we choose the broad side direction $\theta = 90°$ (Figure 6.2) in which the field components E_θ and H_ϕ are maximum and $E_{\text{r}} = 0$.

Figure 6.3 shows the normalized field amplitudes of E_θ and H_ϕ in dB versus the normalized distance $\beta r\ (= 2\pi r/\lambda)$ from a uniformly excited current element source. The field quantities are normalized with respect to their values at $\beta r = 1$, that is,

$$\begin{aligned} |E_\theta| &= 20 \log_{10} \frac{|E_\theta(\beta r)|}{|E_\theta(\beta r = 1)|} \text{ dB} \\ |H_\phi| &= 20 \log_{10} \frac{|H_\phi(\beta r)|}{|H_\phi(\beta r = 1)|} \text{ dB}, \end{aligned}$$

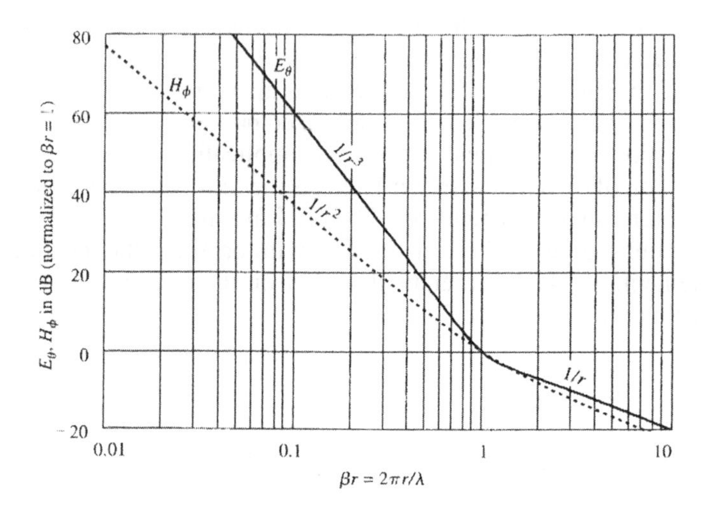

Figure 6.3 Plots of $|E_\theta|$ and $|H_\phi|$ versus normalized distance in the near and far zones. (Source: From [10], p. 30.)

where E_θ and H_θ are obtained from (6.28) and (6.27), respectively. The transition distance ($\beta r = 1$) at which the nature of the field variation changes is clearly evident in Figure 6.3, which indicates that in the reactive near field region (also called the *induction field region*), $|E_\theta|$ decays off as $1/r^3$ (or 60 dB/decade of distance) and $|H_\phi|$ decays off as $1/r^2$ (or 4.0 dB/decade of distance). In the far field region ($\beta r > 1$) both $|E_\theta|$ and $|H_\phi|$ fall off as $1/r$ (or 20 dB per decade of distance).

As discussed earlier, in the far field region E_θ and H_ϕ together form a special TEM wave propagating away from in the radial directions from the source. It is important to note that for sufficiently large distance, the wavefront at the field point P will locally appear as that of a plane wave. This approximation is made in many practical cases. Observe also that E_r component consists of $1/(\beta r)^2$ and $1/(\beta r)^3$ terms (see (6.28)) and hence are neglected in the far zone. For a current element with $d\ell = 1$ m, $I = 1$ A (with triangular currant distribution), the magnitude of the total electric field in the broadside direction versus the distance is shown in Figure 6.4 for three frequencies 100 kHz ($\lambda = 3000$ m), 1 MHz ($\lambda = 300$ m), and 10 MHz ($\lambda = 30$ m). Note that the assumption $d\ell \ll \lambda$ is satisfied.

Again, the near and far zone type of variations of field are evident in Figure 6.4. Note that for this case the fields for 100 kHz are entirely in the near zone within the range of r shown. Also in the near field region the field strength is greater at lower frequencies, and it increases with increase of frequencies in the far field region.

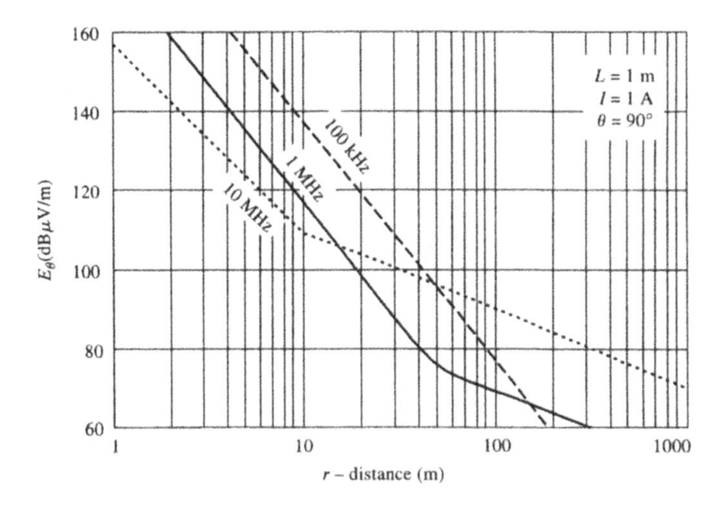

Figure 6.4 Total $|E_\theta|$ versus distance from a short dipole. (Source: From [10], p. 31.)

6.3.4 Radiated Power and Radiation Pattern

In an unbounded medium, the total power radiated by the current element is given by

$$P_{\text{rad}} = \oint_S \mathbf{S}_{\text{av}} \cdot d\mathbf{s}, \tag{6.39}$$

where (Figure 6.5) S is a closed spherical surface in the far zone surrounding the antenna, $\mathbf{S}_{\text{av}} = \hat{r}|\mathbf{S}_{\text{av}}|$ is the power flow density or the radiation density produced by the antenna at the location of an elementary vector surface $d\mathbf{s}$ located on S at (θ, ϕ):

$$d\mathbf{s} = \hat{r}\, ds = \hat{r}\, r^2 \sin\theta\, d\theta\, d\phi = \hat{r}\, r^2\, d\Omega$$

with $d\Omega$ as the solid angle subtended by ds at the origin O of S (Figure 6.5). We now rewrite (6.39) as

$$\begin{aligned} P_{\text{rad}} &= \int_0^{2\pi} \int_0^{\pi} r^2 |\mathbf{S}_{\text{av}}| \sin\theta\, d\theta\, d\phi \\ &= \int_0^{2\pi} \int_0^{\pi} U(\theta)\, d\Omega, \end{aligned} \tag{6.40}$$

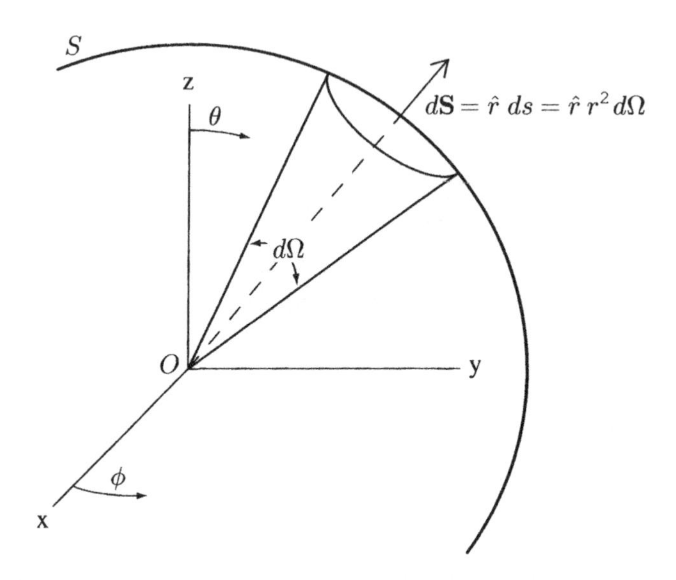

Figure 6.5 Geometry of the spherical surface S with origin at O and the antenna at O.

where

$$U(\theta) = r^2|\mathbf{S}_{\text{av}}| \\ = r^2\frac{|E_\theta|^2}{2\eta} \quad \text{W/solid angle} \tag{6.41}$$

is defined as the radiation intensity of the antenna in the direction $\theta,\ \phi$, which physically represents the power radiated from the antenna per unit solid angle. Note that $U(\theta)$, as defined, is independent of r and in general is a function of θ and ϕ.

Using (6.40) and (6.38), we find in the present case

$$U(\theta) = U_0 \sin^2\theta \tag{6.42}$$

with

$$U_0 = U_{\theta=\pi/2} = \frac{1}{2}\,\frac{\eta\beta^2 I^2\, d\ell^2}{16\pi^2} \tag{6.43}$$

as the maximum radiation intensity produced by the current element in the direction $\theta = \pi/2$.

The distribution of radiation intensity as a function of direction in space is called the *radiation pattern* of the antenna. For example, (6.42) gives the radiation pattern of the current element. Generally, the pattern is normalized with respect to its maximum value, and the normalized (or relative) pattern for any antenna is given by

$$U_n(\theta) = \frac{U(\theta)}{U_0}. \tag{6.44}$$

In the present case for the z-directed current element the normalized pattern is

$$U_n(\theta) = \sin^2\theta. \tag{6.45}$$

Frequently the pattern is expressed in dB (i.e., $10\log_{10} U_n(\theta)$) versus direction in a graphical representation.

In general, the radiation pattern of an antenna is a three-dimensional surface. It is common practice to show planar sections of patterns instead of the complete surface. The most commonly used are the principal E-plane and H-plane patterns. The E-plane pattern is a view of the radiation pattern is a view of the radiation pattern obtained in a plane section containing the maximum value of the radiation field in which the E field lies. Similarly the H-plane pattern is a pattern obtained in a plane section containing the H field and the maximum value of the radiated field. The two principal planes are orthogonal to each other.

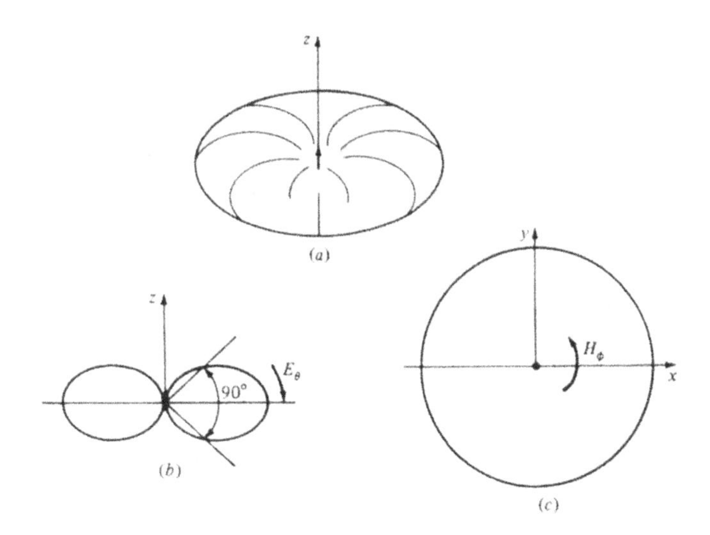

Figure 6.6 Radiation patterns for a short current element: (*a*) Three-dimensional; (*b*) principal E plane; (*c*) principal H plane. (Source: From [7]., p. 26)

The three views of the normalized pattern $U_n(\theta) = \sin^2\theta$ for the $\hat{z}$-directed current element are shown in Figure 6.6, where Figure 6.6*a*, Figure 6.6*b* and Figure 6.6*c* are the three-dimensional, the E-plane, and the H-plane patterns, respectively. Note that in the present case the E-plane pattern is the pattern in any longitudinal (ϕ = constant) plane containing current element and the H-plane pattern is in the xy plane. It should be noted that in the present case, the three-dimensional pattern shown in Figure 6.6*a* may be obtained by rotating the figure of eight E-plane patterns in Figure 6.6*b* around the z axis.

The half power beam width (HPBW) in a plane containing the direction of maximum radiation is the angle between the two directions in which the radiation intensity is one-half (or 3 dB down) the maximum radiation intensity. It is usually given for the E and H planes. For the current element the E-plane HPBW is 90°, and the H-plane pattern in Figure 6.6*c* does not show any HPBW since it is omnidirectional in this plane.

Radiation Resistance

The total power radiated by the current element in an unbounded medium is given by (6.40). After introducing the value of $|\mathbf{S}_{\text{av}}|$ from (6.38), it can be shown that

$$\begin{aligned} P_{\text{rad}} &= \frac{1}{2}\,\frac{I^2\,d\ell^2\,\beta^2\eta}{16\pi^2}\left(\int_0^{2\pi} d\phi\right)\left(\int_0^{\pi}\sin^3\theta\,d\theta\right) \\ &= \frac{I^2\,d\ell^2\,\beta^2\eta}{12\pi}\,. \end{aligned} \tag{6.46}$$

If the medium is assumed to be free space, we have $\beta = \beta_0 = 2\pi/\lambda_0$, λ_0 being the wavelength in free space, and $\eta = \eta_0 = 120\pi$. We have the total power radiated in free space by the current element as

$$P_{\text{rad}} = 40\pi^2 I^2\left(\frac{d\ell}{\lambda_0}\right)^2. \tag{6.47}$$

Equating the total power radiated $I^2R_r/2$ by the antenna to the hypothetical resistance R_r carrying a peak current I, we obtain

$$R_r = \frac{2P_{\text{rad}}}{I^2}\,, \tag{6.48}$$

where R_r is called the *radiation resistance* of the antenna. After substituting (6.46) in (6.47), we obtain the radiation resistance of the current element in free space as

$$R_r = 80\pi^2\left(\frac{d\ell}{\lambda_0}\right)^2. \tag{6.49}$$

For short antennas $d\ell/\lambda_0 \ll 1$; hence R_r is very small. For this reason such antennas are very difficult to excite. However, inspection of (6.49) implies that R_r can be increased by making $d\ell$ comparable to λ_0, which is the case as we will see later.

6.3.5 Wave Impedance

We saw earlier, in Section 6.3.2, that in the far field region of the electric current element the two mutually orthogonal field components E_θ and H_ϕ, radiated by the antenna, propagate as a spherical TEM wave in the radial direction. The wave impedance of this wave is defined as the ratio of the transverse electric field to the transverse magnetic field. Hence, by definition, the wave impedance Z_W is as follows:

$$Z_W = \frac{E_\theta}{H_\phi} . \tag{6.50}$$

To investigate the behavior of Z_W for all distances, we use the complete expressions for E_θ, H_ϕ given by (6.28) and (6.27), respectively, in (6.50) and obtain the following for the wave impedance of an electric dipole

$$Z_W = \eta \frac{\dfrac{j}{\beta r} + \dfrac{j}{(\beta r)^2} - \dfrac{j}{(\beta r)^3}}{\dfrac{j}{\beta r} + \dfrac{j}{(\beta r)^2}} , \tag{6.51}$$

where η is the intrinsic impedance of the medium and $\beta = \frac{2\pi}{\lambda}$ is the phase constant. Equation (6.51) indicates that in the near field region where $\beta r \ll 1$,

$$Z_W \simeq -j\frac{\eta}{\beta r} = \frac{\eta}{\beta r}\angle{-90^\circ} , \tag{6.52a}$$

and in the far field region where $\beta r \gg 1$, (6.51) indicates that

$$Z_W \simeq \eta. \tag{6.52b}$$

Thus in the near field region the wave impedance of an electric dipole is capacitive and its magnitude is larger than the intrinsic impedance of the medium. This is why an electric dipole is referred to as a high impedance source. In the far field region the wave impedance of the electric dipole equals the intrinsic impedance of the medium. Figure 6.7 shows variation of wave impedance as a function of distance from an electric dipole.

6.4 RADIATION FROM A SMALL LOOP OF CURRENT

A small loop of current is another fundamental source of radiation like the electric dipole, and in fact it is frequently referred to as the *magnetic dipole*. The loop is considered to be small if its radius $a \ll \lambda$, where λ is the wavelength in the medium under consideration. Such a loop may also be treated as a point source. The fields produced by a small loop of current placed in a linear, lossless, isotropic, and homogeneous medium of infinite extent are investigated.

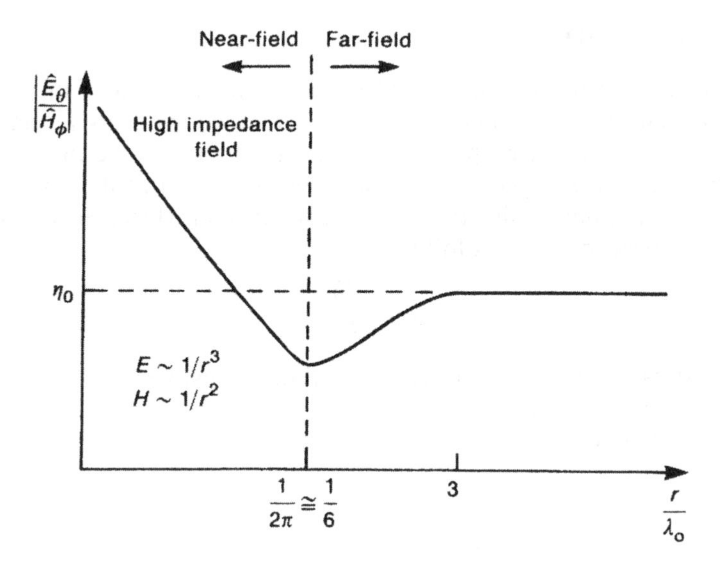

Figure 6.7 Wave impedance $|Z_W|$ versus distance for an electric dipole. (Source: From [11], p. 653.)

6.4.1 Complete Fields

Figure 6.8 shows a conducting circular loop of radius a placed symmetrically in the xy plane with its center located at the origin O. The loop carries a uniform phasor current of peak value I. From (6.20) it is found that the vector potential at the field point $P(r, \theta\phi)$ due to the loop of current is

$$\mathbf{A} = \frac{\mu I}{4\pi} \oint_L \frac{e^{-j\beta R}}{R} \, d\boldsymbol{\ell}', \tag{6.53}$$

where R, $d\boldsymbol{\ell}'$, L are as shown in Figure 6.8 and a is the radius of the loop, μ the permeability of the medium, and β is the phase constant in the medium.

Exact evaluation of (6.53) is difficult, and we approximately evaluate the integral under the assumptions $R, r \gg A$, which along with our small loop assumption $a \ll \lambda$ also implies that $R, r \gg \lambda$. Referring to Figure 6.8, we can now make the approximations

$$R \simeq r - \mathbf{r}' \cdot \hat{r} = r - a\cos\gamma, \tag{6.54a}$$

where γ is the angle between $\mathbf{r}$ and $\mathbf{r}'$ and is given by

$$\cos\gamma = \frac{\mathbf{r}' \cdot \hat{r}}{a} = \sin\theta[\cos\phi\cos\phi' + \sin\phi\sin\phi']. \tag{6.54b}$$

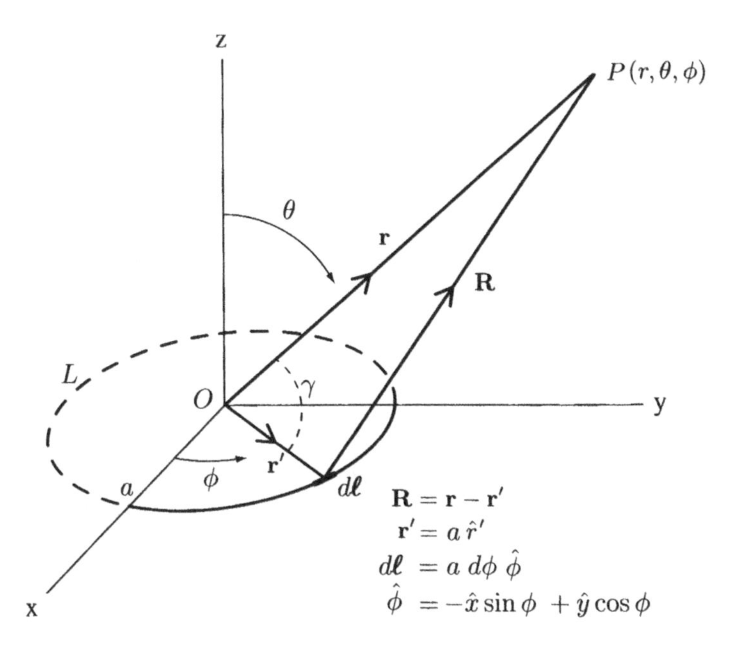

Figure 6.8 Small loop of current in the xy plane and the coordinate system used.

Under the approximations (6.54a) and (6.54b) we now approximate the integrand in (6.53) as

$$\begin{aligned} \frac{e^{-j\beta R}}{R} &\simeq \frac{e^{-j\beta r}}{r}\left[e^{j\beta a \cos r}\right]\left[1 - \frac{a}{r}\cos\gamma\right]^{-1} \\ &\simeq \frac{e^{-j\beta r}}{r}\left[1 + \left(\frac{1}{r} + j\beta\right) a\cos\gamma\right], \end{aligned} \tag{6.55a}$$

where we have retained only terms of the order of βa and a/r in the series expansions of the last two terms in the previous expression and this is consistent with the assumptions $\beta a \ll 1$ and $a/r \ll 1$. Then

$$d\boldsymbol{\ell}' = a\, d\phi'(-\hat{x}\sin\phi' + \hat{y}\cos\phi'). \tag{6.55b}$$

After introducing (6.55a) and (6.55b) in (6.53), we obtain the following for $\mathbf{A}$:

$$\begin{aligned} \mathbf{A} &\simeq \frac{\mu I a}{4\pi}\frac{e^{-j\beta r}}{r} \\ &\times \int_0^{2\pi}\left[1 + \left(\frac{1}{r} + j\beta\right) a\cos\gamma\right]\left[-\hat{x}\sin\phi' + \hat{y}\cos\phi'\right] d\phi', \end{aligned} \tag{6.56}$$

where $\cos\gamma$ is given by (6.54b) . Equation (6.56) can be evaluated, and after rearrangement of terms we obtain

$$\mathbf{A} = \hat{\phi}\,\frac{\mu\pi a^2 I}{4\pi}\left(\frac{1}{r} + j\beta\right)\frac{e^{-j\beta r}}{r}\sin\theta, \tag{6.57}$$

where we have used $\hat{\phi} = -\hat{x}\sin\phi + \hat{y}\cos\phi$.

We next determine the desired $\mathbf{E}$ and $\mathbf{H}$ fields by using (6.16) and (6.17) with $\mathbf{A}$ given by (6.56). In the present case they are determined by using the following relations:

$$\begin{aligned}\mathbf{E} &= -j\omega A_\phi\hat{\phi}\\ \mathbf{H} &= -\frac{\hat{\theta}}{\mu r}\frac{\partial}{\partial r}(rA_\phi),\end{aligned} \tag{6.58}$$

where A_ϕ is given by (6.56). The complete field components produced by the current loop are

$$E_\phi = -j\frac{\beta^2\omega\mu(\pi a^2)I}{4\pi}\sin\theta\left[\frac{j}{\beta r} + \frac{1}{(\beta r)^2}\right]e^{-j\beta r} \tag{6.59}$$

$$E_r = E_\theta \equiv 0 =$$

$$H_r = \frac{2\beta^2\omega\mu(\pi a^2)}{4\pi\eta}\cos\theta\left[\frac{j}{(\beta r)^2} + \frac{1}{(\beta r)^3}\right]e^{-j\beta r} \tag{6.60}$$

$$H_\theta = \frac{\beta^2\omega\mu(\pi a^2)I}{4\pi\eta}\sin\theta\left[-\frac{1}{\beta r} + \frac{j}{(\beta r)^2} + \frac{1}{(\beta r)^3}\right]e^{-j\beta r} \tag{6.61}$$

$$H_\phi = 0.$$

A current loop where fields are given by (6.59)–(6.61) is frequently referred to as a *magnetic dipole of moment* $\mathbf{m} = (\pi a^2)I\hat{z}$ placed at the center of the loop where $\mathbf{m}$ and $I\hat{\phi}$ are related by the right-hand rule as shown in Figure 6.9.

It should be noted that the field of a small magnetic dipole is the dual of that due to a Hertzian dipole on small current element. Comparison of (6.59)–(6.61) with (6.27) and (6.28) indicates that the roles of electric and magnetic fields are interchanged; that is, the $\hat{z}$-directed magnetic dipole has E_ϕ, H_r, and H_θ components with $E_r = E_\theta = H_\phi \equiv 0$ components, whereas the electric dipole has E_r, E_θ, and H_ϕ components with $E_\phi = H_r = H_\theta \equiv 0$. The duality principle states that with proper normalization of the dipole moments the fields of an electric dipole can be obtained from those of the magnetic dipole, and vice versa. Using Table 6.4.1 for the transformation of parameters will help achieve the desired field transformation. By replacing $\mathbf{m}$ by $I\,d\ell$, and

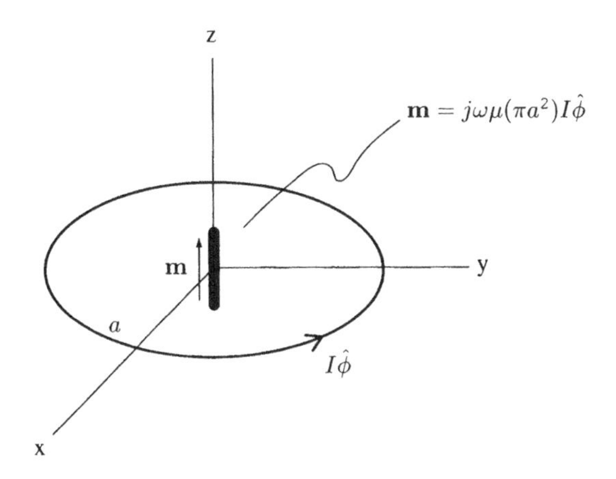

Figure 6.9 Current loop and its equivalent magnetic dipole.

Table 6.1 Duality Transformation

Electric Dipole	Magnetic Dipole
$I\, d\boldsymbol{\ell} = I\, dl\, \hat{z}$	$j\omega\mu(\pi a^2)I\hat{\phi} = \mathbf{m}$
$\dfrac{I}{j\omega}\, d\boldsymbol{\ell} = \mathbf{b}$	$j\omega\mu\mathbf{m} = j\omega\mu m\hat{z}$
$\mathbf{H}$	$-\mathbf{E}$
$\mathbf{E}$	$\mathbf{H}$
ϵ	μ

$-E_\phi$ by H_ϕ in the magnetic dipole field component E_ϕ in (6.59), we obtain the H_ϕ component of the electric dipole field given by (6.27), and similarly for the other components.

6.4.2 Far Zone Fields

Depending on the distance from the source, the fields of the magnetic dipole (6.59)–(6.60) can be classified into near zone and far zone fields. Here we give only the far zone fields of a magnetic dipole as

$$\mathbf{E} = \frac{\beta\omega\mu(\pi a^2)I}{4\pi}\sin\theta\,\frac{e^{-j\beta r}}{r}\,\hat{\phi} \tag{6.62}$$

$$\mathbf{H} = -\frac{\beta\omega\mu(\pi a^2)I}{4\pi\eta}\sin\theta\,\frac{e^{-j\beta r}}{r}\,\hat{\theta}. \tag{6.63}$$

Note that (6.62) and (6.63) together form a spherical TEM wave carrying power in the radial $\hat{r}$ direction as it should be.

The time dependent electric field is from (6.62),

$$\begin{aligned}\mathbf{E}(\mathbf{r},t) &= \mathrm{Re}[\mathbf{E}e^{j\omega t}] \\ &= \frac{\beta\omega\mu(\pi a^2)I}{4\pi}\frac{\sin\theta}{r}\cos(\omega t-\beta r)\,\hat{\phi},\end{aligned} \tag{6.64}$$

which indicates that the time dependant far electric field is polarized parallel to the source current direction which should be compared with the observation made from (6.37a) for the electric dipole field. The time dependant magnetic field $\mathcal{H}(\mathbf{r},t)$ can be similarly obtained, if so desired.

6.4.3 Radiated Power

The time average power flow for the radiated wave is given by

$$\mathbf{S}_{\mathrm{av}} = \mathrm{Re}\left[\frac{\mathbf{E}\times\mathbf{H}^*}{2}\right],$$

where $\mathbf{E}$ and $\mathbf{H}$ are given by (6.62) and (6.63). In the present case we obtain

$$\mathbf{S}_{\mathrm{av}} = \eta\frac{(\beta a)^4}{32}I^2\frac{\sin^2\theta}{r^2}\hat{r} \quad \mathrm{W/m^2}. \tag{6.65}$$

Since the dipole is in unbounded medium, the total power radiated by the dipole is

$$P_{\mathrm{rad}} = \oint_S \mathbf{S}_{\mathrm{av}}\cdot\hat{r}\,ds = \eta\left(\frac{\pi}{12}\right)(\beta a)^4 I^2. \tag{6.66}$$

Equating P_{rad} to the effective power radiated $I^2R_r/2$, we obtain the radiation resistance of the magnetic dipole as

$$R_r = \eta\left(\frac{\pi}{6}\right)(\beta a)^4. \tag{6.67a}$$

The area of the loop is $A = \pi a^2$. If the medium is free space, we have $\eta = \eta_0 = 120\pi\ \Omega$. Then the radiation resistance of a small current loop in free space is

$$R_r \simeq 31171\left(\frac{A}{\lambda^2}\right)\quad \Omega. \tag{6.67b}$$

As an example consider a loop with $a = 1$ cm at $f = 300$ MHz for $\lambda = 1$ m. We have for this loop $R \simeq 3.08\times 10^{-3}\ \Omega$. The current I needed to radiate 1 mW of power is about 0.8 A.

6.4.4 Wave Impedance

Using the complete fields given by (6.59) and (6.61) we obtain the wave impedance for the magnetic dipole case

$$\begin{aligned} Z_\omega &= -\frac{E_\phi}{H_\theta} \\ &= \eta \frac{j\left[\dfrac{1}{(\beta r)^2} + \dfrac{j}{\beta r}\right]}{\dfrac{1}{(\beta r)^3} + \dfrac{1}{(\beta r)^2} - \dfrac{1}{\beta r}} . \end{aligned} \tag{6.68}$$

In the near field region, we obtain

$$Z_\omega \simeq j\eta\beta r, \qquad \beta r \ll 1, \tag{6.69}$$

which is inductive and its magnitude is less than the intrinsic impedance of the medium. The magnetic dipole is therefore a low impedance source in the near field region.

In the far field region we obtain

$$Z_\omega \simeq \eta, \quad \beta r \gg 1, \tag{6.70}$$

which is real and equal to the intrinsic impedance of the medium. Figure 6.10 gives the wave impedance for the magnetic dipole as a function of distance from the dipole.

6.5 FUNDAMENTAL ANTENNA PARAMETERS

In previous sections we discussed certain characteristics of fields produced by electric and magnetic dipoles. Frequently these are considered as point sources, and as such, they are idealistic and impractical. However, many of their radiation field characteristics also apply to practical antennas in general. In this section we generalize those concepts to introduce a few fundamental parameters that are used to describe the performance of all antennas. Wherever possible we utilize the dipole results given earlier as an illustration. Our discussion will be brief; more detailed description can be found in textbooks on antennas, for example, in [1–6].

6.5.1 Radiation Intensity

We will assume that a given antenna is radiating in a linear, lossless, isotropic, and homogeneous medium of infinite extent. We will further assume that the

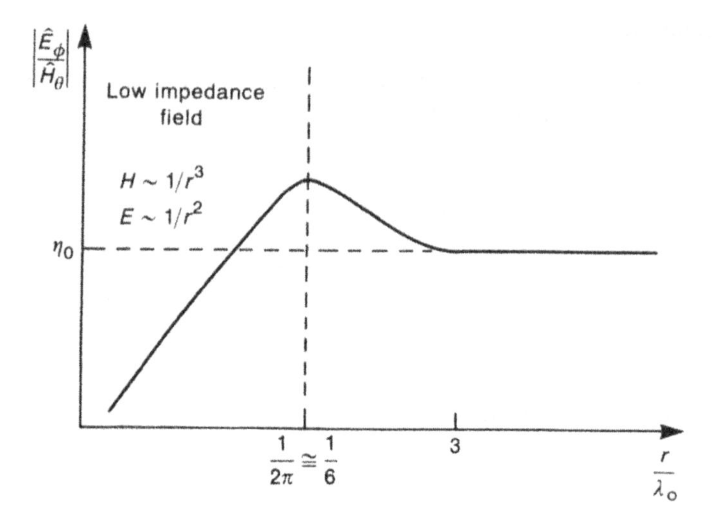

Figure 6.10 Wave impedance $|Z_W|$ for a magnetic dipole as a function of distance. (Source: From [11], p. 653.)

antenna is radiating TEM spherical waves and its phase center, assumed to be somewhere in the antenna, is located at the origin O of a spherical coordinate system as shown in Figure 6.5. The far zone radiation fields **E** and **H** of the wave produced by the antenna are orthogonal to each other, and they are each orthogonal to the radiation direction from an antenna that is also the direction of propagation of the waves. Thus, most generally, the radiation fields of the antenna can be written as

$$\begin{aligned}\mathbf{E} &= E_\theta\hat{\theta} + E_\phi\hat{\phi} \\ \mathbf{H} &= H_\theta\hat{\theta} + H_\phi\hat{\phi}\end{aligned} \tag{6.71}$$

with

$$\frac{|E_\theta|}{|H_\phi|} = -\frac{|E_\phi|}{|H_\theta|} = \eta. \tag{6.72}$$

With the fields given above, the time average power flow for the wave is

$$\begin{aligned}\mathbf{S}_{\text{av}} &= \text{Re}\left[\frac{\mathbf{E}\times\mathbf{H}^*}{2}\right] \\ &= \text{Re}\left[\tfrac{1}{2}(E_\theta H_\phi^* - E_\phi H_\theta^*)\hat{r}\right].\end{aligned} \tag{6.73}$$

In view of (6.72) we obtain from (6.73)

$$\mathbf{S}_{\text{av}} = \frac{1}{2\eta}(|E_\theta|^2 + |E_\phi|^2)\hat{r}$$
$$= \frac{\eta}{2}(|H_\theta|^2 + |H_\phi|^2)\hat{r} \quad \text{W/m}^2. \tag{6.74}$$

The power flow in the radial direction (θ, ϕ) and through an elementary vector surface ds located on a large spherical surface S having its origin at O (see Figure 6.5) is

$$dP(\theta, \phi) = \mathbf{S}_{\text{av}} \cdot d\mathbf{S}$$
$$= |\mathbf{S}_{\text{av}}| r^2 \sin\theta \, d\theta \, d\phi$$
$$= |\mathbf{S}_{\text{av}}| r^2 \, d\Omega, \tag{6.75}$$

where $d\Omega$ is the solid angle included by the surface dS as shown in Figure 6.5.

We now define the radiation intensity $U(\theta, \phi)$ in a direction (θ, ϕ) as the power radiated in that direction per unit solid angle and has the units of watts per square radian (or steradian). Thus symbolically the radiation intensity is given by

$$U(\theta, \phi) = \frac{dP(\theta, \phi)}{d\Omega}$$
$$= r^2 |\mathbf{S}_{\text{av}}|, \quad \text{W/steradian} \tag{6.76}$$

where we have used (6.75). Note that since for a spherical wave $|\mathbf{S}_{\text{av}}| \propto (1/r^2)$, $U(\theta, \phi)$ is independent of distance r, but in general, it is a function of both θ and ϕ.

$U(\theta, \phi)$ versus (θ, ϕ) is called the *power pattern* of the antenna. It is common practice to normalize the pattern so that its maximum value is unity. This is done by expressing the radiation intensity as

$$\frac{U(\theta, \phi)}{U_0} = |F(\theta, \phi)|^2, \tag{6.77}$$

where U_0 is the maximum value of $U(\theta, \phi)$ in the direction (θ_0, ϕ_0) and $|F(\theta, \phi)|^2$ has the maximum value of unity in the direction (θ_0, ϕ_0). $|F(\theta, \phi)|^2$ versus (θ, ϕ) is referred to as the *normalized power pattern* of the antenna. It should be mentioned here that $|F(\theta, \phi)|$ versus (θ, ϕ) is the field pattern for the antenna. In practice, the patterns are expressed in dB; that is, they are represented as $10 \log_{10}(U(\theta, \phi)/U_0)$ or $20 \log_{10} |F(\theta, \phi)|$ versus (θ, ϕ).

It is not always convenient to present the three-dimensional views of the patterns. Instead, most commonly the patterns are presented in two orthogonal planes called the E and H planes or the principal planes that were described earlier in this section, and will not be represented here.

As an illustration for the short current element, from (6.38) we have

$$r^2|\mathbf{S}_{\text{av}}| = \frac{1}{2}\frac{\eta\beta^2 I^2\, d\ell^2}{16\pi^2}\sin^2\theta,$$

and hence

$$U_0 = \frac{1}{2}\frac{\eta\beta^2 I^2\, d\ell^2}{16\pi^2} \quad \text{and} \quad |F(\theta)|^2 = \sin^2\theta.$$

6.5.2 Directivity and Gain

Directivity

The directivity of an antenna in a given direction is defined as the radiation intensity in that direction divided by the average radiation intensity. Thus, by definition, the directivity $D(\theta, \phi)$ is

$$\begin{aligned} D(\theta,\phi) &= \frac{\text{radiation intensity in the direction } (\theta,\phi)}{\text{average radiation intensity}} \\ &= \frac{U(\theta,\phi)}{U_{\text{av}}}, \end{aligned} \tag{6.78}$$

where $U(\theta, \phi)$ is identified by (6.77) and U_{av} is the average radiation intensity given by

$$U_{\text{av}} = \frac{\text{total radiated power}}{4\pi} = \frac{P_{\text{rad}}}{4\pi} \tag{6.79}$$

with P_{rad} obtained from (6.76) as

$$P_{\text{rad}} = \oint_S U(\theta,\phi)\, d\Omega \tag{6.80}$$

and $d\Omega = \sin\theta\, d\theta\, d\phi$. Now, using (6.77) and (6.79)–(6.80), it can be shown that (6.78) reduces to

$$D(\theta,\phi) = \frac{|F(\theta,\phi)|^2}{\dfrac{1}{4\pi}\displaystyle\int_0^{2\pi}\int_0^{\pi} |F(\theta,\phi)|^2 \sin\theta\, d\theta\, d\phi}, \tag{6.81}$$

where we have shown the integration variables explicitly. Equation (6.81) gives the directivity of the given antenna in any direction (θ, ϕ) obtained from its pattern function. It is convenient to define a quantity called the *beam angle* Ω_A as

$$\Omega_A = \int_0^{2\pi}\int_0^{\pi} |F(\theta,\phi)|^2 \sin\theta\, d\theta\, d\phi, \tag{6.82}$$

in terms of which $D(\theta, \phi)$ is given by

$$D(\theta, \phi) = 4\pi \frac{|F(\theta, \phi)|^2}{\Omega_A} . \tag{6.83}$$

Equations (6.81) and (6.83) show that the directivity of an antenna is entirely determined by its pattern function $|F(\theta, \phi)|^2$, meaning by its shape. Note that in view of (6.77) and (6.82), (6.80) reduces to

$$P_{\text{rad}} = U_0 \Omega_A, \tag{6.84}$$

which implies that all the power would be radiated through an equivalent solid angle Ω_A if the power radiated per unit angle or the radiation intensity equals the maximum value beam area [2].

It is appropriate to mention here that when directivity (also radiation intensity) is quoted as a single number without reference to a direction, it implies the maximum directivity (in other words, the old terminology of directive gain or maximum directivity is discarded [6, 13, 14]). Thus the maximum directivity is given by (from (6.84))

$$D = \frac{U_0}{U_{\text{av}}} = \frac{4\pi}{\Omega_A} . \tag{6.85}$$

Finally, using the definition of $D(\theta, \phi)$ given by (6.78), we can relate it to the maximum directivity and the pattern function as

$$D(\theta, \phi) = \frac{U(\theta, \phi)}{U_{\text{av}}} = \frac{U_0 |F(\theta, \phi)|^2}{U_{\text{av}}} = D|F(\theta, \phi)|^2, \tag{6.86}$$

where it should be noted that the maximum value of $|F(\theta, \phi)|^2$ is unity, and it occurs in the same direction as that of D.

Gain

Directivity of an antenna is entirely determined by its radiation pattern and its definition does not include any loss associated with the antenna. For practical reasons the gain of an antenna, to be defined, is a more realistic parameter because it includes the losses associated with the antenna. Power gain or simply the gain $G(\theta, \phi)$ of an antenna is defined as 4π times the ratio of the radiation intensity in a given direction to the net power input to the antenna (not P_{rad}), that is,

$$G(\theta, \phi) = \frac{4\pi\, U(\theta, \phi)}{\text{input power}} = \frac{4\pi\, U(\theta, \phi)}{P_{\text{in}}} . \tag{6.87}$$

The net power input P_{in} is not all radiated by the antenna as P_{rad}, since there is loss associated. Thus we have

$$P_{\text{rad}} = \alpha P_{\text{in}}, \tag{6.88}$$

where α is the radiation efficiency of the antenna. Note that $0 \leq \alpha \leq 1$. Using (6.87) in (6.87), we obtain the relationship between the gain and directivity of an antenna as

$$G(\theta, \phi) = \alpha \; D(\theta, \phi), \tag{6.89}$$

and we have the maximum gain related to the maximum directivity by

$$G = \alpha D. \tag{6.90}$$

Both (6.89) and (6.90) imply that the gain of an antenna is equal to its purely directional characteristics multiplied by its radiation efficiency.

EIRP (Effective Isotropically Radiated Power)

From (6.87) we obtain the following relation in the direction of maximum gain:

$$GP_{\text{in}} = 4\pi U = \text{EIRP}. \tag{6.91}$$

For any antenna the relation above can be interpreted as the product of maximum gain and the input power is equivalent to effective isotropically radiated power (EIRP) with maximum radiation intensity. The term EIRP is used in satellite communications area during the design phase for the choice of antennas. Equation (6.91) indicates that for the same EIRP the required input power to an antenna can be reduced by using a higher gain antenna.

Example 6.1

Radiation properties of a Hertzian dipole in free space.

Solution

We know its radiation intensity $U(Q, \phi)$ is

$$U(\theta, \phi) = \frac{\eta^2 I^2 \beta_0^2 \, d\ell^2}{32\pi^2} \sin^2\theta = U_0 \sin^2\theta$$

[see Section 6.5.1] with

$$U_0 = \frac{\eta^2 I^2 \beta_0^2 \, d\ell^2}{32\pi^2} \qquad \text{at } \theta = \frac{\pi}{2}\,.$$

Total radiated power P_{rad} is

$$P_{\mathrm{rad}} = \oint_S U(\theta,\phi)\, d\Omega$$
$$= U_0 \int_0^{2\pi} \int_0^{\pi} \sin^3\theta \, d\theta \, d\phi$$
$$= U_0 2\pi \left[\frac{4}{3}\right] = U_0 \frac{8\pi}{3}$$

$$\text{therefore,} \quad D(\theta,\phi) = 4\pi \frac{U(\theta,\phi)}{P_{\mathrm{rad}}} = \frac{3}{2}\sin^2\theta.$$

Next we have

$$\text{maximum directivity } D = 1.5 \quad \text{at } \theta = \frac{\pi}{2}$$
$$\text{maximum directivity in dB} = 10\log_{10} 1.5 = 1.76 \text{ dB}.$$

Frequently antenna directivity is expressed in terms of an isotropic radiator

$$U_{\mathrm{i}}(\theta,\phi) = \frac{P_{\mathrm{rad}}}{4\pi},$$

where "i" signifies an isotropic radiator. Therefore,

$$D_{\mathrm{i}}(\theta,\phi) = 4\pi \frac{U(\theta,\phi)}{P_{\mathrm{rad}}} = 1.$$

Hence for an isotropic radiator we have

$$D_{\mathrm{i}} = 1 \text{ or } 0 \text{ dB}.$$

We write the maximum directivity of a dipole with respect to an isotropic radiator as

$$\frac{D_{\mathrm{dipole}}}{D_{\mathrm{i}}} = \frac{1.5}{1} = 1.5 \text{ or } 1.76 \text{ dB}. \tag{6.90}$$

It is commonly referred to that the maximum directivity of a Hertzian dipole is 1.76 dBi, where the "i" means "as compared with an isotropic radiator." Antenna directivities are generally expressed in dBs compared with an isotropic radiator or with a Hertzian dipole.

6.6 FAR FIELDS OF ARBITRARY CURRENT DISTRIBUTIONS

In this section we discuss a general method to determine the far fields of known but arbitrary current distributions of finite extent placed in a lossless, linear, isotropic, and homogeneous medium of infinite extent. The method described is particularly advantageous to apply to linear antennas having a variety of configurations.

6.6.1 The Radiation Vector and the Far Fields

The geometry of the problem is shown in Figure 6.11. We start from (6.20), which gives the exact vector potential produced by the given source at the field point P. In Section 6.3, for point sources, we made certain approximations to evaluate (6.20) and classified the entire space surrounding the source into near and far zones by zones for which $r < \lambda/2\pi$ and $r > \lambda/2\pi$, respectively, where λ is the wavelength in the medium and r is the distance from the source. In view of the finite extent of the source, the approximations made to evaluate (6.20) for the present configuration are to be modified. We will discuss later in more detail the precise definitions of various zones of fields for an extended radiating system. For the present, we assume that the far or the radiation field region involves the following assumptions:

$$\begin{aligned} &r \gg r'_{\max} \\ &r \gg \lambda \text{ or } \quad \beta r \gg 1, \quad \text{as } r \to \infty, \end{aligned} \tag{6.92}$$

where it should be noted that the extent of the source is not assumed to be small, meaning $r'_{\max}$ is not small compared with λ, and in fact many practical antennas have $r'_{\max}$ comparable to or larger than λ.

Under the assumptions of (6.92), we approximate (6.20) by

$$\mathbf{A} \simeq \frac{\mu}{4\pi} \frac{e^{-j\beta r}}{r} \int_{V'} \mathbf{J}(\mathbf{r}')\, e^{+j\beta \mathbf{r}' \cdot \hat{r}}\, dv', \tag{6.93}$$

where we have made the following approximations in the integral (6.20):

$$R \simeq r \text{ in the denominator} \tag{6.94a}$$

$$\begin{aligned} R &\simeq r - \mathbf{r}' \cdot \hat{r} \\ &= r - r' \cos\gamma \text{ in the exponent} \end{aligned} \tag{6.94b}$$

with

$$\cos\gamma = \hat{r}' \cdot \hat{r} = \cos\theta' \cos\theta + \sin\theta' \sin\theta \cos(\phi' - \phi). \tag{6.94c}$$

In antenna theory (6.93) is referred to as the *far zone* or *far field approximation* to (6.20) and is extensively used to determine the far zone fields. The approximation implies that waves originating from different parts of the source travel along parallel rays to the far point as shown in Figure 6.11.

We now rewrite (6.93) as

$$\mathbf{A} = \frac{\mu}{4\pi} \frac{e^{-j\beta r}}{r} \mathbf{N}(\theta, \phi) \tag{6.95}$$

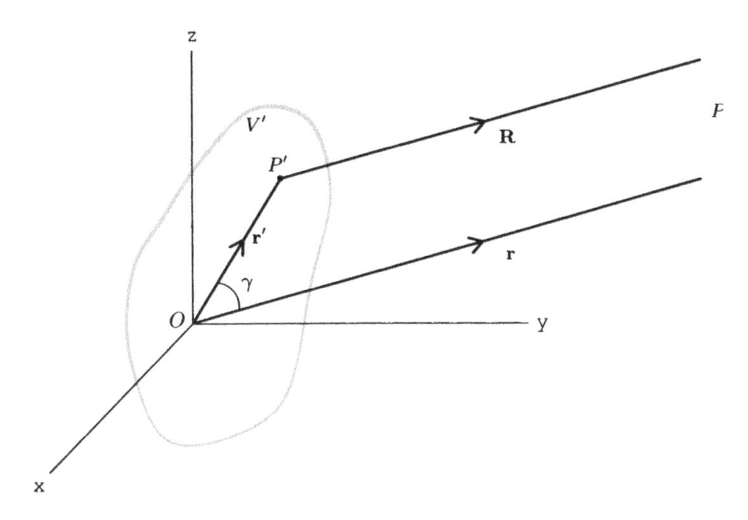

Figure 6.11 Geometry of the far zone approximation: $\mathbf{R} \parallel \mathbf{r}$ for all P'.

with

$$\mathbf{N}(\theta, \phi) \equiv \mathbf{N} = \int_{v'} \mathbf{J}(\mathbf{r}')\, e^{+j\beta\mathbf{r}' \cdot \hat{r}}, \tag{6.96}$$

where it should be noted that $\mathbf{N}$ is independent of r and is in general a function of θ and ϕ.

With $\mathbf{A}$ given by (6.95) the fields outside the source region, where $\mathbf{J} = 0$, can be obtained from the following:

$$\mathbf{H} = \frac{1}{\mu} \nabla \times \mathbf{A} = \frac{1}{4\pi} \nabla \times \left(\frac{e^{-j\beta r}}{r} \mathbf{N} \right) \tag{6.97}$$

$$\mathbf{E} = \frac{1}{j\omega\epsilon} (\nabla \times \mathbf{H}), \tag{6.98}$$

where μ and ϵ are the permeability and permittivity of the medium.

Calculation of the far fields (whose amplitudes vary as $1/r$) is considerably simplified in (6.97) and (6.98) if we make the following (far field) approximation to the curl operation involved (see Example 6.2):

$$\begin{aligned} \nabla \times (\cdot) &\simeq (-j\beta\hat{r}) \times (\cdot) \\ \text{or} \quad \nabla &\simeq -j\beta\hat{r}, \end{aligned} \tag{6.99}$$

where the term $(\cdot)$ contains the spherical wave function $e^{-j\beta r}/r$ multiplied by $\mathbf{N}(\theta,\phi)$. Applying (6.99) in (6.97), we obtain

$$\begin{aligned}\mathbf{H} &= \frac{1}{4\pi}(-j\beta\hat{r}) \times \left(\frac{e^{-j\beta r}}{r}\mathbf{N}\right) \\ &= \frac{-j\beta e^{-j\beta r}}{4\pi r}(\hat{r} \times \mathbf{N}) \\ &= \frac{-j\beta e^{-j\beta r}}{4\pi r}(\hat{r} \times \mathbf{N}_{\mathrm{t}}),\end{aligned} \tag{6.100}$$

where

$$\mathbf{N}_{\mathrm{t}} = \mathbf{N} - N_{\mathrm{r}}\hat{r} \tag{6.101}$$

represents the component of $\mathbf{N} \equiv \mathbf{N}(\theta,\phi)$ which is entirely transverse to the $\hat{r}$ direction. The vector $\mathbf{N}_{\mathrm{t}} = \mathbf{N}_{\mathrm{t}}(\theta,\phi)$ is called the *radiation vector*, due to Schelkunoff [3], of the current distribution.

Example 6.2

Given that $\Phi = e^{-j\beta r}/r$ and $\mathbf{N} = \mathbf{N}(\theta,\phi)$, show that

$$\nabla \times (\Phi\mathbf{N}) \simeq (-j\beta\hat{r}) \times (\Phi\mathbf{N})$$

if terms of order higher than $1/r$ are neglected.

Solution

We start from the vector identity (Appendix A)

$$\nabla \times \Phi\mathbf{N} = \nabla\Phi \times \mathbf{N} + \nabla\Phi \times \mathbf{N}.$$

Now

$$\begin{aligned}\nabla \times \Phi\mathbf{N} &= \nabla\left(\frac{e^{-j\beta r}}{r}\right) \times \mathbf{N} + \frac{e^{-j\beta r}}{r}\nabla \times \mathbf{N} \\ &= \left(1 + \frac{j}{\beta r}\right)\frac{e^{-j\beta r}}{r}(-j\beta)\hat{r} \times \mathbf{N} + \frac{e^{-j\beta r}}{r}\nabla \times \mathbf{N}.\end{aligned}$$

It can be shown that

$$\begin{aligned}\nabla \times \mathbf{N} &= \frac{1}{r^2\sin\theta}\begin{vmatrix}\hat{r} & r\hat{\theta} & r\sin\theta\,\hat{\phi} \\ \dfrac{\partial}{\partial r} & \dfrac{\partial}{\partial\phi} & \dfrac{\partial}{\partial\phi} \\ N_r & rN_\theta & r\sin\theta N_\phi\end{vmatrix} \\ &= O\left(\frac{1}{r}\right),\end{aligned}$$

meaning it is of the order of $1/r$. Therefore

$$\frac{e^{-j\beta r}}{r} \nabla \times \mathbf{N} \quad \text{is of the order of } 1/r^2,$$

meaning $O(1/r^2)$. Hence,

$$\begin{aligned}
\nabla \times (\phi N) &= \left(1 + \frac{j}{\beta r}\right) \frac{e^{-j\beta r}}{r} (-j\beta)\hat{r} \times \mathbf{N} + \frac{e^{-j\beta r}}{r} \nabla \times \mathbf{N} \\
&= \frac{e^{-j\beta r}}{r} (-j\beta)\hat{r} \times \mathbf{N} + O\left(\frac{1}{r^2}\right) + O\left(\frac{1}{r^2}\right) \\
&\simeq \frac{e^{-j\beta r}}{r} (-j\beta)\hat{r} \times \mathbf{N} \\
&= (-j\beta\hat{r}) \times \phi\mathbf{N}
\end{aligned}$$

after the higher order terms are neglected.

After using (6.98) along with (6.97) and (6.99), we obtain

$$\begin{aligned}
\mathbf{E} &\simeq \frac{1}{j\omega\epsilon} (-j\beta\hat{r}) \times \mathbf{H} \\
&= \frac{j\beta\eta}{4\pi} \frac{e^{-j\beta r}}{r} [\hat{r} \times (\hat{r} \times \mathbf{N}_t)] \\
&= -\frac{j\beta\eta e^{-j\beta r}}{4\pi r} \mathbf{N}_t,
\end{aligned} \tag{6.102}$$

where $\eta = \sqrt{\mu/\epsilon}$ is the intrinsic impedance of the medium.

Note that the far fields $\mathbf{E}$ and $\mathbf{H}$ given by (6.102) and (6.100) satisfy the following conditions required of a spherical TEM wave propagating in the $\hat{r}$ direction:

$$\mathbf{H} = \frac{\hat{r} \times \mathbf{E}}{\eta}, \qquad \mathbf{E} = -(\hat{r} \times \mathbf{H})\eta \tag{6.103}$$

and that $\hat{r}$, $\mathbf{E}$, and $\mathbf{H}$ are orthogonal to each other.

6.6.2 Vector Effective Length of an Antenna

The effective length (or height) of an antenna is a useful concept used to specify the performance of a transmitting antenna and also to specify the performance when the antenna is receiving plane waves of arbitrary linear polarization [3]. Using the radiation vector $\mathbf{N}_t$, Sinclair [12] introduced a fundamental antenna parameter called the *vector effective length* $\mathbf{h}$ for an antenna that is a generalization of the conventional effective length to include a specification of the polarization of the field radiated by the antenna. Effective length is also useful in calculating the voltage at the terminals of the antenna when it is used to receive plane waves of arbitrary elliptic polarizations.

Inspection of (6.102) for the far zone electric field indicates that $\mathbf{N}_\mathrm{t}$ can be a complex vector, so it is capable of characterizing the elliptically polarized fields. The radiation vector is in fact capable of completely specifying the distant field of an arbitrarily polarized transmission antenna. Also for applications to the receiving case the vector effective length of an antenna is defined [1–4] as

$$\mathbf{h} = -\frac{\mathbf{N}_\mathrm{t}}{I_\mathrm{i}}, \tag{6.104}$$

where I_i is the input current to the antenna. Note that $\mathbf{h}$ is now independent of the current amplitude. With (6.104) we now have the far electric field radiated by the antenna as

$$\mathbf{E} = \frac{j\beta\eta I_\mathrm{i}}{4\pi r}\, e^{-j\beta r}\mathbf{h}, \tag{6.105}$$

which can be compared with (6.102). The negative sign in (6.104) is used so that in the case of a linear antenna the vector length of a linear antenna reduces to the effective length of the antenna [14]. Note that dimensional analysis of (6.104) gives

$$\begin{aligned}
[\mathbf{h}] = \left[-\frac{\mathbf{N}_\mathrm{t}}{I_\mathrm{i}}\right] &= \frac{[\mathbf{N}]}{[I_\mathrm{i}]} \\
&= \frac{\frac{A}{m^2}\, m^3}{A} \qquad \text{from (6.95)} \\
&= [m].
\end{aligned} \tag{6.106}$$

Despite the result (6.106) it is only in the special case of linear antenna that a convenient physical interpretation can be made for the dimension of length. Generally, $\mathbf{h}$ is a function of θ and ϕ, and it is transverse to the direction of propagations $\hat{r}$ of the radiated wave. It is expressed in component form as

$$\mathbf{h} = h_\theta\hat{\theta} + h_\phi\hat{\phi}. \tag{6.107}$$

Since $\mathbf{h}$ is a complex quantity, it is capable of indicating the polarization state of the radiated fields. The definition of $\mathbf{h}$ given by (6.104) indicates that the vector effective length depends on the choice of the terminals of an antenna. To properly account for the polarization effects, the vector effective length for receiving antenna should be used as $\mathbf{h}^*$. The interaction of an incident plane wave with electric field $\mathbf{E}^\mathrm{i}$ with a receiving antenna yields an open-circuit voltage V_{oc} across the terminals of the antenna. It can be shown that V_{oc} is given by

$$V_{oc} = \mathbf{E} \cdot \mathbf{h}^*, \tag{6.108}$$

where the complex conjugate sign is used because of the receiving conditions.

Some Examples

The set of expressions identified by (6.109a)–(6.109d) give the vector effective heights for the selected antennas:

$\hat{z}$-Directed Hertzian dipole or short current element–

$$\mathbf{h} = d\ell \sin\theta\hat{\theta} \tag{6.109a}$$

$\hat{z}$-Directed center-fed linear antenna of length ℓ–

$$\mathbf{h} = \frac{\lambda}{\pi}\frac{\cos(\frac{\beta\ell}{z}) - \cos(\frac{\beta\ell}{z})}{\sin\theta}\hat{\theta} \tag{6.109b}$$

$\hat{z}$-Directed center-fed $\lambda/2$ dipole–

$$\mathbf{h} = \frac{\lambda}{\pi}\frac{\cos(\frac{\pi}{z}\cos\theta)}{\sin\theta}\hat{\theta} \tag{6.109c}$$

Circular current loop of radius a placed in the xy plane (or a magnetic dipole oriented along the z axis)–

$$\mathbf{h} = -j\frac{2\pi}{\lambda}(\pi a^2)\sin\theta\hat{\varphi}. \tag{6.109d}$$

6.6.3 Summary

In view of the extensive use of the radiation vector and the vector effective length in variety of antenna problems, we summarize the procedures involved to obtain the far fields produced by arbitrary linear current distributions:

Step 1. Given current distribution

$$\mathbf{J}(\mathbf{r}) \text{ in } V$$

Obtain

$$\mathbf{N}(\theta, \phi) = \mathbf{N} \equiv \int_V \mathbf{J}(\mathbf{r}')\, e^{j\beta\mathbf{r}'\cdot\hat{r}}\, dv' \tag{6.95}$$

Step 2. Obtain

$$\mathbf{N}_{\mathrm{t}} = \mathbf{N} - N_{\mathrm{r}}\hat{r} = N_\theta\hat{\theta} + N_\phi\hat{\phi} \tag{6.100}$$

Step 3. Obtain

$$\mathbf{E} = \frac{-j\beta\eta e^{-j\beta r}}{4\pi r}\mathbf{N}_{\mathrm{t}} \tag{6.101}$$
$$= \frac{j\beta\eta I_{\mathrm{i}}}{4\pi r} e^{-j\beta r}\mathbf{h}$$
$$= E_\theta\hat{\theta} + E_\phi\hat{\phi}$$

Step 4. Obtain

$$\mathbf{H} = \frac{1}{\eta}(\hat{r} \times \mathbf{E}) \tag{6.102}$$

Step 5. Check the power flow is in the direction of $\mathbf{E} \times \mathbf{H}$ (i.e., in the direction of $\mathbf{N}_{\mathrm{t}} \times (\hat{r} \times \mathbf{N}_{\mathrm{t}}) = \hat{r}$).

The various notations used are as explained earlier.

6.7 LINEAR ANTENNAS

We have seen that the resistance of short current elements is very small. Hence they are inefficient and impractical to excite. Because the radiation resistance of short elements is proportional to the square of their electrical length, longer current elements have larger radiation resistance. So most useful linear antennas have lengths comparable to or longer than a wavelength. The radiation from a class of center-fed linear antennas called the *dipole antenna* is discussed in the present section.

6.7.1 Center-Fed Linear Antenna

We consider a center-fed linear antenna, or dipole antenna, of arbitrary length ℓ oriented along the $\hat{z}$ axis as shown in Figure 6.12*a*. The antenna consists of two equal lengths of conducting wire of radius a and length ℓ, separated by a gap of negligible width at the center where a generator is located and supplies current $I(z)$ to the antenna. Knowledge of the current distribution $I(z)$ on the antenna sustained by the source is important in antenna theory. Finding the current distribution involves a difficult electromagnetic boundary value problem, which is described in advanced books on antenna theory [5–7]. We will not attempt to do that here because the linear antennas we consider sustain current distributions that are sinusoidal in nature.

Approximate justification for the assumption above is provided by the two-wire transmission line problem applied to the dipole antennas. From this viewpoint, consider a segment of an open-circuited transmission line of length $\ell/2$

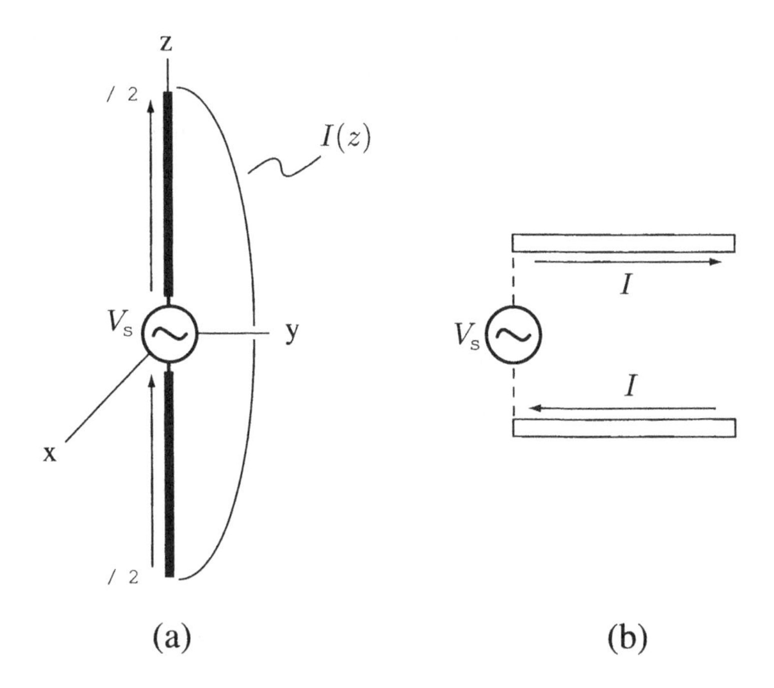

Figure 6.12 (*a*) Center-fed linear antenna of length ℓ fed by a generator and (*b*) open-circuited two-wire line of length $\ell/2$ fed by a generator.

excited by a generator similar to that of the dipole shown in Figure 6.12*b*. It can be seen that if the two lines are folded out in the manner shown, the configuration resembles the dipole under consideration. From transmission line theory (Chapter 5) we know that there will be a standing wave current on the line, and hence a sinusoidal distribution of current having zeros at the two ends of the antenna seems to be plausible. It is appropriate to remark here that under thin wire approximations, the rigorous theory predicts a sinusoidal distribution of current [3]. For an antenna with length ℓ and the element wire radius a, the thin wire approximations are

$$\ell \gg a, \qquad a \ll \lambda, \text{ and} \tag{6.110}$$

$$\text{thickness parameter } \Omega = 2\ln\frac{\ell}{a} > 12,$$

where λ is the wavelength in the medium. We now assume that the current distributions on the dipole of length under thin wire approximation is

$$I(z') = \begin{cases} I_0 \sin\beta\left(\frac{\ell}{2} - z'\right), & 0 \leq z' \leq \frac{\ell}{2} \\ I_0 \sin\beta\left(\frac{\ell}{2} + z'\right), & -\frac{\ell}{2} \leq z' \leq 0, \end{cases} \tag{6.111}$$

where I_0 is the peak value of the current, $\beta = 2\pi/\lambda$, λ being the wavelength in the medium, and z' refers to the coordinate on the antenna. The current distributions (6.111) or dipoles of length $\ell = \lambda/4$, $\lambda/2$, λ, $3\lambda/2$, and 2λ are shown in Figure 6.13 with the currents noted at the feed point and the two end points.

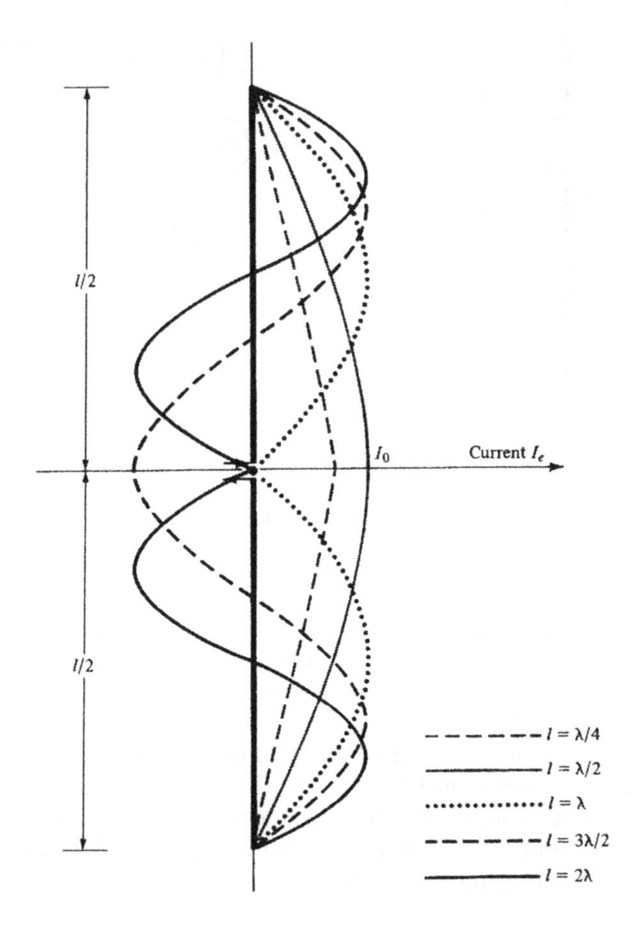

Figure 6.13 Current distributions along the length of a linear antenna for various ℓ. (Source: From [6], p. 186.)

6.7.2 Far Fields of a Dipole of Length ℓ

We assume that the far fields of the dipole Figure 6.12*a* are the same as those produced by a filamentary current $I(z)$ given by (6.110). The geometry of the equivalent problem and the coordinate system used are as shown in Figure 6.14 for far field approximations.

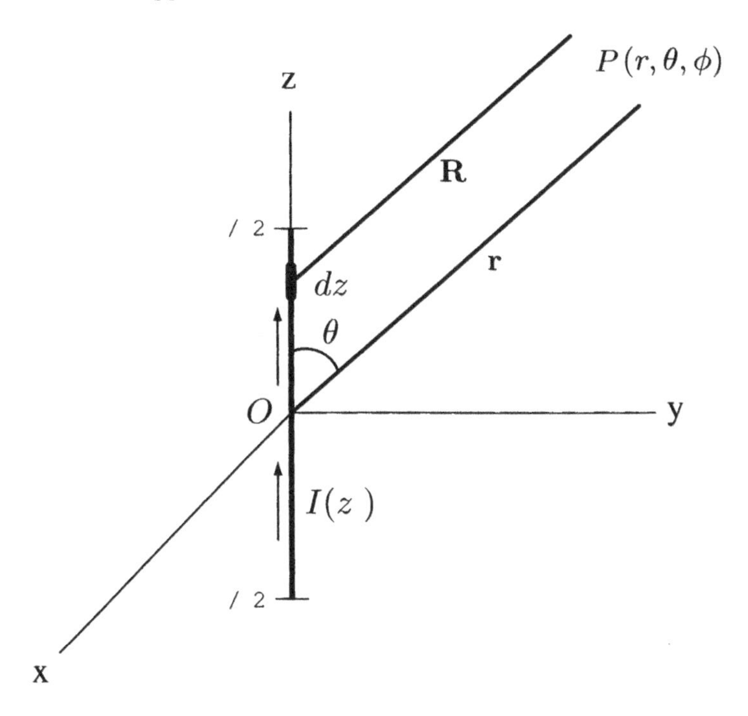

Figure 6.14 Filamentary current $I(z')$ oriented along the z axis. The arrangement shown is for far field approximations.

We now follow the procedure given in Section 6.6.3 to determine the fields. We evaluate **N** given by (6.96) appropriate for the antenna. In the present case we have $\mathbf{J}(\mathbf{r}')\,dv' = I(z')\,dz'\,\hat{z}$. Therefore

$$\mathbf{N} = \hat{z}\int_{-\ell/2}^{\ell/2} I(z')\,e^{j\beta z'\cos\theta}\,dz' \tag{6.112}$$

with $I(z')$ given by (6.111). After performing the integration in (6.112), we obtain

$$\mathbf{N} = \hat{z}\,\frac{2I_0}{\beta}\left[\frac{\cos\left(\frac{\beta\ell}{2}\cos\theta\right) - \cos\left(\frac{\beta\ell}{2}\right)}{\sin^2\theta}\right]. \tag{6.113}$$

Using the vector relationship $\hat{z} = \hat{r}\cos\theta - \hat{\theta}\sin\theta$, we obtain the transverse (i.e., transverse to the radial direction $\hat{r}$) component of **N**. This is the radiation vector $\mathbf{N}_t$ for the antenna:

$$\mathbf{N}_t = N_\theta \hat{\theta} = -\frac{2I_0}{\beta}\left[\frac{\cos\left(\frac{\beta\ell}{2}\cos\theta\right) - \cos\left(\frac{\beta\ell}{2}\right)}{\sin\theta}\right]. \tag{6.114}$$

Using (6.102) and (6.103), we obtain the far electric and magnetic fields produced dipole of length ℓ as

$$\begin{aligned}\mathbf{E} &= E_\theta \hat{\theta} \\ &= \frac{j\eta I_0}{2\pi}\frac{e^{-j\beta r}}{r}\left[\frac{\cos\left(\frac{\beta\ell}{2}\cos\theta\right) - \cos\left(\frac{\beta\ell}{2}\right)}{\sin\theta}\right]\hat{\theta}\end{aligned} \tag{6.115}$$

$$\mathbf{H} = \frac{\hat{r}\times\mathbf{E}}{\eta} = \frac{E_\theta}{\eta}\hat{\phi} = H_\phi \hat{\phi}, \tag{6.116}$$

where $\eta = \sqrt{\mu/\epsilon}$.

6.7.3 Radiated Power and Directivity

The average power flow from the antenna is

$$\begin{aligned}\mathbf{S}_{\text{av}} &= \frac{1}{2}\,\text{Re}[\mathbf{E}\times\mathbf{H}^*] \\ &= \frac{\eta|I_0|^2}{8\pi^2 r^2}\left[\frac{\cos\left(\frac{\beta\ell}{2}\cos\theta\right) - \cos\left(\frac{\beta\ell}{2}\right)}{\sin\theta}\right]^2 \hat{r}\quad \text{W/m}^2.\end{aligned} \tag{6.117}$$

The radiation intensity produced by the antenna at the field point $P(r,\theta,\phi)$ is, from (6.76) and (6.117),

$$\begin{aligned}U(\theta,\phi) &= r^2|\mathbf{S}_{\text{av}}| \\ &= \frac{\eta|I_0|^2}{8\pi^2}\left[\frac{\cos\left(\frac{\beta\ell}{2}\cos\theta\right) - \cos\left(\frac{\beta\ell}{2}\right)}{\sin\theta}\right]^2.\end{aligned} \tag{6.118}$$

The radiation intensity produced by the antenna at the far field point $P(r,\theta,\phi)$ from (6.76) and (6.117) is

$$U(\theta,\phi) = r^2|\mathbf{S}_{\text{av}}| = U_0|F(\theta)|^2, \tag{6.119}$$

where

$$U_0 = \frac{\eta |I_0|^2}{8\pi^2} = \text{maximum radiation intensity} \tag{6.120}$$

and

$$F(\theta,\phi) = F(\theta) = \frac{\cos\left(\frac{\beta\ell}{2}\cos\theta\right) - \cos\left(\frac{\beta\ell}{2}\right)}{\sin\theta} \tag{6.121}$$

is the normalized field pattern. The total power radiated by the antenna from (6.80) and (6.119) is

$$P_{\text{rad}} = 2\pi U_0 \int_0^{\pi} |F(\theta)|^2 \sin\theta \, d\theta, \tag{6.122}$$

with U_0 given by (6.120). Now we obtain the following expression for the radiation resistance of the antenna:

$$\begin{aligned} R_r &= \frac{2P_{\text{rad}}}{|I_0|^2} \\ &= \frac{\eta}{2\pi} \int_0^{\pi} |F(\theta)|^2 \sin\theta \, d\theta, \end{aligned} \tag{6.123}$$

where $F(\theta)$ is given by (6.121).

From the definition of the directivity given by (6.78) and (6.79), and after making use of (6.119) and (6.122), we obtain the following expression for the directivity of the antenna:

$$\begin{aligned} D(\theta,\phi) = D(\theta) &= \frac{2|F(\theta)|^2}{\int_0^{\pi} |F(\theta)|^2 \sin\theta \, d\theta} \\ &= D|F(\theta)|^2, \end{aligned} \tag{6.124}$$

where $F(\theta)$ is given by (6.121) and D is the maximum directivity of the antenna. Note that the determination of the radiation resistance and directivity of the antenna requires knowledge of the radiation pattern of the antenna. The integrals of the pattern function involved in (6.123) and (6.124) can be evaluated numerically. However, they can be expressed in terms of known integrals, which are available in numerical forms. We discuss these integrals in the next section.

6.7.4 Cosine, Sine, and Modified Cosine Integrals

These integrals are discussed in many advanced books on antennas [1,6] their numerical values are available in mathematical tables [15, 16].

The cosine integral is defined as

$$\begin{aligned} \mathrm{Ci} &= \int_{\infty}^{x} \frac{\cos u}{u}\, du \\ &= -\int_{x}^{\infty} \frac{\cos u}{u}\, du, \end{aligned} \tag{6.125}$$

and the modified cosine integral is defined as

$$\mathrm{Cin}(x) = \int_{0}^{x} \frac{1-\cos u}{u}\, du. \tag{6.126}$$

Note that [1] defines $\mathrm{Cin}(x)$ as $S_1(x)$. It can be shown that [1]

$$\begin{aligned} \mathrm{Cin}(x) &= \ln \gamma x - C_{\mathrm{i}}(x) \\ &= C + \ln x - C_{\mathrm{i}}(x), \end{aligned} \tag{6.127}$$

where

$$\gamma = e^{C} = 1.781$$
$C = \ln \gamma = 0.5772$ is called *Euler's constant.*

Sometimes a compact notation is used [1, 15] to represent the cosine, sine, and associated cosine integrals by

$$\begin{aligned} L(x) &= \int_{0}^{x} \frac{1-e^{-ju}}{u}\, du \\ &= \int_{0}^{x} \frac{1-\cos u}{u}\, du + j\int_{0}^{u} \frac{\sin u}{u}\, du \\ &= \mathrm{Cin}(x) + j\mathrm{Si}(x), \end{aligned} \tag{6.128}$$

where the sine integral $\mathrm{Si}(x)$ is defined by

$$\mathrm{Si}(x) = \int_{0}^{x} \frac{\sin u}{u}\, du. \tag{6.129}$$

Series representations of the functions above are available [1,6], and they are

$$\mathrm{Ci}(x) = \ln \gamma x - \frac{x^2}{2!2} + \frac{x^4}{4!4} - \frac{x^6}{6!6} + \cdots \tag{6.130a}$$

$$\mathrm{Cin}(x) = \frac{x^2}{2!2} - \frac{x^4}{4!4} + \frac{x^6}{6!6} - \cdots \tag{6.130b}$$

$$\mathrm{Si}(x) = x - \frac{x^3}{3!3} + \frac{x^5}{5!5} - \cdots . \tag{6.130c}$$

From the definitions and the series expansions given above, we can obtain both small argument and large argument values for these functions. The following values of the functions are of importance in linear antenna impedance and directivity calculations:

$$\begin{aligned}
&\mathrm{Cin}(2\pi) \simeq 2.437, \quad \mathrm{Si}(\infty) = \frac{\pi}{2} \\
&\mathrm{Si}(x) \simeq x \text{ as } x \to 0 \\
&\mathrm{Si}(x) \simeq \frac{\pi}{2} - \frac{\cos x}{x} \text{ as } x \to \infty \\
&\mathrm{Ci}(x) \simeq \ln \gamma x = 0.577 + \ln x \text{ as } x \to 0 \\
&\mathrm{Ci}(x) \simeq \frac{\sin x}{x} \text{ as } x \to \infty.
\end{aligned} \tag{6.131}$$

Further information regarding the above functions are available in the references cited.

6.7.5 The Half-Wave Dipole

The half-wave dipole is a center-fed linear antenna of length $\ell = \lambda/2$. Thus all the pertinent results for the antenna can be obtained from the results described in Section 6.7.1. The explicit expressions for the far zone electric and magnetic fields, the radiation resistance and the directivity of the antenna are now given by

$$\mathbf{E} = E_\theta \hat{\theta} = \frac{j\eta I_0}{2\pi} \frac{e^{-j\beta r}}{r} \frac{\cos\left(\frac{\pi}{2}\cos\theta\right)}{\sin\theta} \hat{\theta} \tag{6.132}$$

$$\mathbf{H} = H_\phi \hat{\phi} = \frac{j I_0}{2\pi} \frac{e^{-j\beta r}}{r} \frac{\cos\left(\frac{\pi}{2}\cos\theta\right)}{\sin\theta} \hat{\phi} \tag{6.133}$$

$$R_{\mathrm{r}} = \frac{\eta}{2\pi} \int_0^\pi \frac{\cos^2\left(\frac{\pi}{2}\cos\theta\right)}{\sin\theta} \, d\theta \tag{6.134}$$

$$D(\theta, \phi) = D(\theta) = \frac{2\cos^2\left(\frac{\pi}{2}\cos\theta\right)/\sin^2\theta}{\displaystyle\int_0^\pi \frac{\cos^2\left(\frac{\pi}{2}\cos\theta\right)}{\sin\theta} \, d\theta}. \tag{6.135}$$

The integrals (6.134) and (6.135) can be evaluated in terms of the known sine and cosine integrals.

Radiation Resistance

After the substitutions $x = \cos\theta$ are made in (6.134), it can be shown that

$$R_r = \frac{\eta}{2\pi}\left[\frac{1}{2}\int_0^1 \frac{1+\cos\pi x}{1-x}\,dx + \frac{1}{2}\int_0^1 \frac{1+\cos\pi x}{1+x}\,dx\right]. \qquad (6.136a)$$

Using the substitutions $1 - x = u/\pi$ and $1 + x = v/\pi$ in (6.136a), we can express R_r as

$$R_r = \frac{\eta}{2\pi}\left[\frac{1}{2}\int_0^{2\pi} \frac{1-\cos u}{u}\,du\right], \qquad (6.136b)$$

which can be evaluated, by using (6.128), as

$$R_r = \frac{\eta}{4\pi}\,\mathrm{Cin}(2\pi). \qquad (6.137)$$

For dipole in free space, $\eta = \eta_0 = 120\pi$, and we obtain the radiation resistance of a half wave dipole as

$$R_r = 30\,\mathrm{Cin}(2\pi) = 30 \times 2.437 \approx 73\ \Omega.$$

With the current distribution given by (6.111) for a center-fed dipole antenna of length ℓ the input current, or $I(z')$ at $z' = 0$, is

$$I_{in} = I_0 \sin\frac{\beta\ell}{2}, \qquad (6.138)$$

where I_0 is the maximum amplitude of the current. Equating the input power to the antenna to the total radiated power and assuming that the antenna is lossless, it can be shown that the input resistance of the antenna of length ℓ is

$$R_{in} = \frac{1}{\sin^2(\beta\ell/2)}\,R_r, \qquad (6.139)$$

where R_r is the radiation resistance of the antenna. (6.135) indicates that the input resistance of the half-wave dipole is

$$R_{in} = R_r = 73\ \Omega.$$

It should be mentioned that for a full wave dipole antenna $\ell = \lambda$ and (6.138) and (6.139) show that $I_{in} = 0$ and $R_{in} = \infty$; however, it is known from (6.123) that $R_r \neq 0$. This means that the antenna cannot be fed from the center. In such cases it should be fed off center at a point suitable to achieve the desired input resistance.

It can be shown [2] that the input impedance of a lossless half-wave dipole is

$$\begin{aligned} Z_{\rm in} &= R_{\rm in} + jX_{\rm in} \\ &= R_r + jX_{\rm in} \\ &\simeq 73 + j42.5 \quad \Omega, \end{aligned}$$

and

$$\begin{aligned} R_{\rm r} = 30{\rm Cin}(2\pi) &= 30 \times 2.437 \\ &\simeq 73\ \Omega \text{ as before,} \end{aligned}$$

where the ohmic loss associated with the antenna is neglected; the evaluations of $X_{\rm in}$ are discussed in [1,2,6,7]. Note that a $\lambda/2$ dipole is slightly inductive. For ℓ slightly few percent less than $\lambda/2$ [2] and under that condition the antennas becomes truly resonant, its input impedance is resistive and is

$$Z_{\rm in} = R_{\rm in} \simeq 70\ \Omega,$$

where the reduction in length reduces the resistance [2].

Directivity

By the value of the integral in (6.135) used in (6.137), it can be shown that the directivity of the half-wave dipoles is

$$\begin{aligned} D(\theta,\phi) &= \frac{4}{{\rm Cin}(2\pi)} \frac{\cos^2\left(\frac{\pi}{2}\cos\theta\right)}{\sin^2\theta} \\ &= D\,\frac{\cos^2\left(\frac{\pi}{2}\cos\theta\right)}{\sin^2\theta}, \end{aligned} \tag{6.140}$$

where

$$D = \frac{4}{{\rm Cin}(2\pi)} = \frac{4}{2.437} = 1.64 = 2.17\ {\rm dB} \tag{6.141}$$

is the maximum directivity of the antenna. We have seen earlier that for a Hertzian dipole $D = 1.54 = 1.64$ dB. Thus a half-wave dipole provides 0.53 dB improvement in maximum directivity; however, its radiation resistance is drastically improved to a value of 73 Ω.

The Radiation Pattern

Figure 6.15 gives the E-plane radiation of a $\lambda/2$ dipole and, for comparison, that of a Hertzian dipole is also shown. The half-power beamwidth of a

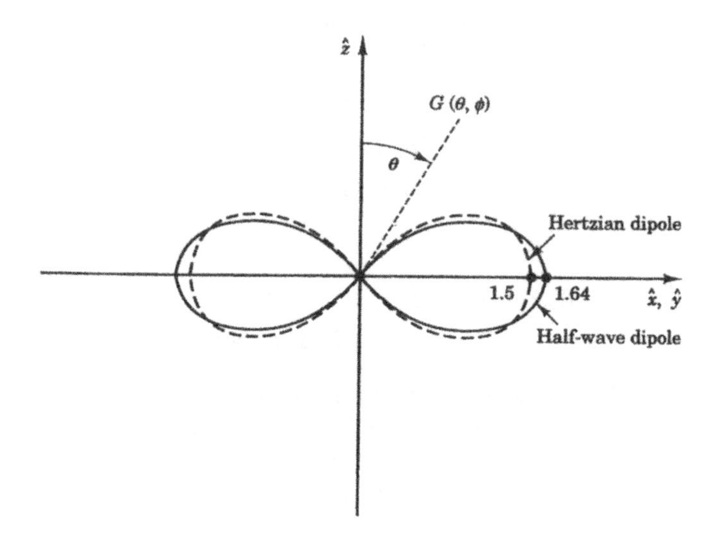

Figure 6.15 Gain of half-wave and Hertzian dipoles. (Source: From [8], p. 415.)

$\lambda/2$ dipole is 78°, which is slightly less than 90° for a Hertzian dipole. It is found that the directive properties of a half-wave dipole are not much different from those of a Hertzian dipole. The most important difference between a $\lambda/2$ dipole and the Hertzian dipole is that the former has a longer radiation resistance which makes it practical to excite, as mentioned earlier.

6.8 NEAR FIELD AND FAR FIELD REGIONS

The near and far field regions of an antenna and their physical extent in space play important roles in variety of electromagnetic radiation and propagation problems. We have already encountered such regions for point source or small antennas in Sections 6.3 and 6.4. In the present section we generalize the field region concepts so that they are applicable to cases when the size of the antenna is comparable to or larger than the operating wavelength. Detailed descriptions of the various aspects of the topic are given in [2, 4, 6].

6.8.1 Basic Assumptions

The determination of the fields produced by a given current distributions begins with that of its vector potential $\mathbf{A}(r, \theta, \phi) \equiv \mathbf{A}$ at any point in space as given by (6.20). The geometry and the appropriate spherical coordinate system are as sketched in Figures 6.1 and 6.11. Exact evaluation of (6.20) for all cases is very difficult. We will make a number of assumptions based on the location of the point $P(r, \theta, \phi)$, the maximum linear dimension of the radiating source $r'_{\rm max}$ and the operating wavelength λ to evaluate (6.20) approximately and thereby the fields, produced. The fields so obtained then pertain to different regions of space associated with the source. As we will see, there may be three such regions in general: (1) reactive and nonradiating near field, (2) radiating near field or Fresnel field, and (3) radiating far field or Fraunhofer field.

Under the assumption $r \gg r'$, the quantity R in (6.20) can be written as

$$R = |\mathbf{r} - \mathbf{r}'| = r\left[1 - \frac{\mathbf{r}' \cdot \hat{r}}{r} + \left(\frac{r'}{r}\right)^2\right]^{1/2}$$

$$\simeq r - r' \cos\gamma + \frac{1}{2\gamma}(r'^2 - r'^2 \cos^2\gamma), \tag{6.142a}$$

where

$$\mathbf{r}' \cdot \hat{r} = r' \cos\gamma \tag{6.142b}$$

and $\cos\gamma$ is given by (6.94c). We now approximate the factor $e^{-j\beta R}/R$ in (6.20) by

$$\frac{e^{-j\beta R}}{R} \simeq \frac{e^{-j\beta r}}{r}\left(1 + \frac{r'}{r}\cos\gamma\right)$$
$$\times\, e^{j\beta\left[r'\cos\gamma - (1/2\gamma)(r'^2 - r'^2\cos^2\gamma)\right]}, \tag{6.143}$$

where we have used the assumption $r \ll r'$. With the approximations (6.143) we obtain (6.20) as

$$\mathbf{A} \simeq \frac{\mu}{4\pi}\frac{e^{-j\beta r}}{r}\int_v \mathbf{J}(\mathbf{r}')\left(1 + \frac{r'}{r}\cos\gamma\right)$$
$$\times\, e^{j\beta\left[r'\cos\gamma - (1/2\gamma)(r'^2 - r'^2\cos^2\gamma)\right]}\, dv'. \tag{6.144}$$

The various field regions are defined on the basis the terms neglected in the integrand of (6.144).

6.8.2 Point or Small Sources

When the maximum linear dimension of the source is small compared with the operating wave length, meaning $r' \ll \lambda$ or $\beta r' \ll 1$, we approximate (6.144) as

$$\mathbf{A} \simeq \frac{\mu I e^{-j\beta r}}{4\pi r}\, d\boldsymbol{\ell}, \tag{6.145}$$

where we have used the relationship $\int_v \mathbf{J}(\mathbf{r}')\, dv' = I\, d\boldsymbol{\ell}$. For a $\hat{z}$-directed short current element (6.145) is the same as that given by (6.25). However, the approximation (6.145) fails for the small current loop or the magnetic dipole for which $\int_V \mathbf{J}(r')\, dv' = \oint_L I\, d\boldsymbol{\ell}'$, where L is the circumference of the loop. This is the reason we retained the term $(1 + (r'/r)\cos\gamma)$ in (6.144). To obtain the appropriate expression for the vector potential produced by a magnetic dipole, we neglect the r'^2 terms in the exponent and approximate $e^{j\beta r' \cos\gamma} \simeq 1 + j\beta\gamma' \cos\gamma$ (since $\beta r' \ll 1$) in (6.144) and obtain

$$\mathbf{A} \simeq \frac{\mu I}{4\pi} \frac{e^{-j\beta r}}{r} \oint_L \left[1 + \left(j\beta + \frac{1}{r}\right) r' \cos\gamma\right] d\boldsymbol{\ell}', \tag{6.146}$$

where L is the circumference of the loop. With $r' < a$, and the loop placed at the origin and symmetrically in the xy plane, we obtain $d\boldsymbol{\ell}' = a\, d\phi'\, \hat{\phi}$; (6.146) reduces to the vector potential for the magnetic dipole given by (6.56) or (6.57).

For the electric and magnetic dipoles we have seen that the nature of the fields produced are different for $\beta r < 1$ and $\beta r > 1$. The two regions so obtained are called the *near field* and *radiative far field regions* (or zones) of the source, respectively. The common boundary is arbitrarily chosen to be at $\lambda/2\pi$ from the source, as we have seen earlier.

6.8.3 Extended Sources

When the maximum linear dimension of the source is comparable to or larger than the operating wavelength the approximation $\beta r' < 1$ is not acceptable. Under this condition the far field region cannot be assumed to start at $r = \lambda/2\pi$. The IEEE standard [12, 13] specifies that for extended sources the far field region starts at the value of r such that

$$\frac{r_{\max}}{2r} = \frac{\lambda}{16} \tag{6.147a}$$

$$\text{i.e.,} \quad r = \frac{8r_{\max}^2}{\lambda}. \tag{6.147b}$$

Assuming D to be the maximum linear dimension of the source and $r'_{\max} = D/2$, we define the far field region of an extended source as

$$r \geq \frac{2D^2}{\lambda}. \tag{6.148}$$

Under the condition (6.147b), and $r \gg r'$, for far field calculations, we approximate (6.144) by

$$\mathbf{A} \simeq \frac{\mu I}{4\pi} \frac{e^{-j\beta r}}{r} \int_V \mathbf{J}(r')\, e^{j\beta r' \cos\gamma}\, dv', \tag{6.149}$$

which we have used for linear antennas in Section 6.6. Note that to obtain (6.149) from (6.144), we have neglected the term

$$\beta\left(\frac{r'^2\cos^2\gamma}{2r} - \frac{r'^2}{2r}\right)$$

in the exponent, which indicates that there will be a maximum phase error of $\beta r'^2_{\mathrm{max}}/2r$ for $r'_{\mathrm{max}} = D/2$, $\gamma = \pi/2$ and $r = 2D^2/\lambda$. The maximum phase error would be $\pi/8$, and for $r > 2D^2/\lambda$ the additional terms neglected in (6.144) will not introduce extra phase angle large enough to cause the various contributions to the integral due to extra terms to give significantly different values of the integral (6.149) [4,6]. Equation (6.149), which does not have any extra phase (error) terms, is sometimes referred to as the *radiation integral* for the radiator.

The radiating near field region also called the *Fresnel field region* of the physical space between the near field and the far field or Fraunhofer regions. The approximations that are made to define the Fresnel region are $r \gg r'$, $\beta r' \gg 1$, but with r, r', and λ such that the terms in r'^2 must be retained when approximating (6.144). Thus the essential difference between the Fraunhofer and Fresnel regions approximations is that the phase term in (6.149) contains extra terms. The Fresnel region to be significant, the extra term $\beta r'^2/2r$ representing the phase error in (6.149) must assume values as large as serial π radians [6]. Values for the Fresnel region boundaries may be determined from diffraction or Fresnel zone considerations, and the fields within the Fresnel region are obtained in terms of Fresnel integrals. For $D \gg \lambda$, the region is defined as [6] $D^2/4\lambda \le r \le 2D^2/\lambda$. It should be mentioned that if D is not sufficiently large, the radiating near field region may not exist.

6.8.4 Definitions of Various Regions

In the context of the above discussions we present here the definitions of various field regions of space surrounding or antennas. These definitions are taken from the *IEEE Dictionary* [12,13].

Reactive Near Field Region

"It is that region of an antenna wherein the reactive field predominates. Note: for most antennas the outer boundary of the region is commonly taken to exist at a distance $\lambda/2\pi$ from the antenna surface." [12]

Radiating Near Field Region

"The region of the field of an antenna between the reactive near field region and the predominate and wherein the angular field distribution is dependent upon the distance from the antenna. If the antenna has a maximum overall dimension which is not large compared with the wavelength, this field region may not exist. For an antenna focused at infinity, the radiating near field region is referred to as the *Fresnel region* on the basis of analogy to optical technology," [13]

Far Field Region

"The region of the field of an antenna where the angular field distribution is essentially independent of the distance has a maximum overall dimension D that is large compared with the wavelength, the far field region is commonly taken to exist at distances greater than $2D^2/\lambda$ from the antennas, λ being the wavelength. For an antenna focused at infinity the far-field region is sometimes referred to as the *Fraunhofer region* on the basis of analogy to optical terminology." [13]

Figures 6.16 gives the sketches of the various field regions of small and extended radiating sources, respectively.

6.8.5 Specific Values of the Region Boundaries

The preceding discussion indicates that there is no general agreement on the exact boundaries of field regions surrounding a radiating system. In particular, boundaries are somewhat arbitrary for the Fresnel field region, and the region is sometimes defined on the basis of included phase errors specific to a problem. Although the boundaries between the nonradiating and radiating near zones may be variable, the criterion for the Fraunhofer region, $r \geq 2D^2/\lambda$, is almost universally accepted. Analytically obtained values for the region boundaries are given in [4, 6], and we quote them as follows:

For antennas having $D \gg \lambda$, the values are

Region	Distance from the Antenna
Reactive near field	0 to $0.62\sqrt{D^3/\lambda}$
Radiating near field	$0.62\sqrt{D^3/\lambda}$ to $2D^2/\lambda$
Radiating far field	$2D^2/\lambda$ to ∞

For point source or small antennas ($D \ll \lambda$), the values are

Region	Distance from the Antenna
Reactive near field	0 to $\lambda/2\pi$
Radiating far field	$\lambda/2\pi$ to ∞

In these cases the Fresnel field region does not exist.

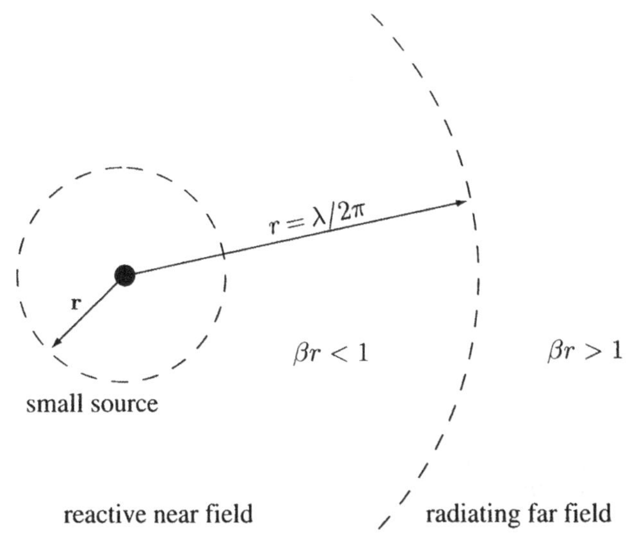

(a) Field regions of a point or small source

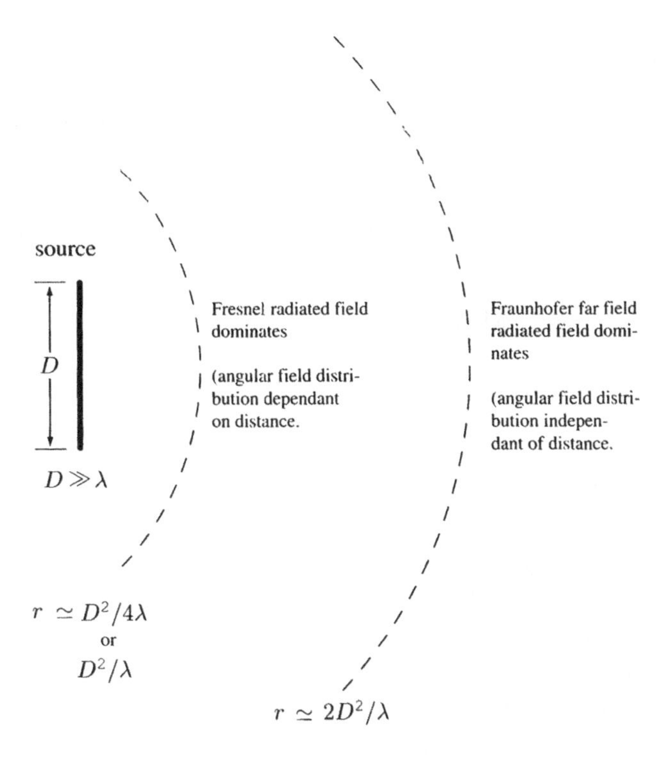

(b) Field regions of an extended radiator.

Figure 6.16 Near and far field regions.

Example 6.3

Estimate the RF emission field regions of a microwave oven.

Solution

The oven operates at $f = 2.45$ GHz, $\lambda = 0.122$ m. We assume that the oven emissions are at 2.45 GHz and a typical effective radiating length $D = 0.5$ m.

We have $2D^2/\lambda \simeq 4.2$ m, $0.62\sqrt{D^3/\lambda} \simeq 0.5$ m. Thus the three field regions surrounding the oven are approximately as follows:

$$\frac{2D^2}{\lambda} \simeq 4.2 \text{ m}, \qquad 0.62\sqrt{\frac{D^3}{\lambda}} \simeq 0.5 \text{ m}.$$

reactive near zone:	$0 \leq r \leq 0.5$ m
radiating near zone (Fresnel):	$0.5 \leq r \leq 4.2$ m
far field (Fraunhofer)	$r \geq 4.2$ m

Note that the large D/λ precludes the beginning of the far field region from $r = \lambda/2\pi \simeq 2$ cm.

Example 6.4

Estimate the RF emission field regions of a personal computer (PC).

Solution

We assume an effective radiating length $D = 1.5$ m. We will estimate it for two frequencies.

Case 1. Assume $f = 150$ MHz, $\lambda = 2$ m.

We have

$$\frac{2D^2}{\lambda} \simeq 2.25 \text{ m}, \quad 0.62\sqrt{\frac{D^3}{\lambda}} \simeq 0.81 \text{ m}.$$

Note that $\lambda/2\pi \simeq 0.32$ m. The three regions are

reactive near field:	$0 \leq r \leq 0.81$ m
radiating near field:	$0.81 \leq r \leq 2.25$ m
far field:	$r \geq 2.25$ m.

Case 2. Assume $f = 1.5$ MHz, $\lambda = 200$ m (this may originate from the switching power supply).

In this case $D/\lambda = 0.0075$, meaning $D/\lambda \ll 1$.

We have

$$\frac{2D^2}{\lambda} \simeq 22.5 \text{ mm}, \quad 0.62\sqrt{\frac{D^3}{\lambda}} \simeq 15\text{mm}$$

$$\frac{\lambda}{2\pi} = 31.8 \text{ m}.$$

For emissions at this frequency, the PC behaves like a point source:

$$\begin{array}{ll} \text{reactive near zone:} & 0 \leq r \leq 31.8 \text{ m} \\ \text{radiating far zone:} & r \geq \lambda/2\pi = 31.8 \text{ m}, \end{array}$$

and the Fresnel zone does not exist.

6.9 EQUIVALENT CIRCUITS OF ANTENNAS

Design of the input and output circuits of transmitting and receiving antennas is facilitated by the use of their equivalent circuits. Detailed discussions of the evolution of such circuits are given in books on antennas [1–8]. We discuss here the main aspects. We will assume that whether or not the antenna is transmitting or receiving, it has a pair of well-defined terminals and its (input) impedance is defined as the ratio of the voltage and current at the terminals.

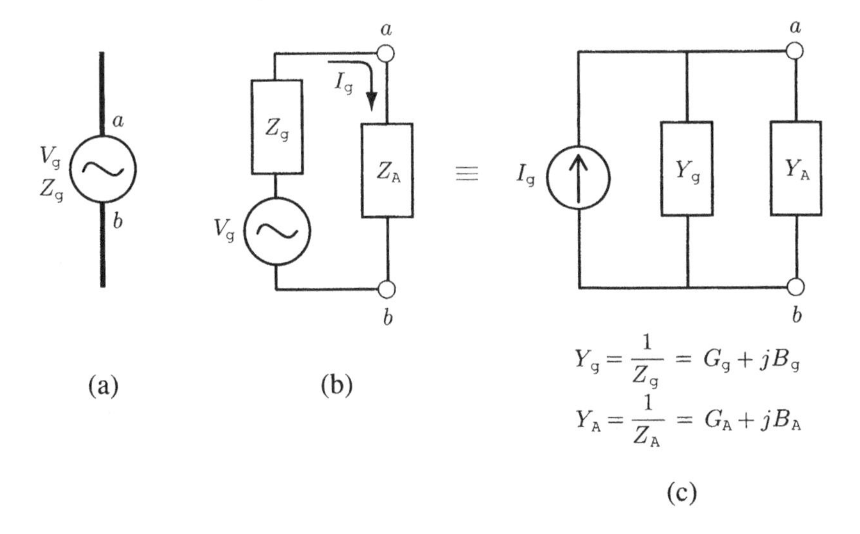

Figure 6.17 Transmitting antenna (*a*) and its Thévenin (*b*) and Norton (*c*) equivalent circuits.

6.9.1 Transmitting Antenna

Consider a transmitting antenna fed by a generator with internal impedance Z_g and supplying a voltage V_g at its terminals a and b, as shown in Figure 6.17*a*. The Thévenin equivalent circuit for the antenna can be represented by Figure 6.17*b* where the generator used in Figure 6.17*a* serves as the Thévenin generator terminated by the antenna input impedance Z_A in the transmitting mode, as shown in Figure 6.17*b*. In general, the input impedance of an antenna is a complex quantity and can be written as

$$Z_A = R_A + jX_A, \tag{6.150}$$

where R_A is the antenna resistance between terminals a and b, and X_A is the antenna reactance between terminals a and b.

We saw earlier (Section 6.7.5) that the antenna resistance consists of its radiation resistance R_r and its loss resistance R_ℓ associated with the finite conductivity of the antenna element. So we write R_A as

$$R_A = R_r + R_\ell. \tag{6.151}$$

In most antennas $R_\ell \ll R_r$; however, we retain it here for generality. The voltage generator supplying a voltage V_g to excite the antenna has an internal impedance given by

$$Z_g = R_g + jX_g, \tag{6.152}$$

where R_g and X_g are respectively the internal resistance and reactance of the generator. Figure 6.17*c* shows an alternative equivalent circuit, called the *Norton equivalent circuit*, for the antenna. It consists of a current generator supplying a current I_g, which is the same current supplied by the voltage generator is the Thévenin circuit:

$$\begin{aligned} I_g &= \frac{V_g}{Z_A + Z_g} \\ &= \frac{V_g}{(R_g + R_A) + j(X_g + X_A)}, \end{aligned} \tag{6.153}$$

The other parameters are as defined before and explained in Figure 6.17*c*. Both equivalent circuits are used to design the input circuits for a given antenna.

We will use the Thévenin circuit for our discussion of power calculations for the antenna. Circuit theory considerations applied to Figure 6.17*b* indicate that for maximum power transfer from the generator to the load (in this case the antenna), we must have

$$X_A = -X_g \tag{6.154a}$$

$$R_A = R_g, \tag{6.154b}$$

meaning the generator and the antenna are complex conjugate impedance matched. We will also refer to (6.154a) and (6.154b) as *tuning* and *matching* the load, respectively. If the system is tuned, (6.153) gives

$$I_{\mathrm{g}} = \frac{V_{\mathrm{g}}}{R_{\mathrm{g}} + R_{\mathrm{A}}}, \tag{6.155}$$

from which we can determine various power in the circuit. Thus we obtain the following powers supplied by the generator V_{g}:

1. Power dissipated in the generator resistance

$$P_{\mathrm{g}} = \frac{|V_{\mathrm{g}}|^2}{2} \frac{R_{\mathrm{g}}}{(R_{\mathrm{g}} + R_{\mathrm{A}})^2}. \tag{6.156}$$

2. Power dissipated in the antenna element as Joule heat

$$P_{\ell} = \frac{|V_{\mathrm{g}}|^2}{2} \frac{R_{\ell}}{(R_{\mathrm{g}} + R_{\mathrm{A}})^2}. \tag{6.157}$$

3. Power radiated by the antenna

$$P_{\mathrm{r}} = \frac{|V_{\mathrm{g}}|^2}{2} \frac{R_{\mathrm{r}}}{(R_{\mathrm{g}} + R_{\mathrm{A}})^2}. \tag{6.158}$$

Note that the power supplied to the antenna is

$$P_{\mathrm{A}} = P_{\mathrm{r}} + P_{\ell} = \frac{|V_{\mathrm{g}}|^2}{2} \frac{R_{\mathrm{A}}}{(R_{\mathrm{g}} + R_{\mathrm{A}})^2} \tag{6.159}$$

of which only P_{r} is radiated by the antenna.

Finally, the power supplied by the generator or the input power under tuned condition is

$$P_{\mathrm{in}} = \frac{|V_{\mathrm{g}}|^2}{2} \frac{1}{(R_{\mathrm{g}} + R_{\mathrm{A}})}. \tag{6.160}$$

Under matched conditions, $R_g = R_A$ and the powers above are

$$P_g = \frac{|V_g|^2}{8}\frac{1}{R_A} \tag{6.161a}$$

$$P_\ell = \frac{|V_g|^2}{2}\frac{R_\ell}{4R_A^2} \tag{6.161b}$$

$$P_r = \frac{|V_g|^2}{2}\frac{R_r}{4R_A^2} \tag{6.161c}$$

$$P_A = P_\ell + P_r = \frac{|V_g|^2}{8}\frac{1}{R_A} \tag{6.161d}$$

$$P_{in} = \frac{|V_g|^2}{4}\frac{1}{R_A}\,. \tag{6.161e}$$

Equations (6.161a), (6.161c), and (6.161e) indicate that under conjugate match conditions, half of the power supplied by the generator is dissipated in the internal resistance of the generator and the other half is supplied to the antenna. The power supplied to the antenna accounts for the power radiated by the antenna and the power dissipated in its ohmic resistance in the following ratio:

$$\frac{P_r}{P_\ell} = \frac{R_r}{R_\ell}\,, \tag{6.162}$$

which indicates that $\mathbf{E}^i$ should be as small as possible. In other words, it is desirable to make antenna elements out of conducting materials.

6.9.2 Receiving Antennas

The general theory of receiving antennas and their equivalent circuits is described in [3,4,7,8]. In the present section we discuss the Thévenin equivalent circuit for a receiving antennas from the viewpoint of power received by the load when the antenna is illuminated by an incident plane electromagnetic wave.

Assume that a plane wave with electric field $\mathbf{E}^i$ is incident at an angle θ on a receiving (linear) antennas oriented along the z axis as shown in Figure 6.18*a*. Observe that $\mathbf{E}^i$ is perpendicular to the incident direction. The load Z_L is assumed to be connected across the terminals a–b of the antenna. The Thévenin equivalent circuit for the antenna is shown in Figure 6.18*b* where V_{oc} is the Thévenin generator voltage that is induced by the incident wave across the terminals a–b under open-circuit conditions. Note that under the open-circuit condition, V_{oc} is essentially applied across the antenna. Hence the induced generator with voltage V_{oc} is assumed to have an internal impedance Z_A of the antenna under transmitting mode.

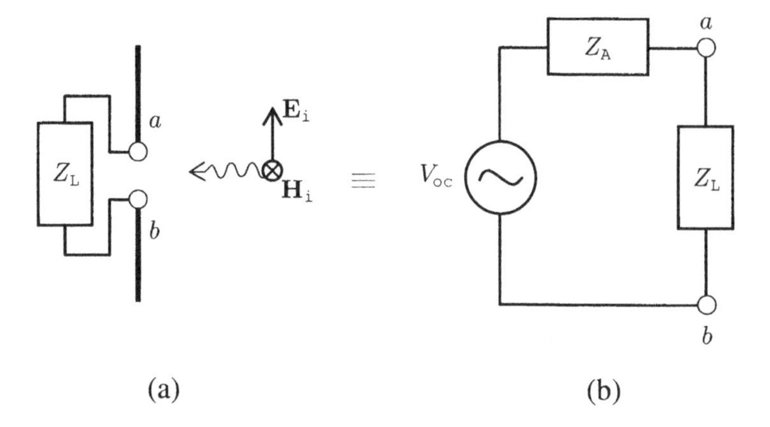

Figure 6.18 (*a*) Receiving antenna illuminated by an incident field **E** and (*b*) the Thévenin equivalent circuit.

Assuming for the time being that V_{oc} is known, we can determine the real power dissipated in the load, meaning the received power, in a manner similar to transmitting case. Instead of describing the detailed step-by-step procedures, we only present the various results below.

Using the equivalent circuit Figure 6.18*b* we obtain the current I_L as

$$I_L = \frac{V_{oc}}{(R_A + R_L) + j(x_A + x_L)}. \tag{6.163}$$

Under conjugate match, $x_A = -x_L$ and $R_A = R_L$, it can be shown that real power collected from the incident wave is

$$P_C = \frac{|V_{oc}|^2}{4} \frac{1}{R_L}. \tag{6.164}$$

The power dissipated or collected by the load is

$$P_L = \frac{|V_{oc}|^2}{8} \frac{1}{R_L}. \tag{6.165}$$

The power lost in Z_A is

$$P_A = \frac{|V_{oc}|^2}{8} \frac{1}{R_A}. \tag{6.166}$$

Note that $P_A = P_r + P_\ell$, where, according to the equivalent circuit, P_r is the power lost in R_r and P_ℓ is the power lost in R_ℓ (i.e., the copper loss in the antenna). Thus,

$$P_r = \frac{|V_{oc}|^2}{8} \frac{R_r}{4R_A^2} \tag{6.167}$$

$$P_\ell = \frac{|V_{oc}|^2}{8} \frac{R_\ell}{4R_A^2} \,. \tag{6.168}$$

Equations (6.165)–(6.168) indicate that under conjugate matching, half of the power collected by the antenna is delivered (or collected) to the load and the other half is delivered to R_A (i.e., antenna resistance). The power delivered to the load is of practical significance and it can be maximized by using the equivalent circuit. It is important to mention that similar practical significance may not be attributed to the power dissipated in the internal impedance of the equivalent circuit. We quote here from [9]: "In the present case with $R_\ell = 0$, it is tempting to interpret P_r as the power re-radiated or scattered by currents on the receiving antenna, and one would conclude that as much power is scattered under condition of perfect match as is absorbed in the load. This condition is not true, except in special cases, where the current distributions may be the same for reception and transmission."

The generator voltage V_{oc} can be obtained from a knowledge of the vector effective height (length) $\mathbf{h}$ of the antenna and the incident field $\mathbf{E}^i$ [5–7]: It is given by

$$V_{oc} = \mathbf{h}^* \cdot \mathbf{E}^i, \tag{6.169}$$

where complex conjugate sign is used because $\mathbf{h}$ is associated with the transmitting case and (6.169) is a receiving relationship.

6.9.3 Equivalent Area

The equivalent area or aperture, or effective area, is a useful concept used to estimate the power delivered to the load by a receiving antenna. It is also used to represent the effective power radiated by a transmitting antenna. Under matched conditions the incident power density at the antennas multiplied by its effective area gives the power received by the load. We have seen earlier that under matched conditions, the power received by the load is (see (6.165))

$$P_L = \frac{|V_{oc}|^2}{8R_r} \,, \tag{6.170}$$

where it is assumed that $R_L = R_A \simeq R_r$, R_r being the relation resistance of the antenna (i.e., we are assuming that the antenna is lossless, and V_{oc} is given by (6.169)).

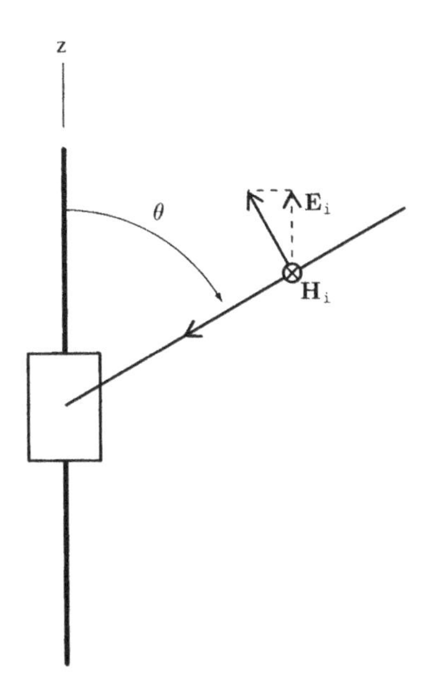

Figure 6.19 Small antenna of length $d\ell$ illuminated by an incident field.

We assume a plane wave incident on an elementary receiving antenna $I\, d\boldsymbol{\ell}$ oriented along the z axis and receiving antenna oriented along the $\hat{z}$ axis and receiving, an incident linearly polarized field $\mathbf{E}^{\mathrm{i}}$ as shown in Figure 6.19.

We have in this case the open-circuit induced voltage for (6.169),

$$V_{\mathrm{oc}} = \mathbf{h}^{*} \cdot \mathbf{E}^{\mathrm{i}} = E_0 \sin\theta \, d\ell \tag{6.171}$$

and

$$P_{\mathrm{L}} = \frac{E_0^2 \sin^2\theta \, d\ell^2}{8R_{\mathrm{r}}} \,. \tag{6.172}$$

By definition, the equivalent area is

$$\begin{aligned} A_{\mathrm{e}}(\theta,\phi) &= \frac{\text{power received by the load}}{\text{incident power density at the antenna}} \\ &= \frac{P_{\mathrm{L}}}{P_{\mathrm{in}}} = \frac{P_{\mathrm{L}}}{E_0^2/2\eta} \,. \end{aligned} \tag{6.173}$$

Note that the effective area generally depends on the incident field direction. From (6.172) and (6.173) we obtain

$$A_{\mathrm{e}}(\theta,\phi) = \frac{\eta \, d\ell^2}{4R_{\mathrm{r}}} \sin^2\theta. \tag{6.174}$$

We know from (6.47) that for a current element $I\ d\ell$ radiating in an infinite medium of intense impedance η, $R_\mathrm{r} = \beta^2\ d\ell^2\ \eta/6\pi$. Substituting this in (6.174), we obtain the following for the effective aperture for an elementary receiving antenna:

$$A_\mathrm{e}(\theta, \phi) = \frac{3\lambda^2}{8\pi} \sin^2\theta. \tag{6.175}$$

The maximum aperture occurs for $\theta = \pi/2$. It occurs when the antenna is polarization patched, and we obtain

$$A_\mathrm{e} = \frac{3\lambda^2}{8\pi}\ . \tag{6.176}$$

We know the directivity of a Hertzian dipole is $D = 3/2$, and we find from (6.176),

$$D = \frac{4\pi A_\mathrm{e}}{\lambda^2}\ . \tag{6.177}$$

Although we have used the Hertzian dipole or an elementary antenna to obtain (6.177), this relationship is universal and applies to all antennas; it is a fundamental relationship in antenna theory. In terms of the gain of the antenna we have

$$G = \alpha D = \alpha\, \frac{4\pi A_\mathrm{e}}{\lambda^2}, \tag{6.178}$$

where α is the efficiency of the antenna. Now, using (6.173) and (6.174), we obtain a relationship between the received power and the incident power density as

$$P_\mathrm{L} = A_\mathrm{e}(\theta, \phi)\ P_\mathrm{in} = \frac{\lambda^2\ G(\theta, \phi)}{4\pi\alpha}\ P_\mathrm{in}. \tag{6.179}$$

It is of significant practical interest to modify (6.179) when the antenna is not impedance matched and its polarization is not matched with that of the incident field. We assume that the impedance mismatch is such that the reflection coefficient at the terminals (a–b) of the antenna is Γ. We define a polarization mismatch parameter p such that

$$p = \frac{|\mathbf{h}^* \cdot \mathbf{E}^\mathrm{i}|^2}{|\mathbf{h}^*||\mathbf{E}^\mathrm{i}|} \tag{6.180}$$

and $0 \le p \le 1$.

Thus, with an impedance and polarization mismatched antenna, it can be shown that

$$P_\mathrm{L} = (1 - |\Gamma|^2)p\, \frac{\lambda^2}{4\pi\alpha}\ G(\theta, \phi)\ P_\mathrm{in}, \tag{6.181}$$

where Γ is the reflection coefficient at the antenna terminal and it accounts for the impedance mismatch, and the other parameters are as explained earlier.

6.10 ANTENNA ARRAYS

An antenna array consists of a number of radiating elements suitably configured and phased such that together they provide high gain and sometimes controllable radiating beam. There exist a variety of antenna arrays having variety of applications. The literature on antenna arrays is vast; however, their basic principles of design and operations are described in almost any textbook on antennas [1–9, 11]. In the present section we discuss only the fundamental procedures used in the design of antenna arrays. This will enable us to determine the combined fields produced in space by a number of arbitrarily arranged and excited radiators. The present discussion is motivated from the viewpoint of EMC, where we are frequently interested in the estimation of undesired fields outside by the radiating components and elements within that system.

6.10.1 General Considerations

Consider a radiator located at a point P_n in a finite region V around the origin O of a rectangular coordinate system shown in Figure 6.20. The spherical coordinates of the source (or the radiating) point P_n and the field point P with respect to the fixed coordinate system are (r_n, θ_n, ϕ_n) and (r, θ, ϕ), respectively, as shown in Figure 6.20 where the associated rectangular axes are xyz.

The field from the radiator P_n will at first be represented in a local spherical coordinate system having its origin at P_n where the corresponding rectangular axes x', y', z' are parallel to the fixed system with origin at O. In the local coordinate system the coordinates of the field point P are (R_n, θ', ϕ') as shown in Figure 6.20. We assume that the medium is free space. At P_n a linear current element of arbitrary length and carry a current $I_n = |I_n|e^{j\alpha_n}$ is placed along the $\hat{z}^{\mathrm{i}} = \hat{z}$ axis, where the amplitude and phase of the complex current are denoted by $|I_n|$ and α_n, respectively. In terms of the local coordinate system, the far fields at P due to the radiators at P_n can be written as

$$\mathbf{E}_n(P) = \hat{\theta}' K I_n \frac{e^{-j\beta R_n}}{R_n} f_n(\hat{\theta}'), \qquad (6.182)$$

where K is a constant to be specified later and $f_n(\theta')$ is the pattern function of the radiator, assumed to be independent of ϕ'. Note that the choice of the two coordinate systems is such that in the far field region $(r \to \infty)$, we have

$$\begin{aligned} \mathbf{R}_n &\parallel \mathbf{r} \\ \hat{\theta}' &= \hat{\theta} \\ \theta' &= \theta. \end{aligned} \qquad (6.183)$$

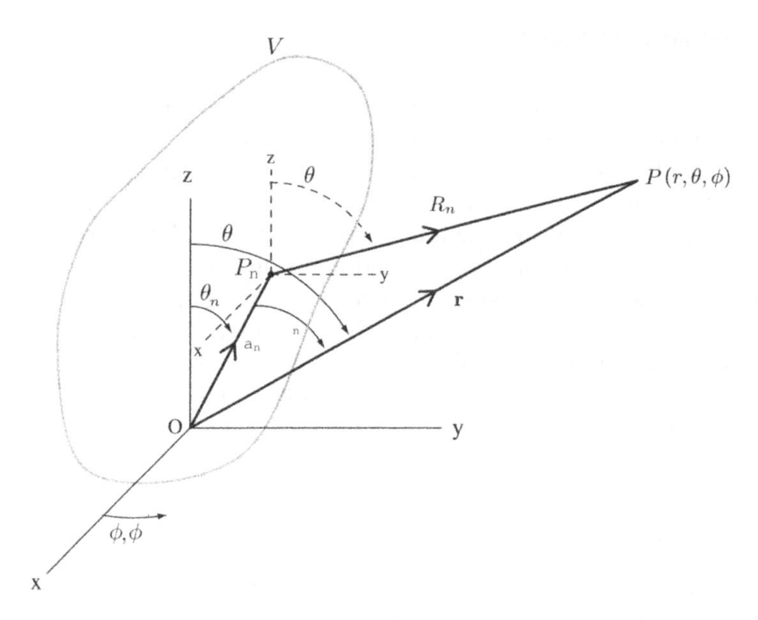

Figure 6.20 Radiator at R_n within V and the coordinate system.

We now express $\mathbf{E}_n(P)$ given by (6.182) in the fixed coordinate system. For this purpose we make the usual far zone approximations to the $e^{-j\beta R_n}/R_n$ term in (6.182). That is, under the assumptions $r,\ R_n \gg \max a_n$ and $r,\ R_n \to \infty$, we approximate

$$\begin{aligned} R_n &\simeq r \text{ in the denominator} \\ R_n &\simeq r - a_n \cos\psi_n \text{ in the exponent} \end{aligned} \tag{6.184}$$

of $e^{-j\beta R_n}/R_n$, where

$$\cos\psi_n = \cos\theta\cos\theta_n + \sin\theta\sin\theta_n\cos(\phi - \phi_n). \tag{6.185}$$

Now using (6.183)–(6.185), we obtain from (6.182)

$$E_n(P) = \hat{\theta}\, K\, \frac{e^{-j\beta r}}{r} \times |I_n| e^{j(\beta a_n \cos\psi_n + \alpha_n)}\, f_n(\theta), \tag{6.186}$$

which expresses the far field at the point P produced by a $\hat{z}$-directed linear radiator at P_n in terms of the fixed coordinate system with origin at O. With N numbers of radiators located within an arbitrary volume V (such that $R_n,\ r \gg$

$\max a_n$) the far field at P is given by

$$\mathbf{E}(P) = \sum_{n=1}^{N} \mathbf{E}_n(P)$$

$$= \hat{\theta}\, K \,\frac{e^{-j\beta r}}{r} \sum_{n=1}^{N} |I_n| e^{j(\beta a_n \cos\psi_n + \alpha_n)}\, f_n(\theta). \tag{6.187}$$

If the individual radiators are different, meaning if $f_n(\theta)$'s are different, then in general (6.187) must be evaluated numerically to obtain $E(P)$. Under the assumption that all the radiators are similar and similarly oriented, or $f_n(\theta) = f(\theta)$, we can write (6.187) as

$$\mathbf{E}(P) = \hat{\theta}\, K \,\frac{e^{-j\beta r}}{r}\, A(\theta, \phi), \tag{6.188}$$

where $A(\theta, \phi)$ represents the far field pattern of the complete array. We define $A(\theta, \phi)$ as the array pattern given by

$$A(\theta, \phi) = f(\theta) \sum_{n=1}^{N} |I_n| e^{j(\beta a_n \cos\psi_n + \alpha_n)}. \tag{6.189}$$

The quantity $f(\theta)$ is the field pattern of individual radiators and is commonly referred to as the *element pattern*. We define here another term called the *array factor*, denoted by AF, which is given by

$$\mathrm{AF} = \sum_{n=1}^{N} |I_n| e^{j(\beta a_n \cos\psi_n + \alpha_n)}. \tag{6.190}$$

It can be seen that (6.190) represents the field pattern of similarly excited N isotropic radiators placed at the phase centers of the original radiators. Thus we can rewrite (6.190) in words as

$$\text{array pattern} = \text{element pattern} \times \text{array factor}. \tag{6.191}$$

Equation (6.191) is the principle of pattern multiplication whose mathematical statement is (6.189). It should be noted that for the principle of pattern multiplication to hold it is necessary that the elements have the same pattern, which is not correct unless the mutual effects between the elements are neglected. For completeness, we include below the appropriate expressions for the parameters K and $f(\theta)$ for three selected linear radiators oriented along the z axis.

Hertzian Dipole
Length dl with constant current of peak value $|I_0|$

$$K = \frac{j\beta\eta|I_0|\,d\ell}{4\pi}, \quad f(\theta) = \sin\theta. \tag{6.192a}$$

Center-Fed Half-Wave Dipole
Length $\lambda/2$carrying sinusoidal current of peak value $|I_0|$

$$K = \frac{j\eta|I_0|}{2\pi}, \quad f(\theta) = \frac{\cos\left(\frac{\pi}{2}\cos\theta\right)}{\sin\theta}. \tag{6.192b}$$

Center-Fed Dipole
Length ℓ carrying sinusoidal current of peak value I_0

$$K = \frac{j\eta|I_0|}{2\pi}, \quad f(\theta) = \frac{\cos\left(\frac{\beta\ell}{2}\cos\theta\right) - \cos\left(\frac{\beta\ell}{2}\right)}{\sin\theta}, \tag{6.192c}$$

where $\beta = 2\pi/\lambda$, λ is the wavelength in the medium, and $\eta = \sqrt{\mu/\epsilon}$ is the intrinsic impedance of the medium. Note that for free space $\eta = \eta_0 = 120\pi\ \Omega$.

6.10.2 A Two-Element Array

We will apply the theory described in the previous section to a two-element antenna array. The array consists of two z-directed half-wave dipoles separated by a distance a and placed symmetrically with respect to the z axis and along the x axis of a coordinate system shown in Figure 6.21. As shown, the dipoles 1 and 2 are excited by currents $I_0\angle 0$ and $I_0\angle\alpha$, respectively.

As shown in Figure 6.21, we have

$$\begin{aligned}
|I_1| = I_0,\ \ \alpha_1 = 0,\ \ a_1 &= -\frac{a}{2} \\
|I_2| = I_0,\ \ \alpha_2 = \alpha,\ \ a_2 &= \frac{a}{2} \\
\theta_1 = \frac{\pi}{2},\ \ \phi_1 &= 0 \\
\theta_2 = \frac{\pi}{2},\ \ \phi_2 &= \pi.
\end{aligned}$$

From (6.185) we have

$$\cos\psi_1 = \sin\cos\phi, \quad \cos\psi_2 = -\sin\theta\cos\phi,$$

where (θ, ϕ) specify the far field point direction.

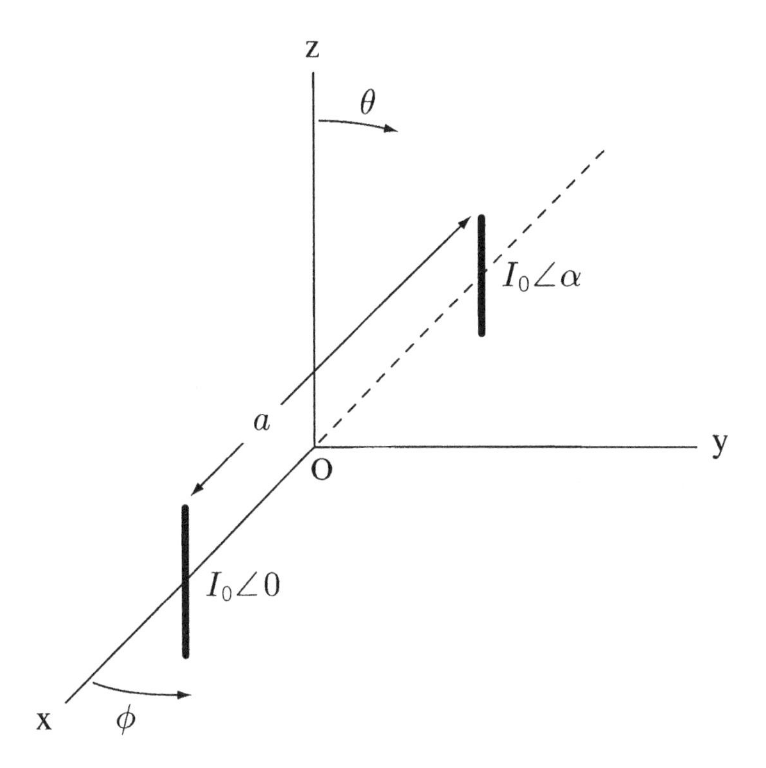

Figure 6.21 Two-element $\lambda/2$-dipole array: Dipole oriented along the z axis (array plane xy).

With the parameters above we obtain the un-normalized array factors from (6.190) as

$$\mathrm{AF} = 2I_0 e^{j\alpha/2} \cos\left(\frac{\beta a}{2}\sin\theta\cos\phi - \frac{\alpha}{2}\right). \tag{6.193}$$

The complete expression for the far field produced by the antennas array is now obtained by using (6.188) and (6.189) as

$$\begin{aligned} E_\theta = {} & \frac{j\eta I_0}{\pi}\, e^{+j\alpha/2}\, \frac{e^{-j\beta r}}{r} \\ & \times \frac{\cos\left(\frac{\pi}{2}\cos\theta\right)}{\sin\theta} \cos\left(\frac{\beta a}{2}\sin\theta\cos\phi - \frac{\alpha}{2}\right). \end{aligned} \tag{6.194}$$

Note that in (6.194) the normalized element and array factors field patterns for the antenna array are

$$f(\theta) = \frac{\cos\left(\frac{\pi}{2}\cos\theta\right)}{\sin\theta} \tag{6.195a}$$

$$\mathrm{AF} = \cos\left(\frac{\beta a}{2}\sin\theta\cos\phi - \frac{\alpha}{2}\right), \tag{6.195b}$$

and the normalized far field pattern of the array is given by

$$A(\theta,\phi) = f(\theta)\ \mathrm{AF} \tag{6.195c}$$

with $f(\theta)$ and AF given by (6.195a) and (6.195b). It is clear now that the pattern of the complete array can be controlled by the choice of a and α. We consider a few choices.

Case 1. Assume that the two dipoles are excited in phase and they are separated by a half-wave length. We then have $\alpha = 0,\ \alpha = \lambda/2$, and

$$f(\theta) = \frac{\cos\left(\frac{\pi}{2}\sin\theta\right)}{\sin\theta}$$

$$\mathrm{AF} = \cos\left(\frac{\pi}{2}\sin\theta\cos\phi\right).$$

The patterns in the xz plane ($\phi = 0$) or in the E plane are sketched in Figure 6.22*a*.

The H-plane patterns are obtained for $\theta = \pi/2$ for which we have

$$f(\theta) = 1 : \mathrm{AF} = \cos\left(\pi 2\cos\phi\right),$$

and they are sketched in Figure 6.22*b*.

Case 2. Assume that the two dipoles are separated by $\lambda/4$ and there is phase difference of $\pi/2$ between them. Further assume $a = \lambda/4,\ \alpha_1 = 0$ and $\alpha_2 = \pi/2$, meaning dipole 2 leads dipole 1 by $\pi/2$.

We have

$$f(\theta) = \frac{\cos\left(\frac{\pi}{2}\cos\theta\right)}{\sin\theta}$$

$$\mathrm{AF} = \cos\frac{\pi}{4}(1 - \sin\theta\cos\phi).$$

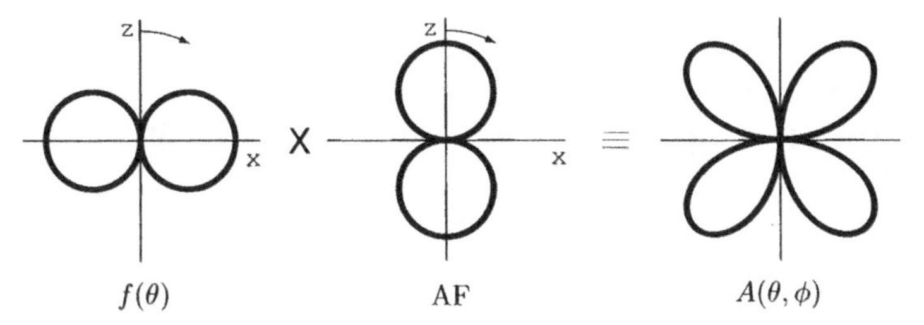

(a) E-plane pattern for the array, $\alpha = 0, a = \lambda/2$.

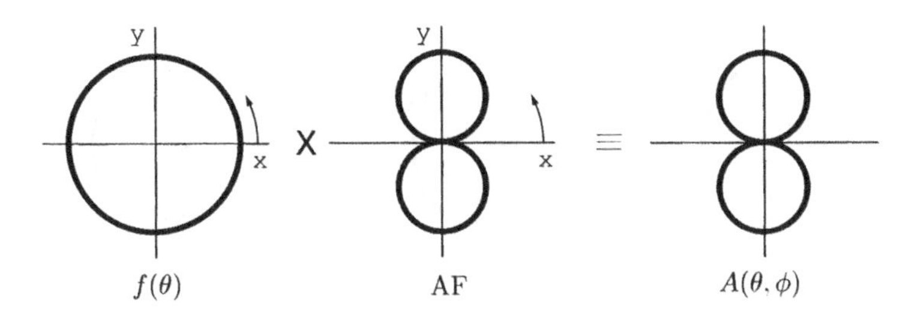

(b) H-plane pattern for the array, $\alpha = 0, a = \lambda/2$.

Figure 6.22 E-plane and H-plane array patterns.

Thus in the x–z plane, $\phi = 0$, we have

$$\mathrm{AF} = \cos\frac{\pi}{4}(1 - \sin\theta)$$

and $f(\theta)$ as above. The E-plane pattern is as sketched in Figure 6.23a.

In the H-plane

$$f(\theta) = 1, \quad \mathrm{AF} = \cos\frac{\pi}{4}(1 - \cos\phi).$$

Then in the H plane, or xy plane, $\theta = \pi/2$,

$$\mathrm{AF} = \cos\frac{\pi}{4}(1 - \sin\theta)$$

The patterns are shown in Figure 6.23b. In this case the pattern is entirely determined by the array factors.

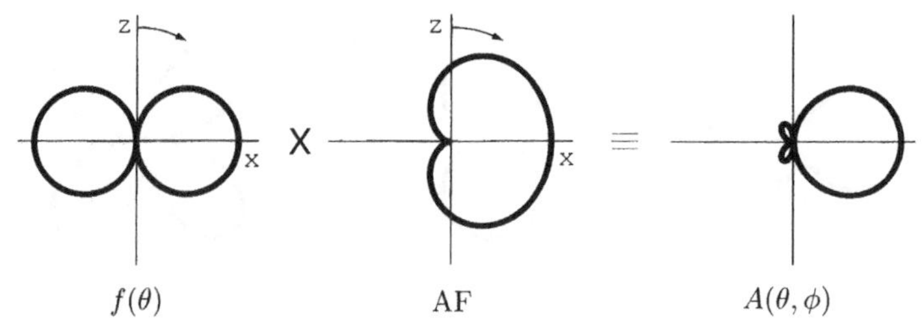

(a) E-plane patterns for the 2-element array: $a = \lambda/4, \alpha_1 = 0, \alpha_2 = \pi/2$.

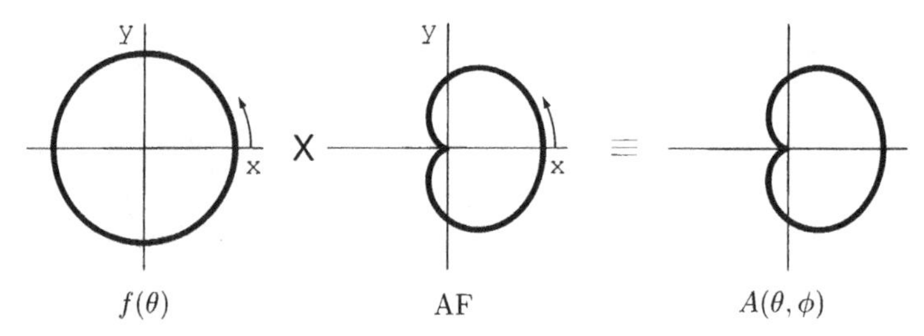

(b) H-plane patterns for the 2-element array: $a = \lambda/4, \alpha_1 = 0, \alpha_2 = \pi/2$.

Figure 6.23 E-plane and H-plane 2-element array patterns.

It is interesting to note that the double-dipole configuration, with $\lambda/4$ spacing and phase difference of $\pi/2$,considered confines most of the energy in the half-plane $x > 0$, that is, in the direction decreasing of phase of the elements along the array. In the example considered we find a maximum in the $+x$ direction. The dipole 2 current leads the dipole 1 current by $\pi/2$, with the separation distance being $\lambda/4$ (equivalent to a phase change of $\pi/2$ due to path difference). Thus in the $+x$ direction the fields from 1 and 2 are in phase and hence will add, whereas those in the $-x$ direction differ by 180° and hence cancel. The simple examples considered here illustrate that the spacing between the element a_n and the phasing (i.e., α_n) provides considerable flexibility in obtaining specific radiation patterns. The third parameter for pattern control is the amplitude of the current $|I_n|$, which we have not considered.

6.11 ANTENNAS ABOVE GROUND

So far we have considered antennas operating in a simple medium of infinite extent, like free space or a similar medium having a dielectric constant different from unity. The impedance and radiation properties of an antenna are generally affected by the presence of material objects in its vicinity. In this chapter we describe briefly how the presumed ground affects the radiation from an antenna.

6.11.1 Ground and Ground Plane

We will make a subtle distinction between ground and ground plane, although in antenna literature the two words are often used interchangeably. We will use the term "ground plane" when it is an integral part of an antenna. An ideal or perfect ground plane is generally infinite in extent and is perfectly conducting. However, in practice ground planes are finite in size and their conductivity is very large, but not infinite. Good conducting materials like copper and aluminum are considered suitable for a practical ground plane. For satisfactory performance, the minimum size of a ground plane must be larger than the radiating length of the antenna, which often implies that it is suitably large compared with the operating wavelength. The effect of the finiteness of an antenna's ground plane on its performance is difficult to estimate analytically; it is usually estimated by experimental and numerical studies, which we will not consider here.

Almost all antennas operate on or above the surface of the earth, which we will refer to as ground, and for our discussion we will consider the important case where the ground surface is planar and infinite in extent. The ground constants μ, ϵ, and σ vary considerably over the electromagnetic spectrum. Ground is a lossy medium whose effective conductivity $\sigma/\omega\epsilon_0$ increases with decrease of frequency. For example, below 50 MHz ground may be considered a perfect conductor, while above 50 MHz it behaves like a lossy dielectric and at increasingly higher frequencies it behaves like a perfect dielectric [1,6,7]. For our discussion we will assume that the ground is an isotropic, homogeneous, and nonmagnetic medium having the following characteristics:

1. Perfectly conducting: $\sigma = \infty, \ \epsilon = \epsilon_0, \ \mu = \mu_0$
2. Lossy dielectric: σ finite, $\epsilon = \epsilon_0\epsilon_r, \ \mu = \mu_0$
3. Perfect dielectric: $\sigma = 0, \ \epsilon = \epsilon_0\epsilon_r, \ \mu = \mu_0$.

Frequently ground is represented by a medium having a complex permittivity $\epsilon_c = \epsilon' - j\epsilon''$ and $\mu = \mu_0$. In the following sections we briefly describe the performance of linear antennas above ground. Detailed discussions of ground effects on antenna performance are given in [1,6,7,10].

6.11.2 Image Theory

Consider a plane interface (at $z = 0$) between free space and a perfectly conducting ground represented by a rectangular coordinate system with origin at O as shown in Figure 6.24*a*.

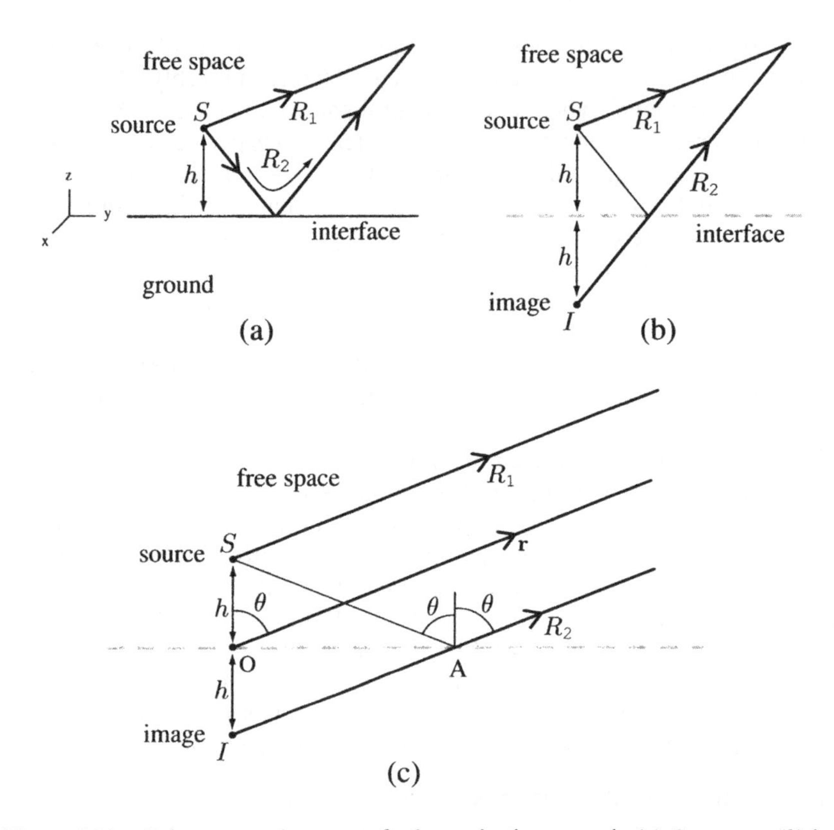

Figure 6.24 Point source above a perfectly conducting ground: (*a*) Geometry; (*b*) image configuration; (*c*) far field configuration.

With a source S located at height h above the ground, the field at P can be determined as the sum of the direct field from S and an indirect field reflected off the ground plane (i.e., the interface) as shown in Figure 6.24*a*. The geometry of the reflection in Figure 6.24*b* indicates that when P is above the ground plane (i.e., above the interface and in the free space region), the field at P can be determined by using the equivalent configurations Figure 6.24*b* entirely in free space based on the image principle. It can be seen from Figure 6.24*b* that this field at P is the sum of the direct field and the field from a virtual (or image) source located directly under the original source and at a distance h below the interface. The image principle holds if the strength of the image (source) is chosen such that the fields satisfy the appropriate boundary conditions of the interface. With proper choice of the image, the fields on the free space side can also be determined by using the image principle. Frequently the far fields produced by a source are of interest. For such cases we use the configuration shown in Figure 6.24c, where it can be seen that the image field can be obtained from a knowledge of the plane wave reflection coefficient at A on the interface for the appropriate field. Note that in the far field region the direct and reflected rays at P are parallel to each other.

6.11.3 Images of Electric Current Elements above Perfect Ground

For differently oriented current elements or Hertzian dipoles of moment $I\,d\ell$ above a perfect ground, the appropriate images and their orientation are shown in Figure 6.25.

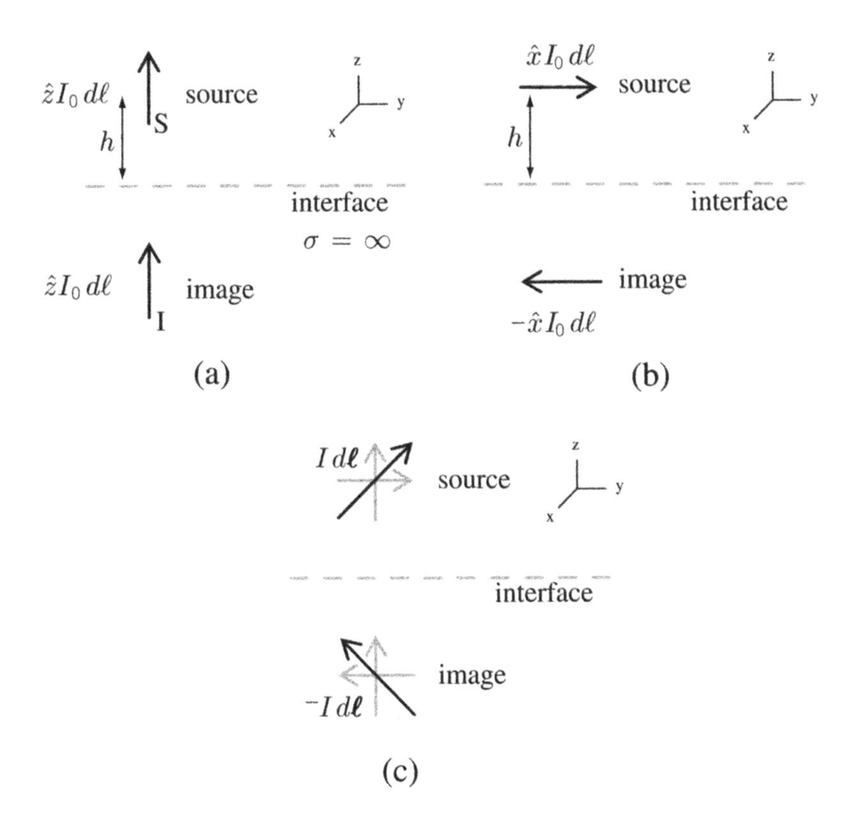

Figure 6.25 Hertzian dipole above a perfect ground and their images.

The image dipole locations and orientations are determined by the fact that the tangential E fields at $z = 0^+$ and $z = 0^-$ produced by the dipole and its image are continuous, and in this case each field equals 0 since $\sigma = \infty$. Note that the vertical dipole and its image are of the same strength and phase; the horizontal dipole and its image are of the same strength but they are oppositely directed. In the case of imperfect ground, the medium below the interface can be considered as a lossy dielectric. The evaluation of image fields can be complicated. However, for far field calculations the image fields are usually determined by making use of the plane wave reflection considerations [1,6,7].

6.11.4 Dipoles above Ground

For the purpose of determining the far fields due to a vertically oriented Hertzian dipole above a flat ground, we use the geometry and configuration shown in Figure 6.24*c* with the origin of coordinate system at Q; the vertically oriented dipole of moment $\hat{z}I_0 d\ell$ and its image are located at S and I as sketched in Figure 6.25*a*. Since the electric field due to the dipoles is rotationally symmetrical around the vertical z direction (i.e., it is ϕ-independent), we can choose any vertical plane through the dipole (e.g., the yz plane) to determine the fields at P.

The total field at P is given by

$$E_\theta(P) = E_\theta^D(P) + E_\theta^r(P)$$
$$= \frac{j\eta I_0\, d\ell}{4\pi} \sin\theta \left[\frac{e^{-j\beta R_1}}{R_1} + \Gamma_\mathrm{v} \frac{e^{-j\beta R_2}}{R_2} \right], \tag{6.196}$$

where $E_\theta^\mathrm{r}(P) = \Gamma_\mathrm{v} E_\theta^D(P)$ is the reflected field at P, Γ_v is the plane wave reflection coefficient for the electric field in the plane of incidence (i.e., for vertical polarization) and is given by (3.296b), which we rewrite in terms of the incident angle θ (instead of the grazing angle ψ in (3.296b)) as

$$\Gamma_\mathrm{v}(=\Gamma_\parallel) = \frac{(\epsilon_\mathrm{r} - j\chi)\cos\theta - \sqrt{(\epsilon_\mathrm{r} - j\chi) - \sin^2\theta}}{(\epsilon_\mathrm{r} - j\chi)\cos\theta + \sqrt{(\epsilon_\mathrm{r} - j\chi) - \sin^2\theta}}, \tag{6.197}$$

with $\chi = \sigma/\omega\epsilon_0$ and σ and ϵ_r being the conductivity and dielectric constant of ground.

Now under the usual far field region approximation, we approximate $R_1 \simeq r - h\cos\theta$ and $R_2 \simeq r + h\cos\theta$ in the exponents and $R_1 = R_2 \simeq r$ in the denominators of the two terms within the brackets (6.197) and obtain

$$E_\theta(P) = \frac{j\beta I_0\, d\ell}{4\pi} \frac{e^{-j\beta r}}{r} \sin\theta \left[e^{j\beta h\cos\theta} + \Gamma_\mathrm{v} e^{j\beta h\cos\theta} \right]. \tag{6.198}$$

If the ground is perfectly conducting, $\Gamma_\text{v} = +1$, and we obtain

$$E_\theta(P) = \frac{j\beta I_0\, d\ell}{4\pi}\,\frac{e^{-j\beta r}}{r}\sin\theta\,[2\cos(\beta h\cos\theta)]. \tag{6.199}$$

Note that the field above ground given by (6.198) can be interpreted as that produced by a 2-element array (dipole and its image); the last two factors on the right-hand side of (6.189) may be identified as the element pattern and the array factors, respectively. In this sense, (6.198) can also be used to determine the field of a $\lambda/2$-dipole above ground provided we use the appropriate element pattern $f(\theta)$ and array factor AF. For future reference we quote the following:

$$\begin{aligned} f(\theta) &= \sin\theta \quad \text{for Hertzian dipole} \\ &= \frac{\cos\left(\frac{\pi}{2}\cos\theta\right)}{\sin\theta} \quad \text{for } \lambda/2\text{-dipole} \end{aligned} \tag{6.200}$$

$$\begin{aligned} \text{AF} &= [2\cos(\beta h\cos\theta)] \quad \text{for perfect ground} \\ &= [e^{j\beta h\cos\theta} + \Gamma_\text{v} e^{-j\beta h\cos\theta}] \quad \text{for imperfect ground.} \end{aligned} \tag{6.201}$$

For perfect ground, (6.199) indicates the far field amplitudes at P will have maxima and minimal nulls at the following angles:

$$\begin{aligned} \theta_\text{max} &= \cos^{-1}\left(\frac{n\lambda}{2h}\right) \\ \theta_\text{min} &= \cos^{-1}\left(\frac{2n+1}{4h}\lambda\right), \end{aligned} \tag{6.202}$$

where $n = 0, 1, 2, \ldots$. Observe that the field amplitude is maximum when P is at the ground surface (i.e., $\theta_\text{max} = \pi/2$ for $n = 0$).

Horizontal Electric Dipole

The analysis of a horizontally oriented Hertzian electric dipole above a plane earth is similar. The geometry and the coordinate system used for the dipole of moment $\hat{y} I_0\, d\ell$ are shown in Figure 6.26. In the present case, the dipole fields not being symmetrical around the z axis (i.e., they are ϕ dependent), it is convenient to express the fields in terms of the angle ϕ measured from the y axis toward the observation point P. Thus the incident and reflected fields at P can be expressed as

$$E_\psi^P(P) = \frac{j\eta I_0 \beta\, d\ell}{4\pi}\,\frac{e^{-j\beta R_1}}{R_1}\sin\psi, \tag{6.203}$$

where $\cos\psi = \hat{r} \cdot \hat{y} = \sin\theta \sin\phi$. The direct and reflected fields at P can be expressed by

$$E_\psi(P) = \frac{j\eta I_0 \beta \, d\ell}{4\pi} \left[\frac{e^{-j\beta R_1}}{R_1} + R_\mathrm{h} \frac{e^{-j\beta R_2}}{R_2} \right], \tag{6.204}$$

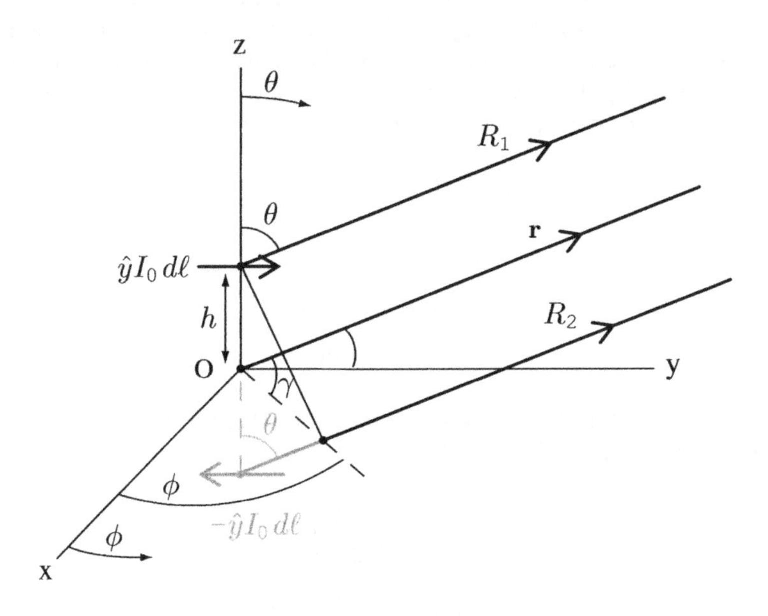

Figure 6.26 Geometry and the coordinate system used for the horizontal dipole above flat ground.

where R_h is the plane wave reflection coefficient appropriate for the horizontal dipole case. After applying the far zone approximations in (6.204), we obtain the far field in any longitudinal plane as

$$\begin{aligned} E_\psi(P) = {} & \frac{j\eta I_0 \beta \, d\ell}{4\pi} \frac{e^{-j\beta r}}{r} \\ & \times \sqrt{1 - \sin^2\theta \sin^2\phi} \\ & \times \left[e^{j\beta h \cos\theta} + R_\mathrm{h} e^{j\beta h \cos\theta} \right], \end{aligned} \tag{6.205}$$

where

$$R_\mathrm{h} = \begin{cases} \Gamma_\perp & \text{for } \phi = 0 \text{ or } 180^\circ, \quad xz \text{ plane} \\ \Gamma_\parallel & \text{for } \phi = 90^\circ \text{ or } 270^\circ, \quad yz \text{ plane.} \end{cases}$$

The reflection coefficient $\Gamma_\perp$ is given by (3.296a), which we rewrite in terms of the incident angle θ (instead of the grazing angle ψ) as

$$\Gamma_\perp = \frac{\cos\theta - \sqrt{(\epsilon_r - j\chi) - \sin^2\theta}}{\cos\theta + \sqrt{(\epsilon_r - j\chi) - \sin^2\theta}}. \tag{6.206}$$

For the perfectly conducting ground plane $R_h = -1$, and we obtain from (6.206)

$$\begin{aligned} E_\psi(P) &= \frac{j\eta I_0 \beta \, d\ell}{4\pi} \frac{e^{-j\beta r}}{r} \\ &\times \sqrt{1 - \sin^2\theta \sin^2\phi}\, [2j \sin(\beta h \cos\phi)], \end{aligned} \tag{6.207}$$

which indicates that the maxima and minima (= 0) of the field occur at

$$\begin{aligned} \theta_{\text{max}} &= \cos^{-1}\left(\frac{2n+1}{4h}\lambda\right) \\ \theta_{\text{min}} &= \cos^{-1}\left(\frac{n\lambda}{2h}\right) \\ n &= 0, 1, 2, \ldots . \end{aligned} \tag{6.208}$$

In this case a minimum occurs at $\theta = \pi/2$, which is when P is on the ground surface as opposed to the vertical dipole case that produces a maximum at $\theta = \pi/2$.

Patterns of Dipoles

The normalized elevation plane patterns of a Hertzian dipole oriented vertically above a flat and horizontal ground surface of infinite extent are shown in Figure 6.27 for the cases when ground is imperfect (σ is finite $< \infty$) and perfect (σ/∞) [6]. The patterns are obtained numerically from (6.198) and (6.199), respectively. Comparison of the two patterns indicates that the imperfect ground tends to increase the field strength toward the vertical direction and it significantly reduces the field near the horizon. In fact the field vanishes along the horizon $\theta = \pi/2$, which can be accounted for by the fact that $\Gamma_\parallel = -1$ along $\theta = \pi/2$.

Similar results for the horizontal dipole ground are shown in Figure 6.28, which gives the elevation plane ($\phi = \pi/2$) patterns of the dipole obtained numerically from (6.205) and (6.207). In this case the two patterns in Figure 6.28 are not found to be significantly different from each other. For the imperfect ground, again, it can be seen that field vanishes at $\theta = \pi/2$ because $\Gamma_\parallel = -1$ in that direction.

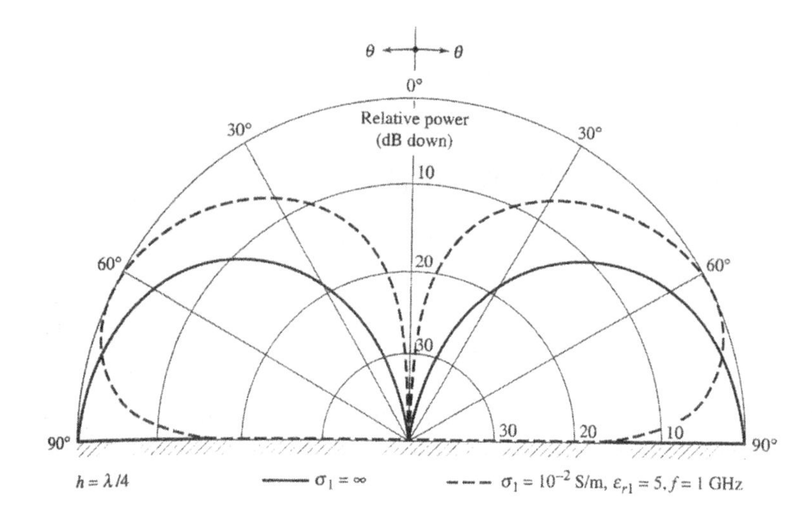

Figure 6.27 Elevation plane antenna pattern of an infinitesimal vertical dipole above a perfect electric conductor ($\sigma_1 = \infty$) and effect of earth ($\sigma_1 = 0.01,\ \epsilon_r = 5,\ f = 1$ GHz). (Source: From [6], p. 183.)

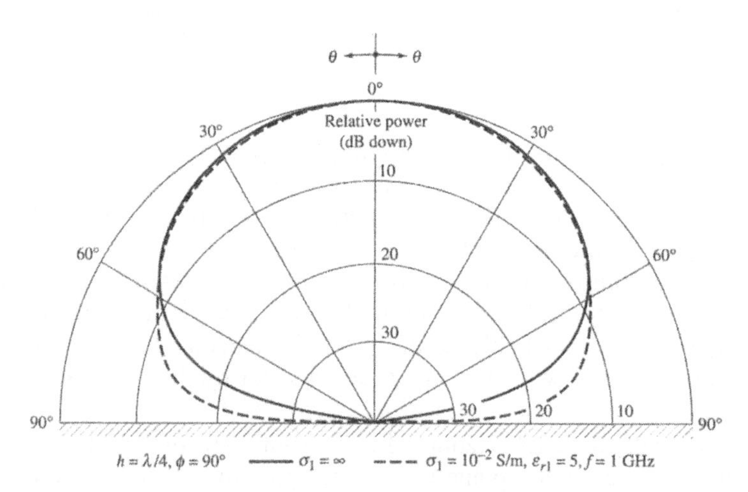

Figure 6.28 Elevation plane antenna pattern of an infinitesimal horizontal dipole above a perfect electric conductor ($\sigma_1 = \infty$) and effect of earth ($\sigma_1 = 0.01,\ \epsilon_r = 5,\ f = 1$ GHz). (Source: From [6], p. 188.)

6.11.5 Monopole Antennas

Monopole antennas and their variations are extensively used in variety of applications. They are used as whip antennas in automobiles to receive AM-FM signals, for land–mobile communications, and frequently as mast antennas to transmit AM broadcast signals.

Normally a monopole has a length of $\lambda/4$ and requires a ground plane for its operation, and it is usually vertically polarized. It is easy to visualize the evolution of a monopole antenna from a vertically oriented center-fed half-wave dipole shown in Figure 6.29.

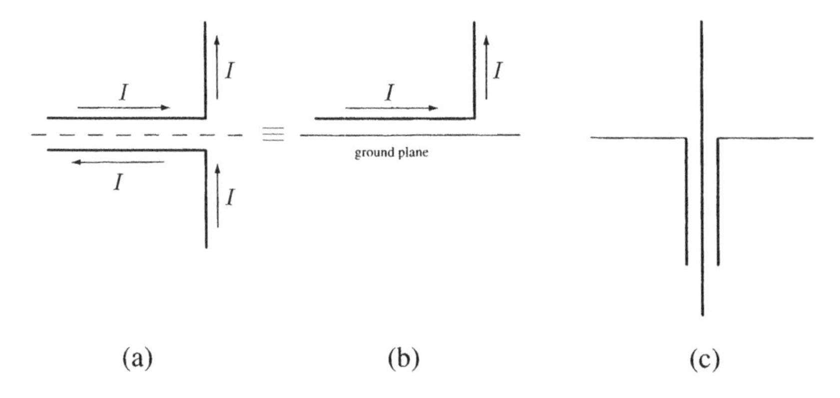

Figure 6.29 Evolution of a monopole antenna: (*a*) Center-fed $\lambda/2$-dipole; (*b*) two-wire transmission line type of monopole antenna; (*c*) coaxial-fed monopole antenna.

An ideal ground plane can be placed symmetrically through the feed lines to divide the dipole into two halves without disturbing the fields and currents, as shown in Figure 6.29*a*. Application of the image principle to the twin line version of the monopole shown in Figure 6.29*b* indicates that it is equivalent to the half-wave dipole in free space. Finally, the coaxial fed version shown in Figure 6.29*c* is the standard monopole above a ground plane. Note that the electric fields in the gap regions of Figures 6.29*a* and 6.29*b* are same, although the gap width of monopole is half that of the dipole. Thus the gap voltage of the monopole is half that of the dipole, and both carry the same current. Hence impedance of the monopole is related to that of the dipole by

$$Z_{A\text{ monopole}} = \frac{Z_{A\text{ dipole}}}{2}. \tag{6.209}$$

Thus

$$Z_{A\text{ monopole}} = 36.05 + j21.3\ \Omega.$$

The monopole and dipole antennas have the same radiation patterns in their upper half regions of space, and their directivity is twice that of the half-wave dipole. A vertically oriented monopole antenna has an omnidirectional pattern in the horizontal plane, and in transmit mode it concentrates most energy near the horizon. This is the reason for its use in broadcasting. Further discussions of monopole of its variations and applications are given in [2, 6, 7].

6.12 BICONICAL ANTENNA

The biconical antenna provides moderately broadband performance and it is typically used in the frequency range 30 to 200 MHz. The antenna can be considered as a finite section of the well-known biconical transmission line described in [1–3, 7].

6.12.1 Biconical Transmission Line

A biconical transmission line consists of two semi-infinite conducting cones of half angle θ_0 placed symmetrically along the z axis with a gap at the center O, which is also the origin of a coordinate system shown in Figure 6.30. The line is excited at the gap. It is known that such a line supports the principal or dominant spherical TEM propagating mode having only E_θ and H_ϕ components. This mode is analogous to the TEM mode in a conventional two-wire or coaxial line. Detailed modal analysis of biconical transmission lines may be found in the references cited. For an understanding of the characteristics of its principal mode we begin with the source-free Maxwell's equations

$$\nabla \times \mathbf{E} = -j\omega\mu\mathbf{H} \tag{6.210}$$
$$\nabla \times \mathbf{H} = j\omega\epsilon\mathbf{E}, \tag{6.211}$$

and in conformity with the geometry we will use a spherical coordinate system.

For the principal mode fields we assume the following:

$$\mathbf{E} = \hat{\theta} E_\theta, \quad \mathbf{H} = \hat{\phi} H_\phi, \quad \begin{array}{l} E_{\mathrm{r}} = E_\phi = 0 \\ H_{\mathrm{r}} = H_\theta = 0 \end{array} \tag{6.212}$$

and $\partial/\partial\phi \equiv 0$, meaning all the field quantities are independent of ϕ. Of course, the fields must satisfy the appropriate boundary conditions.

To obtain explicit equations for E_θ and H_ϕ from (6.201) and (6.202), we need the spherical component expansions for $\nabla = ti\mathbf{E}$ and $\nabla \times \mathbf{H}$ on the left-hand sides of those equations. We do this by using the explicit expression

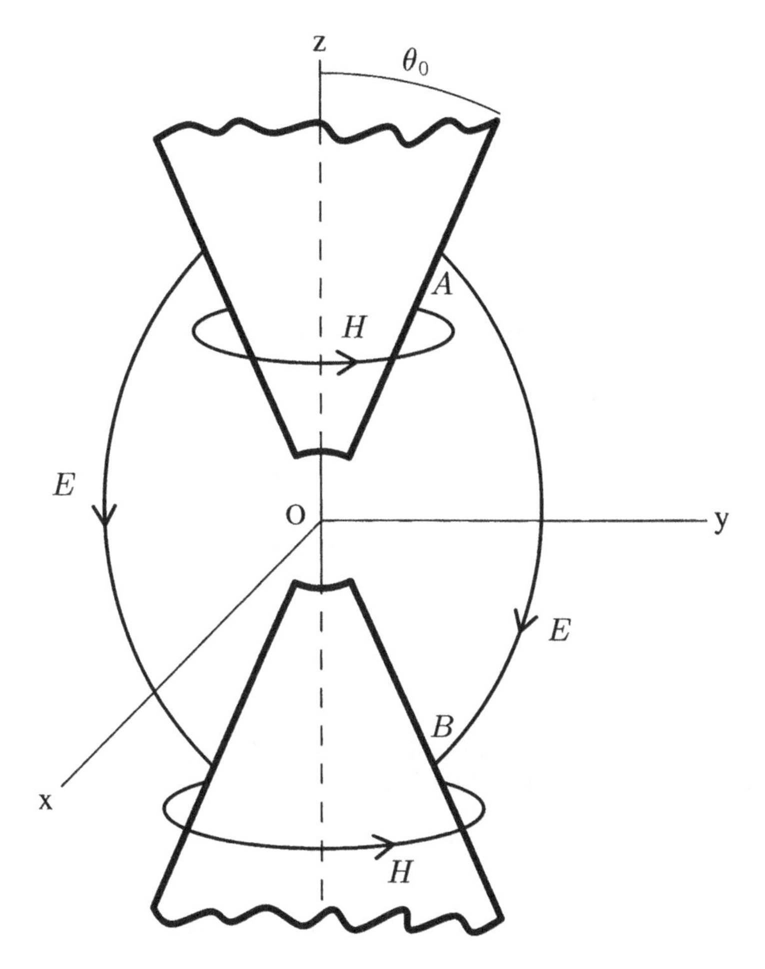

Figure 6.30 Biconical transmission line showing the principal mode field components E_θ and H_ϕ.

for $\nabla \times \mathbf{A}$, for any arbitrary vector $\mathbf{A}$ (given in Appendix A) repeated below:

$$\nabla \times \mathbf{A} = \frac{1}{r^2 \sin\theta} \begin{vmatrix} \hat{r} & r\hat{\theta} & r\sin\theta\,\hat{\phi} \\ \dfrac{\partial}{\partial r} & \dfrac{\partial}{\partial \theta} & \dfrac{\partial}{\partial \phi} \\ A_{\mathrm{r}} & rA_\theta & r\sin\theta\, A_\phi \end{vmatrix}, \tag{6.213}$$

where A_{r}, A_θ, and A_ϕ are the spherical components of an arbitrary vector $\mathbf{A}$.

Under the assumptions (6.212) and making use of (6.213), we obtain the following relations from (6.210) and (6.211):

$$\frac{\hat{\phi}}{r}\frac{\partial}{\partial r}(rE_\theta) = -j\omega\mu H_\phi \hat{\phi} \tag{6.214a}$$

$$\frac{\hat{r}}{r\sin\theta}\frac{\partial}{\partial\theta}(\sin\theta\, H_\phi) - \frac{\hat{\phi}}{r}\frac{\partial}{\partial r}(rH_\phi) = j\omega\epsilon E_\theta\hat{\theta}, \tag{6.214b}$$

which implies that

$$\frac{\partial}{\partial r}(rE_\theta) = -j\omega\mu(rH_\phi) \tag{6.215a}$$

$$\frac{\partial}{\partial r}(rH_\phi) = -j\omega\epsilon(rE_\theta) \tag{6.215b}$$

and

$$\frac{\partial}{\partial\theta}(\sin\theta\, H_\phi) = 0.$$

So the implied θ-dependence of both E_θ and H_ϕ is $1/\sin\theta$. After differentiating (6.215a) with respect to r and using (6.215b) to eliminate (rH_ϕ), we obtain the following equations for rE_θ:

$$\frac{\partial^2(rE_\theta)}{\partial r^2} + \beta^2(rE_\theta) \tag{6.216}$$

with $\beta = \omega\sqrt{\mu\epsilon}$. It can be shown that (rH_ϕ) also satisfies an equation similar to (6.216). In view of the $1/\sin\theta$ dependence of E_θ we can write the solution of (6.216) as

$$E_\theta = \frac{A}{\sin\theta}\frac{e^{-j\beta r}}{r} + \frac{B}{\sin\theta}\frac{e^{j\beta r}}{r}, \tag{6.217}$$

where A and B are arbitrary constants. Now, using (6.217) and (6.215a), we obtain

$$H_\phi = \frac{A}{\eta\sin\theta}\frac{e^{-j\beta\gamma}}{r} - \frac{B}{\eta\sin\theta}\frac{e^{j\beta\gamma}}{r}, \tag{6.218}$$

where $\eta = \sqrt{\mu/\epsilon}$ is the intrinsic impedance of the medium.

Observe that the first terms in (6.217) and (6.213) represent a wave moving away from the center (source) and the second terms represent waves approaching the center (source), meaning they are outgoing and incoming waves. For our semi-infinite biconical line we assume that there is no incoming wave and thus we have the following:

$$E_\theta = \frac{A}{\sin\theta}\frac{e^{-j\beta r}}{r} \tag{6.219a}$$

$$H_\phi = \frac{A}{\eta\sin\theta}\frac{e^{-j\beta r}}{r}, \tag{6.219b}$$

and $E_\phi = \eta H_\phi$.

The voltage between the two conductors, namely between points A and B in Figure 6.30, is given by

$$V(r) = -\int_{\pi-\theta_0}^{\theta_0} \mathbf{E}\cdot d\boldsymbol{\ell} = \int_{\theta_0}^{\pi-\theta} E_\theta r\, d\theta,$$

which after using (6.17) can be written as

$$V(r) = V^A e^{-j\beta\gamma} + V^B e^{+j\beta\gamma} \tag{6.220a}$$

with

$$V_A = 2A\ln\cot\left(\frac{\theta_0}{2}\right)$$
$$V_B = 2B\ln\cot\left(\frac{\theta_0}{2}\right). \tag{6.220b}$$

The current density on the upper cone surface $\mathbf{J}_s(r)$ is

$$\mathbf{J}_s(r) = \hat{n}\times\mathbf{H}_s = \hat{r}H_\phi|_{\theta=\theta_0}, \tag{6.221}$$

where $\hat{n} = \hat{\theta}$ is the outward down unit normal on any point on the upper surface and $\mathbf{H}_s$ is the magnetic field at the upper surface at that point. Using (6.218) in (6.221), we obtain

$$\mathbf{J}_s(r) = \left(\frac{A}{\eta}\frac{e^{-j\beta r}}{r\sin\theta_0} - \frac{B}{\eta}\frac{e^{j\beta r}}{r\sin\theta_0}\right)\hat{r}. \tag{6.222}$$

The total current at $r = r$ on the upper surface, i.e., the axial current at $I(r)$ (see Figure 6.30) is given by Ampère's law:

$$\oint_r \mathbf{H}\cdot d\boldsymbol{\ell} = I_{\text{enclosed}} = I(r), \tag{6.223}$$

which implies

$$I(r) = \int_0^{2\pi} H_\phi \hat{\phi} \cdot \hat{\phi} r \sin\theta_0 \, d\phi.$$

After substituting the value of H_ϕ from (6.218) in the above, we obtain

$$I(r) = \frac{2\pi A}{\eta} e^{-j\beta\gamma} - \frac{2\pi B}{\eta} e^{j\beta r},$$

which by virtue of (6.220b) can be written as

$$I(r) = \frac{1}{\frac{\eta}{\pi} \ln\cot\left(\frac{\theta_0}{2}\right)} \left[V^A e^{-j\beta r} - V^B e^{+j\beta r}\right] \tag{6.224}$$

For a semi-infinite biconical line $V^B \equiv 0$ and we obtain the voltage and current along the same line as

$$V(r) = V^A e^{-j\beta r} \tag{6.225}$$

$$I(r) = \frac{1}{\frac{\eta}{\pi} \ln\cot\left(\frac{\theta_0}{2}\right)} e^{-j\beta r},$$

which give the outgoing travelling wave on the line. The characteristic impedance of the line is given by

$$Z_C = \frac{V(r)}{I(r)} = \frac{\eta}{\pi} \ln\cot\left(\frac{\theta_0}{2}\right) \tag{6.226}$$

and is independent of r, that is, the line has uniform characteristic impedance. For free space, $\eta = 120\pi$, and we have the characteristic impedance of a biconical transmission line of the half-cone angle θ_0 in free space:

$$Z_C = 120 \ln\cot\left(\frac{\theta_0}{2}\right). \tag{6.227a}$$

For a small cone angle, $\theta_0 \lessapprox 20°$,

$$Z_C = 120 \ln\left(\frac{2}{\theta_0}\right). \tag{6.227b}$$

The input impedance Z_i of an infinite biconical antenna with TEM wave is

$$Z_i = \left.\frac{V(r)}{I(r)}\right|_{r\to 0} = Z_C \tag{6.228}$$

and is also independent of r, when Z_C is as given above. Note Z_i as given above is purely resistive, where Z_c is given by (6.227b). The input impedance Z_i of an infinite biconical antenna and of a single (semi-infinite) cone above a perfect ground plane as functions of the cone half angle are shown in Figure 6.31. Note that when the lower cone of the infinite biconical antenna is replaced by a large ground plane, its input resistance is half that of the former.

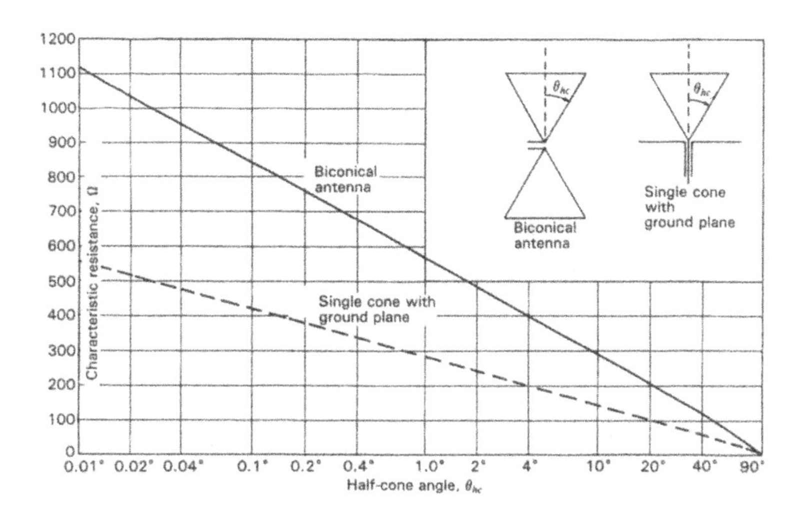

Figure 6.31 Input resistance of an infinite biconical antenna and a single semi-infinite cone above a large (infinite) ground plane as functions of the cone's half-angle. (Source: From [2], p. 347.)

6.12.2 Finite Biconical Antenna

A finite biconical antenna is obtained by using the two finite cones of length $r = \ell$ in Figure 6.30. If we assume that an ideal open circuit exists at $r = \ell$, then using the boundary condition $I(\ell) = 0$ we obtain by using (6.224) that the current distribution on the antenna is

$$I(r) = j\frac{2V^A}{Z_C} e^{-j\beta\ell} \sin\beta(\ell - r), \tag{6.229}$$

that is, for the TEM mode assumed on the antenna, the current distribution is sinusoidal under ideal open-circuit condition. For a finite biconical antenna Schelkunoff [3] defines a boundary sphere of radian l as shown in Figure 6.32.

Within the sphere, namely for $r \leq \ell$, TEM and higher order modes exist; outside the sphere, namely for $r \geq \ell$, only higher order modes exist. The

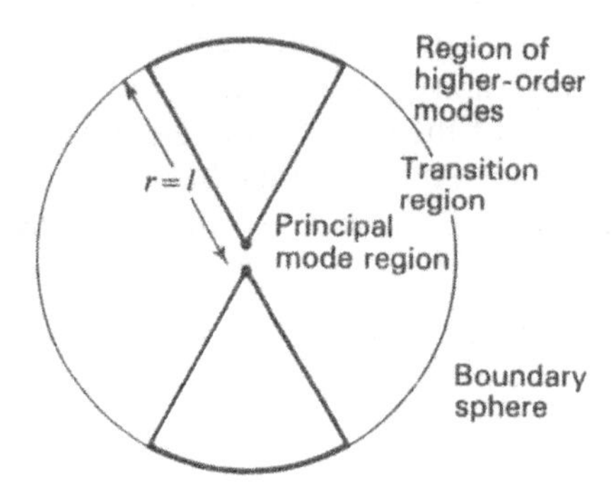

Figure 6.32 Finite biconical antenna with boundary sphere replaced by shell of magnetic material. (Source: From [2], p. 348.)

higher order modes within $r \leq \ell$ are excited at the boundary $r = l$, and they decay in amplitude as r decreases. So at the input $r = r_{\mathrm{i}}$ they produce small effects. The overall effect of these higher order modes within $r \leq \ell$ and of the modes outside ($r > \ell$) on the TEM mode is to produce an effective terminating impedance Z_{t} and $r = \ell$, which differ from the ideal open circuit. Near the input region, the higher order mode effects are small enough so that the antenna input impedance can be obtained approximately by

$$Z_{\mathrm{i}} = Z_{\mathrm{c}} \frac{Z_{\mathrm{t}} + jZ_{\mathrm{c}} \tan \beta\ell}{Z_{\mathrm{c}} + jZ_{\mathrm{t}} \tan \beta\ell}, \tag{6.230}$$

where Z_{t} is the terminating impedance introduced by Schelkunoff.

For $\ell = \lambda/4$ we obtained from (6.230)

$$Z_{\mathrm{i}} = \frac{Z_{\mathrm{c}}^2}{Z_{\mathrm{t}}} = \frac{\eta^2 \left(\ln \cot \dfrac{\theta_0}{2} \right)^2}{\pi^2 Z_{\mathrm{t}}}. \tag{6.231}$$

Schelkunoff showed [3] that for $\theta_0 \to 0$ the terminating impedance Z_{t} grows very large at a rate faster than the logarithmic growth of Z_{L} in the limit of $\theta_0 \to 0$. Thus, for an infinitely thin biconical antenna, the ideal open-circuit condition is achieved. Schelkunoff's biconical antenna theory thus provides a basis for assuming sinusoidal current on thin wire antenna. Distribution of the terminating impedance Z_{t} is discussed by Schelkunoff in [3] and a simpler version of it is given in [2].

A practical method of construction for biconical antennas is to use wires to approximate the cone surfaces [2, 6, 7]. The half-angle θ_0 for a biconical

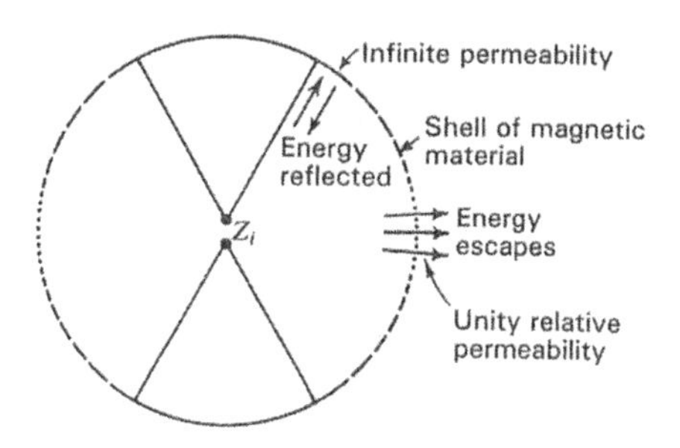

Figure 6.33 Finite biconical antenna with boundary sphere replaced by shell of magnetic material. (Source: From [2], p. 349.)

antenna is usually chosen such that the characteristic impedance of the biconical configuration is nearly the same as that of the transmission line feeding the antenna.

Small-angle antennas are not very practical, but wide-angle configurations $30^\circ \leq \theta_0 \leq 60^\circ$ are frequently used as broadband antennas. For linearly polarized waves incident on the antenna from the broadside ($\theta = 90^\circ$) direction, the antenna responds to the component of $\mathbf{E}$ parallel to its axis; it can be used to perform both vertical and horizontal field measurements required for regulatory compliance verification for EMC.

REFERENCES

1. E. C. Jordan and K. G. Balmain, *Electromagnetic Waves and Radiating Systems*, Prentice-Hall, Inc., Englewood Cliffs, NJ, 1968.

2. J. D. Kraus, *Antennas*, 2nd ed., McGraw-Hill, New York, 1988.

3. S. A. Schelkunoff and H. T. Friis, *Antennas: Theory and Practice*, Wiley, New York, 1952.

4. S. Silver, ed., *Microwave Antenna Theory and Design*, M.I.T. Radiation Laboratory Series, McGraw-Hill, New York, 1949.

5. W. L. Stutzman and G. A. Thiel, *Antenna Theory and Design*, 2nd ed., Wiley, New York, 1998.

6. C. A. Balanis, *Antenna Theory: Analysis and Design*, 2nd ed., Wiley, New York, 1997.

7. R. E. Collin, *Antennas and Radio Wave Propagation*, McGraw-Hill, New York, 1985.

8. D. H. Staelin, A. W. Morgenthaler and J. A. Kong, *Electromagnetic Waves*, Prentice-Hall, Englewood Cliffs, NJ, 1994.

9. S. Ramo, J. R. Whinnery and T. Van Duzer, *Fields and Waves in Communication Electronics*, 3rd ed., New York, 1994.

10. A. A. Smith Jr., *Radio Frequency Principles and Applications*, IEEE Press/ Chapman & Hall, New York, 1998.

11. C. R. Paul, *Introduction to Electromagnetic Compatibility*, Wiley, Inc., New York, 1992.

12. IEEE, *IEEE Standard 100– 1972, IEEE Standard Dictionary of Electrical and Electronics Terms*, Wiley-Interscience, New York, 1972.

13. IEEE, 1983 version of 12.

14. G. Sinclair, "The Transmission and Reception of Elliptically Polarized Waves," *Proc. IRE, 38 (February 1950): 148–151.*

15. E. J. Jahnke and F. Emde, *Tables of Functions*, 4th ed., Dover, New York, 1945.

16. M. Abramowitz and I. A. Stegun, *Handbook of Mathematical Functions*, Applied Mathematics Series 55, National Bureau of Standards, Washington, DC, June 1969.

PROBLEMS

6.1 The power density radiated by an antenna in free space is given by

$$\mathbf{S}_{\text{av}} = A_0 \frac{\sin\theta}{r^2}\hat{r} \quad \text{W/m}^2,$$

where A_0 is the peak value of the power density and the other notations are as described in the text. Determine the following for the antenna: **(a)** the total power radiated (obtain it by using $\mathbf{S}_{\text{av}}$ and also by using the concept of the radiation intensity), **(b)** the directivity, and **(c)** the maximum directivity in absolute value and in dB.

Answer: **(a)** $\pi^2 A_0$; **(b)** $D(\theta,\phi) = \frac{4}{\pi}\sin\theta - D\sin\theta$; **(c)** $D = \frac{4}{\pi} = 1.27$ dB.

6.2 The maximum radiation intensity of a 90% efficient antenna is 200 mW/steradian (or unit solid angle) in free space. Find the maximum directivity and gain in absolute value and in dB when **(a)** the input power is 40π mW and **(b)** the radiated power is 40π mW.

Answer: **(a)** $D = 22.22 = 13.47$ dB, $G = 20 = 13.01$ dB, **(b)** $D = 20 = 13.01$ dB, $G = 0.9\,(20) = 18$ dB.

6.3 The normalized radiation intensity of a given antenna in free space is

$$P(\theta,\phi) = \begin{cases} \sin\theta\,\sin\phi \quad \text{W/steradian} & \text{for } 0 \le \theta \le \pi,\ \ 0 \le \phi \le \pi \\ 0 & \text{otherwise.} \end{cases}$$

Determine its **(a)** maximum directivity in absolute value and in dB, and **(b)** azimuthal and elevation plane half-power beam width (HPBW) in degrees.

Answer: **(a)** $D = 4 = 6.02$ dB; **(b)** HPBW in azimuth $= 2 \times 60 = 120°$, HPBW in elevation $= 2 \times 60 = 120°$.

6.4

(a) The radiation intensity function for an antenna in free space is given by

$$P(\theta, \phi) = \begin{cases} P_0 \cos^2 \theta \quad \text{W/steradian} & \text{for } 0 \leq \theta \leq \pi/2 \\ \pi/2 \leq \phi \leq \pi & \text{otherwise} \end{cases}$$

for all ϕ.

Determine following parameters for the antenna: **(a)** $D(\theta, \phi)$, D in absolute values, and **(b)** D relative to an isotropic radiator (i) and Hertzian dipole (HD).

Answer: **(a)** $D(\theta, \phi) = 6 \cos^2 \theta, D = 6$; **(b)** $D = \frac{6}{1} = 6 \equiv 7.78$ dBi, $D = \frac{6}{1.5} = 4 \equiv 6.04\ \text{dB}_{\text{HD}}$.

(b) Determine the same parameters for the antenna if its radiation intensity is $P(\theta, \phi) = P_0 \cos^2 \theta$ in the range $0 \leq \theta \leq \pi$ for all ϕ.

Answer: **(a)** $D(\theta, \phi) = 3 \cos^2 \theta, / D = 3$; **(b)** $D = \frac{3}{1} = 3 \equiv 4.77$ dBi, $D = \frac{3}{1.5} = 2 \equiv 3.01\ \text{dB}_{\text{HD}}$.

6.5 The input impedance of a center-fed $\lambda/2$-dipole is equal to $73 + j\,42.5\,\Omega$. Assume the dipole to be lossless. Determine the following: **(a)** the input impedance, assuming that the feed terminals are shifted to a point on the dipole which is $\lambda/8$ from one end, **(b)** the capacitive or inductive reactance that must be placed across the new input terminals in (a) so that the dipole is self-resonant, and **(c)** the VSWR at the input terminals when the self-resonant dipole of (b) is connected to a transmission line having a characteristic impedance of $300\,\Omega$.

Answer: **(a)** $Z_{\text{in}} = 146 + j\,85\,\Omega$; **(b)** required reactance $= -j\,335.776\,\Omega$ (capacitive); **(c)** VSWR $= 1.53$.

6.6 A linear $\lambda/2$-dipole is operating at a frequency of 1 GHz. Determine the capacitance or inductance that must be placed across the input terminals of the dipole so that the antenna becomes resonant. What is the VSWR when it is connected to a $50\,\Omega$ line?

Answer: Required capacitance $C_{\text{in}} = 0.95$ pF; VSWR $S = 1.96$.

6.7 A $\lambda/2$-dipole is radiating in free space. The coordinate system is such that the origin is at the center of the dipole and the dipole is oriented along the z axis. Input power to the dipole is 100 W. Assuming 50% efficiency. determine the radiated power density at $r = 500$ m, $\theta = 60°$ and $\phi = 0°$.

Answer: 1.741×10^{-5} W/m^2.

6.8

(a) An FM broadcast radio station has a 3 dB gain antenna system and 100 KW transmitted power. What is the EIRP?

Answer: EIRP $= 2G = 200$ KW.

b) Use proper assumption(s) and reasoning(s) to obtain the effective area of the following antennas: an isotropic radiator, a short current element, a $\lambda/2$ resonant dipole and a small current loop.

Hint: Use $A_e = D = \lambda^2 G/4\pi, \lambda^2/4\pi, 0.12\lambda^2, 0.13\lambda^2, 0.12\lambda^2$.

6.9 A plane wave in free space with electric field $\mathbf{E} = E_0 e^{j\beta r}\hat{t}$, with $E_0 =$ 5 μV/m peak, is incident on a z-directed $\lambda/2$-dipole from the direction $\theta =$ 60°. Find the open circuit voltage at the antenna terminals when $\lambda = 10$ m.

Answer: 13 μV.

6.10 Determine the magnetic field intensity at a distance of 10 Km from an antenna having a directivity of 6 dB and radiating a total power of 25 KW in free space.

Answer: $|H| = 6.5$ mA/m; $|E| = \eta_0 H = 2.44$ V/m.

6.11 The sensitivity of the receiver onboard a geosynchronous satellite approximately 36,000 Km away from a $\lambda/2$-dipole antenna source is 2 μV/m. Assuming free space conditions **(a)** determine the minimum antenna input current that would produce fields sufficient for detection by the satellite and **(b)** calculate the approximate power radiated by the $\lambda/2$-dipole.

Answer: **(a)** $I_{\text{min}} = 1.2$ A; **(b)** $P_{\text{rad}} = 52.56$ W.

6.12 Derive the expressions given by (6.109a)–(6.109d).

CHAPTER 7

BEHAVIOR OF CIRCUIT COMPONENTS

7.1 INTRODUCTION

The resistance (R), inductance (L) and capacitance (C) are the basic elements used in electronic circuits. They play important roles not only in engineering design and circuit performance but also in the analysis of various kinds of electromagnetic interactions produced by a circuit. In all circuits these elements are interconnected generally by conducting paths that also form an integral part of the transmission line system. Investigation of circuits is based fundamentally on Kirchoff's and Ohm's laws, which provide accurate results only at sufficiently low frequencies. We define a passive device as a resistor, inductor, or capacitor if its primary purpose is to introduce ideal resistance, inductance, or capacitance to a circuit, respectively. From the electromagnetic fields point of view, the lumped circuit parameters R, L, and C are looked upon as energy dissipative, magnetic energy storage and electric energy storage elements. Application of Maxwell's equations to an elementary RLC circuit indicates that certain idealizations are required for the validity of Kirchoff's and Ohm's

Applied Electromagnetics and Electromagnetic Compatibility. By D. L. Sengupta, V. V. Liepa
ISBN 0-471-16549-2

laws. We will see that in general, when the linear dimensions of the circuit and the circuit components are comparable to the operating wavelength, these idealizations break down and the circuit may start to radiate. Under these conditions the behavior of the circuit and its components will be considerably different from operation at dc or frequencies below 100 kHz.

In this chapter the behavior of circuits and circuit components are discussed in the context of electromagnetic field theory.

7.2 THE SERIES RLC CIRCUIT

Field theory descriptions of fundamental circuit relations and concepts are given in most textbooks on electromagnetic field theory, for example in [1–5]. We present a discussion of Kirchoff's voltage law applied to a series RLC circuit fed by a time-varying voltage source shown in Figure 7.1. It is important to note that we are not using phasor notation here; all relevant fields, voltages and currents are time-varying quantities and are real. At any instant of time the

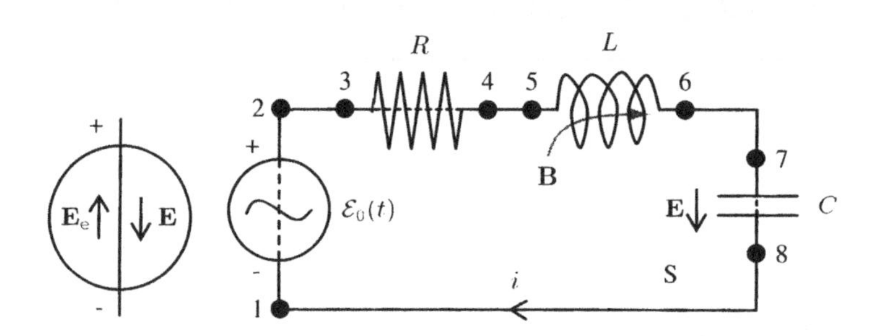

Figure 7.1 A series RLC circuit fed by a time-varying voltage or producing a current i through the circuit.

polarity of V is as shown in Figure 7.1. It is assumed that the emf generating electric field $\mathbf{E}_e$ is confined within the source region only, and it is directed opposite to the induced $\mathbf{E}$ as shown in the inset. The induced currents and charges in the system produce the electric field $\mathbf{E}$ and (possibly) the magnetic flux density $\mathbf{B}$ everywhere as shown. It is assumed that the connecting leads 2–3, 4–5, 6–7, 8–1, the inductance wire, the other leads used to connect the source, the inductance, and the capacitance are all perfectly conducting.

We now apply Faraday's law—or the integral form of Maxwell's equation (3.24b)—using the loop $\ell = 123456781$ defined by the filamentary conducting paths, including the path provided by L and the dashed paths through R, C, and the source enclosing the surface S (which also includes the surface enclosed by the loop, shown in Figure 7.1, and obtain

$$-\oint_{\ell} \mathbf{E} \cdot d\boldsymbol{\ell} = \frac{d}{dt} \int_S \mathbf{B} \cdot ds. \tag{7.1}$$

It should be noted that the direction of integration along the closed loop on the left-hand side and the positive unit normal to the surface S on the right-hand side of (7.1) are related by the right-hand rule. The E-fields along the conducting paths being zero, we obtain from (7.1)

$$-\int_1^2 \mathbf{E} \cdot d\boldsymbol{\ell} - \int_3^4 \mathbf{E} \cdot d\boldsymbol{\ell} - \int_7^8 \mathbf{E} \cdot d\boldsymbol{\ell} = \frac{d}{dt} \int_S \mathbf{B} \cdot ds. \tag{7.2}$$

For convenience, we rearrange the terms and rewrite (7.2) as

$$-\int_1^2 \mathbf{E} \cdot d\boldsymbol{\ell} = \int_3^4 \mathbf{E} \cdot d\boldsymbol{\ell} + \frac{d}{dt} \int_S \mathbf{B} \cdot ds + \int_7^8 \mathbf{E} \cdot d\boldsymbol{\ell}. \tag{7.3}$$

The left-hand side of (7.3) can be identified as the applied emf or the source voltage $\mathcal{E}(t)$ as follows:

$$-\int_1^2 \mathbf{E} \cdot d\boldsymbol{\ell} = \int_1^2 \mathbf{E}_e \cdot d\boldsymbol{\ell} = \mathcal{E}_0(t), \tag{7.4a}$$

where it is assumed that under open circuit condition the induced $\mathbf{E}$ and the emf generating field $\mathbf{E}_e$ are equal but oppositely directed [1,4]. The three terms on the right-hand side of (7.3) are associated with R, L, and C, respectively.

We assume that the resistance is in the form of cylinder of length h and cross section a, and is made of a material with conductivity σ. It is assumed that a uniform $\mathbf{E}$ field directed along its length sustains a uniform current density in the same direction and is given by $\mathbf{J} = \sigma \mathbf{E}$. It can then be shown that

$$\begin{aligned} \int_3^4 \mathbf{E} \cdot d\boldsymbol{\ell} &= \int_3^4 \frac{\mathbf{J} \cdot d\boldsymbol{\ell}}{\sigma} = \frac{Jh}{\sigma} \\ &= \frac{ih}{\sigma a} = iR \end{aligned} \tag{7.4b}$$

with $R = h/\sigma a$ as the resistance.

We assume now that the $\mathbf{B}$-field is confined only within the inductance coil. Thus we can write

$$\begin{aligned} \frac{d}{dt} \int_S \mathbf{B} \cdot ds &= \frac{d\psi}{dt} = \frac{d}{dt}(Li) \\ &= L\frac{di}{dt}, \end{aligned} \tag{7.4c}$$

where ψ is the flux linkage through the coil surface S, L is the inductance of the coil given by $L = \psi/i$, i being the current carried by the coil.

Finally, assume that the capacitance is made of two parallel conducting plates each of area A and separated by a distance d; the space between the plates in filled with a dielectric material with permittivity ϵ. It is assumed that the uniform electric field is entirely confined within the capacitor and directed as shown in Figure 7.1. Under these assumptions, it can be shown that

$$\begin{aligned}\int_7^8 \mathbf{E} \cdot d\boldsymbol{\ell} &= \int_7^8 \frac{\mathbf{D}}{\epsilon} \cdot d\boldsymbol{\ell} = \int_7^8 \frac{\rho_s \, d\ell}{\epsilon} \\ &= \frac{\rho_s d}{\epsilon} = \frac{Qd}{A\epsilon} = \frac{Q}{\frac{\epsilon A}{d}} \\ &= \frac{Q}{C} = \frac{\int i \, dt}{C}, \end{aligned} \tag{7.4d}$$

where $\mathbf{D} = \epsilon \mathbf{E}$ is the electric flux density within C, ρ_s is the electric charge density magnitude or each plate, $Q = \int i \, dt$ is the total charge magnitude on each plate, $C = \frac{\epsilon A}{d}$ is the capacitance of the elements.

Now, combining (7.4a)–(7.4d), we obtain from (7.3)

$$\mathcal{E}(t) = iR + L\,\frac{di}{dt} + \frac{\int i \, dt}{C}, \tag{7.5}$$

which is the well-known Kirchoff's voltage law for a series circuit, expressing that at any time the applied voltage in a series circuit is the sum of the voltage drops across the individual circuit elements.

We have just described the equivalence of circuit and field theory. However, it is important to realize that the field theory clearly brings out the limitations of the circuit theory, and most important, it suggests certain conditions that must be met for the satisfactory application of ordinary circuit theory to be valid. They are:

1. Any displacement current is confined within the capacitor, which implies that capacitance is confined within the capacitor.

2. Any magnetic flux is confined within the inductor, which implies that inductance is confined within the inductor.

3. Any Jule loss is confined within the resistor, which implies that resistance is confined within the resistor.

4. At any instant of time the current is the same at all points on the circuit, which implies that the disturbance travels (propagates) instantaneously around the circuit.

Most important, from the field theory point of view, condition (4) requires that any linear dimension of the circuit and its components must be very small compared with the operating wave length (note that $f = c/\lambda$, c being the velocity of light in free space). Any violation of the conditions above requires that ordinary circuit theory must be modified in the light of field theory.

7.3 DEFINITIONS OF LUMPED CIRCUIT PARAMETERS R, L, AND C

The basic parameters R, L, and C are described in this section from the viewpoints of circuit and field theories.

7.3.1 Circuit Theory Description

Figure 7.2 gives the circuit theory representations of R, L, and C along with a legend summarizing the basic circuit relations and definitions and where the notations used are standard, and Ψ represents the flux linkage by the current through the inductor coil. Note that each element can have two alternative representations between its terminals 1 and 2. We assume that the current I flows from terminal 1 to 2 (i.e., V is the voltage drop between the positive terminal 1 and the negative terminal 2). Also note that Ψ is the flux linkage the associated with the current I through the inductor coil and that $+q$ and $-q$ are the positive and negative charges on the capacitor plates. It should be noted that for convenience of discussion we have assumed instantaneous polarities of the terminals as indicated in Figure 7.2.

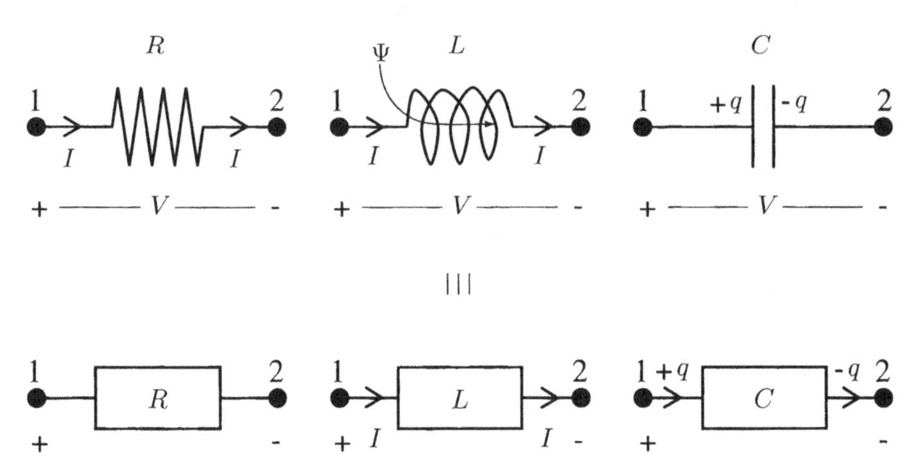

Figure 7.2 Circuit symbols and defining relations for R, L, and C.

We now give the circuit definitions of the lumped parameters as

Resistance

$$R = \frac{V}{I} = \frac{\text{voltage drop}}{\text{current}} \quad \Omega \tag{7.6a}$$

Inductance

$$L = \frac{\Psi}{I} = \frac{\text{flux linkage}}{\text{current}} \quad \text{H} \tag{7.6b}$$

Capacitance

$$C = \frac{q}{V} = \frac{\text{magnitude charge on each plate}}{\text{voltage}} \quad \text{F} \tag{7.6c}$$

It should be noted that the definitions of the circuit elements given by (7.6a)–(7.6c) are entirely determined by the terminal values of V, I, q, and so on, of each element, and all three are directly measurable.

7.3.2 Field Theory Description

The definitions given in Section 7.2.1 do not provide any guidance to analytically determine the values of R, L, and C for given configurations. For this purpose the electric and magnetic fields produced in the excited element are necessary so that field theory concepts can be applied to obtain the necessary terminal relations for the element. We now consider the following configurations of R, L, C as shown in Figure 7.3. The resistor R consists of a conducting material of arbitrary shape and carrying a current I as shown in Figure 7.3*a*. $\mathbf{E}$, $\mathbf{J}$ are the electric field and current density within the resistance, generated by I.

Figure 7.3*b* shows a one-turn loop inductor with inductance L. Current I carried by the loop produces magnetic field $\mathbf{H}$. The flux Ψ generated is linked with the current as shown. The inductor is assumed to be in a lossless homogeneous medium $\epsilon = \epsilon_0, \ \mu = \mu_0\mu_r$.

Figure 7.3*c* shows a capacitor consisting of two oppositely charged conductors of arbitrary shape and separated a certain distance. The medium is assumed to be a lossless dielectric having $\epsilon = \epsilon_0\epsilon_r$ and $\mu = \mu_0$.. The $\mathbf{E}$ field produced by the charged conductor is sketched in Figure 7.3*c*.

It is straightforward now to express (7.6a) in terms of the fields within the resistor shown in Figure 7.3*b* as

$$R = \frac{-\int_1^2 \mathbf{E} \cdot d\boldsymbol{\ell}}{\int_S \mathbf{J} \cdot d\mathbf{s}}, \tag{7.7}$$

where we have assumed $\mathbf{J} = \sigma\mathbf{E}$, σ being the conductivity of the material of R, and S is the cross section of the resistance.

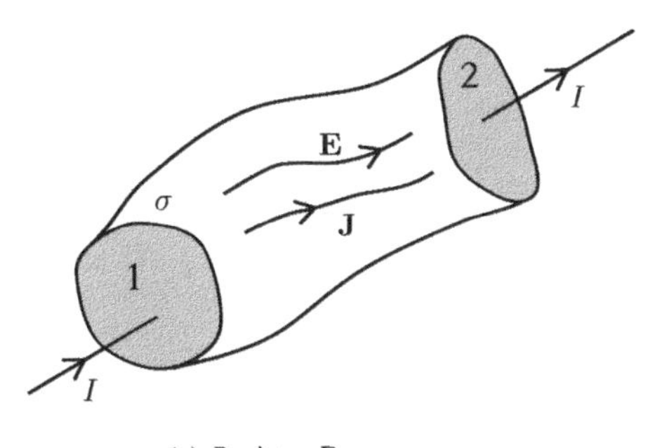

(a) Resistor R

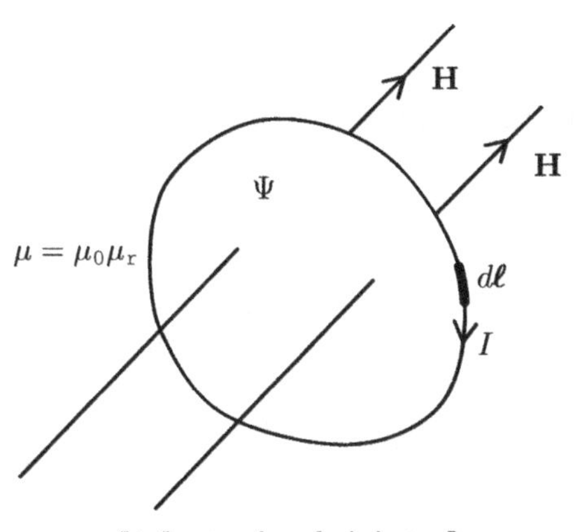

(b) One-turn loop for inductor L

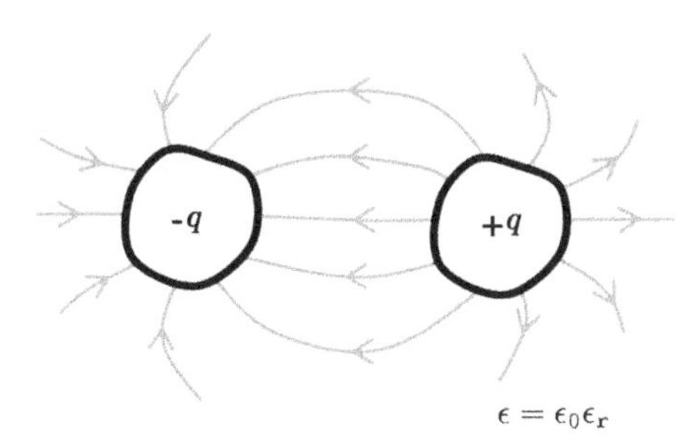

(c) Capacitor C consisting of two conductors

Figure 7.3 R,L, and C configurations.

Using Figure 7.3*b*, we can write the inductance of the loop given by (7.6b) as

$$L = \frac{\psi}{I} = \frac{\int_S \mathbf{B} \cdot d\mathbf{s}}{I}$$
$$= \frac{\oint_c \mathbf{A} \cdot d\boldsymbol{\ell}}{I} = \frac{\int_S \nabla \times \mathbf{A} \cdot d\mathbf{s}}{I}, \tag{7.8}$$

where S is the surface enclosed by the loop contour c, and $\mathbf{A}$ is the vector potential produced by the loop current at the location of $d\boldsymbol{\ell}$.

Using (7.6c) and Figure 7.3*c*, we obtain the following for the capacitance C:

$$C = \frac{\oint_{S_1} \epsilon \mathbf{E} \cdot d\mathbf{s}}{-\int_2^1 \mathbf{E} \cdot d\boldsymbol{\ell}}, \tag{7.9}$$

where S is the total surface area of the positively charged conductor and the integration in the denomination is carried along any $\mathbf{E}$ line from 1 to 2.

Energy Definitions

Frequently it is found convenient to define circuit parameters from the viewpoint of the energy associated with them. For example, the resistance R indicates a measure of the energy lost as heat; inductance indicates a measure of the magnetic energy stored in the element; the capacitance indicates a measure of the electric energy stored in the element.

The power P_{L} (i.e., the energy lost per second) due to when a current I (causing $\mathbf{J}$ and $\mathbf{E}$ within R) passes through the resistance is

$$P_{\mathrm{L}} = \int_\tau \mathbf{J} \cdot \mathbf{E} \, d\tau = I^2 R, \tag{7.10}$$

where τ is the volume of the resistor. Thus we can define R as

$$R = \frac{P_{\mathrm{L}}}{I^2} = \frac{\int_\tau \mathbf{J} \cdot \mathbf{E} \, d\tau}{I^2}, \tag{7.11}$$

The magnetic energy stored within the volume τ of an inductor is [4, 5]:

$$W_{\mathrm{m}} = \int_\tau \frac{\mathbf{H} \cdot \mathbf{B}}{2} \, d\tau = \frac{1}{2} L|I|^2, \tag{7.12}$$

from which we obtain

$$L = \frac{2W_{\mathrm{m}}}{I^2} = \frac{\int_\tau \mathbf{H} \cdot \mathbf{B} \, d\tau}{I^2}, \tag{7.13}$$

where $\mathbf{H}$, $\mathbf{B}$ are the magnetic field intensity and flux density within the inductance and $\mathbf{B} = \mu\mathbf{H}$, and where

$$I = \int_S \mathbf{J} \cdot d\mathbf{s}.$$

Note that the volume integral in (7.13) extends over the entire region occupied by the fields. The portion of the inductance associated with energy of the fields inside a conductor is called its *internal (self) inductance*, and that associated with energy outside is called the *external (self) inductance*. Thus the total inductance is the summation of the two.

The stored electric energy within the volume of the 2-conductor capacitor is [4, 5]

$$W_{\mathrm{E}} = \int_\tau \frac{\mathbf{E} \cdot \mathbf{D}}{2}\, d\tau = \frac{1}{2} C V^2, \tag{7.14}$$

from which we obtain:

$$C = \frac{2W_{\mathrm{E}}}{V^2} = \frac{\int_\tau \mathbf{E} \cdot \mathbf{D}\, d\tau}{V^2} \tag{7.15}$$

with

$$V = -\int_2^1 \mathbf{E} \cdot d\boldsymbol{\ell}$$

and $\mathbf{D} = \epsilon\mathbf{E}$.

7.4 ROUND WIRES

Round wires of conducting material are used to provide interconnection between electronic components and systems. Frequently the main purpose is to transfer signals without loss and distortion. A given length of wire will have resistance, inductance, and sometimes even capacitance associated with it; these RLC effects can manifest significantly at higher frequencies thereby introducing loss as well as distortion to the signal passing through.

Conducing paths of non-circular cross section are also in use for the same purpose. However, it is often possible to define an equivalent round cross section for such cases. Therefore it is sufficient to discuss the circuit properties of wires with circular cross section.

We will consider two types of wires:

1. A single solid cylindrical conducting wire.

2. A standard wire consisting of several strands of solid cylindrical wires placed parallel to each other in the form of a round cylinder.

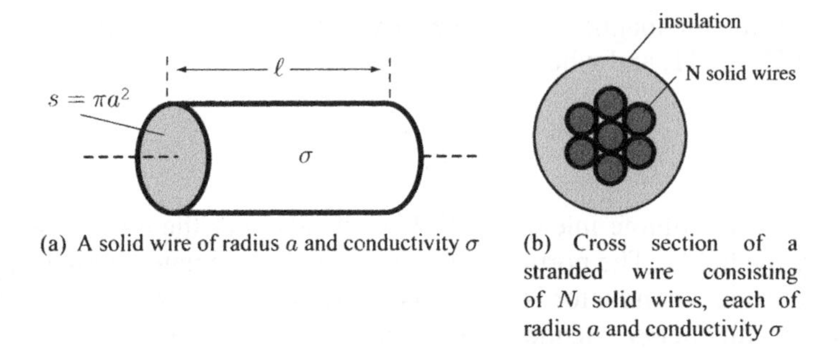

(a) A solid wire of radius a and conductivity σ

(b) Cross section of a stranded wire consisting of N solid wires, each of radius a and conductivity σ

Figure 7.4 A solid wire a and a standard wire b.

The solid wires as well as the stranded wire are coated with some dielectric material for insulation. The two kinds of wires are shown in Figure 7.4.

Wire dimensions and other information are given in handbooks [6]; similar information appropriate for our purpose are given in [7]. The solid wires are assigned a number according the American Wire Gauge (i.e., an AWG number) such that each AWG number is associated with specific wire diameter, cross section, and other relevant information. The stranded wires are also assigned an AWG number for solid wires such that the diameter of the bundled wire approximates that corresponding to the equivalent solid wire. Some selected [6, 7] wire gauges and wire diameters are shown below.

AWG Wire Gauges

Wire Gauge (AWG)	Solid Wire Diameter (in mils)	Stranded Wire Diameter (in mils)
10	101.9	116.0 (105 × 30)
20	32.0	36.0 (41 × 36)
30	10.0	12.0 (7 × 38)

Note that diameters are given in mils where 1 mil = 0.001 inch = 2.54×10^{-5} m. The conversion should be used to convert mils to meters, as required by the SI system.

For solid wires the AWG number gives the diameter of the wire as given. The numbers given for stranded case need some explanation. For example, 116.0 (105 × 30) for the stranded wire AWG 10 means that there are 105 wires each of gauge AWG 30 are in the bundle and 116.0 mils is the approximate diameter of the standard wire. The equivalent diameter of the bundled wire is 101.9 mils; hence it is assigned the gauge AWG 10. The information above can be used to calculate the values of the circuit parameters for a given length of solid or stranded wire. It should also be mentioned here that for resistance and internal inductance calculations for a stranded wire consisting of N strands, it is generally assumed that the component wires or the strands are electrically connected in parallel. Thus the resistance and internal inductance of the stranded wire are obtained by dividing by N the respective values of the single solid wire (i.e., of the strand).

7.4.1 Resistance

The dc resistance of the solid wire shown in Figure 7.4*a* is

$$R_{\mathrm{dc}} = \frac{\ell}{\sigma s} \quad \Omega. \tag{7.16}$$

We define the resistance per unit length as

$$r_{\mathrm{dc}} = \frac{1}{\sigma s} = \frac{1}{\pi a^2 \sigma} \quad \Omega/\mathrm{m}, \tag{7.17}$$

where a is the radius of the wire. Similarly the per unit length resistance $r_{\mathrm{dc}}^{\mathrm{s}}$ of a stranded wire consisting of N strands of single wires each having a radius a is

$$r_{\mathrm{dc}}^{\mathrm{s}} = \frac{r_{\mathrm{dc}}}{N} = \frac{1}{\pi a^2 \sigma N}\,. \tag{7.18}$$

We have seen earlier that at higher frequencies the skin effect phenomena confine the fields and currents within a conductor. In this case the current is assumed to be confined around the surface as shown in Figure 7.5.

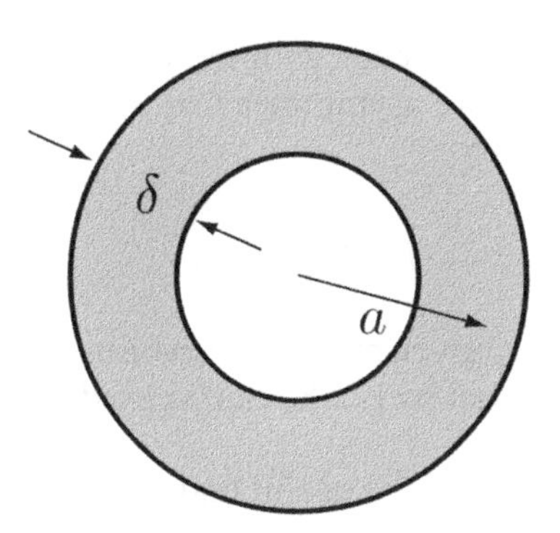

Figure 7.5 Current confined to a ring of width S around the surface.

Thus the appropriate cross section for the high-frequency resistance is $s = 2\pi a\delta$. Using the value above for the cross section, it can be shown that the high frequency resistance per unit length for a solid wire of radius a and conductivity σ is given by

$$r_{\mathrm{HF}} = r_{\mathrm{dc}} \frac{a}{2\delta} \quad \Omega/\mathrm{m}$$
$$= \frac{1}{2a}\sqrt{\frac{\mu_0}{\pi\sigma}}\sqrt{f} \quad \Omega/\mathrm{m}, \tag{7.19}$$

which indicates that $r_{\mathrm{HF}} > r_{\mathrm{dc}}$, and increases as $\sqrt{f}$. Thus the high-frequency resistance increases at a rate of 10 dB per decade increase of frequency, similarly for stranded wires. Figure 7.6 gives a plot of r versus a/δ, where the

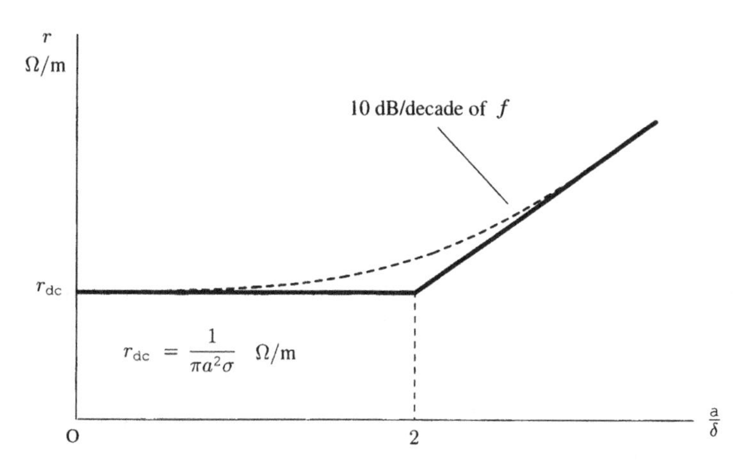

Figure 7.6 Variations of per unit length resistance of a solid wire versus a/δ.

dotted curve merges at the two ends with the solid lines giving the calculated values; the two solid lines are asymptotes that meet at $a = 2\delta$, which occurs at the frequency f_{a} given by

$$f_{\mathrm{a}} = \frac{4}{\pi\mu_0\sigma a^2}\ . \tag{7.20}$$

Approximate dc values r_{dc} given by (7.17) can be used for frequencies $f \leq f_{\mathrm{a}}$ and the HF values r_{HF} given by (7.18) can be used for the $f > f_{\mathrm{a}}$ for rough estimation.

The preceding estimates will be useful finding the resistance of component leads.

7.4.2 Internal Inductance

We will use (7.13) to determine the internal inductance of a solid wire of length ℓ, radius a, and permeability μ. The wire is assumed to carry a total longitudinal current i distributed uniformly over its cross section. After identifying τ in (7.13) as the internal volume of the length of the wire, it can be shown that the internal inductance of the wire is (from (7.13))

$$\begin{aligned} L_{\mathrm{i}} &= \frac{2W_{\mathrm{m}}}{I^2} = \frac{2}{I^2}\int_{\tau}\frac{\mathbf{B}\cdot\mathbf{H}}{2}\,d\tau \\ &= \frac{2}{I^2}\cdot\frac{1}{2}\int_0^{\ell}\int_0^{2\pi}\int_0^{a}\frac{\mu I\rho}{2\pi a^2}\cdot\frac{I\rho}{2\pi a^2}\,\rho\,d\rho\,d\phi\,dz \\ &= \frac{\mu\ell}{8\pi}\,. \end{aligned} \tag{7.21a}$$

From (7.21a) we obtain the internal inductance per unit length as

$$\ell_{\mathrm{i}}^{\mathrm{dc}} = \frac{L_{\mathrm{i}}}{\ell} = \frac{\mu}{8\pi} \quad \mathrm{H/m}, \tag{7.21b}$$

and we note that it is independent of the length and radius of the wire.

With $\mu = \mu_0 = 4\pi \times 10^{-7}$, the internal inductance of the wire is $\ell_{\mathrm{i}}^{\mathrm{dc}} \simeq$ 50 nH/m or 1.27 nH/inch. The calculation assumes that the current in the wire is uniformly distributed across its cross section, which is true at dc (or zero frequency) and may be approximately true at low frequencies. At higher frequencies this is no longer the case, and the results must be modified for skin effects.

At higher frequencies the current is confined within a skin depth δ of the wire surface. We can use the surface impedance concept (see Section 3.6.7) for a conducting surface to obtain the internal inductive reactance in the present case as

$$\omega L_{\mathrm{i}}^{\mathrm{HF}} = \frac{\ell}{\omega}\,X_{\mathrm{s}}, \tag{7.22a}$$

where $\omega = 2\pi a$, $X_{\mathrm{s}} = 1/\sigma\delta$ is the surface reactance (inductive) of the conducting surface. From (7.22a) we obtain the internal inductance per unit length as

$$\begin{aligned} \ell_{\mathrm{i}}^{\mathrm{HF}} &= \frac{1}{2\pi a\sigma\delta} = \ell_{\mathrm{i}}^{\mathrm{dc}}\,\frac{2\delta}{a} \\ &= \frac{1}{4\pi a}\sqrt{\frac{\mu}{\pi\sigma}}\,\frac{1}{\sqrt{f}}\,. \end{aligned} \tag{7.22b}$$

A plot of ℓ_{i} versus a/δ is given in Figure 7.7, where it can be seen that the high-frequency internal inductance per unit length decreases asymptotically at a rate of 10 dB per decade of frequency, which of course assumes that $a \gg \delta$. Note that $\ell_{\mathrm{i}}^{\mathrm{dc}}$ is obtained for $a \ll \delta$. Again, we see that breakdown in the curve occurs at $a/\delta = 2$, meaning at the frequency f_{a} given by (7.20).

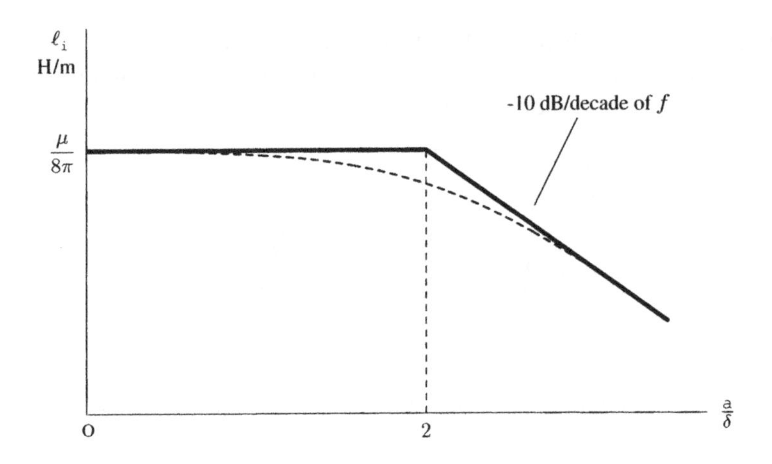

Figure 7.7 Variations of per unit length internal inductance of a solid wire versus a/δ.

7.5 EXTERNAL INDUCTANCE OF ROUND WIRE CONFIGURATIONS

In this section we discuss the external inductances of closed loops of arbitrary shape and made of conducting solid wires.

7.5.1 General Relations

Consider a loop of arbitrary shape carrying a filamentary current I as shown in Figure 7.8. In the figure it is clear that the magnetic flux generated by the current loop will link with itself. Therefore it is convenient to use the flux linkage concept to determine the self-inductance of the loop in the present case. The total flux linked with the current I is given by

$$\begin{aligned}\psi &= \int_S \mathbf{B} \cdot ds = \int_S \nabla \times \mathbf{A} \cdot ds \\ &= \oint_C \mathbf{A} \cdot d\boldsymbol{\ell}_1,\end{aligned} \tag{7.23}$$

where $\mathbf{B}$ is the magnetic flux density produced by I at the locations of ds on the surface S enclosed by the contour C of the loop, and $\mathbf{A}$ is the vector potential at the location of the current element $d\boldsymbol{\ell}_1$, produced by the entire current. It is given by

$$\mathbf{A} = \frac{\mu I}{4\pi} \oint_C \frac{d\boldsymbol{\ell}_2}{R_{12}}, \tag{7.24}$$

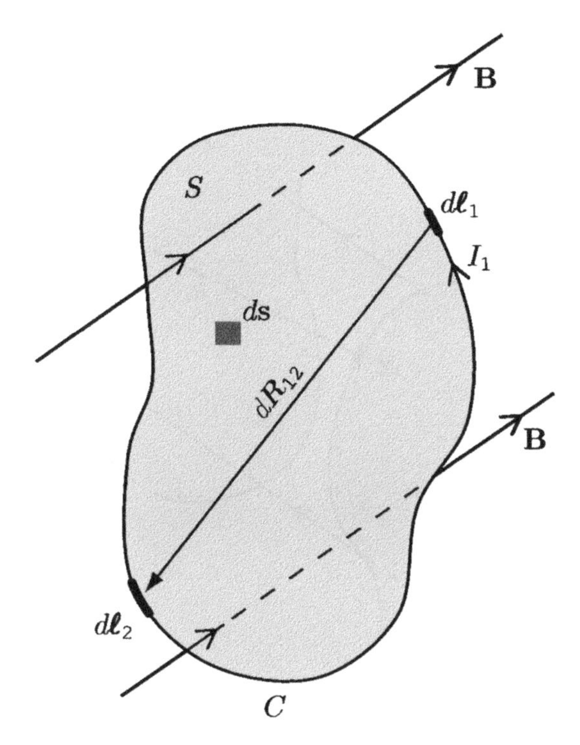

Figure 7.8 Loop of filamentary current I.

where $d\boldsymbol{\ell}_2$ is a current element on the same loop different from $d\boldsymbol{\ell}_1$ as indicated in Figure 7.8. Thus, combining (7.23) and (7.24), we obtain the following for the total flux linkage with I as

$$\begin{aligned} \psi &= \oint_C \mathbf{A} \cdot d\boldsymbol{\ell}_1 \\ &= \frac{\mu I}{4\pi} \oint_C \oint_C \frac{d\boldsymbol{\ell}_1 \cdot d\boldsymbol{\ell}_2}{R_{12}} . \end{aligned} \tag{7.25}$$

The self-inductance of the loop is now obtained by using (7.25) as

$$L = \frac{\psi}{I} = \frac{\mu}{4\pi} \oint_C \oint_C \frac{d\boldsymbol{\ell}_1 \cdot d\boldsymbol{\ell}_2}{R_{12}} . \tag{7.26}$$

With two filamentary current loops carrying currents I_1 and I_2, the flux generated by I_1 and linked with I_2 can be written as

$$\psi_{12} = \oint_{C_2} \mathbf{A}_1 \cdot d\boldsymbol{\ell}_2, \tag{7.27}$$

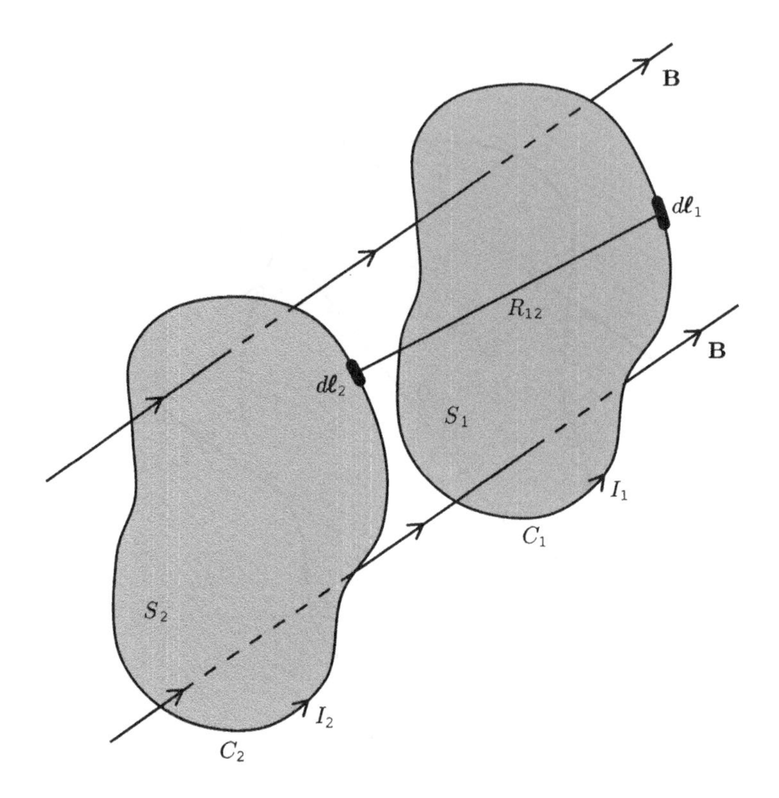

Figure 7.9 Two filamentary current loops.

where $\mathbf{A}_1$ is the vector potential produced by current I_1 at the location of $d\boldsymbol{\ell}_2$ on the loop 2 of contour C_2 and R_{12} is as shown in Figure 7.9. In the present case $\mathbf{A}_1$ is given by

$$\mathbf{A}_1 = \frac{\mu I_1}{4\pi} \oint_{C_1} \frac{d\boldsymbol{\ell}_1}{R_{12}} . \tag{7.28}$$

Combining (7.27) and (7.28), we obtain the flux generated by current I_1 and linked with current I_2 as

$$\psi_{12} = \frac{\mu I_1}{4\pi} \oint_{C_1} \oint_{C_2} \frac{d\boldsymbol{\ell}_1 \cdot d\boldsymbol{\ell}_2}{R_{12}} . \tag{7.29}$$

We now obtain from (7.29) the mutual inductance between loops 1 and 2 as

$$\begin{aligned} M_{12} = \frac{\psi_{12}}{I_1} &= \frac{\text{flux due to } I_1 \text{ linked with } I_2}{\text{current } I_1} \\ &= \frac{\mu}{4\pi} \oint_{C_1} \oint_{C_2} \frac{d\boldsymbol{\ell}_1 \cdot d\boldsymbol{\ell}_2}{R_{12}} . \end{aligned} \tag{7.30}$$

Note that when $C_1 = C_2$, $M_{11} = L_{11} = L$ = self-inductance of one loop given by (7.26). Also note that $M_{12} = M_{21}$, provided that the medium characterized by μ is reciprocal.

The expressions (7.26) and (7.30) are known as Neumann's relations for self- and mutual inductances of conducting loops. For arbitrary shape of the loops it is not possible to carry out the integrals analytically. In such cases the inductances can be obtained by numerical means. For both numerical and analytical use of Neumann's relations, it is important to observe that the parameter R_{12} can be zero in (7.26), whereas it is nonzero in (7.30). This implies that for self-inductance calculations the radius of the wire plays a significant role. Inductances of circular loops can be determined analytically as described in the next section.

7.5.2 Circular Loops

We will apply Neumann's relations to determine the mutual inductance of two parallel circular loops and apply the results to determine the self inductance of a single loop.

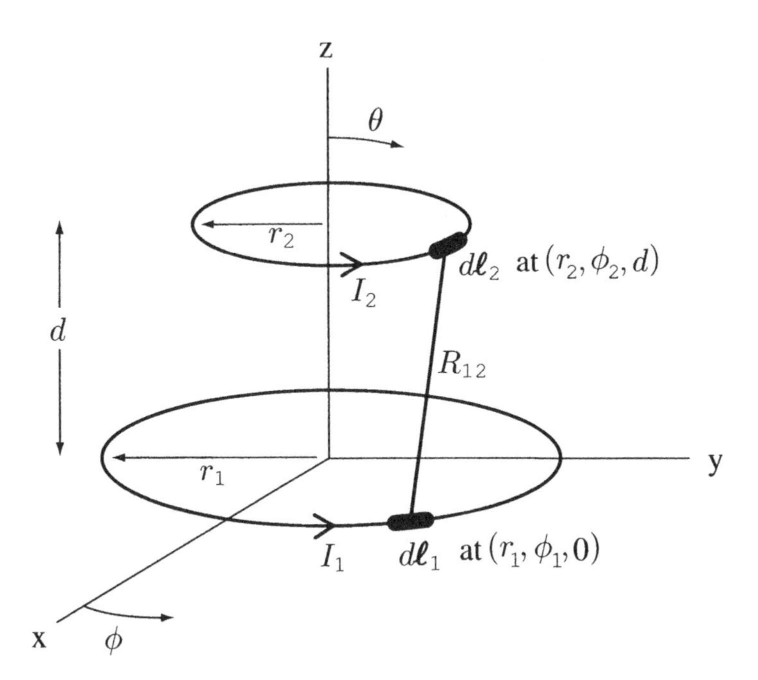

Figure 7.10 Two parallel filamentary current loops with the geometry and coordinate system.

Mutual Inductance of Two Loops

Consider two parallel circular loops separated by a distance d and oriented symmetrically along the z axis of a cylindrical coordinate system (r, ϕ, z) as shown in Figure 7.10. We will use (7.30) to determine M_{12}. It can be shown, by using the appropriate parameters given in Figure 7.10, that

$$d\boldsymbol{\ell}_1 \cdot d\boldsymbol{\ell}_2 = r_1 r_2 \cos(\phi_1 - \phi_2)\, d\phi_1\, d\phi_2$$

$$R_{12} = [r_1^2 + r_2^2 - 2r_1 r_2 \cos(\phi_1 - \phi_2) + d^2]^{1/2}.$$

Introducing the above, we obtain from (7.30)

$$M_{12} = \frac{\mu r_1 r_2}{4\pi} \int_0^{2\pi} \int_0^{2\pi} \frac{\cos(\phi_1 - \phi_2)\, d\phi_1\, d\phi_2}{[r_1^2 + r_2^2 - 2r_1 r_2 \cos(\phi_1 - \phi_2) + d^2]^{1/2}} \,. \tag{7.31}$$

Mutual inductance M_{12} given by (7.31) can be expressed in terms of known integrals. Since this is not obvious, we outline the key steps involved without going into the lengthy process.

We first use the substitution $\phi_1 - \phi_2 = \alpha$ in (7.31), which allows us to carry out the ϕ_2 integral and obtain

$$M_{12} = \frac{\mu r_1 r_2}{2} \int_0^{2\pi} \frac{\cos\alpha\, d\alpha}{[r_1^2 + r_0^2 - 2r_1 r_2 \cos\alpha + d^2]^{1/2}} \,. \tag{7.32}$$

We use another substitution in (7.32) as $\alpha = \pi - 2\beta$ and obtain

$$M_{12} = \frac{2\mu r_1 r_2}{[(r_1 + r_2)^2 + d^2]^{1/2}} \cdot \int_0^{\pi/2} \frac{(2\sin^2\beta - 1)}{\left[1 - \dfrac{4r_1 r_2}{(r_1 + r_2)^2 + d^2} \sin^2\beta\right]^{1/2}}\, d\beta$$

$$= \mu\sqrt{r_1 r_2}\, k \int_0^{\pi/2} \frac{(2\sin^2\beta - 1)}{\sqrt{1 - k^2 \sin^2\beta}}\, d\beta, \tag{7.33}$$

where

$$k^2 = \frac{4r_1 r_2}{(r_1 + r_2)^2 + d^2} \,. \tag{7.34}$$

After some algebraic manipulations, it can be shown that M_{12} between the two loops is given by

$$M_{12} = \mu\sqrt{r_1 r_2} \left[\left(\frac{2}{k} - k\right) K - \frac{2}{k} E\right], \tag{7.35}$$

where

$$K = K(k) = \int_0^{\pi/2} \frac{d\beta}{\sqrt{1 - k^2 \sin^2\beta}}\, d\beta \tag{7.36}$$

is the complete elliptic integral of the first kind [8] and

$$E = E(k) = \int_0^{\pi/2} \sqrt{1 - k^2 \sin^2 \beta}\, d\beta \tag{7.37}$$

is the complete elliptic integral of the second kind [8]. Complete elliptic integrals are discussed in tabular forms in [9, 10].

Self-inductance of a Circular Loop

Consider a conducting wire of radius a bent into a circular loop of mean radius r_0 and placed in a medium having a constant permeability μ, as shown in Figure 7.11.

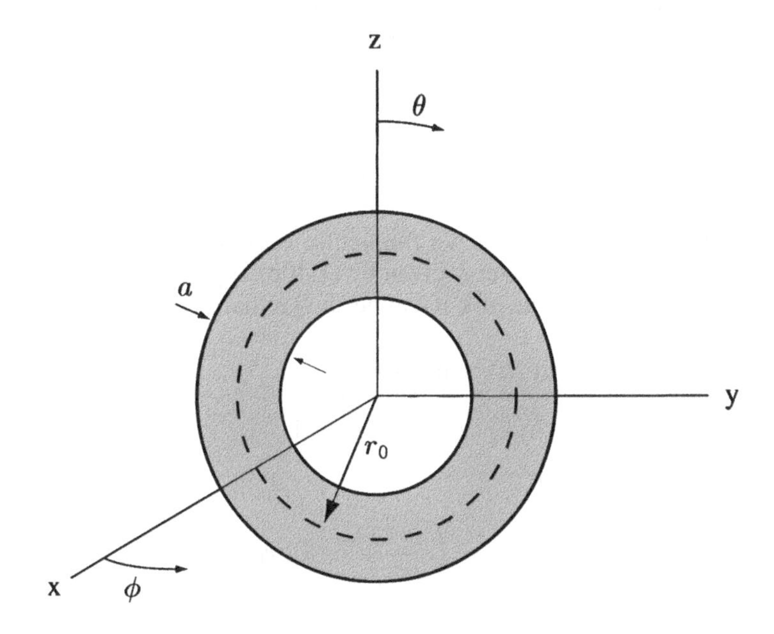

Figure 7.11 Loop of mean radius r made of conducting wire of radius a.

The total inductance L of the loop can be expressed as

$$L = L_{\text{i}} + L_{\text{ext}}, \tag{7.38}$$

where L_{i} is the internal inductance of the loop obtained from (7.21a) as

$$L_{\text{i}} = \mu_0 \frac{2\pi r_0}{8\pi} = \frac{\mu_0 r_0}{4} \quad \text{H}, \tag{7.39}$$

where μ_0 is the permeability of the conducting wire and L_{ext} is the external inductance of the loop to be determined. As was mentioned earlier, Neumann's relation (7.26) cannot be used directly to determine L_{ext} unless the finite radius of the loop wire is taken into account. We will modify Figure 7.10 by placing two filamentary current loops I_1 and I_2 in the x–y plane of Figure 7.11 identified as C_1 and C_2, respectively. The two current loops configuration can now be identified with those of Figure 7.11, provided that we make the following modifications:

$$\begin{aligned} r_1 &= r_0 - a \quad \text{for loop } C_1 \\ r_2 &= r_0 \quad\quad\;\; \text{for loop } C_2 \end{aligned} \tag{7.40a}$$

$$k^2 = \frac{4r_0(r_0 - a)}{(2r_0 - a)^2}\,. \tag{7.40b}$$

With use of (7.40a) and (7.40b) we obtain the M_{12} appropriate for the two-loop configurations of Figure 7.11 and identify it as its external inductance. Thus the external inductance of the circular loop is, from (7.35),

$$L_{\text{ext}} = \mu\sqrt{r_0(r_0 - a)}\left[\left(\frac{2}{k} - k\right)K - \frac{2}{k}E\right], \tag{7.41}$$

where k is given by (7.40b) and the complete elliptic integrals K, E are given by (7.36) and (7.37), respectively.

For practical cases of interest we frequently encounter situations where $r_0 \gg a$, for which the parameter k given by (7.40b) approaches the value of unity. With this limiting value of k it can be shown that

$$\begin{aligned} K(k) &\simeq \ln\frac{4}{\sqrt{1-k^2}} \\ E(k) &\simeq 1, \quad \text{as } k \to 1. \end{aligned} \tag{7.42a}$$

From (7.40b) it is found that

$$1 - k^2 = \frac{a^2}{(2r_0 - a)^2} \simeq \left(\frac{a}{2r_0}\right)^2 \quad \text{for } r_0 \gg a. \tag{7.42b}$$

Using the conditions (7.42a) and (7.42b), we obtain from (7.41)

$$L_{\text{ext}} \simeq \mu r_0\left[\ln\left(\frac{8r_0}{a}\right) - 2\right]. \tag{7.43}$$

Thus the total inductance of the circular loop is

$$\begin{aligned} L &= L_{\text{ext}} + L_{\text{i}} \\ &= \mu r_0\left[\ln\left(\frac{8r_0}{a}\right) - 1.75\right] \quad \text{H}. \end{aligned} \tag{7.44}$$

Equation (7.44) is commonly used to calculate the inductance of conducting loops.

7.6 INDUCTANCE OF STRAIGHT WIRES

Wire, inductance Practical lumped circuit components use metallic leads; similar leads are used to interconnect systems and components, and to transfer signals from one place to another. Ideal leads are lossless and reactance free. We have seen earlier how to determine the resistance and internal inductance of conducting wires.

In the present section we describe the determination of external inductance of finite lengths of conducting wires. Although Neumann's relations (7.26) and (7.30) can be used for this purpose, provided that the loop contours are appropriately modified, we describe here a slightly different method. External inductance of a finite length of wire can be approximately determined by considering it as a part of a closed loop consisting of a finite number of sections and then utilizing the concept of partial inductance of wires [7, 11, 12]. In the following we describe the basic steps involved in the determinations of partial inductances of conducting wires.

7.6.1 Partial Inductance

The external inductance of a closed loop C of filamentary current I shown in Figure 7.8 is given by

$$L_{\text{ext}} = \frac{\int_S \mathbf{B} \cdot d\mathbf{s}}{I} = \frac{\int_C \mathbf{A} \cdot d\boldsymbol{\ell}}{I}, \tag{7.45}$$

where $\mathbf{A}$ is the vector potential at the location of the current element $d\boldsymbol{\ell}$ chosen in the direction of the current, and the other notations are as explained in Figure 7.7. We will use the expression containing $\mathbf{A}$ in (7.45) to determine L_{ext}, although the expression containing $\mathbf{B}$ is also used for the same purpose [7, 12].

Assume that the contour C consists of N numbers of linear sections, each of length $\ell_{\rm i}$, so that we can write (7.45) as

$$L_{\text{ext}} = \sum_{i=1}^{N} \frac{\int_{\ell_{\rm i}} \mathbf{A}_{\rm i} \cdot d\boldsymbol{\ell}_{\rm i}}{I}, \tag{7.46}$$

where $\mathbf{A}_{\rm i}$ now represents the vector potential at the location of $d\boldsymbol{\ell}_{\rm i}$ produced by all sections of currents including the ith section. Because $\mathbf{A}_{\rm i}$ is the contributions from all the current sections, we have

$$\mathbf{A}_{\rm i} = \sum_{j=1}^{N} \mathbf{A}_{ij}, \tag{7.47}$$

where we interpret $\mathbf{A}_{ij}$ as the vector potential at the location of element $d\boldsymbol{\ell}_{\mathrm{i}}$ due to the current element j. Introducing (7.47) in (7.46), we obtain

$$L_{\text{ext}} = \sum_{i=1}^{N} \left(\sum_{j=1}^{N} \int_{\ell_{\mathrm{i}}} \frac{\mathbf{A}_{ij} \cdot d\boldsymbol{\ell}_{\mathrm{i}}}{I_{\mathrm{j}}} \right), \tag{7.48}$$

where we have identified $I = I_{\mathrm{j}}$, as the current in the jth element. Note that all elements carry the same current.

We now define the mutual partial inductance between the elements i and j by

$$L_{\mathrm{p}_{ij}} = \int_{\ell_{\mathrm{i}}} \frac{\mathbf{A}_{ij} \cdot d\boldsymbol{\ell}_{\mathrm{i}}}{I_j}, \tag{7.49}$$

and the partial self-inductance of the element of the element L_{ij} is obtained from (7.49) by assuming $i = j$ and taking into account the radius of the wire element (to be discussed later).

Finally we write the external inductance (7.48) of the ith current element of

$$L_{\text{ext}} = \sum_{i} L_i, \tag{7.50}$$

where L_i is the total inductance of the current element and is given by

$$L_{\mathrm{i}} = \sum_{j} \pm L_{\mathrm{p}_{ij}}, \tag{7.51a}$$

and $L_{\mathrm{p}_{ij}}$ is defined as the partial mutual inductance between the current elements i and j. The sign of each term is related to the orientation of the currents in the elements i and j. The total inductance of the original loop C can now be obtained from the partial inductances by using

$$L_{\text{loop}} = \sum_{i} L_i, \tag{7.51b}$$

where i indicates the summations over all the segments composing the original loop. Note that although the partial inductance concept was introduced for a closed loop, the partial self-inductance of a straight section so obtained can be used for the inductance of the section in isolation. It should be mentioned here that the reverse of (7.51b) is not true; that is, each partial inductance cannot be determined from a knowledge of L_{loop} alone [7].

Expression for $L_{\mathrm{p}_{ij}}$

Consider two parallel current elements of equal length l and carrying equal currents I in the same direction and oriented with respect to a coordinate system shown in Figure 7.12.

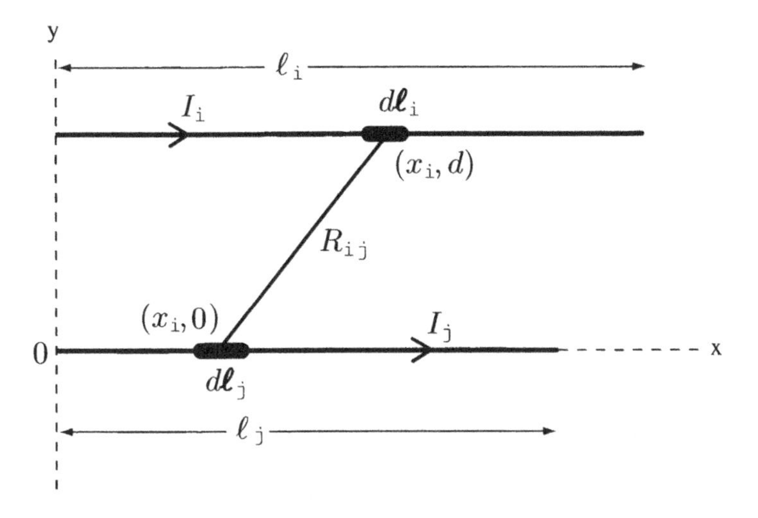

Figure 7.12 Two parallel current elements.

Note that for identification purpose, the current and length of the two elements are identified as I_i, ℓ_i and I_j, ℓ_j, respectively, although $I_i = I_j = I$ and $\ell_i = \ell_j = \ell$.

The vector potential at (x_i, d), meaning at the location of $d\boldsymbol{\ell}_i = \hat{x}\ dx_i$, produced by the current element j is

$$\mathbf{A}_{ij} = \frac{\hat{x}\mu I_j}{4\pi} \int_0^{\ell_j} \frac{dx_j}{R_{ij}}, \tag{7.52}$$

where μ is the permeability of the medium and

$$R_{ij} = [(x_i - x_j)^2 + d^2]^{1/2}. \tag{7.53}$$

Using (7.52) and (7.53) in (7.49), we obtain the following for the mutual partial inductance between the two elements i and j:

$$L_{\mathrm{p}_{ij}} = \frac{\mu}{4\pi} \int_0^{\ell_i} \left[\int_0^{\ell_j} \frac{dx_j}{R_{ij}}\ dx_j \right] dx_i. \tag{7.54}$$

The partial self-inductance for each element can be obtained from (7.54) by assuming $d = a$. Note that we have assumed $\ell_i = \ell_j = \ell$ in the expressions above.

It can be shown that (7.54) gives the following for the mutual partial inductances between two equal length and parallel elements separated by distance

d:

$$L_{\mathrm{p}_{ij}} = \frac{\mu}{2\pi} \ell \left[\ln \left\{ \left(\frac{\ell}{d} \right) + \sqrt{\left(\frac{\ell}{d} \right)^2 + 1} \right\} + \frac{d}{\ell} - \sqrt{1 + \left(\frac{d}{\ell} \right)^2} \right] . \quad (7.55)$$

The partial self-inductance of each element is now obtained from (7.55) by assuming $d = a$:

$$L_{\mathrm{p}_{ii}} = \frac{\mu}{2\pi} \ell \left[\ln \left\{ \frac{\ell}{a} + \sqrt{\left(\frac{\ell}{a} \right)^2 + 1} \right\} + \frac{a}{\ell} - \sqrt{1 + \left(\frac{a}{\ell} \right)^2} \right] . \quad (7.56)$$

For practical cases we can have: $d/\ell \ll 1$ or $\ell/d \gg 1$ and $\ell/a \gg 1$, and the following expressions are of sufficient accuracy:

$$L_{\mathrm{p}_{ij}} \simeq \frac{\mu}{2\pi} \ell \left[\ln \left(\frac{2\ell}{d} \right) - 1 \right], \quad \frac{d}{\ell} \ll 1 \quad (7.57a)$$

$$L_{\mathrm{p}_{ii}} \simeq \frac{\mu}{2\pi} \ell \left[\ln \left(\frac{2\ell}{a} \right) - 1 \right], \quad \frac{a}{\ell} \ll 1. \quad (7.57b)$$

7.6.2 Inductance of a Closed Rectangular Loop

As an application of the partial inductance concept, we consider a rectangular loop made of conducting wire of radians a and excited by voltage source V at one corner such that the loop carries a current I as shown in Figure 7.13. The loop directions are shown in Figure 7.13a. The partial inductances associated with side 1 and the instantaneous voltage along with its polarity across each side are as indicated in Figure 7.13b.

From circuit considerations we find from Figure 7.13b that

$$V_1 = L_{\mathrm{p}11} \frac{dI_1}{dt} + L_{\mathrm{p}12} \frac{dI_2}{dt} + L_{\mathrm{p}13} \frac{dI_3}{dt} + L_{\mathrm{p}14} \frac{dI_4}{dt} , \quad (7.58a)$$

where we have identified currents I_1, I_2, I_3, and I_4 as the currents on each side although they are same. Now, since the currents I_1, I_3 and $(I_2,\ I_4)$ are oppositely directed, we obtain from (7.58a)

$$V_1 = (L_{\mathrm{p}11} - L_{\mathrm{p}13}) \frac{dI}{dt} + (L_{\mathrm{p}12} - L_{\mathrm{p}14}) \frac{dI}{dt} . \quad (7.58b)$$

Since the currents I_1 and I_2, and I_3 and I_4 are orthogonal in space, we have in the present case $L_{\mathrm{p}12} = L_{\mathrm{p}14} \equiv 0$. We then obtain

$$V_1 = (L_{\mathrm{p}11} - L_{\mathrm{p}13}) \frac{dI}{dt} . \quad (7.59)$$

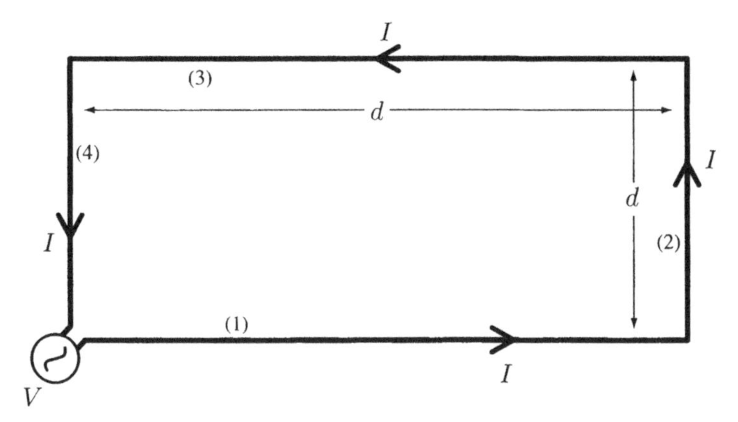

(a) Rectangular loop carrying a current I

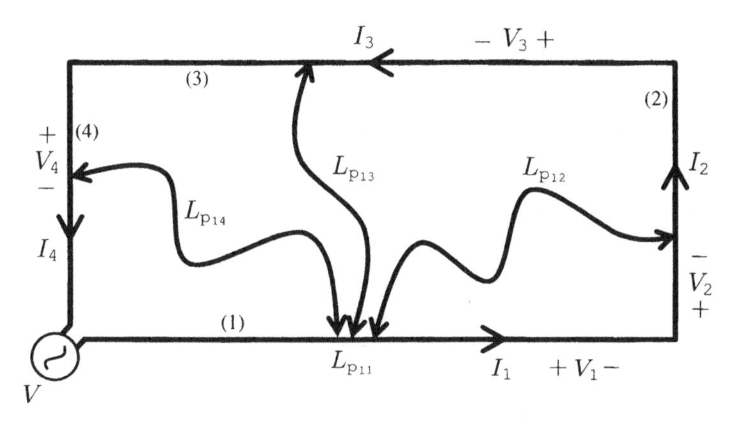

(b) Partial, self, and mutual inductances associated with side (1)

Figure 7.13 A rectangular loop and its inductance calculations.

Similarly we can find V_2, V_3, and V_4 and obtain

$$V = \sum_{i=1}^{4} V_i \tag{7.60}$$

for $i, j = 1, 2, 3, 4$. In the present case $L_{p_{ii}} = L_{p_{jj}}$ and $L_{p_{ij}} = L_{p_{ji}}$. Thus it can be shown from (7.60) that

$$\begin{aligned} V &= j2\omega[(L_{p_{11}} - L_{p_{13}}) - (L_{p_{22}} - L_{p_{24}})]I \\ &= j\omega L I, \end{aligned} \tag{7.61}$$

where we have assumed a harmonic time dependence (of angular frequency ω) for the current, and L is the total inductance of the rectangular loop. The partial inductances in (7.61) can be determined by using the appropriate equations (7.55)–(7.57). During calculations of $L_{\mathrm{p}_{ij}}$'s, the ℓ, d parameters must be appropriately identified with the sides of the rectangle involved.

For ℓ/d large, the loop inductance calculated by using the two-wire transmission line theory approximates closely that obtained from partial inductance theory. Further discussion and an example are given in [7].

7.7 OTHER CONFIGURATIONS

In addition to microstrip and strip lines described in Chapter 5, printed circuit board (PCB) lines are extensively used in electronic circuits. In this section we describe certain circuit aspects of PCB lines and also briefly describe how to determine certain circuit parameters of microstrip and strip lines. Detailed descriptions of these and other aspects of these guiding structures are given in [13–15].

7.7.1 Printed Circuit Board (PCB) Lines

Printed circuit boards are composed of dielectric substrates, typically of glass epoxy with $\epsilon_{\mathrm{r}} \simeq 4.7$, on which planar conducting strips or traces are etched. Figure 7.14 shows one trace on one side of the PCB.

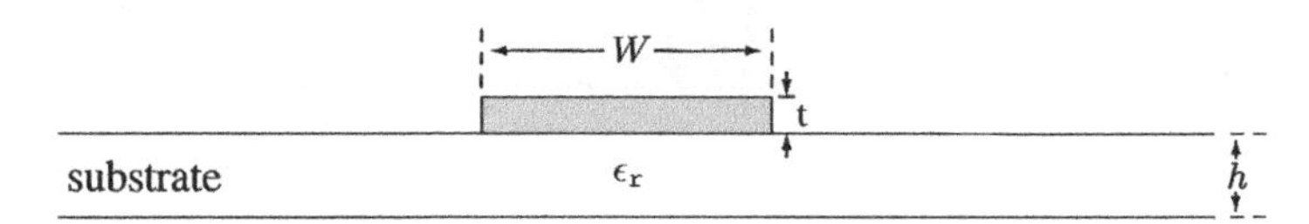

Figure 7.14 Cross section of a PCB line.

A good description of PCB lines and their circuit configurations along with pertinent design information is given in [13–16]; some old but useful information is also given in [7] and [17].

For purposes of discussion, we will assume that typical values of the geometrical parameters of a PCB line are (see Figure 7.14):

$$w = 10 \text{ mils}$$
$$h = 62 \text{ mils}$$
$$t = 3 \text{ mils.}$$

We assume that the trace material is copper. It is to be noted that the skin depth of copper $\delta_{\mathrm{cu}} = 2.6, 0.82$, and 0.26 mils at 1, 10, and 100 MHz, respectively.

The current distribution in the line behaves in a manner similar to that in conducting wires. For dc and low frequencies, RF currents are approximately uniform. At high frequencies, the skin effects confine the current within a skin depth δ of the trace surface which can be used to estimate its circuit parameters. However, because of the trace's rectangular geometry it is difficult to estimate them analytically. They are usually determined numerically.

The dc and low-frequency resistance of a trace is

$$r_{\mathrm{dc}} = r_{\mathrm{LF}} = \frac{1}{\sigma w t} \quad \Omega/\mathrm{m}, \tag{7.62}$$

where σ is the conductivity of the trace and w and t are as indicated in Figure 7.14.

For high frequency, the resistance obtained by making the skin effect approximations to the rectangular cross section is

$$r_{\mathrm{HF}} \simeq \frac{1}{2w\sigma\delta} \quad \Omega/\mathrm{m}. \tag{7.63}$$

Internal inductances of PCB traces are generally small compared with their external inductances at low frequencies.

The concept of partial inductance developed for wires in Section 7.6 applies to PCB lines in a similar manner. The partial self-inductance of a PCB line of length ℓ, trace width w and thickness t is discussed in [16]. The partial inductance of a PCB line with zero thickness is given in [7, 16] as

$$\begin{aligned} L_{\mathrm{p\ trace}} = \frac{\mu_0}{2\pi}\Bigg\{ & \ln(u + \sqrt{u^2+1}) + u \ln\left[\frac{1}{u} + \sqrt{\left(\frac{1}{u}\right)^2 + 1}\right] \\ & + \frac{u^2}{3} + \frac{1}{3u} - \frac{1}{3u}(u^2+1)^{3/2}\Bigg\}, \end{aligned} \tag{7.64}$$

where $u = \ell/w$. For trace lengths ℓ that are long compared with their width, $u = \ell/w \gg 1$. Under this condition (7.64) can be approximated by [7]:

$$\begin{aligned} L_{\mathrm{p\ trace}} &\simeq \frac{\mu_0}{2\pi}\,\ell\left(\ln 2u + 1 + \frac{1}{3u} - \frac{1}{2}\right) \\ &\simeq \frac{\mu_0}{2\pi}\,\ell\left(\ln\left(\frac{2\ell}{w}\right) + 1 + \frac{1}{2}\right). \end{aligned} \tag{7.65}$$

The partial mutual inductance between two parallel traces can be approximated as being that between two filamentary wires if the separation between the two traces is large [11].

7.7.2 Microstrip, Strip, and Coplanar Lines

Microstrip, strip, and coplanar lines using dielectric substrates or PCBs perform a variety of functions in electronic circuits. The first two lines were described in Chapter 5. The coplanar line consists of two conducting parallel strips or traces mounted on the same side of a dielectric substrate or PCB shown in Figure 7.15*a*.

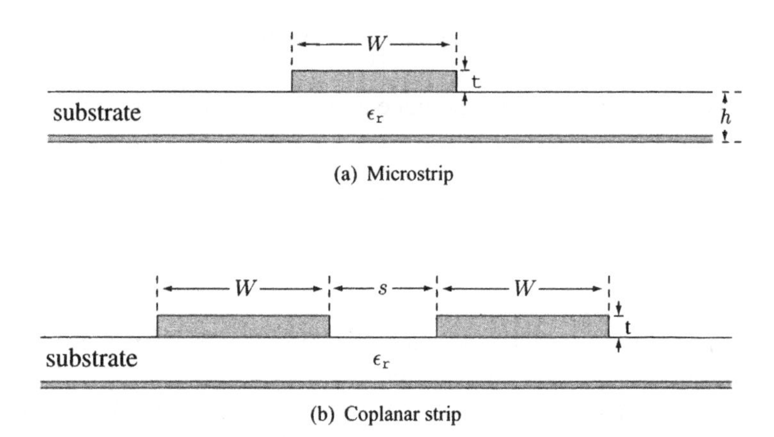

Figure 7.15 Cross-sectional views of strip type transmission lines.

It is not easy to determine analytically the circuit parameters for these lines. However, the analysis can be simplified considerably by considering the fact that these lines support predominantly TEM or quasi-TEM mode of wave propagation. Under this condition static analysis can be carried out to determine the capacitance C per unit length of the line. Thus, TEM considerations and the knowledge of C are sufficient to evaluate the desired circuit parameters. Analytical and numerical investigations of this class of transmission lines are described in [14,15] and compact descriptions of approximate expressions are given in [7, 13]. The characteristic impedance Z_0 and effective dielectric constant ϵ_r' for the microstrip and strip lines were given in Section 5.7.5.

Z_0 and ϵ_r' appropriate for the coplanar line shown in Figure 7.15*b* with trace thickness $t = 0$ are given by [7]

$$Z_0 = \frac{377}{\sqrt{\epsilon_r'}} \begin{cases} \dfrac{\frac{1}{\pi} \ln\left(2 \frac{1+\sqrt{k}}{1-\sqrt{k}}\right)}{\pi} & \text{for } \frac{1}{\sqrt{2}} \leq k < 1 \\ \dfrac{}{\ln\left(2 \dfrac{1+\sqrt{k'}}{1-\sqrt{k'}}\right)} & \text{for } 0 \leq k < \frac{1}{\sqrt{2}} \end{cases}, \qquad (7.66)$$

with

$$k' = \sqrt{1 - k^2} \tag{7.67a}$$

$$k = \frac{s}{s + 2w} \tag{7.67b}$$

and

$$\epsilon_r' = \frac{\epsilon_r + 1}{2} \left\{ \tanh \left[0.775 \ln \left(\frac{h}{w} \right) + 1.75 \right] + \frac{kw}{h} [0.04 - 0.7k + 0.01(1 - 0.1\epsilon_r)(0.25 + k)] \right\}. \tag{7.68}$$

Further discussion about other approximations is given in [7].

We summarize below the basic steps involved in the determination of the circuit parameters of the TEM or TEM-like circuits:

Step 1. Determine the static capacitance per unit length of the line with and without the dielectric present. This is done analytically and numerically. Thus obtain the following:

$$\frac{C}{C_0} = \epsilon_r', \tag{7.69}$$

where C is the per unit length capacitance in the presence of the dielectric substrate, C_0 is the per unit length capacitances in the absence of the dielectric substrate, and ϵ_r' is the effective dielectric constant of the substrate. Note that ϵ_r' is different from the ϵ_r of the substrate.

Step 2. Assuming TEM mode propagating in the structure and assuming that medium is characterized by μ, ϵ, we have

$$\ell_e C = \mu\epsilon, \tag{7.70}$$

where ℓ_e is the external inductance and C is the capacitance appropriate for the line. Since the permeability of the medium is μ_0 everywhere, and since the magnetic field is unaffected by the variations of ϵ, we obtain from (7.70)

$$\ell_e = \frac{\mu_0\epsilon_0}{C_0} = \frac{1}{v_0^2 C_0}, \tag{7.71}$$

where $v_0 = 1/\sqrt{\mu_0\epsilon_0}$is the velocity of light in free space. C_0 is as defined earlier.

Thus we have the information about the parameters ϵ_r', ℓ_e, and C.

Step 3. In the literature, generally approximate expressions are given about Z_c and ϵ_r' appropriate for a configuration. We know that

$$Z_e = \sqrt{\frac{\ell_e}{C}} \tag{7.72a}$$

$$v = \frac{1}{\sqrt{\ell_e C}} = \frac{v_0}{\sqrt{\epsilon_r'}}, \tag{7.72b}$$

from which ℓ_e and C can be determined by using the following:

$$\ell_e = \frac{Z_e}{v} \tag{7.73a}$$

$$C = \frac{1}{v^2 Z_c}. \tag{7.73b}$$

7.8 BEHAVIOR OF CIRCUIT ELEMENTS

Earlier we defined ideal passive circuit elements and their expected behavior. In practice, the fabrications procedure and the required leads for mounting them cause their behavior to detract from the ideal. For example, some fields may exist outside the component and the leads may introduce undesirable inductance, capacitance, and even resistance to a component regardless of whether it is designed to act as a resistor, inductor, or capacitor. Although PCB's use surface mount technology (SMT), such effects cannot be completely eliminated, and non-ideal behavior of a circuit element will occur at sufficiently high frequencies.

Generally, each component behaves ideally over a limited band of frequencies, outside of which it ceases to provide the behavior it is designed for. In the present section we describe the frequency behavior that a resistor, capacitor, and inductor have on an assumed equivalent circuit for each. Although such findings are theoretical, they are extremely helpful for their design, and for the estimations of their expected performance. Findings based on equivalent circuits are confirmed by experimental investigation as described in [7, 17].

7.8.1 Bode Plots

The behavior of passive components is usually judged by their impedance variation with frequency. In the EMC area it is of considerable interest to estimate the field intensities as functions of frequency. The envelope of the maximum variations rather than the finer details of their variations is often sufficient for such estimation. Moreover the range of frequencies involved may be so large that to visualize the variations it is found convenient to use logarithms for representation of such results. For example, instead of representing an impedance ($Z(\omega)$) plot as $|Z(\omega)|$ versus ω on linear scale, it is found convenient to plotu $|Z(\omega)|_{\text{dB}} = 20\log_{10}|Z(\omega)|$ versus $\log\omega$. If one uses a logarithmic representation, then we can use linear scale for $|Z(\omega)|_{\text{dB}}$, and the (relative) frequency ω can be identified on the logarithmic scale. Such a plot is called a *Bode plot*. The Bode plot is extensively used in circuit theory for representations of various results. Detailed description of Bode plots are given in [17] and textbooks on circuit theory.

We shall use Bode plots not only to represent the frequency behavior of passive components but also of many other parameters. For these reasons we will start with the basics of such plots by describing the representations of some selected simple complex functions:

$$\begin{aligned} Z(\omega) &= K \\ Z_1(\omega) &= (j\omega) \text{ or } (j\omega + 1) \\ Z_2(\omega) &= \frac{1}{j\omega} \text{ or } \frac{1}{(j\omega+1)}, \end{aligned} \tag{7.74}$$

where K is a real constant, $j = \sqrt{-1}$, and $\omega = 2\pi f$ may be identified with the radian frequency of interest. We wish to investigate the limiting (or asymptotic) behavior of $Z(\omega)$, $Z_1(\omega)$, and $Z_2(\omega)$ for all frequencies and their representation. We consider three special cases.

Case 1. Let $Z(\omega) = K$, which is a function independent of ω. Thus, assuming $K = |K|e^{j\phi}$, we have

$$|Z(\omega)|_{\text{dB}} = 20\log_{10}|K| = K_1 \text{ (say)} \tag{7.75a}$$

$$\phi = \arg Z(\omega). \tag{7.75b}$$

Now, depending on the value of the real constant K, it follows, from (7.75a) and (7.75b),

$$K_1 \text{ is negative for } |K| < 1$$
$$K_1 \text{ is positive for } |K| > 1$$

and

$$\begin{aligned} \phi &= 0 \quad \text{for } K \text{ positive} \\ &= \pi \quad \text{for } K \text{ negative.} \end{aligned}$$

Thus the Bode plots for $Z(\omega) = K$ are as shown in Figure 7.16.

The frequency can be identified as ω as a relative frequency.

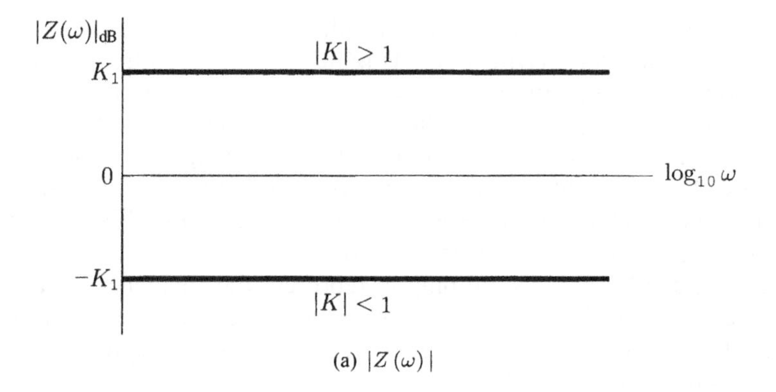

(a) $|Z(\omega)|$

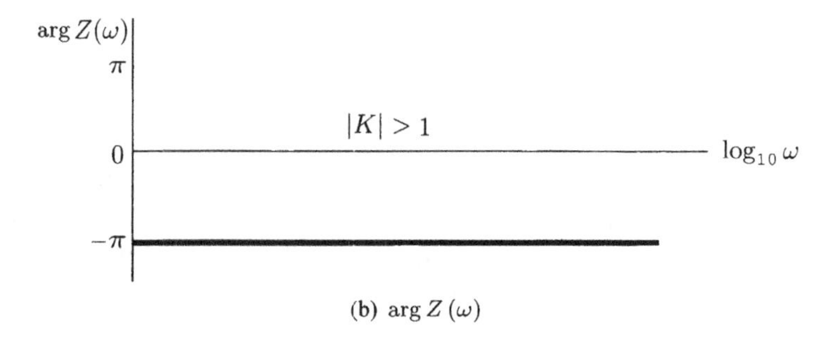

(b) $\arg Z(\omega)$

Figure 7.16 Bode plots for $Z(\omega) = K$.

Case 2. Assume $Z_1(\omega) = (j\omega)$ and $Z_2(\omega) = 1/j\omega$. We have

$$\begin{aligned} |Z_1(\omega)|_{\mathrm{dB}} &= 20\log_{10}\omega \\ \arg Z_1(\omega) &= \frac{\pi}{2} \end{aligned} \tag{7.76a}$$

and

$$\begin{aligned} |Z_2(\omega)|_{\mathrm{dB}} &= -20\log_{10}\omega \\ \arg Z_2(\omega) &= -\frac{\pi}{2}\,. \end{aligned} \tag{7.76b}$$

The Bode plots of (7.76a) and (7.76b) are shown in Figure 7.17.

Note that we have normalized the frequency scale with respect to an arbitrary frequency ω_0.

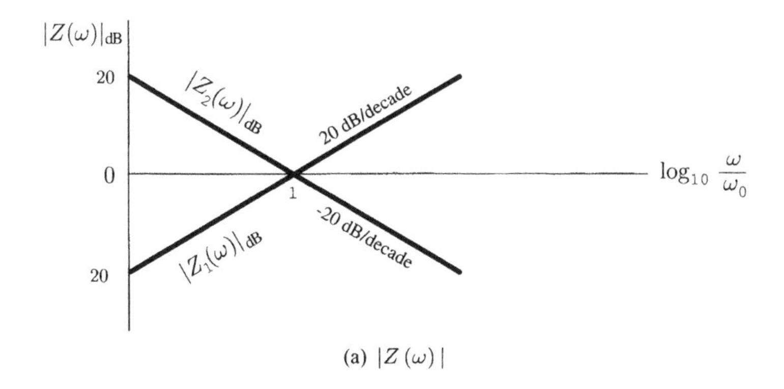

(a) $|Z(\omega)|$

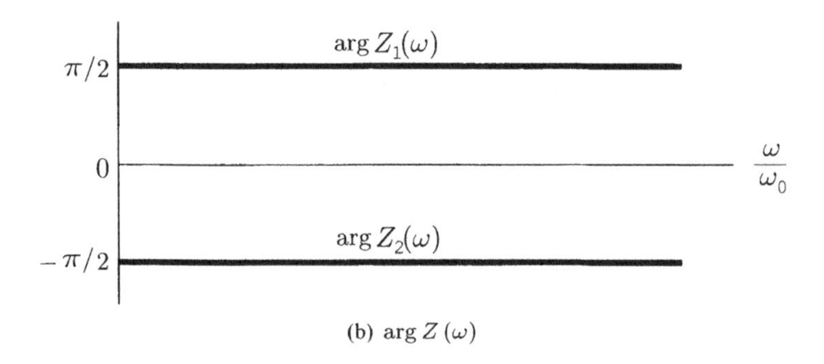

(b) $\arg Z(\omega)$

Figure 7.17 Bode plots for $Z_1(\omega) = (j\omega)$ and $Z_2(\omega) = \frac{1}{j\omega}$.

Case 3. Assume $Z_1(\omega) = (j\omega + 1)$ and $Z_2(\omega) = 1/j\omega + 1$. Here we consider only the amplitude variations. It can be shown that in the present case

$$|Z_1(\omega)|_{\text{dB}} = 20\log_{10}(\omega^2 + 1)^{1/2} \tag{7.77a}$$

$$|Z_2(\omega)|_{\text{dB}} = 20\log_{10}(\omega^2 + 1)^{-1/2}. \tag{7.77b}$$

From (7.77a) and (7.77b) we obtain the following limiting values:

$$\begin{aligned} |Z_1(\omega)|_{\text{dB}} &\simeq 0, && \text{for } \omega \ll 1 \\ &\simeq 20\log_{10}\omega, && \text{for } \omega \gg 1 \end{aligned} \tag{7.78a}$$

and

$$\begin{aligned} |Z_2(\omega)|_{\text{dB}} &\simeq 0, && \text{for } \omega \ll 1 \\ &= -20\log\omega, && \text{for } \omega \gg 1. \end{aligned} \tag{7.78b}$$

Thus in the present case the Bode plots for $Z_1(\omega)$ and $Z_2(\omega)$ are as shown in Figure 7.18.

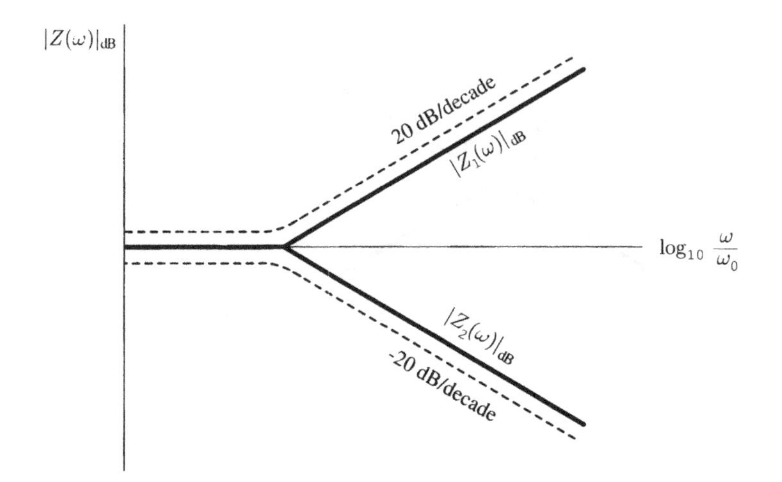

Figure 7.18 Bode plots for $Z_1(\omega) = j\omega + 1$ and $Z_2(\omega) = 1/j\omega + 1$.

Note that the actual variation limits of $|Z_1(\omega)|$, $|Z_2(\omega)|$ may be as shown by the dotted lines. The low- and high-frequency asymptotes meet at $\omega = \omega_0$, which may be defined as the breakpoint. For quick estimate of the magnitude variation of the response, the straight line approximation is an invaluable visual aid. We will have many occasions later to utilize such representations of parameter variations.

7.8.2 Resistors

A symbolic representation of a resistor R is as shown in Figure7.2. The element R ideally behaves as a dissipative element, meaning the current through it provides for the Joule (I^2R) loss. Non-ideal behavior comes about as a result of inherent inductance and capacitance of the package and the material used.

Resistors are available in various forms, such as tubular (high power), leaded, and surface mount (SMT); they are either wire wound, film, or compensation type. An equivalent circuit for a resistor is shown in Figure7.19, where R is the design value of the resistance, C is the equivalent shunt capacitance, and L is the equivalent inductance, which may include the lead inductance and the inductance from the element itself. The values of C and L depend on the resistor type and are described in [7] and [17], where the evaluation of the equivalent circuit shown in Figure7.19 is described.

For a typical SMT resistor, the shunt capacitance C is on the order of $0.1 - 0.5$ pF. The shunt capacitance can be of significance when R is high. The inductance is primarily due to the leads. The leads can contribute to the

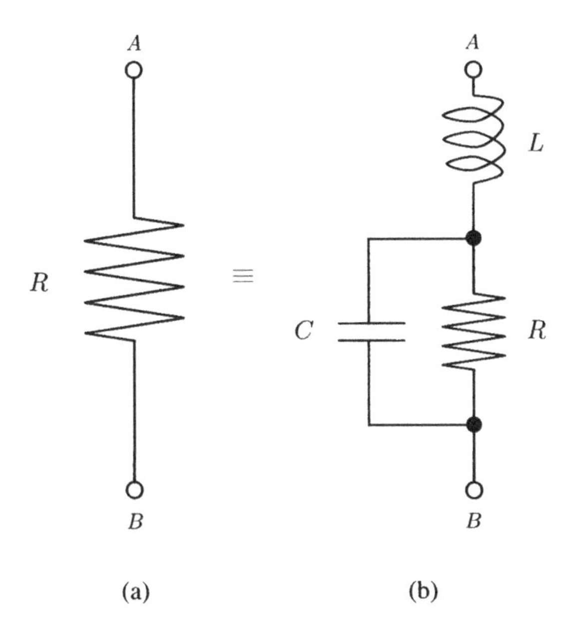

Figure 7.19 A resistor R (*a*) and its equivalent circuit (*b*).

shunt capacitance; also the body of the resistance can contribute in the form of leakage capacitance [7]. The impedance behavior of an ideal resistor $|Z| = R$ versus ω and $\arg Z$ versus ω are shown in Figure 7.20.

The ideal behavior shown in Figure 7.20 is achieved over a limited band of frequencies, not over all frequencies as shown. To investigate the frequency behavior of the resistor we obtain the impedance Z_{A-B} of the resistor by using the equivalent circuit in Figure 7.19 as

$$Z(\omega) = L\,\frac{\omega_2^2 - \omega^2 + j\omega\omega_1}{j\omega + \omega_1}, \tag{7.79}$$

where

$$\omega_1 = \frac{1}{RC}, \quad \omega_2 = \frac{1}{\sqrt{LC}}, \quad \text{and usually } \omega_1 \ll \omega_2.$$

Equation (7.79) indicates the following approximate behavior of $Z(\omega)$:

$$Z(\omega) \simeq R \quad \text{for } \omega \to 0 \tag{7.80a}$$

$$Z(\omega) \simeq \frac{1}{j\omega C} \text{for} \omega_1 \leq \omega \leq \omega_2 \tag{7.80b}$$

$$Z(\omega) \simeq j\omega L \text{ for } \omega \gg \omega_2. \tag{7.80c}$$

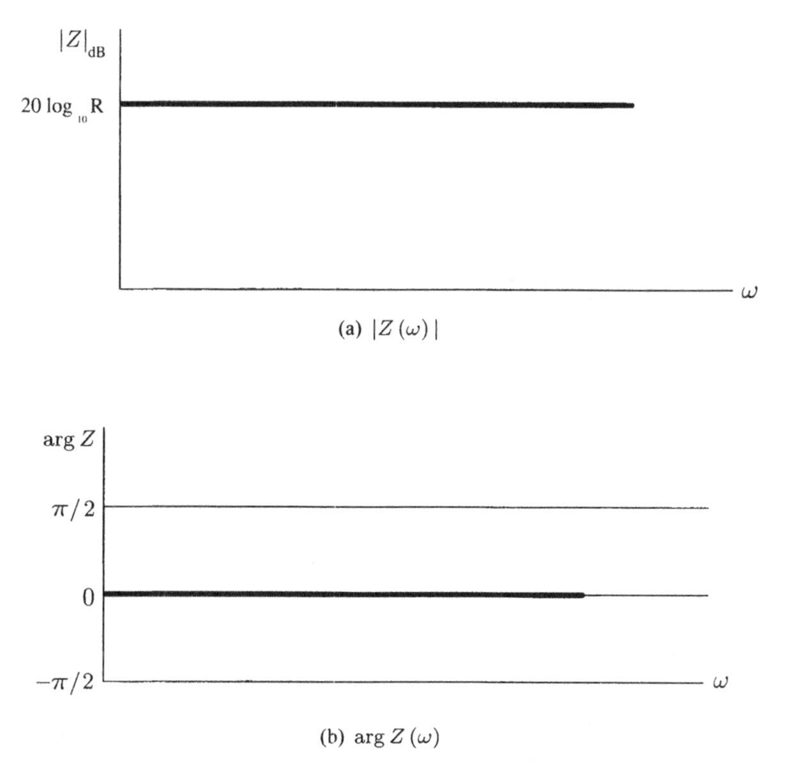

Figure 7.20 Behavior of an ideal resistor $Z(\omega) = R$: (*a*) $|Z(\omega)| = R$ versus ω and (*b*) $\arg Z(\omega)$ versus ω

In view of (7.80a)–(7.80c), the Bode plots for the resistor R whose impedance function obtained from its equivalent circuit is given by (7.79) are shown in Figure 7.21.

It is found from Figure 7.21 that the resistor behaves like an ideal resistance from dc to a frequency $\omega_1 = 1/RC$; it is capacitive in the frequency band $\omega_1 = 1/(RC) \le \omega \le 1/\sqrt{LC}$ and it is inductive for frequencies $\omega > \omega_2 = 1/\sqrt{LC}$. Such observations obtained from a study of the equivalent circuit are found useful for the design and fabrications of the resistor. Measured results confirming such findings are described in [7].

7.8.3 Capacitors

An ideal capacitor C shown symbolically in Figure 7.2 behaves as an element that stores electric energy within itself and does not produce any electric field outside. Again, its non-ideal behavior is caused by the inherent inductance of

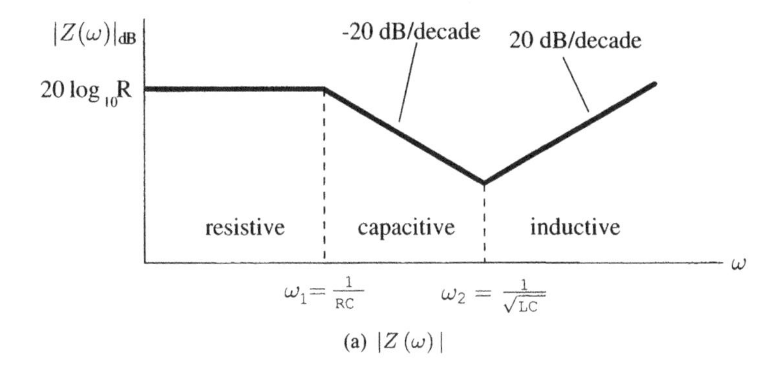

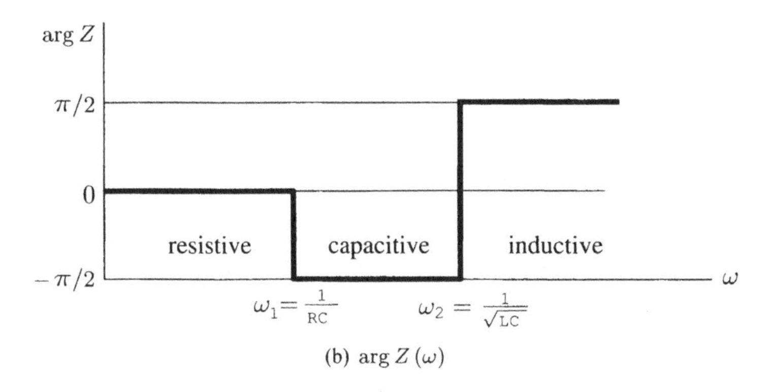

Figure 7.21 Behavior of a non-ideal resistor $Z(\omega) = R$: (*a*) $|Z(\omega)| = R$ versus ω and (*b*) arg $Z(\omega)$ versus ω

the package and dielectric losses. For example, in addition to the necessary leads the losses associated with the dielectric material used, and the finite conductivity of the plates can contribute significantly to its non-ideal behavior. Descriptions of various capacitors are given in [7,17]. An equivalent circuit for a capacitor C is shown in Figure 7.22, where C is the value of the capacitor and L is the lead (package) inductance, R_s is the equivalent series resistance (also abbreviated as ESR), representing the losses associated with the device [7]). Note that we have neglected the loss resistance associated with the dielectric; if need be, it should be placed in parallel with C in the equivalent circuit.

The impedance between the terminals of and ideal capacitor is

$$Z = \frac{1}{j\omega C} = \frac{1}{\omega C} \angle -\pi/2 , \tag{7.81}$$

which we can summarize by the plots shown in Figure 7.23.

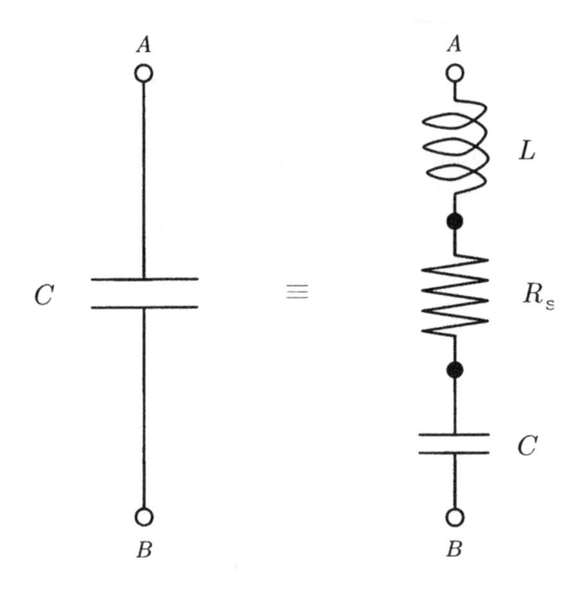

Figure 7.22 Equivalent circuit for a capacitor.

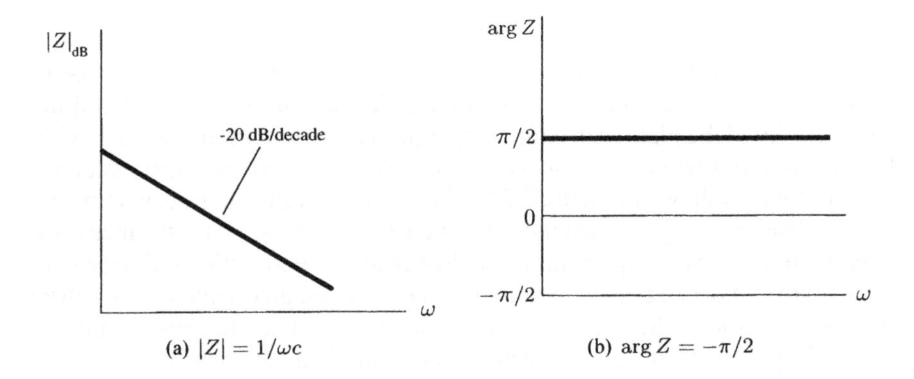

Figure 7.23 Behavior of an ideal C.

The impedance function $Z(\omega)$ appropriate for the capacitor can be obtained from its equivalent circuit shown in Figure 7.22 and is given by

$$Z(\omega) = L\,\frac{\omega_1^2 - \omega^2 + j\omega R_s/L}{j\omega}\,, \tag{7.82}$$

where

$$\omega_1 = \frac{1}{\sqrt{LC}} \tag{7.83}$$

and $R_s/L \ll 1$.

Equation (7.82) indicates the following asymptotic behavior of $Z(\omega)$:

$$Z(\omega) \simeq L\,\frac{\omega_1^2}{j\omega} = \frac{1}{j\omega C} \tag{7.84a}$$

for $\omega \ll \omega_1$

$$Z(\omega) \simeq L\,\frac{-\omega^2}{j\omega} = \frac{1}{j\omega L} \tag{7.84b}$$

for $\omega \gg \omega_1$.

Note that $Z(\omega) = R_s$ at $\omega = \omega_1$. Using (7.84), we obtain the Bode plots for the capacitor C as shown in Figure 7.24.

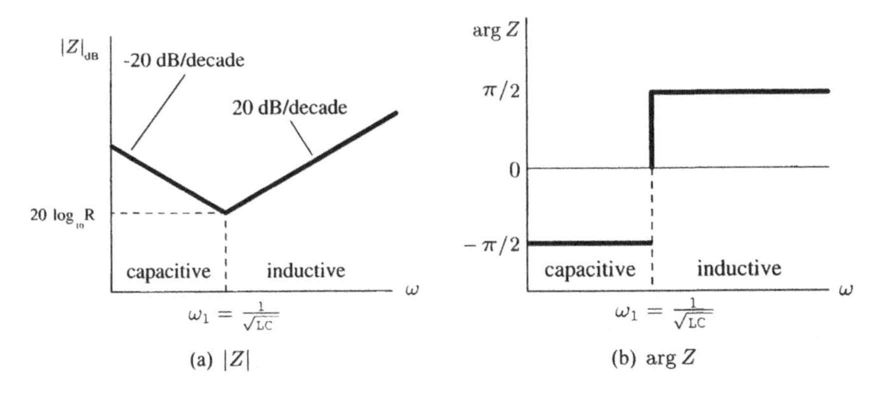

Figure 7.24 Bode plots for the capacitor C.

Thus we find that the capacitor C behaves like a capacitance within the range of frequencies $\omega < \omega_1$, where ω_1 is called the *self-resonant frequency* of the capacitor or the *maximum effective frequency* for a capacitor and is usually limited by the inductance of the leads and the capacitor itself. Descriptions of measured characteristics for typical capacitors are given in [7, 17].

7.8.4 Inductors

An inductor L shown symbolically in Figure 7.2 ideally acts as an element which stores magnetic energy within itself and does not produce any magnetic field outside its boundaries. Non-ideal behavior occurs as a result of packaging. They are leaded and SMT packages. Detailed development of the equivalent circuit for an inductor is given in [7, 17]. A satisfactory equivalent circuit is shown in Figure 7.25 where L is the value of the inductance, C is the

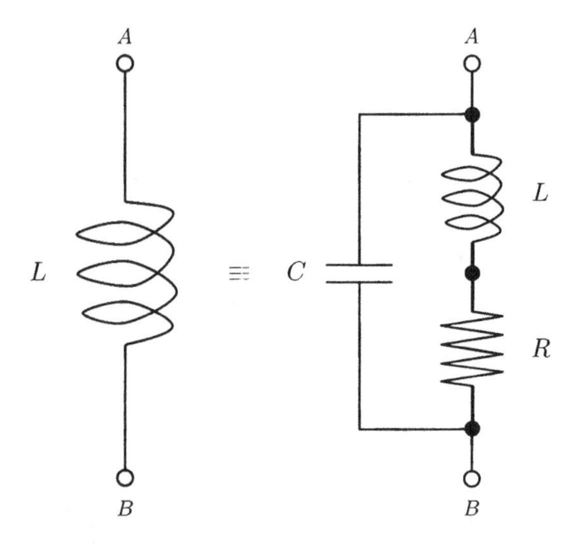

Figure 7.25 Equivalent circuit of an inductor.

parasitic capacitance between the windings of the inductor, and R represents the wire and magnetic material losses. We have assumed here that $L \gg L_{\text{lead}}$ and $C \gg C_{\text{lead}}$, where L_{lead} and C_{lead} are the inductance and capacitance, respectively, of the leads [7].

The impedance behavior of and ideal inductor is given by

$$Z(\omega) = j\omega L = \omega L \angle 90^{\circ}, \tag{7.85}$$

which is represented by the diagrams given in Figure 7.26.

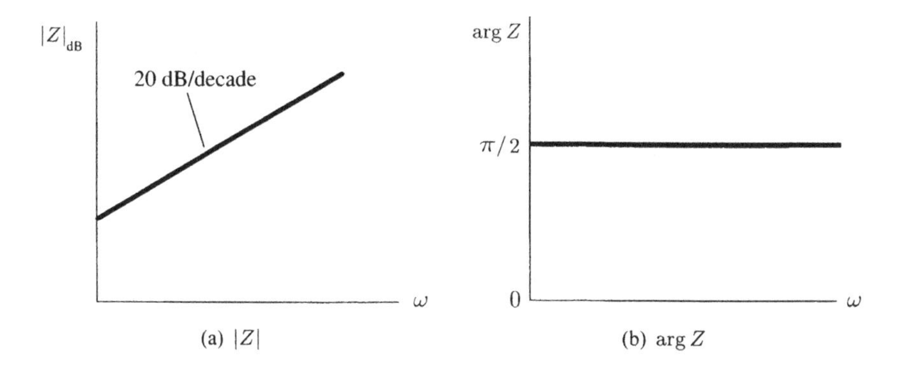

Figure 7.26 Impedance behavior of an ideal inductor.

The impedance function $Z(\omega)$ for the inductor obtained by using the equivalent circuit given in Figure 7.25 is

$$Z(\omega) = R \frac{1 + j\omega/\omega_1}{1 - \dfrac{\omega^2}{\omega_2^2} + j\omega RC}, \tag{7.86}$$

where

$$\omega_1 = \frac{R}{L}, \quad \omega_2 = \frac{1}{\sqrt{LC}} \tag{7.87}$$
$$\omega_1 \leq \omega_2.$$

The asymptotic behavior of $Z(\omega)$ as obtained from (7.86) can be expressed as

$$Z(\omega) \simeq R \tag{7.88a}$$

for $\omega \to 0$

$$Z(\omega) \simeq Rj \frac{\omega}{\omega_1} - j\omega L \tag{7.88b}$$

for $\omega_1 \leq \omega \leq \omega_2$

$$Z(\omega) \simeq Rj \frac{j\omega/\omega_1}{-\omega^2/\omega_2^2} \tag{7.88c}$$
$$= \frac{1}{j\omega C} \quad \text{for } \omega \gg \omega_2.$$

The Bode plots for the inductor are as shown in Figure 7.27.

From the plots in Figure 7.27 it is found that the dc resistance of the inductor dominates up to a frequency $\omega = R/L$. The value of R is generally very low,

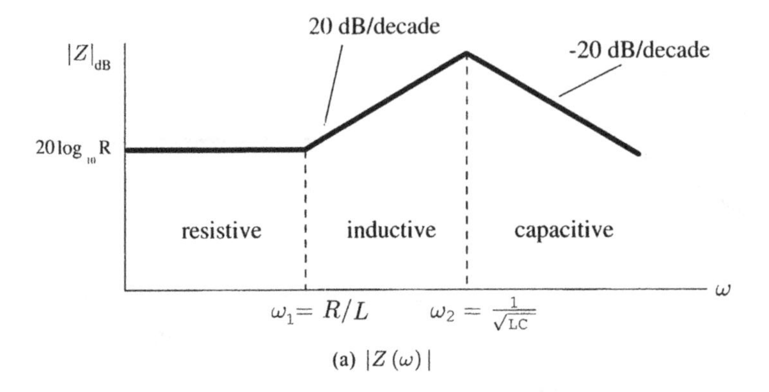

(a) $|Z(\omega)|$

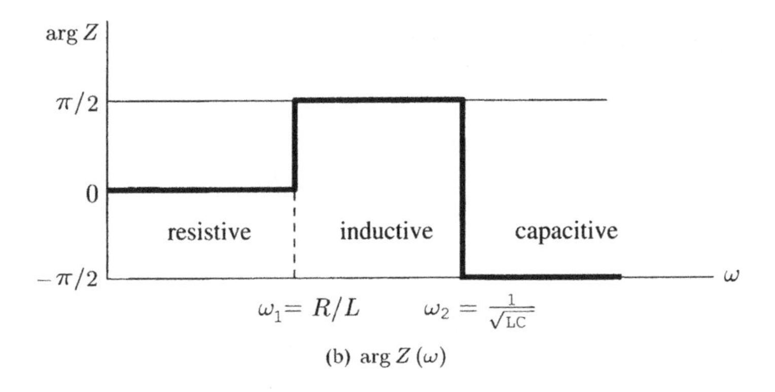

(b) arg $Z(\omega)$

Figure 7.27 Bode plots for an inductor.

typically 1 Ω or less. Thus ω_1 is very close to 0. As the frequency is increased, the inductor performs as it should up to a frequency $\omega_2 = 1/\sqrt{LC}$, called its *self-resonant frequency*, determined by the condition when the inductor reactance equals its parasitic capacitive reactance. At frequencies $\omega > \omega_2$ the inductor behaves like a capacitor, Further discussion of the experimental investigation confirming the results are given in [7].

Selected Topics Related to Inductors

Magnetic Core Inductors

Inductors are generally classified according to the type of cores on which they are wound. Most commonly used cores are as follows:

1. Air core or any non-magnetic core that fits into this group

2. Magnetic core, which can be subdivided depending on whether the core is open or closed as in a toroid.

A toroid, shown in Figure 7.28, using a high permeability magnetic material provides large inductance. Such an inductor is discussed in [5] and many other textbooks on electromagnetics.

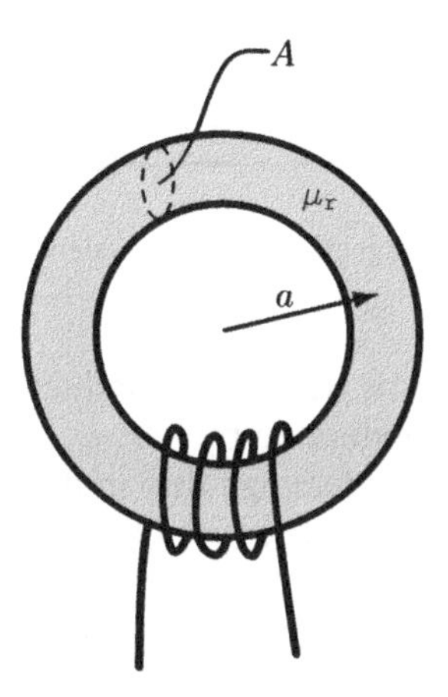

Figure 7.28 Toroidal inductor with ferromagnetic core. Cross section A and mean radius a.

It can be shown that the inductance of the toroid shown in Figure 7.28 is given by [1]

$$L = \frac{\mu_0 \mu_r N^2 A}{\ell} \quad \text{(H)}, \tag{7.89}$$

where N is the number of turns, μ_0 is the permeability of free space, μ_r is the relative permeability of the core material, $\ell = 2\pi a$ is the mean circumference, and A is the cross section of the core.

Equation (7.89) clearly indicates that with large μ_r, a high value of L can be achieved, hence the use of ferromagnetic materials. It should be noted that air core or open magnetic core inductors are most likely to cause interference, since their flux extends a considerable distance from the inductor. Inductors wound on closed magnetic core of high μ_r confine their flux mostly within. This may have implications for coupling and interference problems.

Magnetic Circuit Applications

Inductance configurations using ferromagnetic cores are used in a variety of core toroid inductance. It is appropriate here to briefly outline the basic steps involved in their determination by using the concepts of magnetic circuits which are described in [5, 7]. The magnetic circuit equivalent to the toroid in Figure 7.28 is shown in Figure 7.29 [5],

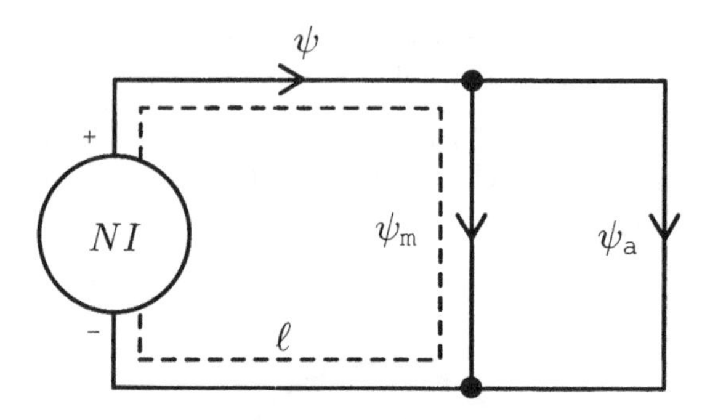

Figure 7.29 Magnetic circuit equivalent to the toroid.

where NI is the Ampère-turns used in the inductor (magnetomotive force), ψ is the total magnetic flux generated by NI, ψ_m is the magnetic flux within the inductor core, and ψ_a is the leakage flux in the air outside.

In analogy with a resistive electric circuit, the magnetomotive force (NI) is equivalent to the voltage applied applied to the electric circuit, and the magnetic flux is equivalent to the current in the electric circuit. In analogy with the resistance of the electric circuit, the *reluctance* $\mathcal{R}$ of the magnetic circuit is defined as [5]

$$\mathcal{R} = \frac{NI}{\pi}, \tag{7.90}$$

where

$$\mathcal{R} = \frac{\ell}{\mu A}, \tag{7.91}$$

ℓ being the length of the magnetic circuit (path) or the magnetic core, A the cross-section of the path, and μ the permeability of the path.

Note that the total path may consist of the magnetic core plus the air. Now, using the magnetic circuit shown in Figure 7.29, it can be shown that

$$\psi_m = \frac{\mathcal{R}_{air}}{\mathcal{R}_{air} + \mathcal{R}_m}, \tag{7.92}$$

where $\mathcal{R}_{air}$ and $\mathcal{R}_m$ are the reluctance of the air path and the magnetic path, respectively. Equations (7.91)–(7.92) indicate that $\mathcal{R}_{air} \ll \mathcal{R}_m$ because of high μ for the core, which implies that the flux is confined in the core. Thus, using (7.92) and (7.90), we obtain

$$\psi_m \simeq \psi = \frac{NI}{\mathcal{R}_m} = \frac{NI\mu A}{\ell}. \tag{7.93}$$

The flux linkage ψ_ℓ with the current I is then

$$\psi_\ell = N\psi_{\mathrm{m}} = \frac{N^2 I \mu A}{\ell} \tag{7.94}$$

from which we obtain the inductance L of the toroid in Figure 7.28 as

$$L = \frac{\psi_\ell}{I} = \frac{N^2 \mu A}{\ell}, \tag{7.95}$$

as given by (7.89).

It should be noted that for ferromagnetic materials the B–H curve is nonlinear, since μ_{r} of the material depends on the value of H (or the exciting current I). Thus the appropriate μ_{r} must be obtained from the slope of the B–H curve at the appropriate value of H (note $H = NI/\ell$ for the toroid). Generally, μ_{r} decreases with increasing value of I.

Common Mode Chokes

Ferrite core inductances or chokes are used to suppress undesirable common mode currents in electric circuits. Such mode currents will be described in greater detail elsewhere. Here we just introduce the concept for the purpose of this section.

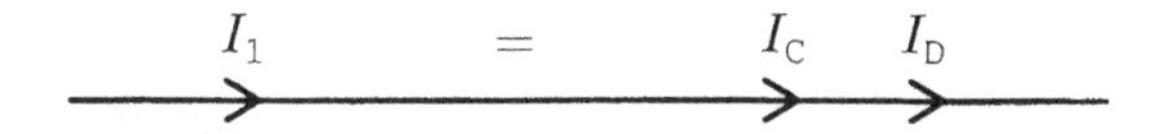

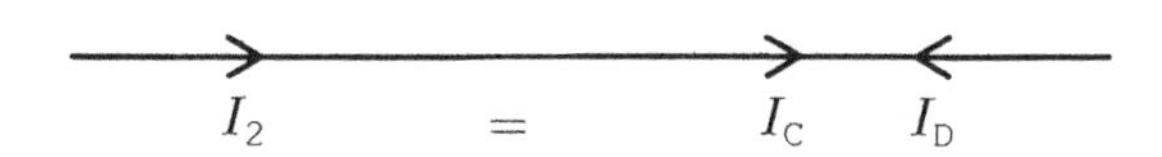

Figure 7.30 Two unequal parallel currents I_1, I_2 decomposed into common and differential currents.

Consider two parallel conducting wires carrying two unequal currents I_1 and I_2 as shown in Figure 7.30. As the figure shows, we can decompose the two currents as

$$I_1 = I_{\mathrm{C}} + I_{\mathrm{D}} \tag{7.96a}$$
$$I_2 = I_{\mathrm{C}} - I_{\mathrm{D}}. \tag{7.96b}$$

From (7.96a) and (7.96b) we obtain

$$I_D = \frac{1}{2}(I_1 - I_2) \tag{7.97a}$$

$$I_C = \frac{1}{2}(I_1 + I_2), \tag{7.97b}$$

where I_D, I_C are called the *differential* and *common mode currents*, respectively. In ordinary circuits we deal with only the differential mode currents (e.g., currents in a balanced two-wire transmission line in free space are in differential mode). Common mode currents occur in a circuit configuration as a result of unbalance in the system. In ordinary circuits common mode currents are undesirable. However, they are of importance in antennas, and this is why they are often referred to as the antenna mode currents.

We now describe how a common mode choke is used to suppress common mode currents in a two-wire line. A common mode choke [7] consists of a toroidal ferromagnetic core having two wires wound on it in a special manner as shown in Figure 7.31. The windings are such that the two currents I_1 and I_2 produce fluxes ψ_1 and ψ_2 in the core that add to each other as shown.

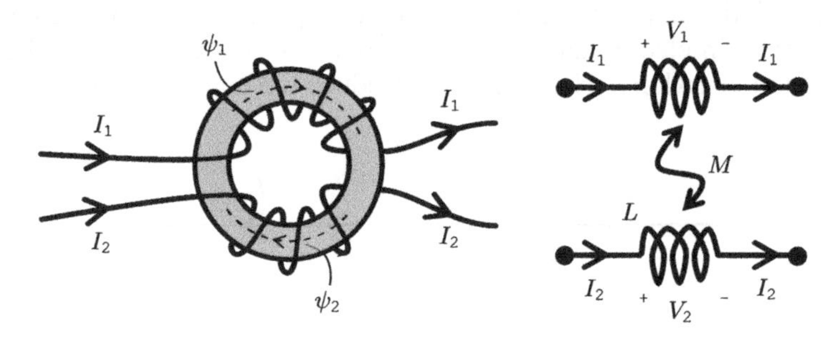

Figure 7.31 Common mode choke and equivalent circuit for the currents.

We assume that the inductance of each winding is L, meaning the windings are identical and the mutual inductance between them is M. Assuming that at any instant the voltages across the two winding are V_1 and V_2 as shown, we can write the impedance Z_1 of the winding carrying I_1 as

$$Z_1 = \frac{V_1}{I_1} = \frac{j\omega L I_1 + j\omega M I_2}{I_1}. \tag{7.98}$$

We now consider the contribution to the series impedance due to I_C and I_D separately.

For the common mode case we assume $I_1 = I_C$ and $I_2 = I_C$ and obtain the series impedance Z_{1C} for the common mode from (7.98) as

$$Z_{1C} = j\omega(L + M). \tag{7.99}$$

It can be shown that the common mode series impedance Z_{2C} of the winding carrying current I_2 will be the same as Z_{1C}.

For the differential mode series impedance Z_{1D} we assume $I_1 = I_D$ and $I_2 = -I_D$ and obtain from (7.99)

$$Z_{1D} = j\omega(L - M), \tag{7.100}$$

where it should be noted that in this case the current being oppositely directed the sign of M is negative in (7.100) and in this case also we have $Z_{1D} = Z_{2D}$.

Now, if the windings are symmetric and all flux remains in the winding, we can have $L = M$, and we obtain

$$Z_{1D} = Z_{2D} \equiv 0. \tag{7.101}$$

Thus in the ideal case, when $L = M$, a common mode choke has no effect on the differential mode currents on a two wire line, but it selectively provides an impedance $Z_{1C} = Z_{2C} = j2\omega L$ (from (7.99) with $L = M$) in series with the common mode currents. The fluxes due to the differential mode currents cancel in the core, which is an advantage. Further discussion of common mode choke is given in [7].

Ferrite Beads

Ferrite materials in the form called *beads* are used to provide selective suppression of high-frequency signals above in the range 10 kHz to 300 MHz. They are commonly used in the EMC area for high-frequency suppression by introducing series loss in the cable. The bead material is characterized by a complex permeability so that the inductor consisting of the bead material as its core has an equivalent circuit consisting of a resistance in series with an inductance. Both the resistive and inductive components of the bead impedance are frequency dependent. More descriptions of the ferrite beads and their applications are given in [7, 17].

REFERENCES

1. S. Ramo, J. R. Whinnery, and T. Van Duzer, *Electromagnetic Fields and Waves in Communication Electronics*, 3rd ed., Wiley, New York, 1994.
2. E. C. Jordan and K. G. Balmain, *Electromagnetic Waves and Radiating Systems*, 2nd ed., Prentice-Hall, Englewood Ciffs, NJ, 1968.
3. J. D. Kraus and D. A. Fleisch, *Electromagnetics with Applications*, 5th ed., WCB/McGraw-Hill, New York, 1999.
4. R. Plonsey and R. E. Collin, *Principles and Applications of Electromagnetic Fields*, McGraw-Hill, New York, 1961.

5. M. N. O. Sadiku, *Elements of Electromagnetics*, 2nd ed., Saunders/Harcourt Brace, Fort Worth, 1994.
6. ITT, *Reference Data for Radio Engineers*, 6th ed., Howard W. Sams, New York, 1975.
7. C. R. Paul, *Introduction to Electromagnetic Compatibility*, Wiley, New York, 1992.
8. H. B. Dwight, *Tables of Integrals*, 4th ed., Macmillan, New York, 1961.
9. M. Abramowitz and I. A. Stegun, eds., *Handbook of Mathematical Functions*, National Bureau of Standards, Applied Mathematics Series 55, Washington, DC, June 1964.
10. E. Jahnke and F. Emde, *Tables of Functions with Formulae and Curves*, 4th ed., Dover, New York, 1945.
11. A. E. Ruehli, "Inductance Calculations in a Complex Integrated Environment," *IBM J. Res. Dev.*, 16 (1972): 470–481, 1972.
12. F. W. Grover, *Inductance Calculations*, Dover, New York, 1946.
13. M. I. Montrose, *Printed Circuit Board Design Techniques for EMC Compliance*, IEEE Press, New York, 1996.
14. K. C. Gupta, R. Garg, and I. J. Bahl, *Microstrip Lines and Slot Lines*, Artech House, Dedham, MA, 1979.
15. K. C. Gupta, R. Garg, and R. Chadha, *Computer-Aided Design of Microwave Circuits*, Artech House, Dedham, MA, 1981.
16. C. Hoer and C. Love, "Exact inductance equations for rectangular conductors with application to more complicated geometries, *J. Res. Natl. Bur. Standards—C, Engineering Instrumentation*, 69C (1965): 127–137.
17. H. W. Ott, *Noise Reduction Techniques in Electronic Systems*, 2nd ed., Wiley, New York, 1988.
18. C. R. Paul, *Analysis of Linear Circuits*, McGraw-Hill, New York, 1989.

PROBLEMS

7.1 A solid metallic wire has length ℓ, cross section s, and conductivity σ. Determine its dc resistance R by applying **(a)** the field definition given by (7.7) and **(b)** the energy definition given by (7.11).

7.2 A parallel plate region is formed by a pair of conducting plates separated by a distance d. Each plate has surface area A, and the region is filled with a homogeneous medium with dielectric constant ε_r. Determine the capacitance C of the configuration by applying **(a)** the field definition (7.9) and **(b)** the energy definition (7.15). Point out the basic assumptions involved in the calculations.

7.3 Derive (7.18) using the surface impedance concept given by (3.118).

7.4 The dimensions for the solid and standard versions of wires AWG 10, 20, and 30 are as given in the text. Determine their dc resistance per unit length.

Answer: AWG 10: 0.0039 Ω/m (solid), 0.0032 Ω/m (standard); AWG 20: 0.033 Ω/m, 0.033 Ω/m; AWG 30: 0.340 Ω/m, 0.304 Ω/m.

7.5 The solid wire AWG 30 has a radius $\alpha = 5$ mils. **(a)** Determine the frequency at which its skin depth δ equals its radius α. **(b)** Determine its resistance per unit length at $f = 1$ MHz.

Answer: **(a)** 1 MHz; **(b)** 0.442 Ω/m, use $r_{\mathrm{HF}} \simeq 1/\pi\,(2a - \delta)\,\sigma\delta$ instead of $\alpha\gamma_{dc}/2\delta$. Why?

7.6 Determine the resistance per unit length for the solid wires of gauges AWG 10, 20, and 30 at 10 MHz.

Answer: AWG 10: 0.1025 Ω/m; AWG 20: 0.3220 Ω/m; AWG 30: 1.037 Ω/m.

7.7 Determine the internal inductance per unit length for the solid and standard versions of the wires of gauges AWG 10, 20, and 30.

Answer: AWG 10: 1.609 nH/m, 0.156 nH/m; AWG 20: 5.13 nH/m, 0.2 nH/m; AWG 30: 16.4 nH/m, 2.93 nH/m.

7.8 Determine the resistance, internal inductance and capacitance of a ribbon cable consisting of two 30 AWG 12.0 (7×30) wires, each 1 m in length, and separated by 50 mils. Assuming free space, determine its characteristic impedance. Assume $f = 100$ MHz.

Answer: 1.17 Ω, $\ell_{\mathrm{iHF}} = 13.00$ nH for each wire; $\ell_{\mathrm{e}} = 0.848\ \mu$H, $c_{\mathrm{e}} = 3.33$ pF, $z_0 = 2.558$ Ω.

7.9 An isolated metallic sphere in a homogeneous dielectric medium is charged to an electric potential of 1 MV. Calculate the minimum radius of the sphere such that dielectric breakdown does not occur for **(a)** air breakdown strength, $E_{\mathrm{B}} = 3$ MV/m, **(b)** oil, $E_{\mathrm{r}} = 2.3$, $E_{\mathrm{B}} = 15$ MV/m, and **(c)** mica, $E_{\mathrm{r}} = 5.4$, $E_{\mathrm{B}} = 200$ MV/m.

Answer: **(a)** 33.3 cm; **(b)** 6.7 cm; **(c)** 5 mm.

7.10 A small loop of metallic wire of radius a lies at a distance d above the center of a larger loop of radius b. The loops are oriented parallel to each other and the medium is free space. **(a)** Assume a current I flowing in the loop a. Determine the total flux ψ_{ab} passing through loop b. **(b)** Now assume that current I flows through the small loop b and the loop a is closed. Assuming loop b as a magnetic dipole, determine the flux ψ_{ba} passing through the loop a. **(c)** Show that $\psi_{ab} = \psi_{ba}$, and hence $M_{ab} = M_{ba} = M$. What is M?

Answer: **(a)** $\psi_{ab} = \frac{\mu_0 I a^2 b^2}{2(b^2+d^2)^{3/2}}$; **(b)** $\psi_{ba} = \psi_{ab}$; **(c)** $M_{ab} = M_{ba} = \frac{\psi_{ab}}{I} = \frac{\mu_0 \pi a^2 b^2}{2(b^2+d^2)^{3/2}} = M$.

7.11 Two parallel and conducting loops of radii r_1, r_2 and separated by distance d are configured in free space as in Figure 7.9. The exact mutual impedance M between the loops is given by (7.35) along with defining equations (7.34), (7.36), and (7.37). Using the assumptions $k \ll 1$, utilize the small argument expansions of the elliptic functions $K(k)$ and $E(k)$ given in [9] to show that

$$M \simeq \frac{\mu_0}{2} \left\{ d^2 + (r_1 + r_2)^2 \right\}^{1/2} \frac{\pi k^4}{16} .$$

7.12 In Problem 7.11, if $r_2 \ll d$, then it follows that $k \ll 1$. Determine the mutual impedance M for this configuration. Compare the result with that obtained in Problem 7.10(c) and comment.

Answer: $M = \frac{\mu_0 r_1^2 r_2^2}{2\left(d^2+r_1^2\right)^{3/2}}$.

7.13 A rectangular loop in Figure 7.12a is made of metal wire AWG 30 with the radius of the wire $a = 5$ mils. Assume that $\ell = 1$ inch and $d = 0.5$ inch. **(a)** Determine the appropriate partial self- and mutual inductances for the loop. Calculate the total loop inductance. **(b)** Determine the total inductance of the loop by using the transmission line analogy, (5.104), and comment on the result.

Answer: **(a)** $Lp_{11} = 23.36$ nH $= Lp_{33}$; $Lp_{22} = 10.92$ nH $= Lp_{44}$; $Lp_{13} = 4.19$ nH $= Lp_{31}$; $Lp_{24} = 0.62$ nH $= Lp_{42}$. $L_{\text{loop}} = 2\,(Lp_{11} - Lp_{12}) + 2\,(Lp_{22} - Lp24) = 5.890$ nH. **(b)** $le = 10.96 \ln(\frac{s}{a}) = 58.07$ nH. Agreement improves if the length $\ell > 1$ inch.

7.14 A coaxial line consists of an inner conductor with outer radius a and an outer conductor with inner radius b. The line is filled with a medium characterized by μ, ε. Determine the inductance L and the capacitance C per unit length of the line by using **(a)** the circuit definitions of L and C by using (7.66) and (7.6c), respectively, and **(b)** the energy definitions of L and C by using (7.13) and (7.15), respectively.

CHAPTER 8

RADIATED EMISSIONS AND SUSCEPTIBILITY

8.1 INTRODUCTION

Fields emitted by a digital electronic device or system are required to be of specified minimum intensity at selected distances from the device; it is also required that the device be not susceptible to degradation of its performance due to exposure to the electromagnetic environment. These requirements will be described at greater length in Chapter 12. In the present chapter we describe the radiated emissions from selected and identifiable components of an electronic device; the same components may also receive ambient fields, thereby making the device susceptible to malfunction. Although the radiated emission and susceptibility requirements for an electronic device are met by actual measurements, the theoretical results obtained by using simplified models will provide useful information for the initial circuit layout design of a system for EMC.

Applied Electromagnetics and Electromagnetic Compatibility. By D. L. Sengupta, V. V. Liepa
ISBN 0-471-16549-2

8.2 MAIN REQUIREMENTS

The compliance requirements for a device, meaning the limits of its radiated emissions and its susceptibility limits to the ambient fields, will be described in Chapter 12. Here we only briefly quote the limits sufficient for the present purpose. In the United States, the FCC requires that over the frequency range 30 MHz ($\lambda = 10$ m) to 40 GHz ($\lambda = 0.75$ cm) the radiated emissions by a digital device be monitored (measured) at distances of 3 and 10 m from the device under test (DUT) for Class B (device used for residential use) and Class A (device used for commercial and industrial applications) devices, respectively, so as to ascertain that the emitted fields are less than some specified limits. Similarly the European organization CISPR recommends measurements over 30 MHz ($\lambda = 10$ m) to 1 GHz ($\lambda = 30$ cm) at a distance of 10 m each for the Class B and Class A devices, respectively. It should be noted that at the specified distances from the DUT, the test antenna may not be in the far zone of the DUT throughout the range of frequencies of interest. With this perspective we describe the far field radiated emissions by some selected circuit parts of a digital electronic device.

8.3 EMISSIONS FROM LINEAR ELEMENTS

In Chapter 6 we described the radiation from a variety of antennas. Here in the context of compliance requirements we further describe the radiation from finite lengths of conducting wires of circular cross section carrying currents of harmonic frequency $\omega = 2\pi f$. Although we consider wires of circular cross sections, the results can be applied to linear sections PCB traces with rectangular cross section provided that we identify a conductive strip of width W and thickness t ($W \gg t$) with a conducting wire of radius a such that $W = 4a$ [1]. We will consider the emissions from a current element of length ℓ, small compared with a wavelength and which carries a current $Ie^{j\omega t}$ with $\omega = 2\pi f$.

The complete electric fields radiated by a small current element have been discussed in Section 6.3. We shall use the same geometry shown in Figure 6.2. except that for convenience we use the z-directed current element as $I\ell\hat{z}$ instead of $I\ d\ell\hat{z}$. The far zone electric field produced by the element is given by (6.35a). We can write the amplitude of the maximum electric field, produced in the broad side direction $\theta = \pi/2$, as

$$E_{\max} = \frac{\beta\eta I\ell}{4\pi r}, \tag{8.1}$$

where all the parameters are as defined earlier. Assuming free space, we obtain

$$E_{\max} = 2\pi \times 10^{-7} \frac{I\ell}{r} f, \tag{8.2}$$

which gives the maximum electric field amplitude produced by a short current element (i.e., $\ell \ll \lambda$). Equation (8.2) can be used to estimate the maximum field emitted by a small wire in a circuit and compare it with FCC requirements. For example, it can be shown by using (8.2) that the emissions from a current I carried by an element of length ℓ would meet the FCC requirement of 40 dBμv ($=$ 100 μv/m) at a distance of 3 m from the current element and over the frequency range 30 to 88 MHz provided that I satisfies the following

$$I \le I_{\max} = \frac{3}{2\pi f_{\mathrm{MHz}} \ell_{\mathrm{m}}}, \tag{8.3}$$

where I or $I_{\max}$ is in mA, f_{MHz} is the frequency in MHz, $\ell_{\mathrm{m}} = \ell$ is length of the current element in meters. It is to be noted that (8.3) should be used when 3 m is outside the far zone distance of the element of length ℓ. Note that (8.2) gives the maximum far field amplitude produced by a sinusoidal current (of length ℓ) of frequency f.

In the area of EMC we are interested in fields produced by nonsinusoidal signals, for example, a clock signal having trapezoidal waveform discussed in Chapter 4. Thus, using (8.2), it can be shown that the amplitude of the maximum field emitted by the element carrying a nonsinusoidal current is given by [2–4]

$$E_{\mathrm{c}}(f) = E_{\max}(f)\; S(f), \tag{8.4}$$

where $E_{\mathrm{c}}(f)$ is the amplitude of the maximum field emitted by the clock signal current, $E_{\max}(f)$ is the amplitude of the maximum field emitted by a sinusoidal current of frequency f, and $S(f)$ is the spectral intensity of the waveform discussed in Chapter 4. $E_{\max}(f)$ is given by (8.2), and it indicates that it is linearly proportional to the frequency f. The spectral intensity $S(f)$ of a similar signal was described earlier in Section 4.4.2.

The Bode plots for $E_{\mathrm{c}}(f)$ pertinent to a clock signal can now be obtained by adding the corresponding plots for $E_{\max}(f)$ and $S(f)$, as shown in Figure 8.1, where we have shown the details of the clock signal waveform in the inset.

It is clear from Figure 8.1 that the maximum occurs in the range of frequencies $1/\pi\tau \le f \le 1/\pi\tau_{\mathrm{r}}$ determined by the choice of τ and τ_{r} of the clock signal waveform. The emissions at $f > 1/\pi\tau_{\mathrm{r}}$ and $f > 1/\pi\tau$ may be neglected.

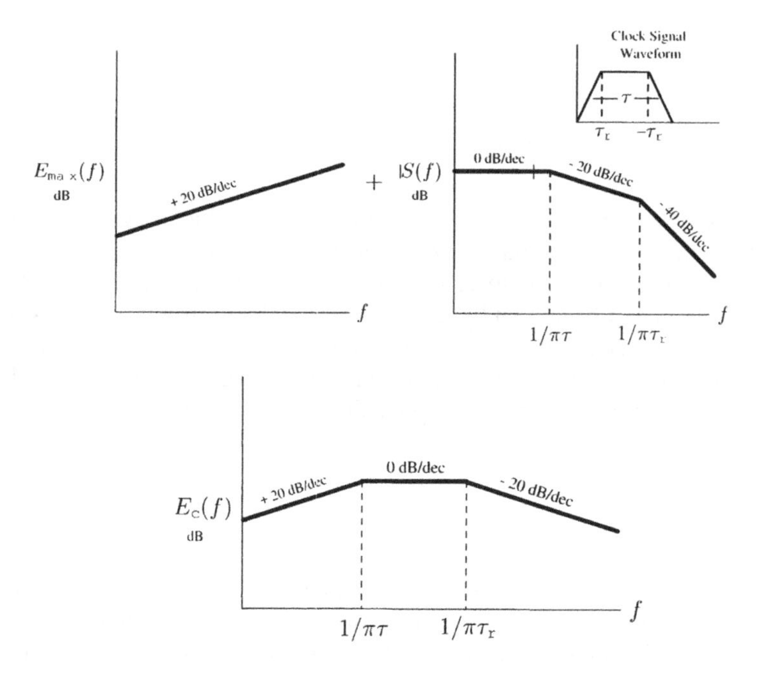

Figure 8.1 Bode plots for the emissions by a linear wire element carrying trapezoidal signal currents.

8.4 TWO PARALLEL CURRENTS

8.4.1 Introduction

Two parallel currents I_1, I_2 passing through two parallel conducting wires and separated by a distance small compared to a wave length occur in electronic circuits. A small length of such configuration may produce significant emissions. As sketched in Figure 7.29, the currents I_1, I_2 can be split into common mode and differential mode currents in the two wires considered together as given by (7.97b) and (7.97a). For the purpose of evaluating the radiation from I_1, I_2 we consider the configuration to be a combination of combination of common and differential mode currents shown in Figure 8.2.

From the far zone electric field patterns of linear current elements discussed in Section 6.3.2, it can be seen that if $d \ll \lambda$ the common mode currents in Figure 8.2*b* will produce electric fields that add and the differential mode currents in Figure 8.2*c* would subtract in the broad side directions. For example, at a far field point P the common mode currents will produce maximum and the differential mode currents will produce minimum (or negligible) radiated fields. This is generally true if the distance between the current elements are

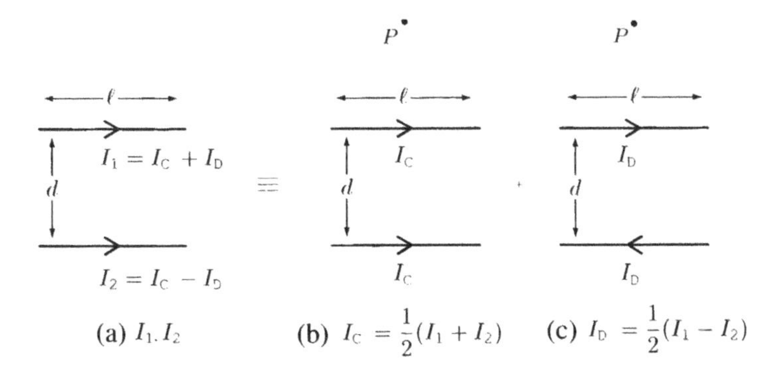

Figure 8.2 Two parallel currents I_1, I_2 (with $I_1 > I_2$) as a combination of similar configurations of common and differential mode currents: (*a*) actual currents; (*b*) common mode; (*c*) differential mode.

small compared to the operating wavelength. In the following sections we investigate in more detail the radiated emissions from such current distributions.

8.4.2 Two Parallel Currents

Consider two parallel current elements $I_1\ d\boldsymbol{\ell} = I_1\ d\ell\ e^{j\alpha}\hat{z}$ and $I_2\ d\boldsymbol{\ell} = I_2\ d\ell\ e^{j\alpha}\hat{z}$ placed in the xz plane and along the x axis at $x = a$ and $x = -a$, respectively, as in Figure 8.3 (i.e., we have assumed $\alpha_1 = 0$ and $\alpha_2 = \alpha$).

Using the results of antenna array theory described in Section 6.10, and in particular applying (6.186) to the configuration in Figure 8.3, we obtain the far electric field $\mathbf{E}(P)$ at any field point $P(r, \theta, \phi)$ produced by the two currents as

$$\mathbf{E}(P) = \hat{\theta}\, K\, \frac{e^{-j\beta_0 r}}{r}\, f(\theta) \times \left[I_1 e^{j\beta a \sin\theta\cos\phi} + I_2 e^{-j(\beta a \sin\theta - \alpha)} \right], \tag{8.5}$$

where $f(\theta) = \sin\theta$ for a short current element

$$K = \frac{j\beta_0\eta_0\, d\ell}{4\pi}, \quad \text{assuming free space.}$$

Assuming $I_1 = I_2 = I$, we obtain from (8.5)

$$\mathbf{E}(p) = \hat{\theta} 2K\, \frac{e^{-j\beta_0 r}}{r}\, I e^{j\alpha/2} \sin\theta \times \cos\left[\beta a \sin\theta\cos\phi - \frac{\alpha}{2} \right], \tag{8.6}$$

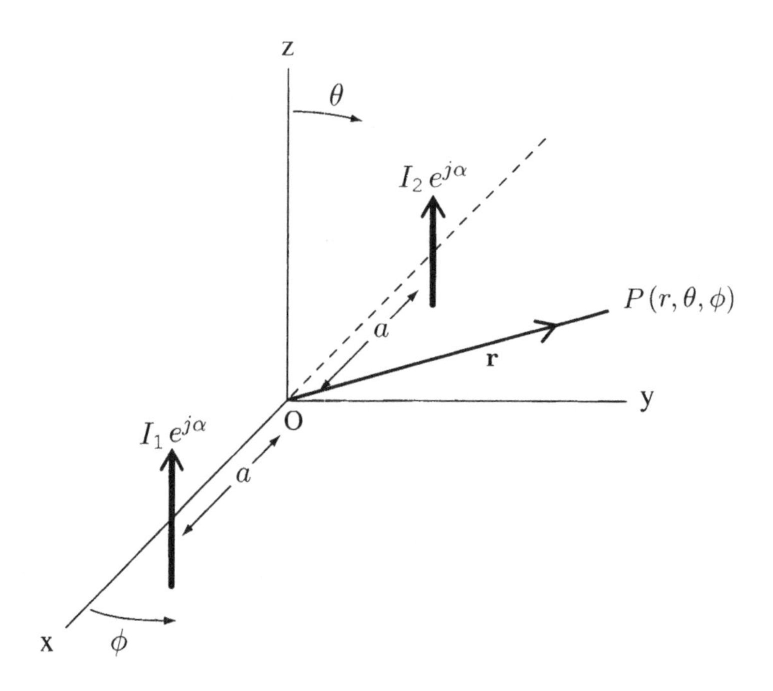

Figure 8.3 Two parallel currents oriented in the z direction and placed in the xz plane.

which gives the far electric field at any far field point P produced by the currents I, $Ie^{j\alpha}$. We now specialize (8.6) to the differential and common mode configuration of currents.

Differential Mode Emissions

For differential mode currents we assume $I = I_D$ and $\alpha = \pi$, and obtain from (8.6) the total field at a far field point $P(r, \theta, \phi)$:

$$E^D(p) = -4\pi \times 10^{-7} I_D(f\ d\ell)\ \frac{e^{-j\beta_0 r}}{r} \sin\theta \sin(\beta_0 a \sin\theta\cos\phi), \quad (8.7)$$

which indicates that maximum electric field occurs at $\theta = \pi/2$ and $\phi = 0$ or π.

Since the spacing between the currents is small compared to the operating wavelength, we can approximate $\sin(\beta_0 a \cos\phi) \simeq \beta a \cos\phi$, and obtain the from (8.7) the following for the magnitude of the maximum radiated electric field at a distance r:

$$|E^D|_{\max} = 4\pi \times 10^{-7} (f\ d\ell)\ \frac{\beta_0 a}{r}\ |I_D|, \quad (8.8)$$

which can be expressed as

$$|E^D|_{\max} = 1.316 \times 10^{-14} \frac{|I_D|}{r} f^2 A \qquad \text{V/m}, \tag{8.9}$$

where $A = 2a\, d\ell$ = cross-sectional area of a loop of dimension $\ell \times 2a$. Note that (8.9) indicates that the maximum radiated emission from the differential current configuration increases as f^2, that is, it increases 40 dB per decade increase of f, and it is proportional to the current loop area A.

Proceeding in a manner similar to the case of a single element carrying clock signal current described in Section 8.3, we obtain the following relations for the emission spectrum of the differential mode current configuration carrying clock signals:

$$E^D_{\text{cl}}(f) = E^D_\theta(f)\, S(f), \tag{8.10}$$

where $E^D_{\text{cl}}(f)$ is the amplitude spectrum of the maximum field intensity emitted by the signal differential mode currents, $E^D_\theta(f)$ is the amplitude of the maximum field emitted by differential mode sinusoidal currents at frequency f and is given by (8.8), $S(f)$ is the spectral intensity of the signal waveform.

The Bode plots for the (8.10) are shown in Figure 8.4. The physical arrangement of the differential current elements is shown in the inset. Note that the results are same at $P(r, 0)$ and $P(r, \pi)$.

It can be seen from Figure 8.4 that the differential mode emissions are mostly confined to the higher frequencies of the regulatory limit by FCC. They are typically above 100 MHz [2]. The maximum differential mode current whose radiated emissions would meet the FCC limit can be estimated by using (8.9). It is evident that such emissions can be reduced by reducing I_D and by reducing the loop area A.

Common Mode Emissions

For common currents we assume $I = I_C$ and $\alpha = 0$ in (8.6), and it can be shown that the radiated field at a far field point $P(r, \theta, \phi)$ is

$$E^C(P) = 2K \frac{e^{-j\beta r}}{r} I_C \sin\theta \cos(\beta_0 a \sin\theta \cos\phi), \tag{8.11}$$

which indicates that $E^C(p)$ is maximum at $\theta = \pi/2$ and $\pi = 0$ or π. Substituting the value of K given earlier, it can be shown, by using (8.11), that the magnitude of the total maximum field radiated by the two common mode currents of sinusoidal frequency f is given by

$$E^C(f) = \frac{1.257 \times 10^{-6}}{r} |I_C| f\, d\ell, \tag{8.12}$$

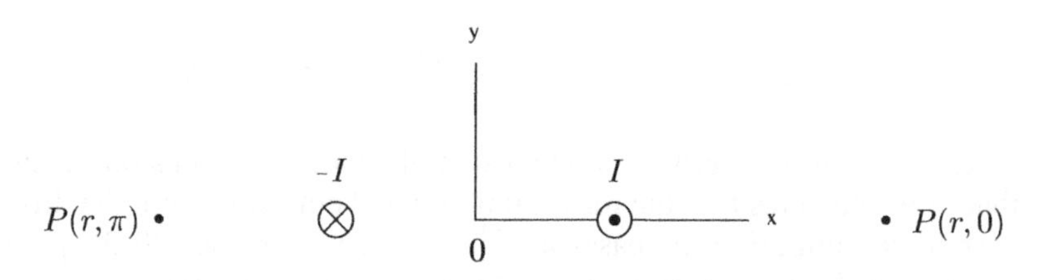

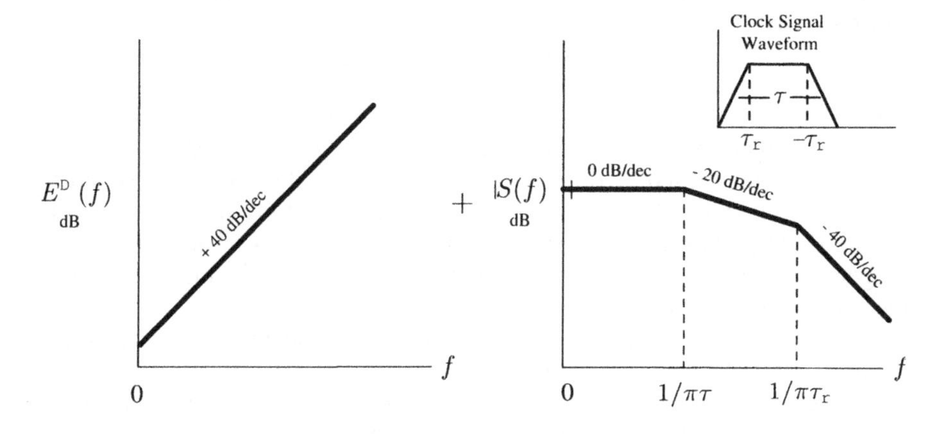

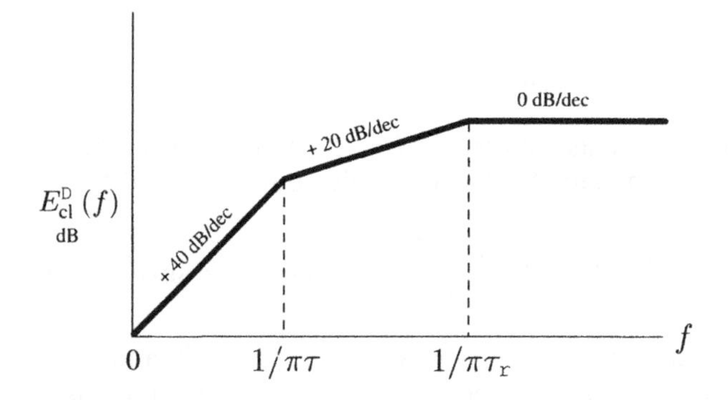

Figure 8.4 Amplitude spectrum for the emissions by a pair of parallel and differential mode trapezoidal signal currents.

where we have used the approximations $\cos(\beta_0 a \cos\phi) \simeq 1$ and the value of K given earlier.

The relation for the emission spectrum of the common mode current configuration carrying clock signals can now be written as

$$E^C_{\text{cl}}(f) = E^C(f)\, S(f), \tag{8.13}$$

where $E^C_{\text{cl}}(f)$ is the amplitude spectrum of the maximum field intensity emitted by the clock signal common mode currents, $E^C(f)$ is the amplitude of the maximum field emitted by the common mode sinusoidal currents at frequency f and is given by (8.12), $S(f)$ is the spectral intensity of the signal waveform.

The Bode plots for the common mode emissions are shown in Figure 8.5. The physical arrangement of the common mode current elements are shown in the insert and note that the results are same at $P(r, 0)$ and $P(r, \pi)$.

Figure 8.5 illustrates that the common mode emissions tend to be confined to the frequencies of the radiated emissions regulatory limits, typically below 200 MHz [2]. The maximum common mode current $|I_C|$ whose radiated emissions would meet the FCC limit can be estimated by using (8.12). Using (8.9) and (8.12), it can be shown that the ratio of maximum common and differential mode currents in the same short parallel current configuration will cause maximum radiated emissions acceptable by FCC as

$$\frac{|I_C|_{\max}}{|I_D|_{\max}} = 1.0475 \times 10^{-5} f_{\text{MHz}} S_{\text{mils}}, \tag{8.14}$$

where f_{MHz} is the frequency in MHz, S_{mills} is the spacing between the currents in the mills. Generally, (8.14) indicates that it takes less common mode current to produce fields larger than the FCC limit [2].

8.5 TRANSMISSION LINE MODELS FOR SUSCEPTIBILITY

8.5.1 Introduction

Analytical investigation of the susceptibility of an electronic device uses finite sections of terminated multi- or two-wire transmission lines illuminated from outside by external electromagnetic fields. The transmission lines are utilized to model certain portions of the internal circuitry of the device. The voltage developed at the terminations can then be used to estimate the susceptibility of the device to such incident fields. Rigorous theoretical and numerical investigation of a variety of such cases is described in [4–6].

In the present section we consider a simplified case of a section of a two-wire line terminated at both ends by arbitrary impedances and illuminated by a plane electromagnetic field. The basic formulation of the problem was given in [7] and subsequently investigated in great detail in [2, 5, 6].

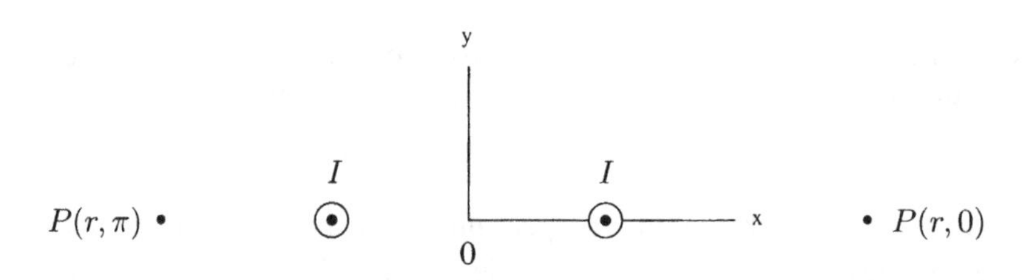

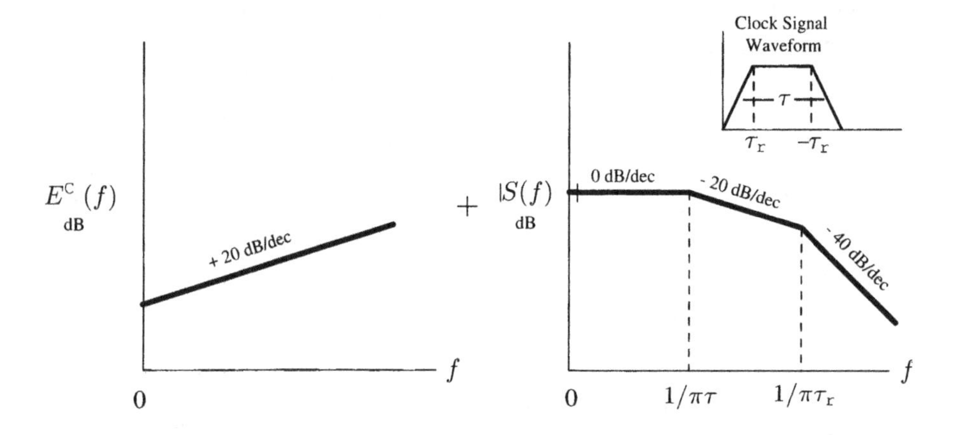

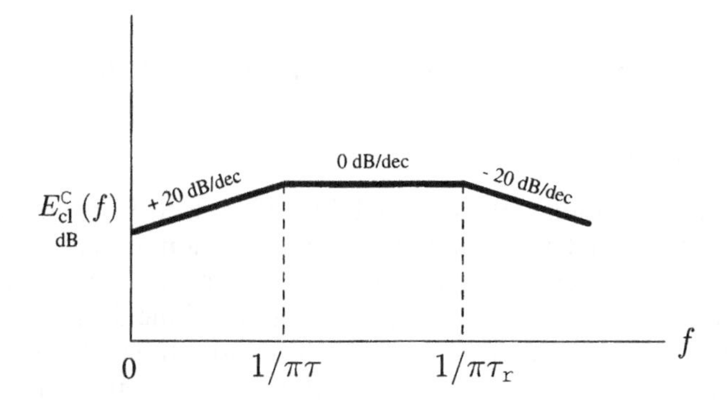

Figure 8.5 Amplitude spectrum for the emissions by a pairs of parallel and common mode trapezoidal signal currents.

8.5.2 Voltage Induced on the Two-Wire Transmission Line

Consider a two-wire transmission line oriented in the xy plane such that the y direction is the longitudinal direction for the line as shown in Figure 8.6, which describes the geometry and the coordinate systems used. We assume that the medium is free space.

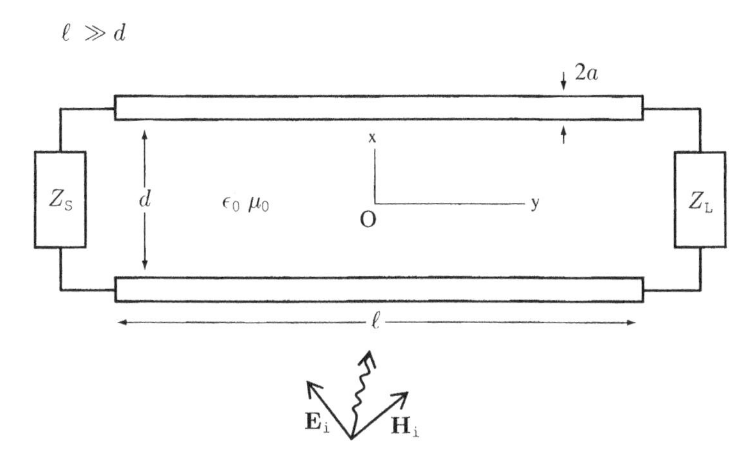

Figure 8.6 Section of a two-wire transmission line illuminated by a plane electromagnetic wave.

As shown, the line of length ℓ is terminated at both ends by arbitrary impedances Z_L, Z_s, called *load* and *sending end impedances*, respectively. A plane electromagnetic wave is incident on the transmission line from outside. We wish to determine the voltages across Z_L and Z_s caused by the incident field. Basic formulations of the problem were described in [7] and its analytical solution in various forms was described in [2,4]– [8].

We assume that the transmission line carries the dominant TEM mode discussed in Chapter 5, and we quote from Section 5.7.2 the following characteristic parameters of the line:

$$L = \frac{\mu_0}{\pi} \ln\left(\frac{d}{a}\right) \quad \text{H/m} \tag{8.15a}$$

$$C = \frac{\pi\epsilon_0}{\ln\left(\frac{d}{a}\right)} \quad \text{F/m} \tag{8.15b}$$

$$Z_C = \sqrt{\frac{L}{C}} = 120 \ln\left(\frac{d}{a}\right) \quad \Omega. \tag{8.15c}$$

The incident plane wave is represented by the electric and magnetic fields $\mathbf{E}^i$ and $\mathbf{H}^i$ such that $\mathbf{E}^i \perp \mathbf{H}^i$ and $\mathbf{E}^i \times \mathbf{H}^i$ describe the direction of incidence (or propagation) of the wave, and as we know, $|\mathbf{E}^{\mathrm{i}}| = \eta_0|\mathbf{H}^{\mathrm{i}}|$, where $\eta_0 = 120\pi$ is the intrinsic impedance of free space.

We now define the following quantities in the $z = 0$ or xy plane:

$\mathbf{E}_{\mathrm{t}}^{\mathrm{i}}$ is the component of the incident electric field $\mathbf{E}^{\mathrm{i}}$ in the plane of the transmission line and transverse to the line conductors (i.e., in the $\hat{x}$ direction).

$\mathbf{H}_{\mathrm{n}}^{\mathrm{i}}$ is the component of the incident magnetic field $\mathbf{H}^{\mathrm{i}}$ that is in the direction of the normal $\hat{n}$ to the transmission line plane and is defined to be in the $\hat{z}$ direction, meaning $\hat{n} = \hat{z}$. The normal component $\mathbf{H}_{\mathrm{n}}^{\mathrm{i}}$ of the incident field and the transverse component $\mathbf{E}_{\mathrm{t}}^{\mathrm{i}}$ will induce a series voltage generator per unit length, and the transverse component $\mathbf{E}_{\mathrm{t}}^{\mathrm{i}}$ will introduce a shunt current generator per unit length along the line.

Induced Voltage and Current Generators

We consider a section of length Δy of the line in Figure 8.6 and represent it as a loop $ABCD$ in Figure 8.7*a*, where the appropriate incident field components in the plane of the loop as well as the coordinate system used are also shown.

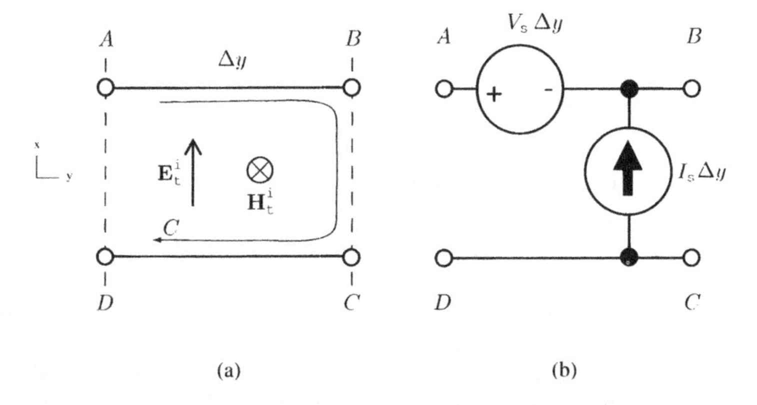

Figure 8.7 Induced voltage and current generators on the line: (*a*) Section of the line and the fields in the plane of the line; (*b*) equivalent voltage and current generators on the line.

Applying Faraday's law along the loop $ABCD$, we obtain the voltage V_{s}' induced in series with the loop as

$$V_{\mathrm{s}}' = -\oint_C \mathbf{E} \cdot d\boldsymbol{\ell} = \frac{d}{dt}\int_S \mathbf{B} \cdot ds, \tag{8.16}$$

where the contour C is as shown in Figure 8.7*a*, along which the integration of the emf generating field $\mathbf{E}$ is carried out, and the other notations are as explained before. It can be shown that (8.16) reduces to

$$\begin{aligned} V_{\mathrm{s}}' &= j\omega\mu_0 \int_S H_{\mathrm{n}}^{\mathrm{i}} \hat{n} \cdot \hat{n}\, ds \\ &\simeq j\omega\mu_0\, \Delta y \int_0^d H_{\mathrm{n}}^{\mathrm{i}}\, dx, \end{aligned} \tag{8.17}$$

where we have assumed that Δy is small enough that $H_{\mathrm{n}}^{\mathrm{i}}$ does not depend on it. Note that it follows from Lenz's law (also from the negative sign on the left-hand side of (8.16)) that the polarity of V_{s}' is such that any induced current through the loop must be anticlockwise. In the present case, where $H_{\mathrm{n}}^{\mathrm{i}}$ is increasing with time, the polarity of v_{s} is such that the induced current would flow opposite to the direction of C. We now define the induced voltage generator per unit length as

$$V_{\mathrm{s}} = \frac{V_{\mathrm{s}}'}{\Delta y} = j\omega\mu_0 \int_0^d H_{\mathrm{n}}^{\mathrm{i}}\, dx \quad \text{V/m}. \tag{8.18}$$

The per unit length induced current generator I_{s} is directed in the direction of $\mathbf{E}_{\mathrm{t}}^{\mathrm{i}}$, which is in the $+x$ direction and is given by

$$I_{\mathrm{s}} = -j\omega C \int_0^d E_{\mathrm{t}}^{\mathrm{i}}\, dx \quad \text{A/m}, \tag{8.19}$$

where C is the per unit length capacitance of the line.

Now, using (8.18) and (8.19), we can introduce the induced generators into the line sections of Figure 8.7a and show it equivalently as Figure 8.7b, where for convenience we have shown the generators as lumped voltage and current generators for the entire length Δy.

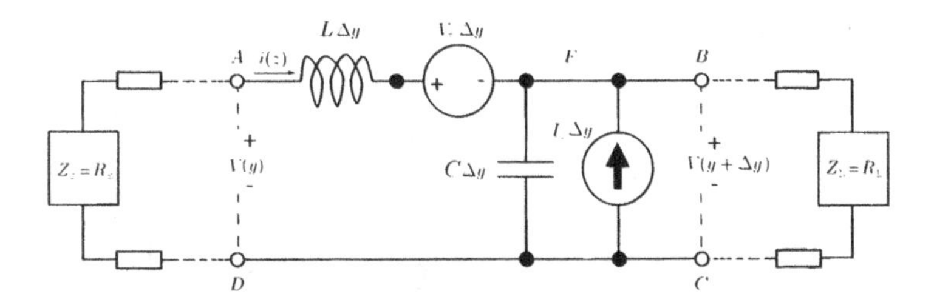

Figure 8.8 Lumped equivalent circuit for the terminated transmission line illuminated by a plane electromagnetic wave.

Transmission Line Equations

To obtain the transmission line equations for the original configuration of Figure 8.6, we now represent its lumped circuit equivalent shown in Figure 8.8, where we have used Figure 8.7*b* for an intermediate section $ABCD$. Applying Kirchoff's voltage law at loop $ABCD$ and current law at the node F, we obtain

$$\frac{dV}{dy} + j\omega LI = -V_\mathrm{s} = -j\omega\mu_0 \int_0^d H_\mathrm{n}^\mathrm{i}\, dx \tag{8.20}$$

$$\frac{dI}{dy} + j\omega CI = I_\mathrm{s} = -j\omega C \int_0^d E_\mathrm{t}^\mathrm{i}\, dx \tag{8.21}$$

which are the appropriate transmission-line type of equations that can be used to determine the voltage $V = V(y)$ and current $I = I(y)$ along a two conductor line illuminated by an incident plane electromagnetic wave. Equations (8.20) and (8.21) apply for lossless lines; they can be generalized for lossy lines by replacing $j\omega L$ and $j\omega C$ by $(R + j\omega L)$ and $(G + j\omega C)$, respectively, in appropriate places.

Rigorous solutions of (8.20) and (8.21) are described in [4–7]. We will describe now a very simplified model for a section of line such that $\ell \ll \lambda$ and $d \ll \lambda$; under these conditions the equivalent lumped parameters $L\ell$ and $C\ell$ in Figure 8.8 are negligible and can be neglected provided the line is neither short circuited nor open circuited at any end [2]. Moreover, under the assumptions above, the appropriate components of the incident field in the plane at the loop can be assumed to be constant. Thus the equivalent circuit of Figure 8.8, reduces to that given in Figure 8.9 [2]:

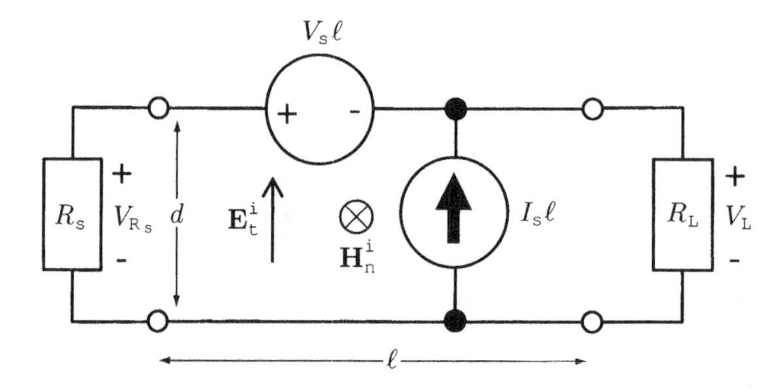

Figure 8.9 Simplified model for a short section of a terminated two-wire line illuminated by a plane electromagnetic wave.

where

$$V_s\ell \simeq j\omega\mu_0 H_n^i S \tag{8.22}$$

$$I_s\ell \simeq -j\omega C E_t^i S \tag{8.23}$$

with $S = d\ell =$ area of the loop.

Applying Kirchoff's law, it can be shown that induced terminal voltages developed across the terminating resistances are

$$V_L = -\frac{R_L}{R_s + R_L}(V_s\ell) + \frac{R_s R_L}{R_s + R_L}(I_s\ell) \tag{8.24}$$

$$V_{R_s} = \frac{R_s}{R_s + R_L}(V_s\ell) + \frac{R_s R_L}{R_s + R_L}(I_s\ell), \tag{8.25}$$

where V_s and I_s are given by (8.22) and (8.23). The simplified model will yield useful estimations of the effects of electromagnetic fields incident on an electrically short two-wire transmission line. Some examples are described in [2].

REFERENCES

1. J. D. Kraus, *Antennas*, 2nd ed., McGraw-Hill, New York, 1988.
2. C. R. Paul, *Introduction to Electromagnetic Compatibility*, Wiley, New York, 1992.
3. H. W. Ott, *Noise Reduction Techniques in Electronic Systems*, 2nd ed., Wiley, New York, 1988.
4. A. A. Smith Jr., *Radio Frequency Principles and Applications*, IEEE Press, New York, 1998.
5. F. M. Tesche, M. V. Ianuz, and T. Karlsson, *EMC Analysis Methods and Computational Models*, Wiley, New York, 1997.
6. C. R. Paul, *Analyses of Multiconductor Transmission Lines*, Wiley, New York, 1994.
7. C. D. Taylor, R. S. Satterwhite, and C. W. Harrison, "The Response of a Terminated Two-Wire Transmission Line Excited by a Non-uniform Electromagnetic Field, *IEEE Trans. Ant. Prop., 13 (1965): 987–989.*
8. C. Christopoulos, *Principles and Techniques of Electromagnetic Compatibility*, CRC Press, Ann Arbor, 1995.

CHAPTER 9

ELECTROMAGNETIC SHIELDING

9.1 INTRODUCTION

Electromagnetic shielding, commonly referred to as shielding, reduces or ideally prevents the coupling of unwanted electromagnetic energy from the outside environment into an electronic device, and it also reduces the transmission of undesirable emissions by the device into its outside environment. Generally, shielding can be achieved by using a suitable conducting or metallic screen or barrier between two regions of space such that it controls or stops transmission of electric and magnetic fields from one side to the other, and vice versa. Field transmissions can occur by conduction and by radiation. In this chapter we will describe only the shielding of radiated emissions. Varying degrees of shielding of electronic devices are required to be effective over a large part of the electromagnetic spectrum. Note that perfect shielding of a device is impractical because of the presence of many required intentional discontinuities in the barrier walls due to the input and output accesses needed by the device. Much of the shielding technologies described here were originally

Applied Electromagnetics and Electromagnetic Compatibility. By D. L. Sengupta, V. V. Liepa
ISBN 0-471-16549-2

developed by Schelkunoff [1]. Discussions of shielding from the view-point of EMC are given in [2–6]. We will briefly describe the basic aspects of the shielding of radiated emissions; more detailed descriptions can be found in the references cited.

9.2 DEFINITIONS

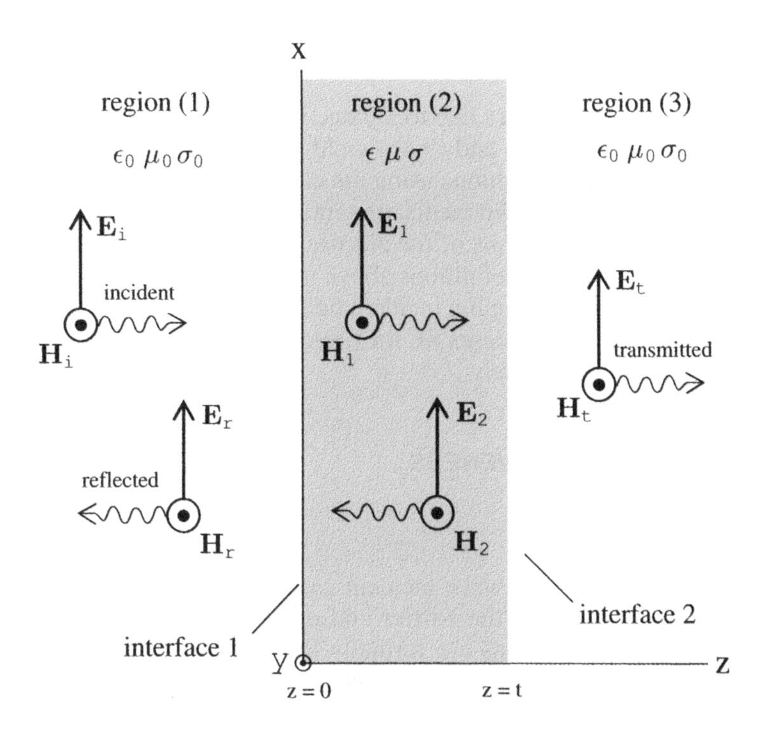

Figure 9.1 Conducting barrier to a normally incident plane TEM wave in free space.

Consider a plane, infinite and with a conducting barrier of thickness t oriented in free space with respect to a rectangular coordinate system as shown in Figure 9.1, which indicates that the entire space is divided into three regions: region 1 and 3 being free space and region 2 being occupied by the conducting barrier. We assume a plane TEM wave incident normally on the barrier from the left as shown. As we saw in Chapter 3 (Section 3.7.1), the barrier will give rise to a reflected wave in the region $z \leq 0$ and a transmitted wave in the region $z \geq t$; in addition the multiple reflection effects at the two interfaces of the barrier can be represented by two waves travelling in the $\pm z$-direction in the region $0 \leq z \leq t$. All these waves are symbolically represented in Figure 9.1.

Assuming that we wish to shield region 3 from region 1, we define the *shielding effectiveness* (SE) as

$$\mathrm{SE} = 20\log_{10}\left|\frac{\mathbf{E}^{\mathrm{i}}}{\mathbf{E}^{\mathrm{t}}}\right| \quad \mathrm{dB} \tag{9.1a}$$

or

$$\mathrm{SE} = 20\log_{10}\left|\frac{\mathbf{H}^{\mathrm{i}}}{\mathbf{H}^{\mathrm{t}}}\right| \quad \mathrm{dB}. \tag{9.1b}$$

Since the incident fields are due to a plane wave and the media of regions 1 and 3 are the same, (9.1a) and (9.1b) yield the same result for SE. If the media are different, the definitions using the electric and magnetic fields give different results. When the two media are same, it is customary to express the shielding effectiveness in terms of the electric fields. It should be noted that practical application of the definitions above implies that the barrier is in the far zone of the source causing the incident field. In many practical situations this may not be true; in such cases SE for E and H fields are not the same, and should be considered separately.

9.3 SHIELDING EFFECTIVENESS

9.3.1 Introduction

We consider now the plane wave incident case shown in Figure 9.1 where the intrinsic impedances of the barrier and other regions are as identified. The fields in the three regions are formally identified in Figure 9.1. As in Section 3.7.1, we write the explicit expressions for the various fields as

$$\begin{aligned}
\mathbf{E}^{\mathrm{i}} &= \hat{x}E^{\mathrm{i}}e^{-j\beta_0 z}, & \mathbf{E}^{\mathrm{r}} &= \hat{x}E^{\mathrm{r}}e^{+j\beta_0 z} \\
\mathbf{H}^{\mathrm{i}} &= \hat{y}\,\frac{E^{\mathrm{i}}}{\eta_0}\,e^{-j\beta_0 z}, & \mathbf{H}^{\mathrm{r}} &= -\hat{y}\,\frac{E^{\mathrm{r}}}{\eta_0}\,e^{-j\beta_0 z} \\
& & &\text{for } z \leq 0,
\end{aligned} \tag{9.2}$$

$$\begin{aligned}
\mathbf{E}_1 &= \hat{x}E_1e^{-\gamma z}, & \mathbf{E}_2 &= \hat{x}E_2e^{\gamma z} \\
\mathbf{H}_1 &= \hat{y}\,\frac{E_1}{\eta}\,e^{-\gamma z}, & \mathbf{H}_2 &= -\hat{y}\,\frac{E_2}{\eta}\,e^{\gamma z} \\
& & &\text{for } 0 \leq z \leq t,
\end{aligned} \tag{9.3}$$

$$\mathbf{E}^{\mathrm{t}} = \hat{x} E^{\mathrm{t}} e^{-j\beta_0 z}$$
$$\mathbf{H}^{\mathrm{t}} = \hat{y}\,\frac{E^{\mathrm{t}}}{\eta_0}\,e^{-j\beta_0 z} \text{for } z \geq t, \tag{9.4}$$

where

$$\beta_0 = \omega\sqrt{\mu_0\epsilon_0}\,,\quad \eta_0 = \sqrt{\frac{\mu_0}{\epsilon_0}}\,,\quad \omega = 2\pi f$$
$$\gamma = \alpha + j\beta = \sqrt{j\omega\mu(\sigma + j\omega\epsilon)} \tag{9.5}$$
$$\eta = \sqrt{\frac{j\omega\mu}{\sigma + j\omega\epsilon}}\,.$$

The various parameters in (9.5) were defined and discussed in Chapter 3 and they will not be further described here.

Analytical expressions for the shielding effectiveness (9.1a) require the knowledge of $\mathbf{E}^{\mathrm{t}}$, which can be obtained by solving a set equations obtained by applying appropriate boundary conditions on the fields given by (9.2)–(9.4). Note that in (9.2)–(9.4) only the parameter $\mathbf{E}^{\mathrm{i}}$ is known and E^{r}, E_1, E_2, and E^{t} are yet unspecified. Application of the boundary conditions on the tangential electric and magnetic fields at the two interfaces 1 and 2 provides us with four equations for four unknowns.

The continuity of the tangential E fields at $z = 0$ and $z = t$ yields

$$E_{\mathrm{i}} + E_{\mathrm{r}} = E_1 + E_2 \tag{9.6a}$$
$$E_1 e^{-\gamma t} + E_2 e^{-\gamma t} = E_{\mathrm{t}} e^{-j\beta_0 t}. \tag{9.6b}$$

Similarly the continuity of the tangential magnetic fields at $z = 0$ and $z = t$ yields

$$\frac{E^{\mathrm{i}}}{\eta_0} - \frac{E^{\mathrm{r}}}{\eta_0} = \frac{E_1}{\eta} - \frac{E_2}{\eta} \tag{9.7a}$$
$$\frac{E_1}{\eta}\,e^{-\gamma t} - \frac{E_2}{\eta}\,e^{-\gamma t} = \frac{E^{\mathrm{t}}}{\eta_0}\,e^{-j\beta_0 t}. \tag{9.7b}$$

From (9.6a), (9.6b) and (9.7a), (9.7b) we obtain the following:

$$\frac{E^{\mathrm{i}}}{E^{\mathrm{t}}} = \frac{(\eta_0 + \eta)^2}{4\eta_0\eta}\left[1 - \left(\frac{\eta_0 - \eta}{\eta_0 + \eta}\right)^2 e^{-2\gamma t}\right] e^{\gamma t} e^{-j\beta_0 t}, \tag{9.8}$$

which is an exact expression.

The propagation constant γ in the conducting barrier is approximated by

$$\gamma = \alpha + j\beta \simeq \frac{1+j}{\delta}, \tag{3.209b}$$

where $\delta = \sqrt{2/\omega\mu\sigma}$ is the skin depth in the conducting medium.

Now using (3.118) and making the approximation $\eta_0 \gg \eta$, we obtain the following from (9.8):

$$\left|\frac{E^{\rm i}}{E^{\rm t}}\right| \simeq \left|\frac{\eta_0}{4\eta}\right| e^{t/\delta}\left|1 - e^{-2t/\delta}e^{-j2t/\delta}\right|. \tag{9.9}$$

Thus we obtain an expression for the shielding effectiveness from (9.9) and express it in the form

$$\begin{aligned}\mathrm{SE}_{\mathrm{dB}} &= 20\log_{10}\left|\frac{\eta_0}{4\eta}\right| + 20\log_{10}(e^{t/\delta}) \\ &\quad + 20\log_{10}\left|1 - e^{-2t/\delta}e^{-j2t/\delta}\right| \\ &= R_{\mathrm{dB}} + A_{\mathrm{dB}} + M_{\mathrm{dB}},\end{aligned} \tag{9.10}$$

where it is common to identify the following: R_{dB} as the reflection loss caused by the interfaces 1 and 2, A_{dB} is the assumption loss suffered by the wave as it proceeds through the barrier, M_{dB} is additional losses suffered by the wave due to multiple reflections at the interfaces. The identification of the three terms in (9.9) as above can be described analytically from detailed investigation of the exact expression (9.8), as in [3].

9.3.2 SE Expressions for Computation

The total loss associated with a shield for plane wave incidence can be calculated for the exact or the approximate expressions given by (9.8) and (9.9). We will recast (9.9) in a form more suitable for computation for practical applications. Using the values of η_0 and η in terms of the medium parameters given earlier, it can be shown from (9.10) that

$$R_{\mathrm{dB}} = 20\log\left(\frac{1}{4}\sqrt{\frac{\sigma}{\omega\mu_{\mathrm{r}}\epsilon_0}}\right), \tag{9.11}$$

where we have assumed that the permeability of the barrier material is $\mu = \mu_0\mu_{\mathrm{r}}$, μ_{r} being the permeability of the barrier material relative to that of free space. It is convenient to refer the conductivity σ of the barrier material to that of copper, meaning $\sigma = \sigma_{\mathrm{cu}}\sigma_{\mathrm{r}}$, where $\sigma_{\mathrm{cu}} = 5.8 \times 10^7$ S/m and σ_{r} is the conductivity of the barrier material relative to that of copper. After substituting $\sigma = \sigma_{cu}\sigma_r$ into (9.11), we obtain the following for the reflection loss:

$$R_{dB} = 168 + 10\log_{10}\left(\frac{\sigma_r}{\mu_r f}\right), \tag{9.12}$$

where f is in Hz and μ_r, σ_r are as defined earlier. Note that (9.12) indicates that reflection loss decreases at a rate of 10 dB per decade increase of frequency; it also indicates that high-conductivity materials provide large loss.
Similarly absorption loss A_{dB} in (9.11) can be expressed in the following alternative forms:

$$A_{dB} = 20\log e^{t/\delta} = 8.6859\frac{t}{\delta} \tag{9.13a}$$

$$= 131\,t\sqrt{f\mu_r\sigma_r} \tag{9.13b}$$

$$= 3.34\,t\sqrt{f\mu_r\sigma_r}\,, \quad t \text{ in m and } f \text{ in inches.}$$

In (9.13b) we have explicitly used the expression for δ, in terms of the barrier material constants given earlier. Equation (9.13a) indicates that for barrier thickness $t = \delta$ and 3δ, f in Hz, $A_{dB} = 8.7$ and 26.1 dB, respectively. Note that the absorption loss increases at the rate of 10 dB per decade increase of frequency. The absorption loss versus frequency is plotted in Figure 9.2 for two thicknesses of copper and steel. This shows the advantage of steel over copper in providing absorption loss.

The multiple reflection loss M_{dB} defined in (9.10) can be written as

$$M_{dB} = 20\log_{10}\left|1 + e^{-4t/d} - 2e^{-2t/d}\cos\left(\frac{2t}{\delta}\right)\right|, \tag{9.14}$$

which indicates that for $t/\delta \gg 1$, meaning for barrier thickness large compared with the skin depth in the barrier, $M \simeq 1$, meaning $M_{dB} \simeq 0$. Generally, multiple reflection loss for thick barriers can be neglected. However, for thin barriers $t/\delta \ll 1$ and (9.14) indicates that a lower bound for M_{dB} is given by

$$M_{dB} \geq 20\log_{10}\left|1 - e^{-2t/\delta}\right| \quad \text{for } \frac{t}{\delta} \ll 1. \tag{9.15}$$

It is clear from (9.15) that M_{dB} is a negative number that now represents a gain, which is a reduction in the loss predicted by (9.10), and this loss will be less than the lower bound given by (9.15). Multiple reflection effects can be neglected for E fields (for all t) and should be taken into account particularly for thin barriers; they can also be neglected for thick barriers for H fields [2, 3]. Figure 9.3 gives a plot of the lower bound $M_{dB} = 20\log_{10}|1 - e^{-2t/\delta}|$ versus t/δ.

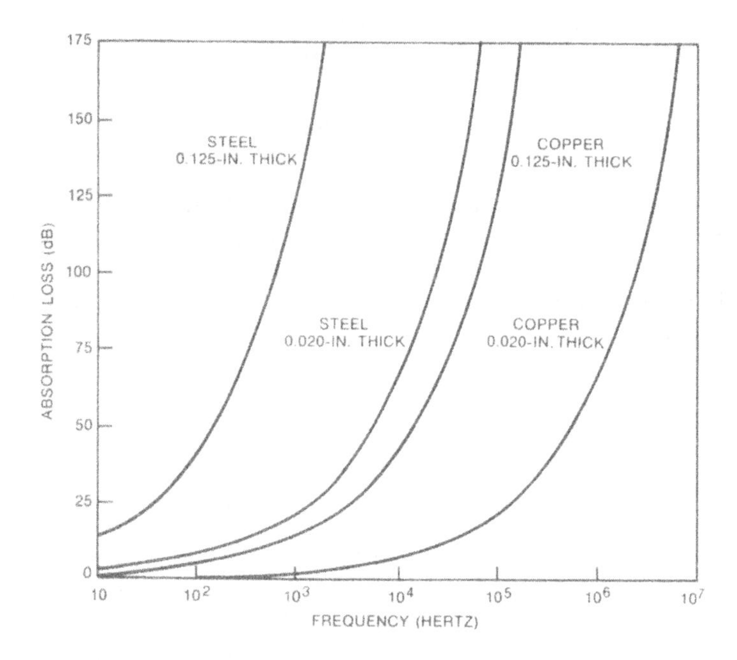

Figure 9.2 Absorption loss versus frequency for two thicknesses of copper and steel. (Source: From [2], p. 168.)

The results clearly indicate that M_{dB} tends to zero for thick barriers. As was mentioned earlier, the results should be applied for magnetic field shielding; but not for electric fields.

We will now describe the total attenuations suffered by a plane wave or the total shielding effectiveness (dB) of copper shield in the far field. For reasons given earlier we will ignore the multiple reflection loss. Thus $SE_{\mathrm{dB}} = R_{\mathrm{dB}} + A_{\mathrm{dB}}$ versus frequency is shown in Figure 9.4 for a copper shield. The reflection loss decreases with increasing frequency; the absorption losses increase with frequency. The minimum shielding effectiveness occurs at some intermediate frequency below 10 kHz. It is clear from the figure that for two planes of different frequencies, reflection loss accounts for most of the attenuation whereas at high frequency most of the attenuation comes from the absorption loss.

9.4 SHIELDING EFFECTIVENESS: NEAR FIELD ILLUMINATION

So far we have considered shields illuminated by plane electromagnetic waves. This implies that the source of the fields to be shielded is sufficiently distant

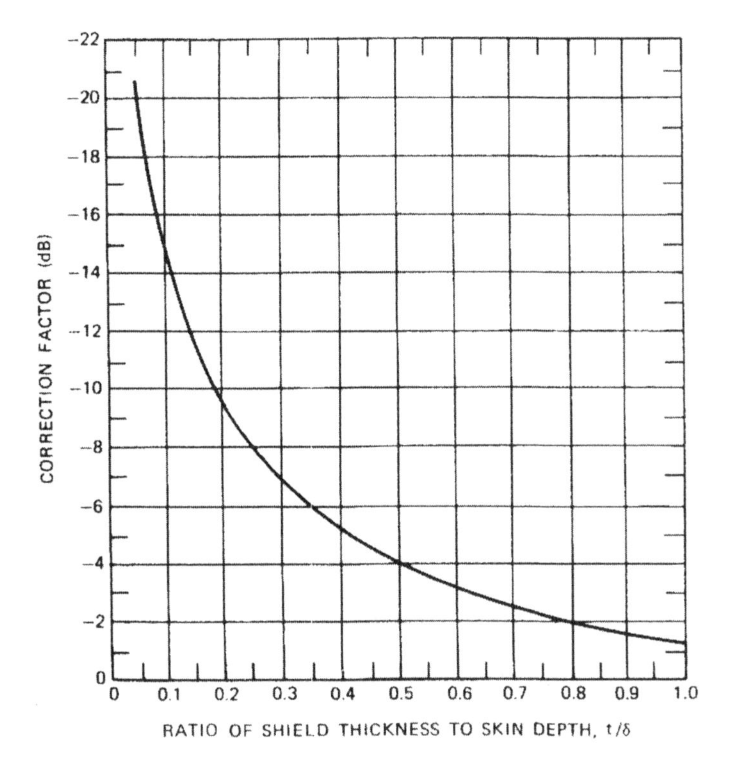

Figure 9.3 Multiple reflection loss lower bound $M_{dB} = 20\log_{10}\left(1 - \exp^{-2t/\delta}\right)$ versus skin depth.

from the shield so that the far zone approximations are valid for the fields incident on the shield.

There are cases where the shield is located within near field region of the source. In the present section we describe the special considerations that must be given when designing shields located in the near shield region of the source. In such cases the design of the shield depends considerably on the nature of the source, that is, whether the source is of electric or magnetic type, as discussed below.

9.4.1 Electric and Magnetic Sources

Basic electric and magnetic sources of radiation are the electric (Hertzian) dipole or a small current element and magnetic dipole or the small loop of current, respectively. The complete fields produced by the sources above,

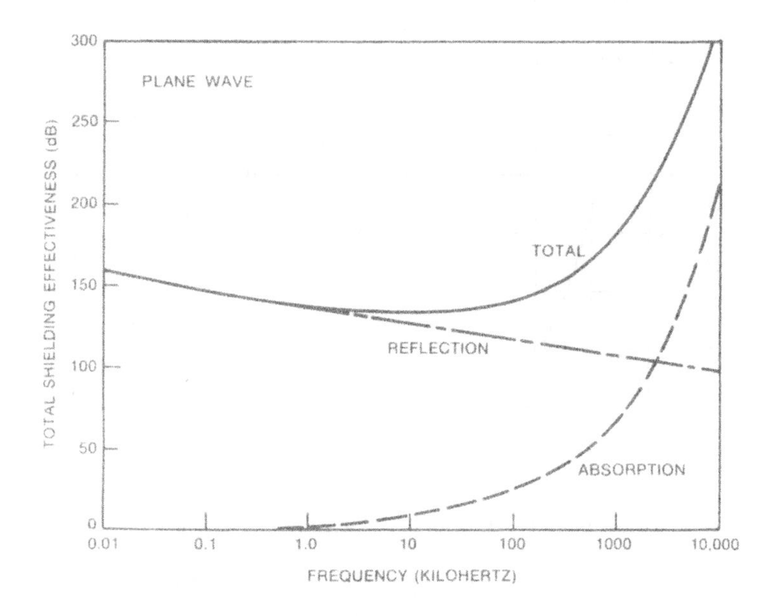

Figure 9.4 Shielding effectiveness of a 0.02 inch thick copper shield in the near field. (Source: From [2], p. 178.)

and the behavior of the fields in various regions of space were described in Sections 6.3 and 6.4.

The far zone fields produced by an electric dipole in free space and located at the origin of a spherical coordinate system, as shown in Figure 6.2, can be written as

$$\begin{aligned} \mathbf{E} &= \hat{\theta} E_\theta \\ \mathbf{H} &= \hat{\phi} H_\phi = \hat{\phi}\,\frac{E_\theta}{\eta_0}\,, \end{aligned} \tag{9.16}$$

where $\eta_0 = \sqrt{\mu_0/\epsilon_0}$ = intrinsic impedance of free space, and E_θ is given by (6.35a). Generally, the wave impedance of the far zone wave given by (9.16) is defined by

$$Z_{\mathrm{w}}^{\mathrm{e}} = \frac{E_\theta}{H_\phi}\,, \tag{9.17}$$

where the superscript "e" indicates it applies to the electric dipole fields. It is clear from (9.17) that in the far zone, the wave impedance for the electric dipole equals the intrinsic impedance of the medium, $Z_W^{\mathrm{e}} = \eta_0$.

Similarly the far zone fields produced by a magnetic dipole (or small current loop) in free space and located at the origin of a spherical coordinate system, as shown in Figure 6.8, can be written as

$$\mathbf{E} = \hat{\phi} E_\phi,$$
$$\mathbf{H} = -\hat{\theta} H_\theta = -\hat{\theta}\,\frac{\hat{\theta} E_\phi}{\eta_0}, \tag{9.18}$$

where E_ϕ is given by (6.61). Thus the wave impedance for the magnetic dipole fields in the far zone is

$$Z_{\mathrm{w}}^{\mathrm{m}} = -\frac{E_\phi}{H_\theta} = \eta_0.$$

It is important to note that in the far field region, the waves impedance for the electric and magnetic dipoles are same. As was discussed in Sections 6.3 and 6.4, in the near field region, the wave impedances for the electric and magnetic fields of dipoles in free space are

$$Z_{\mathrm{w}}^{\mathrm{e}} \simeq -j\frac{\eta_0}{\beta_0 r} = \frac{-j}{\omega\epsilon_0 r} = \frac{1}{j2\pi f\epsilon_0 r} \tag{9.19}$$
$$Z_{\mathrm{w}}^{\mathrm{m}} \simeq j\eta_0\beta_0 r = j\omega\mu_0 r = j2\pi f\mu_0 r \tag{9.20}$$
$$\text{for } \beta_0 r \ll 1,$$

which implies that the two wave impedances are not only unequal but they also depend on the distance from the source. However, as we saw earlier in the far zone,

$$Z_{\mathrm{w}}^{\mathrm{e}} = Z_{\mathrm{w}}^{\mathrm{m}} = \eta_0.$$

For comparison we show in Figure 9.5 the variations of $|Z_{\mathrm{w}}|$ versus distance from the sources for the electric and magnetic dipole fields.

It is clear from (9.19) and (9.20) and from Figure 9.4 that in the near zone

$$|Z_{\mathrm{w}}^{\mathrm{e}}| > \eta_0 \quad \text{and} \quad |Z_{\mathrm{w}}^{\mathrm{m}}| < \eta_0.$$

This is why electric and magnetic dipoles are often referred to as high- and low-impedance sources, respectively.

Fields generated by arcing in a spark gap are of electric type, and hence these are high-impedance sources. The fields generated by some transformer action are predominantly magnetic type, and hence they are low-impedance sources.

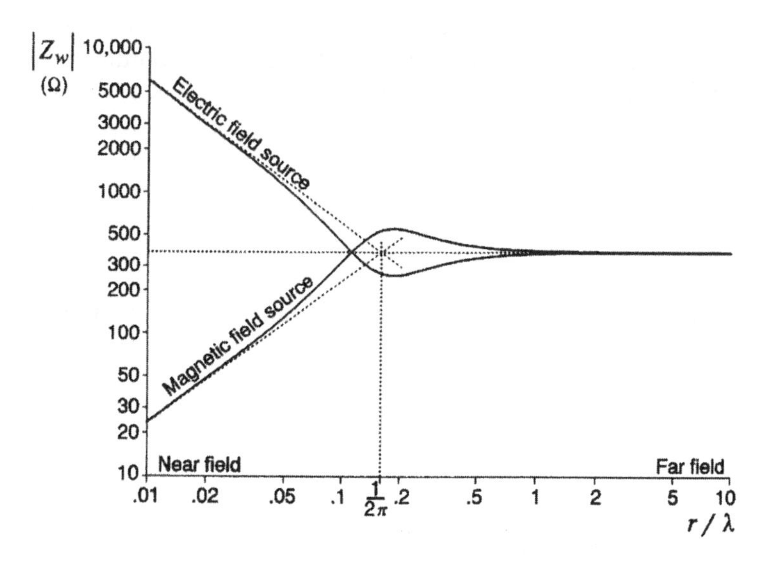

Figure 9.5 Wave impedance versus distance from a radiating source. (Source: From [5], p. 144.)

9.4.2 SE Expressions: Near Zone Considerations

The shields described in the previous section are meant for incident uniform plane electromagnetic waves. Proper considerations must be given to the shield that is located within the near zone region of source producing the incident fields. In such cases we saw earlier (Sections 6.3.5 and 6.4.4) that two types of fields, namely electric and magnetic, may exist for which the wave impedances have been found to be higher and lower, respectively, than the wave far field wave impedance. This requires special treatment for the reflection losses associated with two types of fields. Inspection of (9.8)–(9.10) for the plane wave incident case indicates that the assumption and multiple reflection losses are independent of the wave impedance. However, the reflection loss significantly depends on the nature of the near fields.

From (9.8) we can write the general expression for the reflection loss as

$$R = \frac{(Z_{\mathrm{w}} + \eta)^2}{4Z_{\mathrm{w}}\eta}, \tag{9.21}$$

where Z_{w} is the wave impedance appropriate for the incident field, and η is the intrinsic impedance of the shield material. Thus the reflection loss for high-

and low-impedance fields can be written as

$$R^{e} = \frac{(Z_{w}^{e} + \eta)^2}{4Z_{w}^{e}\eta} \tag{9.22}$$

$$R^{m} = \frac{(Z_{w}^{m} + \eta)^2}{4Z_{w}^{m}\eta}, \tag{9.23}$$

where Z_{w}^{e} and Z_{w}^{m} are as given by (9.19) and (9.20), respectively, and are the wave impedances for electric and magnetic type of fields. It can be shown that $|Z_{w}^{e}|, Z_{w}^{m}| \gg \eta$ for conducting shields, and we can approximate (9.22) and (9.23) by

$$R^{e} \simeq \frac{|Z_{w}^{e}|}{4|\eta|} = \frac{1}{8\pi f \epsilon_0 r|\eta|} \tag{9.24}$$

$$R^{m} \simeq \frac{|Z_{w}^{m}|}{4|\eta|} = \frac{2\pi f \mu_0 r}{4|\eta|}. \tag{9.25}$$

The reflection losses for the fields of electric and magnetic types are formally given by

$$R_{dB}^{e} = 20\log_{10}\frac{Z_{w}^{e}|}{4|\eta|} \tag{9.26}$$

$$R_{dB}^{m} = 20\log_{10}\frac{Z_{w}^{m}|}{4|\eta|}, \tag{9.27}$$

where

$$|\eta| = \sqrt{\frac{\omega\mu_0\mu_r}{\sigma_{cu}\sigma_r}}$$

is the intrinsic impedance of the barrier material, μ_r is its relative permeability and σ_r is its conductivity relative to that of copper as discussed earlier. Equations (9.26) and (9.28) can be expressed in the following convenient forms for computations:

$$R_{dB}^{e} = 142 + 10\log_{10}\left(\frac{\sigma_r}{\mu_r f^3 r^2}\right) \quad \text{dB} \tag{9.28}$$

$$R_{dB}^{m} = 74.6 + 10\log_{10}\left(\frac{f r^2 \sigma_r}{\mu_r}\right) \quad \text{dB}, \tag{9.29}$$

where r is in meters and f is in megahertz.

The near field reflection losses introduced by a copper shield to the electric and magnetic type of shields are shown in Figure 9.6. The plots are shown as functions of frequency and for selected distances between the shield and the source. The near field results shown are valid up to a distance r determined by $\beta_0 r = 1$ such that $r = \lambda_0/2\pi$, where λ_0 is the free space wavelength; this implies that the near field results are valid up to a frequency $f = 3 \times 10^8/r$ Hz with r in meters and f in Hz. The appropriate limiting frequencies for the two distances $r = 1$ and 30 m are as indicated in Figure 9.6.

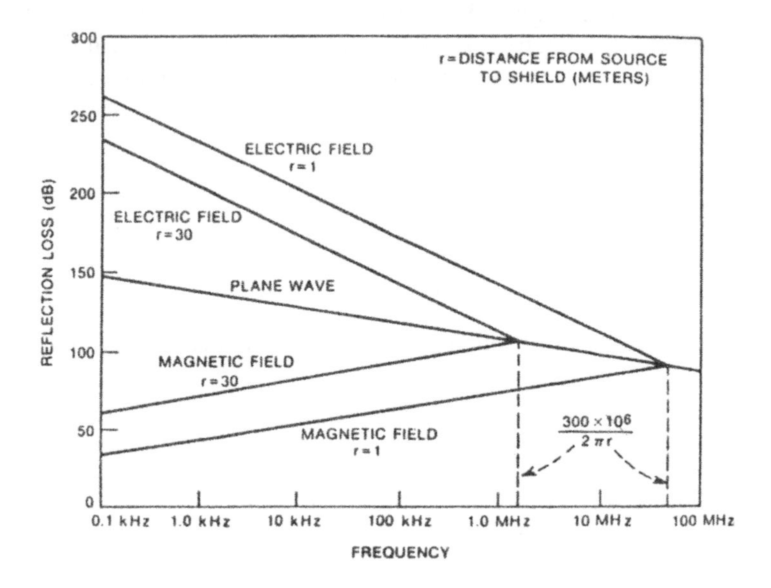

Figure 9.6 Reflection loss of near field electric and magnetic sources. (Source: From [2], p. 173.)

Although the reflection loss results obtained for the plane wave illumination are not valid in the near field region, the asymptotic behavior of the plane wave reflection loss is also indicated in Figure 9.6 for comparison. Note that the results indicate that $R^{\mathrm{e}}_{\mathrm{dB}}$ and $R^{\mathrm{m}}_{\mathrm{dB}}$ are larger and smaller, respectively, from the plane wave case.

9.5 DISCUSSION

The design of a shield for electromagnetic fields depends significantly on whether it is in the near or far zone of the sources of the fields of interest. We summarize here the essential loss mechanisms, discussed earlier, that would determine the shielding effectiveness in a given situation.

9.5.1 Far Zone Fields

With the shield illuminated by uniform plane electromagnetic fields the reflection loss provides the predominant shielding mechanism at low frequencies while the absorption mechanism dominates at higher frequencies (see Figure 9.4). Further details are given in [2, 3].

9.5.2 Near Zone Fields

Electric Type:

The near field situation is mostly unchanged from that of the plane wave case: the reflection loss dominates at low frequencies and the adsorption loss dominates at high frequencies.

Magnetic Type:

The situation is different at low frequencies. Absorption loss is dominant at all frequencies. However, both reflection and absorption losses are small at low frequencies. Thus other more effective methods of shielding against low-frequency fields must be used. For example, the use of high μ_r material to provide a low-reluctance path to the magnetic flux, and the generation of opposing flux via Faraday-Lenz's law for such cases are described in [2, 3].

REFERENCES

1. S. A. Schelkunoff, *Electromagnetic Waves*, Van Nostrand, New York, 1943.
2. H. W. Ott, *Noise Reduction Techniques in Electronics Systems*, 2nd ed., Wiley, New York, 1988.
3. C. R. Paul, *Introduction to Electromagnetic Compatibility*, Wiley, New York, 1992.
4. V. P. Kodali, *Engineering Electromagnetic Compatibility*, IEEE Press, New York, 1996.
5. J. P. Mills, *Electromagnetic Interference Reduction in Electronic Systems*, PTR Prentice-Hall, Englewood Cliffs, NJ, 1993.
6. C. Christopoulos, *Principles and Techniques of Electromagnetic Compatibility*, CRC Press, Ann Arbor, 1995.

CHAPTER 10

COUPLING BETWEEN DEVICES

10.1 INTRODUCTION

Electromagnetic signals, desired or undesired, can couple from one device (circuit) to another in a variety of manners. For example, if the circuit 1 (called the *offending circuit*) or some part of it shares a common path or impedance with another circuit 2 (called the *receptor circuit*), then current flow and/or voltage fluctuations in the common impedance caused by circuit 1 will produce corresponding fluctuations or interference in circuit 2.

If the common impedance is a conductor, then the resultant fluctuations in the receptor are called *conducted interference*. If the impedance is capacitive or inductive, then the interference is called capacitive (electric) or inductive (magnetic), respectively. Note that conductive interference may be associated with the inductive interference when the inductor is directly a part of the common impedance.

Even when there is no physical connections (through a common impedance) between the two circuits, there can be interference between the two circuits:

Applied Electromagnetics and Electromagnetic Compatibility. By D. L. Sengupta, V. V. Liepa
ISBN 0-471-16549-2 ©2005 John Wiley & Sons, Inc.

Case 1. Near Field Coupling:. If the separation between the two circuits is small compared to the wavelength of interest, there can occur coupling of signals from one to the other. This can occur in two ways: (1) capacitive or purely electric type of coupling and (2) inductive or purely magnetic type of coupling. In both cases the near zone E and H fields are of interest, and they generally can be treated separately. This type of coupling can be treated by applying circuit theory concepts.

Case 2. Far Field Coupling:. The affected circuits are separated by a distance large compared with the wavelength of interest. Here the coupling mechanism is truly electromagnetic, and both the far zone E and H fields are involved in the interference effects.

In the present chapter we describe a few selected topics in low-frequency near electric and magnetic field coupling between two circuit components. Many such and more complicated configurations are described in detail in [1–4]. Our description will be brief and illustrative of capacitive and inductive couplings. More details are given in the references cited.

10.2 CAPACITIVE (ELECTRIC) COUPLING [1, 3]

Consider a configuration of two conductors labelled 1 and 2 placed above a conducting ground plane as shown in Figure 10.1. The two conductors whose cross sections are shown may constitute a two-wire transmission line in free space, or they may be components of two different circuits. The capacitance between the conductors 1 and 2 is represented by C_{12}; $C_{1\text{g}}$ and $C_{2\text{g}}$ represent the capacitances between conductor 1 and ground, and conductor 2 and ground, respectively, as shown in Figure 10.1. An arbitrary harmonic voltage V of frequency ω is applied across conductor 1 and ground. We wish to determine the voltage across the resistance R, connected between conductor 2 and ground as in Figure 10.1. Although the problem we are describing is hypothetical, it can be applied to many practical configurations. For example, we may consider that the two conductors, 1 and 2, are each of diameter d and separated by D to form a two-wire transmission line.

It can be shown that with a voltage V_1 applied as shown in Figure 10.1, the voltage developed across R is given by

$$V_{\text{N}} = \frac{j\omega C_{12} R}{1 + j\dfrac{\omega}{\omega_0}} V_1, \tag{10.1}$$

where

$$\omega_0 = \frac{1}{R(C_{12} + C_{2\text{g}})} . \tag{10.2}$$

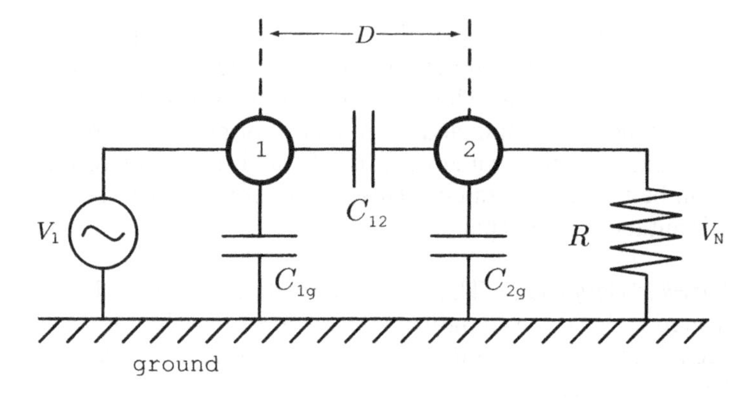

Figure 10.1 Two-conductor configuration illustrating the capacitive (or electric) coupling of signals from one conductor to another.

From (10.1) and (10.2) it can be shown that

$$V_N \simeq j\omega C_{12} R V_1 \tag{10.3}$$

for $\omega \ll \omega_0$, that is, when $R \ll 1/[\omega(G_2 + C_{2g})]$, and

$$V_N \simeq \frac{C_{12}}{C_{12} + C_{2g}} V_1 \tag{10.4}$$

for $\omega \gg \omega_0$, or when $R \gg 1/[\omega(G_2 + C_{2g})]$.

A plot of $|V_N|$ versus ω obtained from (10.1) is shown in Figure 10.2, which gives the frequency response of a capacitively or electrically coupled noise voltage if V_1 is considered to be the noise injected.
Figure 10.2 indicates that for $\omega \ll \omega_0$, or sufficiently small resistance from conductor 2 to ground, the capacitively coupled voltage can be obtained by using a current generator model connected between conductor 2 and ground as shown in Figure 10.3.
Thus (10.3) shows that for $\omega \ll \omega_0$, V_N is directly proportional to the frequency and amplitude of the input (noise) voltage, and to the resistance R and the coupling capacitance C_{12}. It is clear that resistance and the coupling capacitance can be controlled to obtain minimum V_N, if so desired. For $\omega \gg \omega_0$, V_N is given by (10.4), which shows that V_N is independent of frequency and is of larger magnitude than when R is smaller. For most practical cases $\omega \ll \omega_0$ [1].

For $\omega \ll \omega_0$, V_N can be reduced by reducing C_{12}. If the two conductors 1 and 2 can be considered as part of a two-wire transmission line, then C_{12} can be reduced considerably by increasing their separation distance. In other cases C_{12} can be reduced by using a proper shielding around conductor 2. Further discussion on this is given in [1].

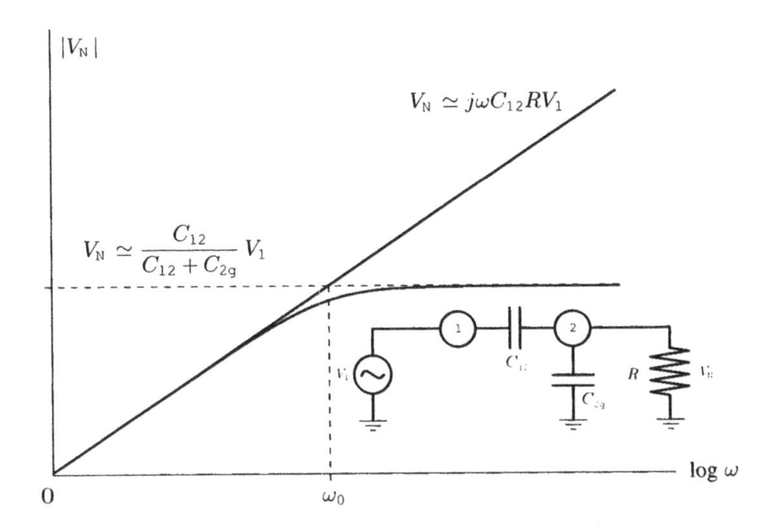

Figure 10.2 Frequency response of the capacitively coupled noise voltage across R. (Source: From [1], p. 33.)

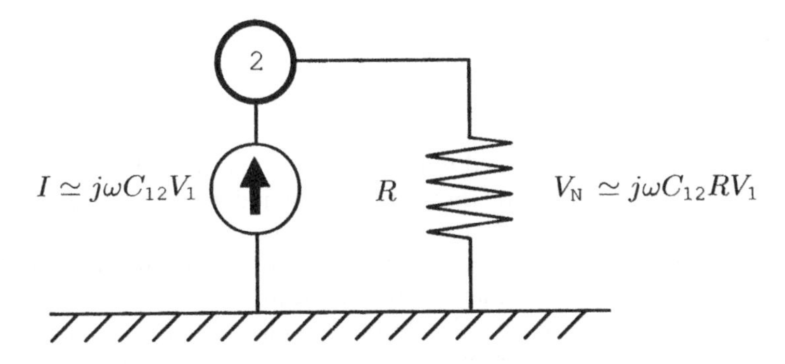

Figure 10.3 Current generator model to obtain the coupled voltage V_N in the circuit of (10.2).

10.3 MAGNETIC (INDUCTIVE) COUPLING

Faraday's law indicates that a current flowing through a closed loop or circuit produces magnetic flux by linking with a second nearby closed loop or circuit which induces a series emf or voltage in the second circuit. This is one way that energy couples magnetically from one circuit to another. The present section describes magnetic or inductive coupling of fields through selected illustrations described in more detail in [1,3,4].

10.3.1 Some Basic Concepts

In Chapter 7 we described mutual and self-inductances from various viewpoints. For the present purpose, we define the self-inductance L of a loop carrying a current I by (see (7.26))

$$L = \frac{\psi}{I}, \tag{7.26}$$

where ψ is the self-generated magnetic flux linked with the current I. Similarly the mutual inductance between two loops carrying currents I_1, I_2 is given by (7.30) and is written as

$$M_{12} = \frac{\psi_{12}}{I_1}, \tag{7.30}$$

where ψ_{12} is the flux generated by circuit 1 and linked with circuit 2. Generally, in a reciprocal medium $M_{12} = M_{21}$.

Faraday's law indicates that a circuit carrying a current I will induce a voltage (V_{in}) in series with a circuit 2 due to flux linkage. The voltage is given by

$$\begin{aligned} V_{\text{in}} &= \frac{d}{dt}(\psi_{12}) = \frac{1}{dt}(M_{12}I_1) \\ &= M_{12}\frac{dI_1}{dt} = j\omega M_{12}I_1. \end{aligned} \tag{10.5}$$

Note that the sign of the induced voltage is such that the induced current in circuit 2 must produce magnetic flux opposing the inducing flux (Lenz's law).

Let us represent symbolically two circuits each consisting of a conducting wire and resistances and connected to ideal ground as in Figure 10.4. Circuit 1 is fed by a voltage V, which may be some undesirable voltage or noise. The two circuits are inductively coupled through their mutual inductance M_{12}. The voltage induced in the circuit 2 is $V_{\text{ind}} = j\omega M_{12}I_1$, and it produces a current I_2 in the direction as shown. Note that inducing flux (due to I_1) in circuit 2 is opposed by that produced by I_2, as indicated in Figure 10.4. The magnetically induced voltage (if undesired) can be reduced by (1) reducing M_{12}, namely, by increasing the separation distance between the loops and/or by adjusting the orientation of the loops, and (2) reducing the area of the loop 2, which would reduce ψ_{12}. It should also be noted that V_{ind} increases with increase of frequency and the current I_1.

From the practical application point of view it is important to observe that in contrast to the electric coupling where the induced current I_{ind} appears as a current generator in shunt with the receptor conductor [1], the magnetic coupling produces an induced voltage V_{ind} in series with the receptor circuit conductor (Figure 10.4).

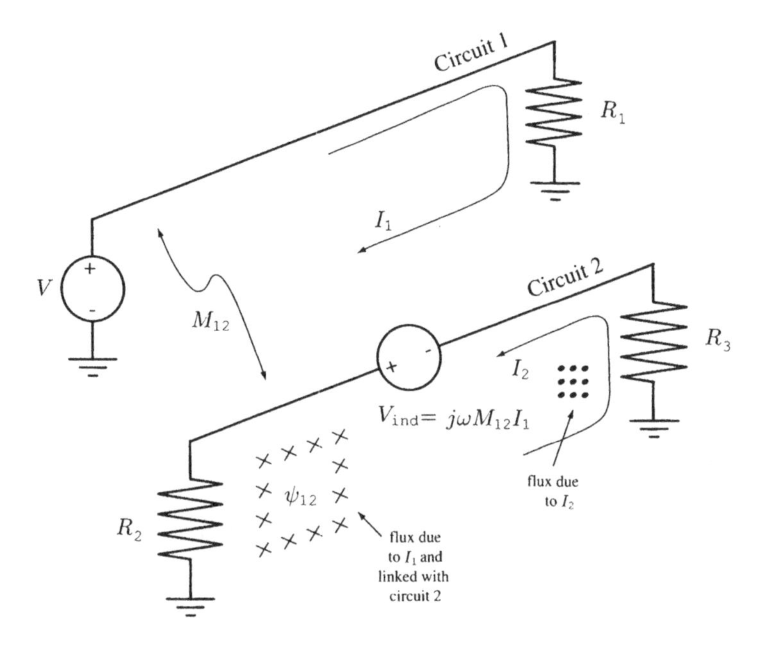

Figure 10.4 Two simple circuits inductively coupled with each other. (Source: From [1], p.39.)

The differences between these two couplings can be demonstrated [1] with the help of the following two equivalent circuits shown in Figure 10.5 for the induced generators for the electric and magnetic coupling cases (also consult Figure 10.1).

For convenience, we have identified A, B as the two ends of the receptor circuit conductor 2 and have assumed that current and voltage generators are at the middle of the receptor conductor. R_v's are variable resistors, and V may be looked upon as the voltage developed across R by electric and magnetic coupling.

With reference to Figure 10.5, if V decreases with decrease of R_v, then the coupling is electric; the coupling is inductive or magnetic if V increases with decrease of R_v.

10.3.2 Shielding of the Receptor Conductor

We now discuss the effects of surrounding the receptor conductor 2 of Figure 10.4 by a cylindrical nonmagnetic shield shown in Figure 10.6, where V_1 is the noise voltage picked up by the circuit 1, M_{1s}, M_{12}, M_{s2} are the mutual inductances between the conductors 1 and the shield, conductor 1 and conductor 2, respectively.

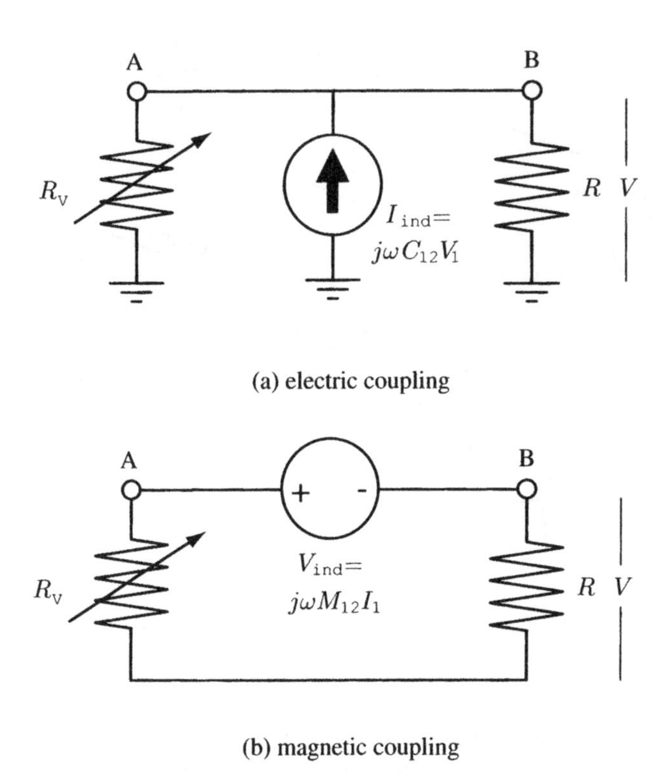

Figure 10.5 Equivalent circuits: (*a*) electric coupling and (*b*) magnetic coupling.

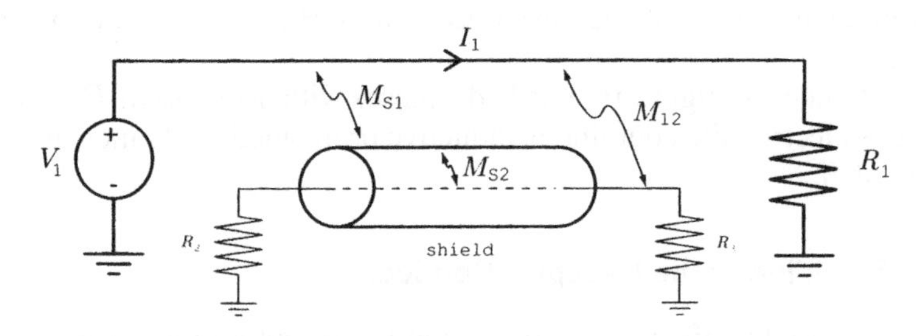

Figure 10.6 Inductively coupled circuits 1 and 2 with the receptor conductor 2 surrounded by a nonmagnetic shield. (Source: From [1], p.43.)

Assuming a uniform axial current along the tubular shield, application of field theory concepts indicates that there will be no magnetic field (flux) within the cavity; the magnetic flux produced by the shield current will be confined outside the shield. It therefore follows that the mutual inductance M_{s2} between the shield and the receptor conductor 2 inside the shield is [1]

$$M_{s2} = L_s, \tag{10.6}$$

where L_s is the inductance of the shield. Also the location and configuration of the shield and the conductor with respect to the conductor 1 are such that

$$M_{1s} = M_{12}. \tag{10.7}$$

We will make use of relations (10.6) and (10.7) for later calculations.

The current I_1 will induce a voltage V_{ind}, in series with the conductor 2, and is given by

$$V_{ind_1} = j = \omega M_{12} I_1. \tag{10.8}$$

There also will be a series voltage V_s induced on the shield by the current I_1, whether the shield is grounded or not, and it is given by

$$V_s = j\omega M_{1s} I_1. \tag{10.9}$$

There cannot be any shield current, if it is ungrounded or grounded at one end only. Under these conditions the shield voltage will not affect the current in conductor 2. However, there will be a shield current I_s when both ends of the shield are grounded, and it will induce an (extra) voltage V_{inds} in series with conductor and thereby affect the current in conductor 2. The induced voltage is given by

$$V_{inds} = j\omega M_{s2} I_s. \tag{10.10}$$

For the purpose of determining the coupled voltages in conductor 2 in the receptor circuit we use the symbolic circuit shown in Figure 10.6 as an equivalent circuit for the two magnetically coupled circuits shown in Figure 10.7 with the shield grounded at both ends, and where all the notations are as explained earlier except L_s and R_s being the inductance and resistance of the shield, respectively.

The polarity of the various voltages and the directions of the appropriate currents should be noted for consistency with Lenz‘s law. From the shield equivalent circuit in Figure 10.7 we have for the shield current

$$I_s = \frac{V_s}{j\omega L_s + R_s}. \tag{10.11}$$

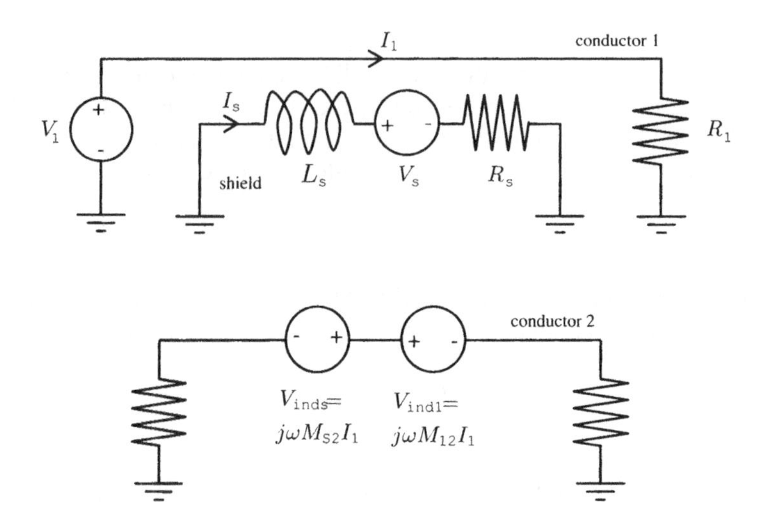

Figure 10.7 Equivalent circuit for the two inductively coupled circuits with the receptor conductor surrounded by a grounded shield. (Source: From [1], p. 48.)

Using (10.10) and (10.11) and in view of (10.6), we obtain the following for the voltage induced on conductor 2 by the shield current I_s:

$$\begin{aligned} V_{\text{inds}} &= j\omega M_{s2} I_s \\ &= \frac{j\omega}{j\omega + \omega_c} V_s, \end{aligned} \tag{10.12}$$

where V_s is given by (10.9) and

$$\omega_c = \frac{R_s}{L_s} \tag{10.13}$$

is called the *shield cut-off frequency*.

V_{inds} versus $\log \omega$ is shown in Figure 10.8. Note that for frequencies larger than the shield cut-off frequency $|V_{\text{inds}}|$ asymptotically approaches the shield voltage. For further discussion see [1].

From the equivalent circuit of the receptor circuit in Figure 10.7, we have for the total induced voltage V_{ind_T} as

$$\begin{aligned} V_{\text{ind}_T} &= V_{\text{ind1}} - V_{\text{inds}} \\ &= j\omega M_{12} I_1 - j\omega M_{s2} I_s, \end{aligned} \tag{10.14}$$

which from the relations given earlier can be written as

$$V_{\text{ind}_T} = j\omega M_{12} \frac{\omega_c}{j\omega + \omega_c} I_1. \tag{10.15}$$

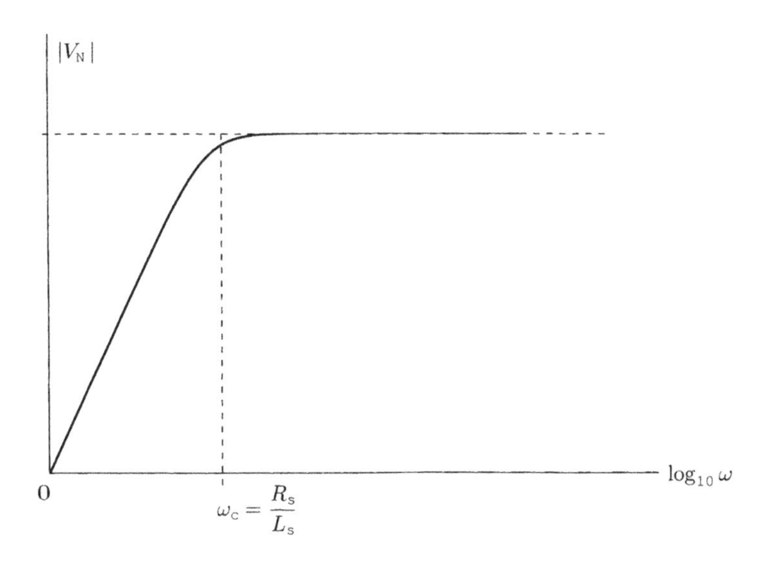

Figure 10.8 Induced voltage on the receptor conductor 2 due to shield current. (Source: From [1], p. 46.)

A plot of $|V_{ind_T}|$ versus $\log \omega$ is shown in Figure 10.8.

At low frequencies $\omega \ll \omega_c$ (10.14) shows that

$$V_{indT} = V_{ind1} = |j\omega M_{12} I_1), \tag{10.16}$$

which is the same as that produced by the unshielded case. But Figure 10.8 indicates that with a grounded shield, for $\omega \gg \omega_c$, the induced voltage remains constant at a value less than that for the under grounded case:

$$\begin{aligned} |V_{ind_T} &\simeq \left| j\omega M_{12} \frac{\omega_e}{j\omega} I_1 \right| \\ &= M_{12} I_1 \frac{R_S}{R_L} . \end{aligned} \tag{10.17}$$

Figure 10.9 gives a plot of (10.15) on a function of ω. It indicates at $\omega < \omega_c$ the noise pickup in the shielded cable is the same as in the unshielded cable; at $\omega > \omega_c$ the pickup voltage in the shielded cable remains constant and below that in the unshielded cable.

Other design considerations for shielding of magnetic field effect are described in [1, 3, 4].

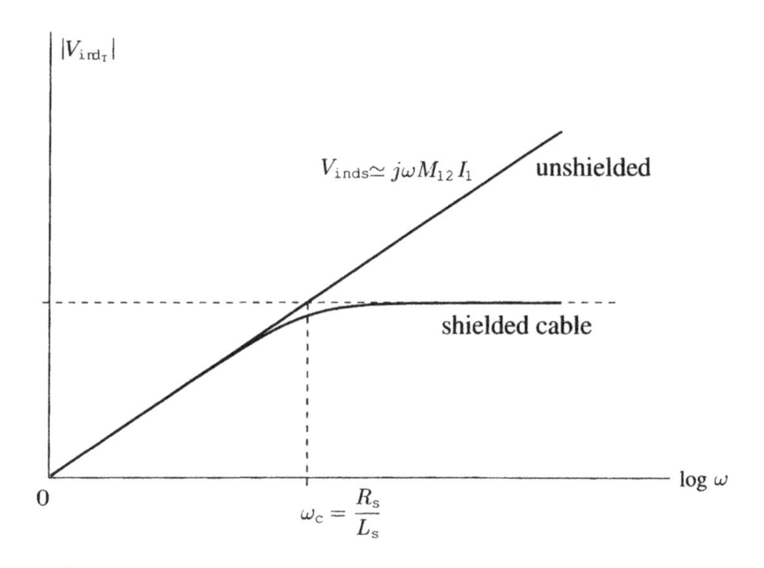

Figure 10.9 Total magnetically coupled induced voltage on the receptor versus frequency for the receptor conductor with a grounded and ungrounded shield. (Source: From [1], p.49.)

REFERENCES

1. H. W. Ott, *Noise Reduction Techniques in Electronic Systems*, 2nd ed., Wiley, New York, 1988.
2. J. P. Mills, *Electromagnetic Interference Reduction in Electronic Systems*, PTR Prentice-Hall, Englewood Cliffs, NJ, 1993.
3. F. M. Tesche, M. V. Ianoz and T. Kasson, *EMC Analysis Methods and Computational Models*, Wiley, New York, 1997.
4. C. R. Paul, *Introduction to Electromagnetic Compatibility*, Wiley, New York, 1992.

CHAPTER 11

ELECTROSTATIC DISCHARGE (ESD)

11.1 INTRODUCTION

Electrostatic discharge (ESD) is a phenomenon in which static electric charges accumulated on one object suddenly discharge to another object having a lower resistance to ground. The *IEEE Guide* [1] defines ESD as the sudden transfer of charge between bodies of different electrostatic potentials. The transfer of charge with resultant discharge currents generates electromagnetic fields over a broad range of frequencies from dc to low gigahertz [1]. The semiconductor integrated circuits and components are especially susceptible to the effects of uncontrolled ESD events in their vicinity. They can affect the performance of an electric device in a variety of ways: by intense electromagnetic fields due to the charge, by breakdown and its associated effects, and by the discharge currents causing conducted and radiated emissions. An ESD event can cause equipment malfunctions in the form of data corruption and equipment lockup, and physical damage in the form of equipment damage and even loss of life [1]. It is therefore important to provide, during the design phase, meaningful ESD

Applied Electromagnetics and Electromagnetic Compatibility. By D. L. Sengupta, V. V. Liepa
ISBN 0-471-16549-2

immunity to a device by taking into account the discharge and associated field configurations that will be encountered by the entire system during its expected lifetime. Various aspects of ESD are described in [1–9]. In the present chapter we describe briefly the ESD phenomenon, its possible effects on an electric device, and the types of interference it may cause. Our discussion is mainly based on the *IEEE Guide* [1] and on [2,3]; more detailed discussion of ESD can be found in the other references cited.

11.2 ACCUMULATION OF STATIC CHARGE ON BODIES

There are a number of ways that a body or bodies can be electrostatically charged where the charging is defined as all processes that produce a separation of positive or negative electric charges on the body under consideration. The main processes relevant to ESD are the following:

- **Triboelectric Charging [1, 2] [4, 5].** Static electricity is generated by contact and/or rubbing of two materials of different dielectric constant. This charging mechanism is referred to as *triboelectric charging* and involves electron or ion transfer upon contact. During the process some materials readily absorb electrons and thereby acquire negative charge while others tend to give up electrons and thereby acquire positive charge. Table 11.1 presents the *triboelectric series* [1,2,5], a listing of materials in the order of their affinity for giving up electrons.

  Note the positive and negative connotations at the top and bottom of the table; this means that when two materials of the table are in contact, the one closer to the beginning of the series will acquire positive charge, while the one closer to the end of the series will acquire negative charge. The farther apart the materials are in the series, the greater the magnitude of the static charge that will develop. For example, rubbing nylon against Teflon can cause electrons to be transferred from the nylon surface to the Teflon surface. As a result nylon may acquire positive charge and Teflon may acquire negative charge. The electrostatic charge that can be developed between two materials depends on their relative position in the triboelectric series, and also in their surface (or volume), respectively [1]. Materials higher in the series can accumulate significant charge because a longer time is required for the charge to decay in such materials (note: decay or relaxation time τ in a material is $\tau = \epsilon/\sigma = \epsilon\rho$, where ρ is the resistivity of the material). Under certain instances even conductive materials such as metals can be charged, as shown in Table 11.1. The order of ranking in a triboelectric series is not fixed [1], and moreover it is not always possible to duplicate the series. Also, the magnitude of the charge transfer depends on a number of factors in addition to the material's position in the series: for example, the

Table 11.1 Triboelectric Series.

	POSITIVE		
1.	Air	18.	Hard rubber
2.	Human skin	19.	Mylar
3.	Asbestos	20.	Epoxy glass
4.	Glass	21.	Nicked, copper
5.	Mica	22.	Brass, silver
6.	Human hair	23.	Gold, platinum
7.	Nylon	24.	Polystyrene foam
8.	Wool	25.	Acrylic
9.	Fur	26.	Polyester
10.	Lead	27.	Celluloid
11.	Silk	28.	Orlon
12.	Aluminum	29.	Polymer transform
13.	Paper	30.	Polyethylene
14.	Cotton	31.	Polypropylene
15.	Wood	32.	PYC (vinyl)
16.	Steel	33.	Silicon
17.	Sealing wax	34.	Teflon
			NEGATIVE

Source: From [2], p. 323.

smoothness of its surface, surface cleanliness, the pressure of contact, and the speed of separation, [1,2].

- **Charging by Induction** Induction or electrostatic induction (an IEEE term) may cause portions of the human body or other items to be differentially charged. If a charged object (e.g., the face of a cathode ray tube) is brought near another object, it will polarize the charges on the object if it is nonconducting (an insulator) or redistribute the charges on the object if it is conductive. Thus, although the object may remain electrically neutral, its charges may be distributed unequally, thereby creating the possibility of ESD from a portion of the person or the body [1].

11.3 CHARGING AND CHARGE SEPARATION

We saw earlier that contact between two materials causes transfer of charge. Because the materials are physically separated, the resultant charge separation creates an electric field E between them. If this E field is sufficiently strong, a voltage breakdown between the objects may occur; this may result in a breakdown of the intervening medium, causing an arc and its associated electromagnetic effects. Consider a simple case where a transfer of charges $+Q$ and $-Q$ has taken place between two bodies. The relationship between the charge amplitude and the voltage between the objects is $V = Q/C$, where C is the capacitance of the system. Observe that as the two bodies are separated the charge Q remains fixed, $Q = VC =$ constant. We know that $C \propto 1/d$, where d is the separation distance between the bodies. Thus for small distances C is large but V is small, as indicated by (10.2). As the objects are separated farther, C decreases; V must therefore increase to keep Q constant.

Example 11.1

Let $C = 10$ pF, $Q = 1\,\mu$C, $V = Q/C = 10^9$ KV for a certain value of d. It is possible that at a certain separation distance the intervening air medium may break down. Note that the breakdown field strength for air is 30 MV/m or 30 KV/cm. Note also that the energy stored in a charged system like the human body or any other object is a function of the capacitance and its voltage, and is given by $W = \frac{1}{2}CV^2$, where C is the capacitance in farads, V is voltage in volts, and W is energy in joules. For example [1], a person walks across a carpet and becomes charged to 10 KV. Assuming a body capacitance of 100 pF, we determine the stored energy as

$$W = \frac{1}{2}\left(100 \times 10^{-12}\right)\left(10 \times 10^{3}\right)^{2} = 5 \text{ mJ}.$$

The charging and separation of accumulated charges and their associated effects are simple to understand when the two objects are insulators. They also would occur when an insulator is separated from a conductor. The degree of charge separation is less than that for two insulators because as the conductor is separated from the insulator the charge tends to redistribute itself over the conductor in relation to the relaxation time ϵ/σ mentioned earlier. Observe also that static charges are stored on the surface of a conducting material; with an insulator the charges tend to remain in the vicinity of the attachment point, whereas for a conducting material the charges tend to redistribute over the surface. Grounding a conductor will bleed off the charge, but grounding an insulator will not. A charged insulating body alone may not pose a problem because the charges on it are not free to move and hence it cannot produce a static discharge. However, a charged insulator can induce charge on a conductor in its vicinity; thus there may be electrostatic discharge associated.

Generally, the ESD is a three-step process: charge is generated on an insulator, the charge is then transferred to a conductor, and the charged conductor comes near a metal object, usually grounded, and a discharge takes place.

Common examples of charging processes are given in [1]. Cathode ray tubes used as display devices and sometimes as input devices for computer terminals, oscilloscopes, and television sets can be sources of charges. Humans are a prime source of electrostatic discharge carrying harmful effects to electronics devices. The next section discusses the human body from this viewpoint.

11.4 HUMAN BODY AS SOURCE OF ESD

Electrostatic charge accumulation takes place as a result of triboelectric effects when a human wearing shoes with soles made of insulating materials (e.g., polyurethane foam) walks on low conductivity materials such as wood or the synthetic materials in carpets [1, 2]. The charged shoe soles cause charge redistribution on the human body by induction. As each step is taken, this charge redistribution increases [1]. The maximum total electrostatic charge that can accumulate is of the order of 10^{-6} coulombs or more, depending on the nature of the carpet and the shoes, the humidity, the distance the person has walked on the carpet, and other factors. This could result in a voltage up to 15 KV [4]. To describe ESD characteristics and variables, we follow the *IEEE Guide* [1] and schematically illustrate the direct ESD event as in Figure 11.1.

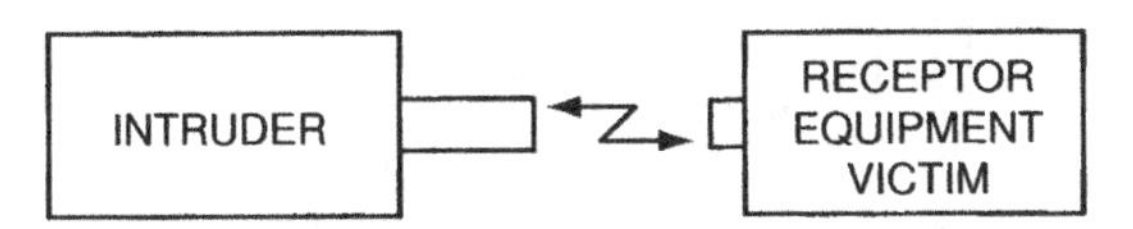

Figure 11.1 Direct ESD event. (Source: From [1], p. 18.)

The intruder is frequently a human hand with or without an associated metal object, but it may be an object such as a chair or equipment cart. The receptor is the stationary object to which the charges are transferred by the intruder where the receptor may be at local electrostatic ground potential. The equipment victim is an electronic device that is often the receptor, but it may also be the intruder or an object that is neither the intruder nor the receptor. In this last case the equipment victim is in close proximity to the ESD event and may be upset or damaged by the electromagnetic fields generated by the discharge between the intruder and the receptor. When either the intruder or the receptor is an equipment victim, the ESD is termed *direct*. When the equipment victim is neither the intruder nor the receptor, the ESD is termed *indirect*.

Examples of direct ESD events of various kinds are shown in Figure 11.2 through 11.8 taken from [1]. The intruder may be human (Figures 11.2 and 11.4), an inanimate object (Figures 11.5; 11.8), or a combination of the two (Figures 11.3, 11.6, 11.7). Table 11.2 lists the various kinds of intruders with the names of the ESD events that result from each.

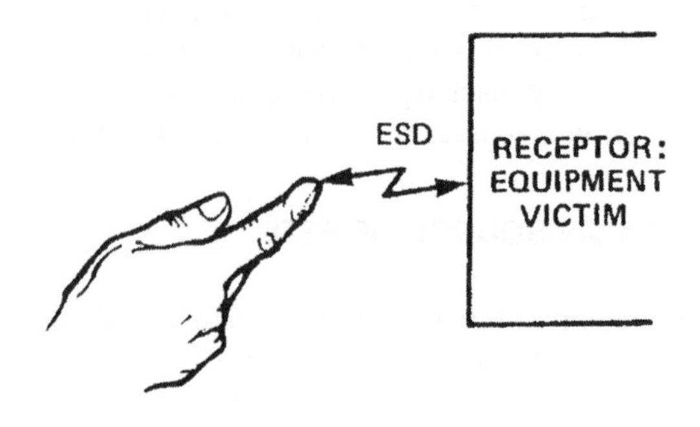

Figure 11.2 Direct body/finger ESD. (Source: From [1], p. 19.)

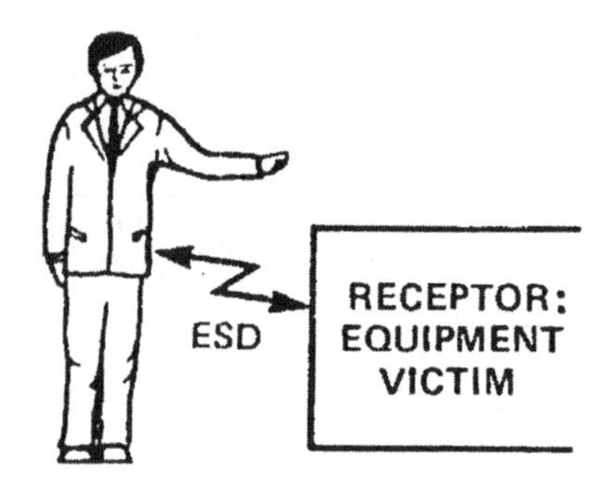

Figure 11.3 Direct hand/metal ESD or body/metal ESD. (Source: From [1], p. 19.)

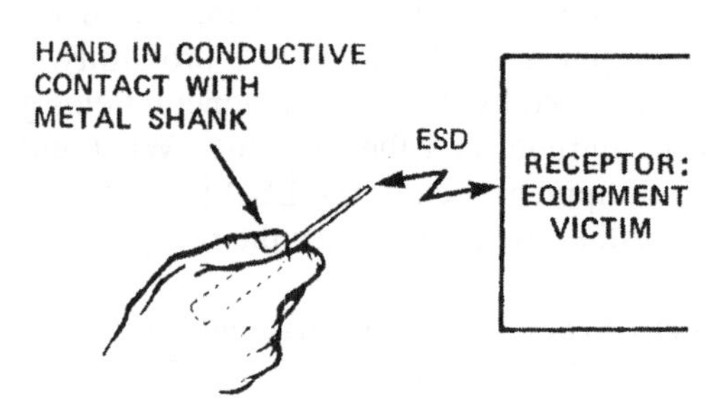

Figure 11.4 Direct brush-by ESD. (Source: From [1], p. 19.)

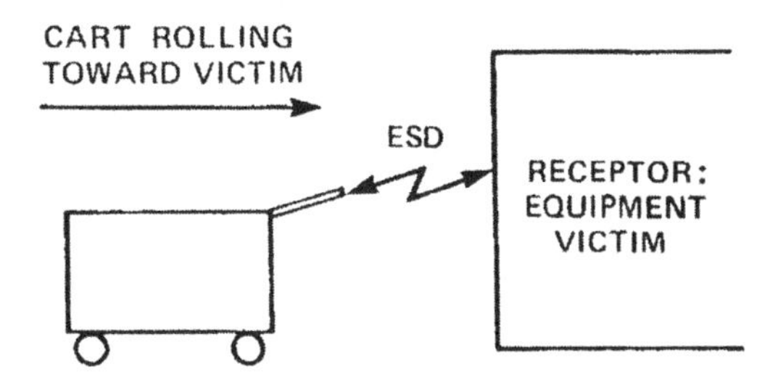

Figure 11.5 Direct furniture ESD—Rolling cart alone. (Source: From [1], p. 20.)

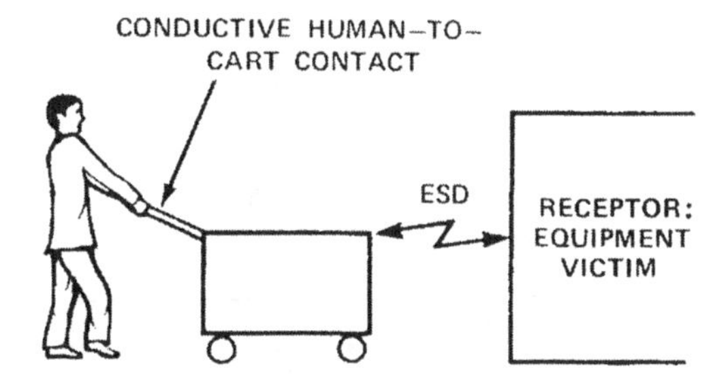

Figure 11.6 Direct furniture ESD—Rolling cart with human. (Source: From [1], p. 20.)

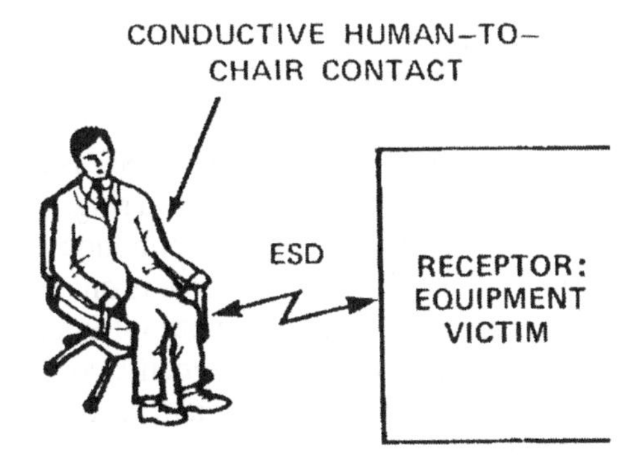

Figure 11.7 Direct furniture ESD—Rolling chair with human. (Source: From [1], p. 20.)

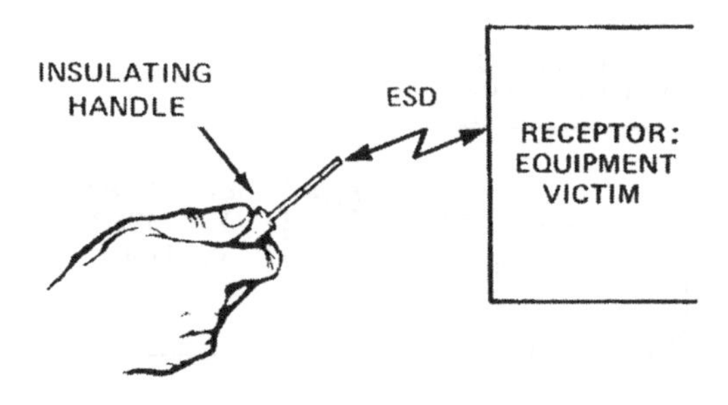

Figure 11.8 Direct furniture ESD—Screwdriver alone. (Source: From [1], p. 21.)

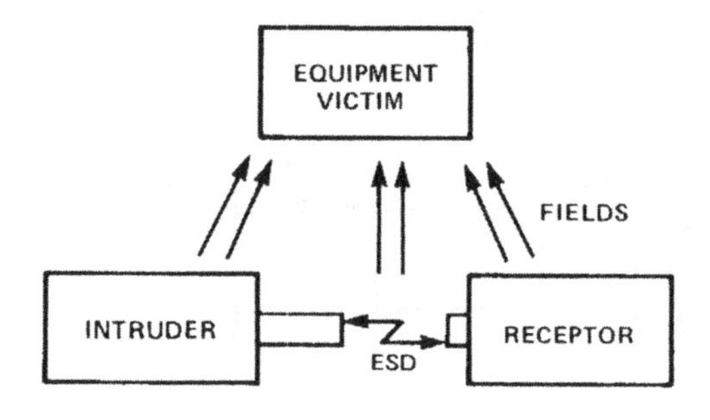

Figure 11.9 Indirect ESD event. (Source: From [1], p. 21.)

In the case of a human intruder, the ESD may be a body/finger ESD that originates directly from the skin (Figure 11.2). It may also be a brush-by that originates from the body (Figure 11.3). In the case of a human contact with a small metallic object, a body/metal or hand/metal ESD is one that originates from a metallic object associated with a hand such as a ring, watch, screwdriver, or key as in Figure 11.4. If the human is not in conductive contact with the small metallic object, the source of the ESD charge is only the metallic object as in Figure 11.8. In the case of a large inanimate intruder the ESD event is called a *furniture ESD*, and it originates from a specific conductive location on the intruder. Inanimate intruders include equipment carts without a human, as in Figure 11.5, and with a human as in Figure 11.6, and a human with a chair as in Figure 11.7. The energy transferred from an intruder to a receptor during ESD depends on the impedance between them and on the energy dissipated in the arc. Figure 11.9 shows an indirect ESD event. The equipment victim is

Table 11.2 ESD Intruders and Resulting ESD

ESD Intruder	Type of Resulting ESD
Human object	Body/finger—from finger without associated metal object Brush-by—from hip, torso, or shoulder
Metal object	Hand-associated—tool or other object not in conductive contact with the hand or body Furniture—not in conductive contact with the hand or body
Human/metal object	Body/metal or hand/metal—tool, ring, watch, or other metal object in conductive contact with the hands or body Furniture and human—in conductive contact with the hand or body; human seated in chair, pushing cart, or vacuum cleaner; holding and in conductive contact with a tool, etc.

(Source: From [1], p. 18.)

irradiated by the fields, where the principal source is a result of the ESD (i.e., the ESD arc) and a function of the the intruder mass and the receptor mass [1].

11.5 ESD WAVEFORMS

Human and furniture ESD current waveforms are described in [1]. Knowledge of these waveforms is important for designing the waveform generators used for ESD testing of electric devices. We describe here a few typical waveforms given in [1]. Current waveforms due to ESD events generally fall into two categories: those with extremely steep initial slopes and those without extremely steep slopes.

Figure 11.10 shows a current waveform originating from a lower voltage (approximately 4 KV) human hand/metal ESD event having a steep initial slope (approximately 1 ns) and an initial spike. Figure 11.11 shows a hand/metal ESD current wave initiated from a higher voltage (or from a sharper point or slower approach). This wave displays a much slower initial slope and has no initial spike or pulse. Furniture ESD events produce current waves with steep initial slope, but they do not always have an initial spike, which

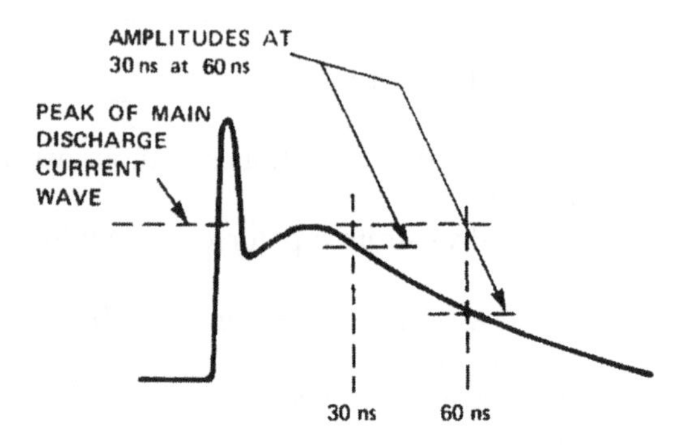

Figure 11.10 Full hand/metal wave. (Source: From [1], p. 27.)

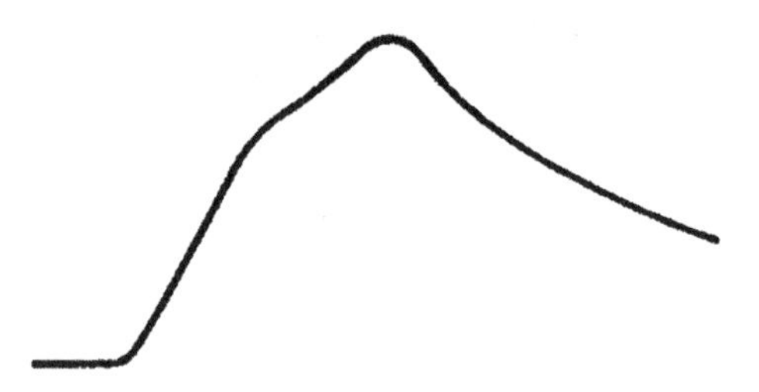

Figure 11.11 Hand/metal wave without initial spike. (Source: From [1], p. 27.)

occurs less frequency with such events. The main discharge wave for furniture ESD is typically oscillatory, since the total impedance in the discharge circuit path is not much more than the resistance; in other words, there is minimal damping in the circuit. Two typical furniture current waves with fast and slow initial slopes are shown in Figures 11.12 and 11.13, respectively. For comparison Figure 11.14 shows an overlay of waveforms in Figure 11.10 and Figure 11.12 for hand/metal ESD and furniture ESD, respectively. Observe that the initial slopes of the two are the same, and also note the oscillatory nature of the full waveform in the furniture case. The transient pulses due to ESD events can cause logic errors, process resets, and even physical damage to semiconductor devices. The fast rise time pulses due to ESD may interfere with clock transitions in digital circuits and thus can give rise to upset and malfunction [9].

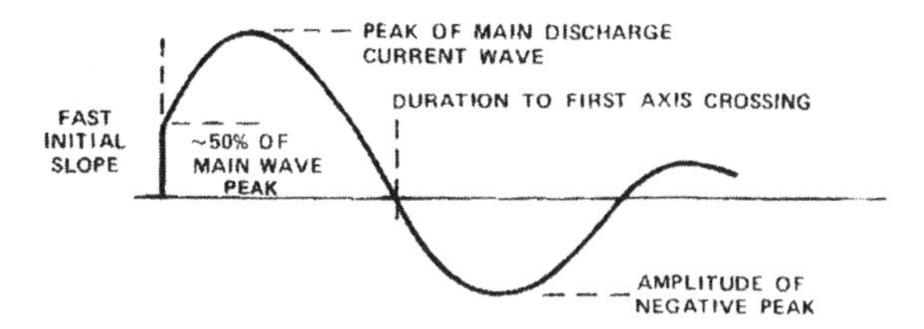

Figure 11.12 Full furniture wave (fast front). (Source: From [1], p. 28.)

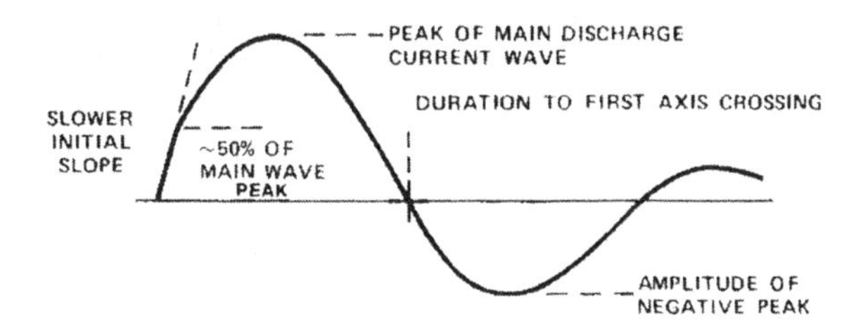

Figure 11.13 Furniture ESD current waves. (Source: From [1], p. 28.)

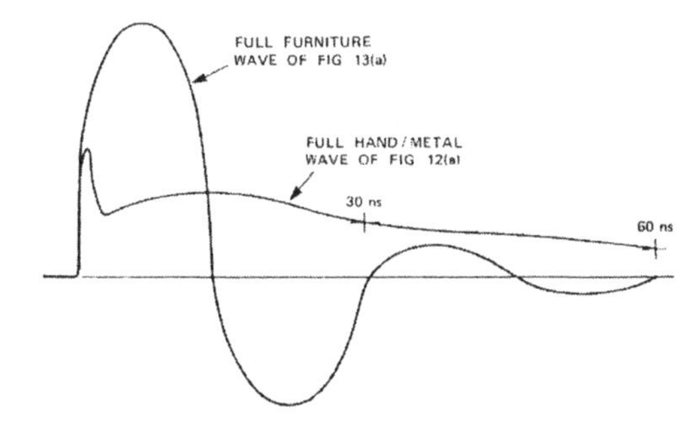

Figure 11.14 Overlaid current waves from hand/metal and furniture ESD current waves. (Source: From [1], p. 29.)

11.6 HUMAN BODY CIRCUIT MODEL

As mentioned earlier, humans are the most common source of ESD to cause undesirable interference to electronic systems in their vicinity. For analytical and/or estimation of interference effects thus produced, it is convenient to use an RLC equivalent circuit to model the path of an ESD involving a human body with its arms and fingers and an object EUT (equipment under test) through which a discharge takes place. Such modeling is described in [1–5]. The three sources of ESD that are of greatest interest to a designer of electronic equipment, and that are therefore the most important to model, are (1) body/finger ESD (discharge from fingertip without an associated metal object; Figure 11.2), (2) body/metal or hand/metal ESD (discharge from a metal object in contact with a hand; Figure 11.3), and (3) furniture or inanimate object ESD (discharge from a cart, chair or other inanimate object with or without a human in conductive contact with the object; Figure 11.8). Equivalent circuit models for the above are described in detail in [1] and also in [2–4, 9]. Approximate equivalent circuit models to study ESD from human body are shown in Figure 11.15, where in the simplified model C_b represents the body capacitance to ground (typically $C_b \sim 150$ pF) and R_b represents the body resistance (typically $R_b \sim 1.5$ KΩ). In the literature one typically finds $C_b \sim 60$–300 pF and $R_b \sim 1.5$–10 KΩ.

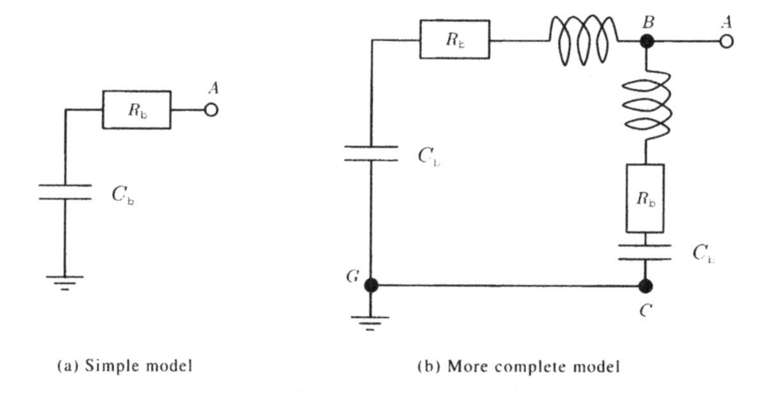

Figure 11.15 Equivalent circuits: (*a*) simple model and (*b*) more complete model for the human body.

When terminal A is brought to contact with a grounded conducting body, the capacitor C_b discharges, causing the flow of current that may result in conducted and radiated interference. Figure 11.15*b* shows an improved model where a small inductance L is added and the branch B–C is a model of the human hand. In this model the discharge current has a fast rise-time component (less than 1 ns) due to branch BC and a slower component (100 ns) due to branch BG. More elaborate models and their descriptions are given in [1] and in the other references cited.

11.7 ESD GENERATOR AND ESD TEST

From the discussions in Section 11.5 it can be seen that voltage/current pulses of desired shape and intensity are required to carry out meaningful ESD testing on equipment and its components. Such pulses can be obtained from an ESD generator. Its discharge electrode for contact discharge has a sharp pointed tip whereas the electrode has a round tip for air discharge tests. The test generator usually has provisions to deliver test voltage/current pulses at various intensity levels. Further details about the generators and their applications are described in [4]. For a given piece of equipment under test (EUT), its immunity to ESD is estimated by carrying out the following tests: (1) air discharge test where a charged electrode is brought close to the device under test so that electrostatic discharge takes place in the form of a spark between the charged electrode and the EUT, and (2) contact discharge test where the charged electrode is held in contact with the EUT and the discharge is initiated by means of a switch in the generator circuit. Further details for the tests are given in [4].

REFERENCES

1. *IEEE Guide on Electrostatic Discharge (ESD): Characterization of the ESD Environment*, IEEE Std. c 62. 47–1992, IEEE Press, New York, 1993.
2. H. W. Ott, *Noise Reduction Techniques in Electronic Systems*, 2nd ed., Wiley, New York, 1988.
3. C. R. Paul, *Introduction to Electromagnetic Compatibility*, Wiley, New York, 1992.
4. V. P. Kodali, *Engineering Electromagnetic Compatibility*, IEEE Press, New York, 1996.
5. N. Boxleitner, *Electrostatic Discharge and Electronic Equipment: A Practical Guide for Designing to Prevent ESD Problems*, IEEE Press, New York, 1989.
6. M. I. Montrose, *Printed Circuits Board Design Techniques for EMC Compliance*, 2nd ed., IEEE Press, New York, 2000.
7. *Electromagnetic Compatibility for Industrial Process Measurement and Control Equipment, Part 2: Electrostatic Discharge Requirements*, International Standard, IEC 801-2, International Technical Commission, Geneva, 1991.
8. *Guide to Electrostatic Discharge Test Methodologies and Criteria for Electronic Equipment*, IEEE c 63. 16, Institute of Electrical and Electronics Engineers, New York, 1991.
9. C. Christopoulos, *Principles and Techniques of Electromagnetic Compatibility*, CRC Press, Ann Arbor, 1995.

CHAPTER 12

EMC STANDARDS

12.1 INTRODUCTION

Most electronic devices, including digital devices, give rise to electromagnetic emissions, which either by radiation through free space (air) or by conduction through some conducting paths may act as unwanted EMI signals or noise to another electric device in their vicinity. It is therefore desirable that such emissions be controlled so that their interference effects may be kept at minimum or acceptable levels. For this reason all electronic devices are required to satisfy certain regulations and standards established by national bodies, trade associations, international organizations, and the like. In most cases these EMC standards are similar, but there are also differences in the specific limits and the methods of measurement specified by various organizations. EMC standards may be classified into three broad categories: (1) those mandated by a national government agency, which must be met, (2) those imposed voluntarily by product manufacturers, which are specific to each manufacturer, and (3) those imposed by military agencies of a country. For example, US military

Applied Electromagnetics and Electromagnetic Compatibility. By D. L. Sengupta, V. V. Liepa
ISBN 0-471-16549-2

agencies require that their own standards must be met by a product for their consideration. US military standards regulate emissions from the product as well as immunity to emissions from other devices, including to ESD. Note that a product that meets these standards may not be immune to interference. Thus, although complying with a country's EMC standards is essential for the marketing of the product in that country, it should not be taken as a substitute for detailed EMC consideration during the design process.

The subject of EMC standards and their evolution is vast. Brief summaries of standards are given in [1–4], and detailed descriptions of US standards and methods of testing are described in [5–7]. In this chapter we will concentrate on the US and European civilian standards and regulations. We will also touch upon the international hierarchy set up to create uniform standards and regulations that apply to the European Union (EU) countries and their equivalence with US regulations. Many countries, including the United States, will eventually accept these standards, known as "harmonized" standards. Finally, these standards and regulations are modified and up-dated regularly; hence one should consult the latest versions before venturing to meet these standards.

12.2 CURRENT US STANDARDS

12.2.1 Introduction

Any electronic device to be marketed in the US must meet certain requirements or standards established by the Federal Communications Commission (FCC) within the country. Historically the rules for nonlicensed use of RF devices were first established by the FCC in 1983 when "the limit applied to the emissions by these early devices was 15 μV/m at a distance equivalent to $\lambda/2\pi$, where λ is the operating wave length [5]." Part of the FCC's responsibility includes creation (or adoption) of regulations that control electromagnetic interference. These regulations are published in the Code of Federal Regulations, Telecommunications 47 (Washington, DC, US National Archives and Records Administration). Regulations specifying limits for electromagnetic emissions for radio frequency devices and equipment (both intentional and unintentional radiators) are covered in Part 15 [5]; Part 18 provides similar information for industrial, scientific, and medical (ISM) equipment. The FCC general procedures for measuring emissions are given in [6]; the FCC also advocates and encourages the use of procedures outlined by the American National Standards Institute (ANSI) [7]. The FCC defines an *RF device* as any "device that in its operation is capable of intentionally or unintentionally emitting radio frequency energy by radiation, conduction or some other means." *RF energy* defined by the FCC is any electromagnetic energy in the frequency range of 9 kHz to 3000 GHz. The purpose of the standards and regulations set by the FCC is that if a given device meets these standards, then it suffices that it does not cause or create unwanted electromagnetic signals or noises that could otherwise affect the performance of another electromagnetic device located beyond a certain minimum distance from the device.

In this chapter we will concentrate on *digital devices*. A digital device is defined by the FCC as "an unintentional radiator (device or system) that generates and uses timing signals or pulses at a rate in excess of 9,000 pulses (cycles) per second and uses digital techniques; inclusive of telephone equipment that uses digital techniques or any device or system that generates and uses radio frequency energy for the purpose of performing data processing functions, such as electronic computations, operations, transformations, recording, filing, sorting, storage, retrieval, or transfer."

Note that an intentional radiator "that contains a digital device is not subject to the standards for digital devices, provided the digital device is used only to enable operation of the radio frequency device and the digital device does not control additional functions or capabilities." Transmitters, such as automobile remote door openers are called intentional radiators, and must meet separate FCC rules and regulations for intentional radiators.

Digital devices may be grouped as Class A and Class B. They are defined in [5] as follows:

Class A Digital Device "A digital device that is marketed for use in a commercial, industrial or business environment, exclusive of a device which is marketed for use by the general public or is intended to be used in the home."

Class B Digital Device "A digital device that is marketed for use in a residential environment notwithstanding use in commercial, business and industrial environments. Examples of such devices include, but are not limited to, personal computers, calculators, and similar electronic devices that are marketed for use by the general public."

12.2.2 FCC Radiated Emission Limits for Digital Devices

The FCC requires that over certain frequency bands the amplitude of the radiated electric field from a device at a specific distance must be equal to or less than certain limiting values and measured following the procedure described in [6]. As an alternative to the FCC's radiated emission limits for digital devices, in an effort to approach the ideal of a harmonized standard, "digital devices may be shown to comply with the standards contained in Third Edition of the International Special Committee on Radio Interference (CISPR), Pub. 22, "Information Technology Equipment–Radio Disturbance Characteristics—Limits and Methods of Measurement," known as CISPR 22. Measurements must be carried out with an antenna horizontally and vertically polarized, meaning both the horizontal and vertical components of the electric field must be measured. Tables 12.1 and 12.2 give the pertinent information for the radiated emissions from Class A and Class B electronic devices.

Table 12.1 FCC and CISPR 22 radiated emission limits for Class A digital devices

	Frequency (MHz)	Field Strength (μV/m)	Field Strength (dB μV/m)	Distance (m)
FCC	30–88	90	39.0	10
	88–216	150	43.5	10
	216–960	210	46.0	10
	>960	300	49.5	10
CISPR	30–230	31.6	30	30
	230–1000	70.8	37	30

Table 12.2 FCC and CISPR 22 radiated emission limits for Class B digital devices

	Frequency (MHz)	Field Strength (μV/m)	Field Strength (dB μV/m)	Distance (m)
FCC	30–88	100	40.0	3
	88–216	150	43.5	3
	216–960	200	46.0	3
	>960	500	54	3
CISPR	30–230	31.6	30	10
	230–1000	70.8	37	10

It is necessary to compare these standards; however, the field limits given above apply to differing distances. To compare them, it is required to express all of them at the same measurement distance, for example, for the Class B distance of 3 m. Assuming that the far zone fields of the source vary as $1/r$, where r is the distance from the source, it can be shown that the relationship between the fields E_1, E_2 at distances d_1, d_2, respectively, from the source is

$$E_1 = \frac{d_2}{d_1} E_2. \tag{12.1a}$$

When the fields are expressed in dBμV/m, the relation is

$$E_1 = 20 \log_{10}\left(\frac{d_2}{d_1}\right) + E_2. \tag{12.1b}$$

Thus, using (12.1b), we add $20 \log_{10}(10/3)$ to the Class A fields given in Table 12.1 to obtain the equivalent fields at a distance of 3 m and then compare them with the corresponding Class B fields given in Table 12.2. Following this procedure, we present in Figure 12.1 Class A and Class B standards at a distance of 3 m.

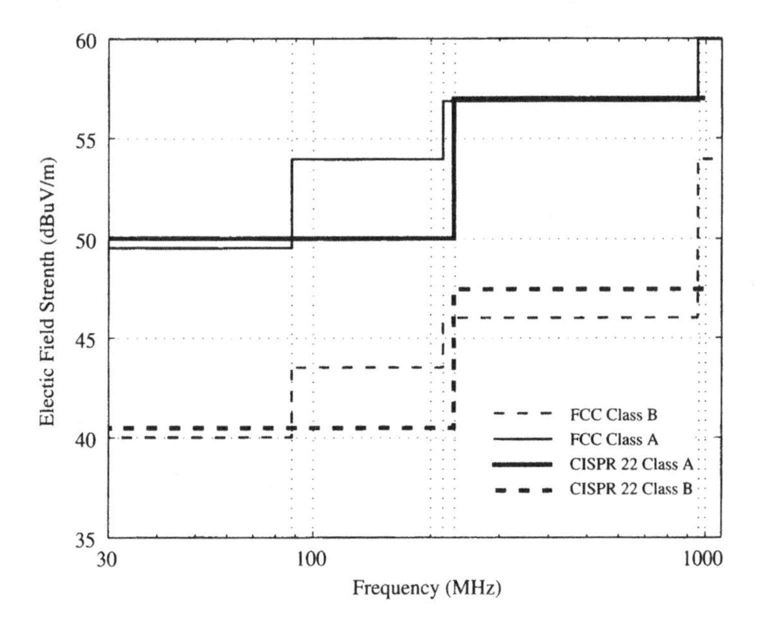

Figure 12.1 FCC and CISPR 22 radiated emission limits extrapolated to a 3 m measurement range.

The results now demonstrate that the Class B standards are more stringent. The above is only approximate. The assumption of far zone behavior of the fields may not be valid at the frequency of interest. For a point radiator (Chapter 6) the far zone distance $d \leq \lambda/2\pi \simeq \lambda/6$. However, in most cases the radiator may be extended; that is, it may have a maximum linear dimension D. In such cases (Chapter 6) the far zone distance is $d \geq 2D^2/\lambda$. Note that for a $\lambda/2$-dipole we have $D \geq 0.5\,\lambda$. Thus the field strength standards may not be in the far field of the device, and, moreover, the measurement antennas and the radiating device may not be in the far zone of each other at the measurement distance for all frequencies tested. This can have significant implications for measurements intended to demonstrate compliance.

12.2.3 FCC Conducted Emission Limits for Digital Devices

The FCC requires that for a "digital device that is designed to be connected to the public utility (AC) power line, the radio frequency voltage that is conducted back onto the AC power line on any frequency or frequencies within the band 150 kHz to 30 MHz shall not exceed the limits in the following table, as measured using a 50 μH/50 ohm line impedance stabilization network (LISN). Compliance with this provision shall be based on the measurement of the radio frequency voltage between the power line and ground at the power terminals [5–7]." Note that the CISPR 22 conducted emission limits have been directly adopted by the FCC.

Table 12.3 FCC and CISPR 22 limits on conducted emission from digital devices

	Frequency of Emission (MHz)	Conducted Limit	
		Quasi-peak (dB μV)	Average (dB μV)
Class A	0.15–0.5	79	66
	0.5–30	73	60
Class B	0.15–0.5	66 to 56[a]	56 to 46[a]
	0.5–5	56	46
	5–30	60	50

[a]Decreases with the logarithm of the frequency.

For quick reference, the conducted emission limits are shown graphically in Figure 12.2.

12.3 EMI/EMC STANDARDS: NON-US COUNTRIES

As mentioned at the beginning of this chapter, most countries, with the exception of the US and Canada, are in the process of adopting European Standards that also include mandatory immunity standards. In the following we give some brief comments about foreign standards; more details and references pertinent to this topic are given in [3].

12.3.1 CISPR Standards

The European-based CISPR (Comité Internationale Spéciale des Perturbations Radio-électroniques) since its founding in 1934 has been developing international standards on EMI/EMC methods of measurements to ensure product

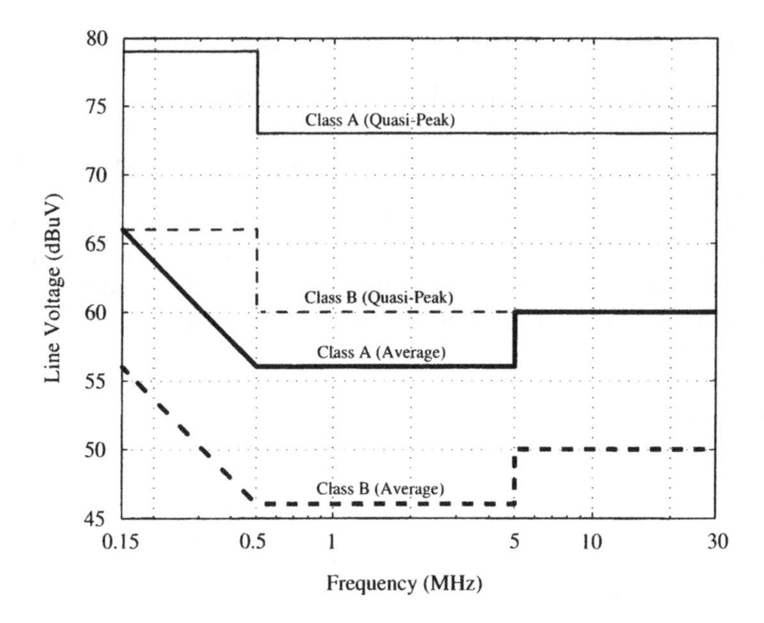

Figure 12.2 FCC and CISPR 22 conducted emission limits for digital devices.

compliance. Among the first agreements in CISPR at its founding, a signal-to-noise ratio of 40 dB was provided as a tolerable limit of interference for a reference field strength of 1 mV/m modulated to a depth of 20 percent [3]. The standards have been published by the IEC (International Electrotechnical Commission). The CISPR/IEC effort is not only on the part of European nations but also other advanced nations such as Australia, Canada, India, Japan, Korea and the United States. CISPR has no regulatory authority, but its standards, when adopted by governments, become national standards. In 1985 CISPR adopted a new set of emission standards (Publication 22) for Information Technology Equipment (ITE) for digital electronics. Many European countries have adopted those requirements as their national standard, and more are expected to do so in the future. The United States, as a voting member of CISPR, voted in favor of the new standards, and is working to merge those standards with its own [2]. The limits given in CISPR Publication 22 with slight modifications are likely to become the international standard. CISPR specifications are described in detail in [1–3].

12.3.2 European Norms

After the formation of the European Common Market and the removal of trade and tariff barriers, a group of countries took the initiative to evolve common EMI/EMC standards in the form of Euro Norms (EN) or European standards [3]. CENELEC (Comité Européen de Normalisation Electrotechnique), representing all concerned European countries, agreed to harmonize and integrate their national standards as Euro Norms that are derived from the related international standards published by CISPR/IEC. Once the Euro Norms are published, the agencies responsible for standardization and regulations in different member countries produce their national standards, which are harmonized with EN standards. Thus identical standards are used in EU member countries. Euro Norms cover not only the emission limits but also the minimum immunity levels for different equipment. More detailed descriptions and guidelines for international aspects of EMI/EMC are given in [3]. Equipment that has been tested and complies with subject European Norms is identified by the letters "CE".

REFERENCES

1. C. R. Paul, *Introduction to Electromagnetic Compatibility,* Wiley, New York, 1992.
2. H. W. Ott, *Noise Reduction Techniques in Electronic Systems,* 2nd ed., Wiley, New York, 1988.
3. V. P. Kodali, *Engineering Electromagnetic Compatibility,* IEEE Press, New York, 1996.
4. C. Christopoulos, *Principles and Techniques of Electromagnetic Compatibility,* CRC Press, Ann Arbor, 1995.
5. FCC General Docket No. 87–389 First Report and Order (Released, April 1987). Revision of part 15 of the Rules Regarding the Operations of Radio Frequency Devices within License, April 18, 1984.
6. FCC, General Docket No. 89–44, Procedure for Measuring Electromagnetic Emissions from Digital Devices, March 7, 1989.
7. ANSI, *Methods of Measurement of Radio Noise Emissions from Low-Voltage Electrical and Electronic Equipment in the Range 9 kHz to 40 GHz,* ANSI. C 63.4–2003, IEEE, New York, 2004.

CHAPTER 13

MEASUREMENTS OF EMISSION

13.1 INTRODUCTION

In the present chapter we discuss the measurement procedures that must be followed to ensure that an electric device meets the recommended electromagnetic emission standards described in Chapter 12. Our discussion in this chapter is confined to the FCC recommended practice for electromagnetic emissions (EME) measurements of Class A and Class B digital devices. The CISPR standards and measurement procedures are similar, although there are certain differences that we will not address here. For convenience, we emphasize here that the emitted radiation is to be measured at the following distances from the device under test (DUT) or the equipment under test (EUT):

FCC requirements for distance d in meters

$d = 3$ from DUT for Class B device
$d = 10$ from DUT for Class A device

Applied Electromagnetics and Electromagnetic Compatibility. By D. L. Sengupta, V. V. Liepa
ISBN 0-471-16549-2

CISPR requirements for distance d in meters

$d = 10$ from DUT for Class B device
$d = 30$ from DUT for Class A device

The appropriate references for the FCC and CISPR recommended measurement procedures are [1–6].

13.2 GENERAL

The FCC specifications require that the radiated and conducted emissions from a DUT must be measured using the complete system described in FCC documents [1,2]. For example, a "minimum configuration of EUT shall be in a manner representative of the equipment as typically used. A digital device marketed to be operated as stand-alone device is to be tested by itself. Conversely, a digital device designed and marketed to be operated as a part of a multi-unit system shall be tested with the device attached to it in a manner that is representative of actual use." Further descriptions of the EUT are given in [1,2,4].

13.3 RADIATED EMISSIONS

13.3.1 Introduction

The FCC recommends that radiated electromagnetic emissions (EME) from the EUT be measured on an open area test site (OATS) meeting the minimum requirements described in FCC OET Bulletin 55 [3]. An alternative site may be used for such measurements provided that the results are corrected satisfactorily to those that would be obtained on an OATS. A shielded room is not acceptable for making radiated EME measurements unless it is properly lined with EME energy-absorbing material to eliminate the reflection effects. It is desirable that the level of both conducted and radiated ambient EME at the test site be at least 6 dB below the FCC limit for the EUT. If the levels are high, guidelines are given in FCC 89–53 [2] for following some corrective procedures. A ground screen is highly recommended but is not mandatory for the measurement of radiated EME. However, OATS are likely to need ground screens when any of the following conditions exists: (1) terrain discontinuity, (2) terrain is subject to extreme seasonal variations in ground conductivity, (3) there are unburied power or control cables on the site, and (4) the site is located on pavement [2].

13.3.2 Receiver

EME measurements can be made with a radio noise meter which conforms to the instrument requirements described in [2]. It is recommended that the measurements at or below 1000 MHz be made with a quasi-peak (QP) detector function on the measuring instrument. Above 1000 MHz measurements are to be made with a peak detector on the measuring instrument. Other instrumentation, such as a spectrum analyzer, may be used for making these measurements. However, the measurements taken with a radio noise meter employing the quasi-peak detector function are preferable to measurements with other instrumentation. The FCC and CISPR test procedures specify that the 6 dB bandwidth of the spectrum analyzer or receiver used to measure radiated EME must be at least 9 kHz for frequencies between 9 and 150 kHz, 100 kHz for frequencies between 150 kHz and 1 GHz, and 1 MHz for frequencies above 1 GHz.

When a spectrum analyzer is used as the measuring instrument, appropriate accessories should be employed to ensure accurate and repeatable measurements of all emissions over the specified frequency ranges. The accessories needed will depend upon specific measurements and may include preamplifiers for improving sensitivity, filters and attenuators for overload protection, and additional quasi-peak circuitry. We note the at this time, most spectrum analyzers with EMC options contain the necessary detection functions.

13.3.3 Antennas

General

The recommended antennas for measuring the radiated EME are as follows:

Frequency range	Antenna type
30–1000 MHz	Tuned half-wave dipole
Above 1000 MHz	Linearly polarized horn

The antennas used must be capable of measuring both horizontal and vertical polarization over the specified frequency. Any other type of linearly polarized antennas may be used to take the measurements between 30 and 1000 MHz provided that the measurements are correlated satisfactorily to a tuned dipole. It is recommended that the test antenna be mounted on a stand that will facilitate raising and lowering the height of the antenna above ground from 1 to 4 m in an even, continuous manner, and there must be provisions made to reorient the antenna from the horizontal to vertical polarization at all heights. The FCC requires [2] that antennas used in testing for radiated EME are to be calibrated for their free space antenna factor (AF) to show tractability to standard dipole antenna of the National Institute of Standards and Technology

(NIST). Antenna factors indicated by the manufacturers may not generally be accurate, so all antennas should be individually calibrated. The radio receiver or the spectrum analyzer used measures the received voltage by the test antenna whereas the standards require the knowledge of the incident field amplitude at the antenna terminals. The antenna factor, specific to the test antenna, provides the incident field amplitude from the measured received voltage. It is important to determine the antenna factor for the test antenna as a function of frequency, that is, to calibrate the antenna as mentioned earlier. This is explained in greater detail in subsequent sections.

Antenna Factor (AF)

The *antenna factor* is defined as the ratio of the amplitude of the incident electric field (or magnetic field) at the test antenna to the voltage received at the antenna terminals:

$$AF = \frac{|\mathbf{E}^{\mathrm{i}}|}{|V_{\mathrm{rec}}|}, \tag{13.1}$$

where $\mathbf{E}^{\mathrm{i}}$ is the incident electric field at the antenna and V_{rec} is the voltage received at the antenna terminals. Figure 13.1 illustrates the antenna receiving the incident field $(\mathbf{E}^{\mathrm{i}}, \mathbf{H}^{\mathrm{i}})$ and its equivalent circuit. In the equivalent circuit Figure 13.1*b*, $V_{\mathrm{oc}} = \mathbf{h}^* \cdot \mathbf{E}^{\mathrm{i}}$ is the open circuit voltage at the antenna terminals induced by the incident E field, and $\mathbf{h}$ is the effective height of the antenna under the transmitting condition, as explained in Section 6.9.2.

Using the definition and the equivalent circuit of Figure 13.1*b*, it can be shown that

$$\mathrm{AF} = \frac{|\mathbf{E}^{\mathrm{i}}||Z_A + R_L|}{|\mathbf{h}^* \cdot \mathbf{E}^{\mathrm{i}}|\, R_L}. \tag{13.2}$$

Now, if it is assumed that the antenna is tuned, $Z_A = R_{\mathrm{r}}$, where R_{r} is the radiation resistance of the antenna and the antenna is assumed to be lossless. Under impedance and polarization matched conditions we have

$$R_{\mathrm{r}} = R_L, \text{ with } \mathbf{h} || \mathbf{E}^{\mathrm{i}} \tag{13.3}$$

and we obtain from (13.2)

$$\mathrm{AF} = \frac{2}{h} \tag{13.4}$$

which indicates that the antenna factor has the dimension of 1/meter. The units of AF are frequently ignored [7], so the dimension is stated in dB. It follows from (13.1) and Figure 13.1 that the antenna factor generally includes balun losses, the effective length h (as we have seen), and impedance mismatches, but usually does not include transmission (cable) losses. Thus the knowledge of AF yields the incident E-field strength $|\mathbf{E}^{\mathrm{i}}|$ as

$$|\mathbf{E}^{\mathrm{i}}| = \mathrm{AF}\, |V_{\mathrm{rec}}| \tag{13.5}$$

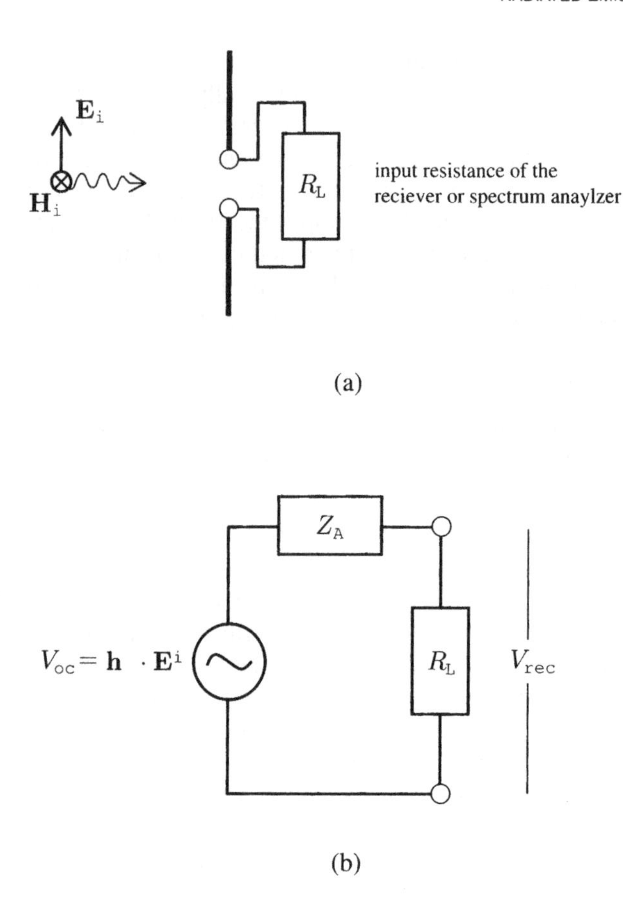

Figure 13.1 Receiving antenna: (*a*) attached to a receiver or spectrum analyzer; (*b*) its equivalent circuit.

which can be expressed in dB as

$$|\mathbf{E}^{\mathrm{i}}|\ (\mathrm{dB}\mu\mathrm{V/m}) = \mathrm{AF}\ (\mathrm{dB}) + |V_{\mathrm{rec}}|\ (\mathrm{dB}\mu\mathrm{V})\,, \tag{13.6}$$

where it is assumed that there no cable loss is included in the AF. With cable loss C_{dB} included, the measured incident E-field intensity (13.6) is

$$|\mathbf{E}^{\mathrm{i}}|\ (\mathrm{dB}\mu\mathrm{V/m}) = \mathrm{AF}\ (\mathrm{dB}) + |V_{\mathrm{rec}}|\ (\mathrm{dB}\mu\mathrm{V}) + C_A\ \mathrm{dB}. \tag{13.7}$$

Measurement of AF

It is now clear that for accurate quantitative measurement of E one requires an accurate knowledge of the AF over the frequencies of interest, in other words, the calibration of the test antenna. There are four methods for the measurement of AF [8]: (1) the standard antenna method (SAM), (2) the standard field method (SFM), (3) the secondary standard antenna method (SSAM), and (4) the standard site method (SSM).

Methods (1) and (2) are suitable for use only by standard laboratories such as the National Institute of Standards and Technology (NIST), formerly the National Bureau of Standards (NBS). Few industrial laboratories have the specialized instrumentation and facilities necessary to use these methods to calibrate antennas. The NIST in Boulder, Colorado, provides antenna calibration to the EMC industry for a fee. Once an antenna is calibrated by the SAM or SFM, it can be used as a secondary standard to calibrate other antennas using the same substitution process employed in SAM [8]. The SSM is the method of choice for many industrial laboratories, since it is easy to implement and requires no special field strength meter or signal generator. The SSM requires site attenuation measurements [9] made in an open field site. Accuracies comparable to those obtained by NIST using SAM and SFM methods can be achieved. Compact descriptions of the four antenna calibration methods are given in [8], and a detailed description of SSM is given in [9].

13.3.4 Some Results

In this section we present measured antenna factor (calibration) curves for a tuned dipole and a biconical antenna, which are commonly used for radiated emission field measurements. Also shown are the measured radiated emissions from a typical digital device.

Figure 13.2 shows a typical measured antenna factor versus frequency for a set of three tuned dipoles and a biconical (bicone) antenna.

Figure 13.3 shows the raw measured radiated emissions from a digital device at a distance of 3 m using a biconical antenna for horizontal and vertical polarizations. Note that the device meets the FCC and CISPR 22 Class A emission limits but fails to meet the Class B limits over the band of frequencies shown.

13.4 CONDUCTED EMISSIONS

13.4.1 Introduction

Conducted EME measurements can be made at any suitable facility including a shielded room provided that the facility otherwise meets the requirements for a proper test environment. A good ground screen is required if tests are performed in laboratory facilities but it is not mandatory at the on-site facility of the EUT. The ground screen is to consist of a floor and at least one vertical earth ground conducting surface, such as the metal floor and metal wall of a shielded test chamber. Each conducting surface is to be at least 2.0×2.0 m in

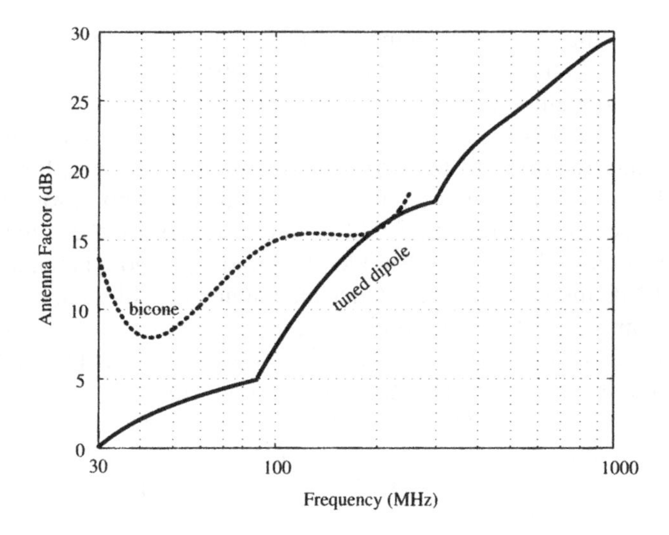

Figure 13.2 Antenna factor (AF) versus frequency for a bicone and a set of three tuned $\lambda/2$ dipoles.

dimension. Part of the ground plane is to be covered by an insulating material 3 mm thick, typically rubber matting or flow tile.

13.4.2 Noise on Power Supply Lines

Generally, low voltage (up to 1000 V) electric power lines are three wire lines. In North America the three wire lines are the the hot line, neutral line, and ground line. The neutral and ground lines are connected to earth ground at each service entrance. The distance between the equipment connected to the power supply line and the actual location of electrical earth is thus limited. In this situation common mode surges and interference will be smaller than the differential mode interference.

13.4.3 Transients on Power Supply Lines

Electrical transients and other disturbances are included in the power supply lines as a consequence of natural electromagnetic phenomena or the operation of a variety of electrical equipment. The most common natural phenomenon is lightning, which can induce transients on overhead power lines either by a direct strike or by way of induction from a strike on a nearby structure. Machine operations such as a local on–off switching of heavy electrical equipment, motor control activation, or welders and industrial cranes can induce substantial electrical transients in the power supply lines. Such transients can appear on

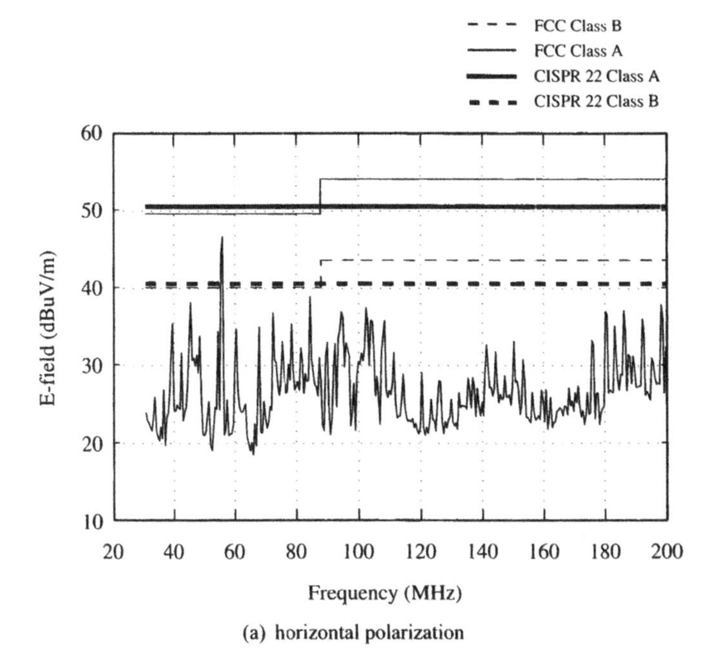

(a) horizontal polarization

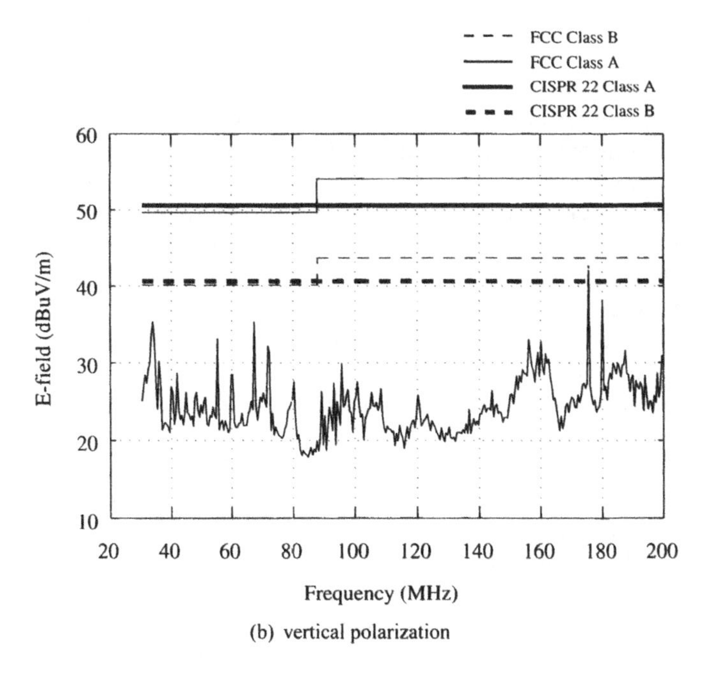

(b) vertical polarization

Figure 13.3 Radiated emissions from a digital device at a distance of 3 m using a biconical antenna.

the ac power lines as a transient voltage difference between the phase and neutral conductors or between the neutral and ground conductors. The FCC specifies that such conducted noise must remain below certain limits over a specified range of frequencies [10].

13.4.4 Conducted Emissions from a DUT

During measurements of conducted emissions from a DUT, it is necessary to make certain that electromagnetic noise and other disturbances from the power line are isolated, so that they do not add to the conducted noise from the DUT. The experimental setup should be such that only the desired emissions from the DUT are measured. During measurement of the conducted emissions from a DUT, a line impedance stabilization network (LISN) is normally connected between the electric power supply mains and the DUT, as shown in Figure 13.4.

Figure 13.4 Block diagram for conducted emissions measurement setup.

The purpose of the LISN is to present a defined, standard impedance to the DUT power input terminals at high frequencies so that the power supply line impedance variations are sufficiently isolated from the DUT. Also the LISN is used to filter out any incoming noise on the main power supply so that only clear power is provided to the DUT. The FCC-specified LISN for conducted emission measurement is shown in Figure 13.5, where the purpose of the 1 μF capacitor between phase and green wire and between neutral and green wire on the commercial power side is to divert external noise on the commercial power line, preventing that noise from flowing through the DUT and thereby contaminating the test data.

Similarly the purpose of the 50 μH inductors is to block out that noise. The 0.1 μF capacitors prevent dc from entering the test receiver input. Note that over the range of test frequencies 150 KHz to 30 MHz the capacitors 0.1 μF and 1 μF are essentially short circuits and the 50 μH inductors present large impedances. The 1 kΩ resistors facilitate the discharge of 0.1 μF capacitors in case the 50 Ω resistors are removed. One of the 50 Ω resistors is the input impedance of the spectrum analyzer or receiver while the other serves as a dummy load to make sure that the impedance between neutral and safety wire is 50 Ω at all times. The voltages between the phase wire and the safety wire, and between the neutral wire and safety wire, V_P and V_N, respectively, shown

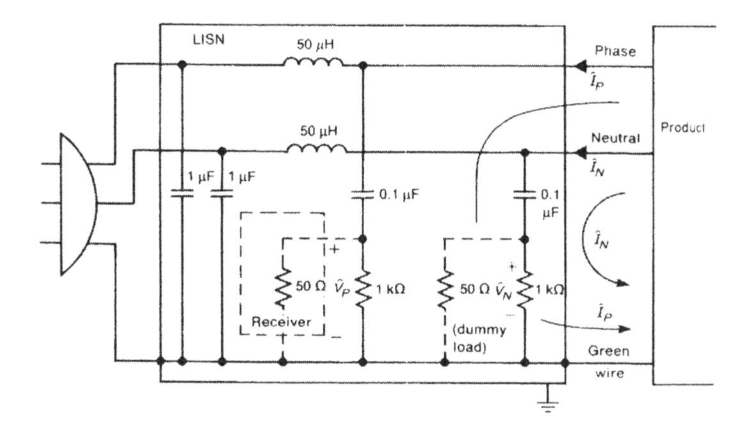

Figure 13.5 LISN circuit for conducted emission measurement.

in Figure 13.5 are to be measured over the specified range of frequencies. These voltages are related to the phase and neutral currents I_{P}, I_{N} by

$$V_{\mathrm{P}} = 50\, I_{\mathrm{P}} \tag{13.8a}$$

$$V_{\mathrm{N}} = 50\, I_{\mathrm{N}}, \tag{13.8b}$$

where we have assumed the capacitors ($1\ \mu$F, $0.1\ \mu$F) and the inductors ($50\ \mu$H) of the LISN are short and open circuits, respectively. Under these ideal conditions the LISN circuit simplifies the measurement considerably. Further description and implications of the test results can be obtained by using the simplified circuit [7].

For example, if the phase and neutral currents I_{P}, I_{N} are decomposed into common and differential mode components of current I_{C}, I_{D}, then by using the simplified circuit described in [7] it can be shown from (13.8) that

$$\begin{aligned} V_{\mathrm{P}} &= 50\ (I_{\mathrm{C}} + I_{\mathrm{D}}) \\ V_{\mathrm{N}} &= 50\ (I_{\mathrm{C}} - I_{\mathrm{D}}), \end{aligned} \tag{13.9}$$

where it is assumed that

$$\begin{aligned} I_{\mathrm{P}} &= I_{\mathrm{C}} + I_{\mathrm{D}} \\ I_{\mathrm{N}} &= I_{\mathrm{C}} - I_{\mathrm{D}}. \end{aligned} \tag{13.10}$$

13.4.5 Some Results

Here we present an example of measured conducted emissions from a typical digital device. Figure 13.6 shows the measured conducted emissions from a digital device as compared with the FCC and CISPR Class B limit.

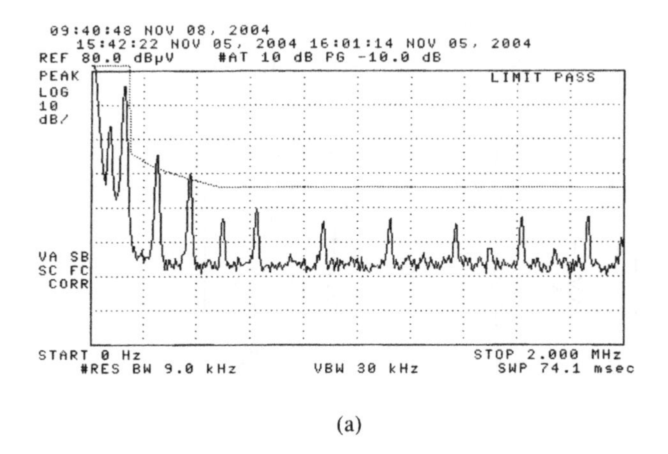

(a)

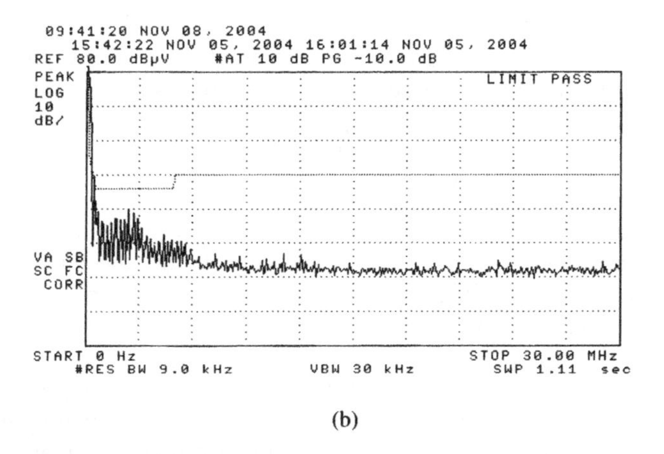

(b)

Figure 13.6 Measured conducted emissions from a digital device as compared with the FCC and CISPR Class B Limit Line: (*a*) 0–2 MHz; (*b*) 0–30 MHz.

REFERENCES

1. FCC *Measurements of Radio Noise Emissions from Computing Devices,* FCC/ OST, MP-4, July 1989.

2. FCC, Updated version of Part 15, *Producers for Measuring Electromagnetic Emissions from Digital Devices,* FCC 89–53, March 7, 1989.

3. FCC, OET Bulletin, *Characteristics of Open Field Test Sites,* FCC Consumer Assistance and Small Business Division, Washington DC.

4. ANSI, *Methods of Measurements of Radio Noise and Electronic Equipment in the Range of 9 kHz to 40 GHz,* ANSI C 63.4-1991, American National Standards Institute, New York, 1991.

5. CISPR Publication 16, *CISPR Specifications for Radio Interference Measuring Apparitions and Measurement Methods,* 2nd ed., 1987.

6. CISPR Publication 22, *Limits and Methods of Measurements of Radio Interference Characteristics of Information Technology,* 1985. (In the USA, CISPR publications are available from the Sales Department of the American National Standards Institute, New York.)

7. C. R. Paul, *Introduction to Electromagnetic Compatibility,* Wiley, New York, 1992.

8. A. A. Smith Jr., *Radio Frequency Principles and Applications,* IEEE Press, New York, 1998.

9. A. A. Smith Jr., "Standard Site Method for Determining Antenna Factor," *IEEE Trans. Electromag. Compat.* 24 (August 1982): 326–322.

10. V. P. Kodali, *Engineering Electromagnetic Compatibility,* IEEE Press, New York, 1996.

APPENDIX A

VECTORS AND VECTOR ANALYSIS

A.1 INTRODUCTION

Vectors and vector fields are used extensively in electromagnetics where it is found convenient to use vector notations and vector analysis in the study of field phenomena. However, it is noteworthy that many electromagnetic concepts were developed before vector analysis was developed. Maxwell in fact formulated his original equations in nonvector form. Oliver Heaviside is credited for first using the vector notation in field analysis.

Vector analysis was developed as a branch of applied mathematics, mainly by theoretical physicists. J. Willard Gibbs [1] established the discipline of vector analysis as we know it today [1–3].

The importance and wide application of vector analysis can hardly be overemphasized. The present appendix reviews aspects of vectors and vector analysis that are of importance for the topics discussed in this book. Much of the derivations of the various relations given here are omitted—more detailed

Applied Electromagnetics and Electromagnetic Compatibility. By D. L. Sengupta, V. V. Liepa
ISBN 0-471-16549-2

derivations and discussions can be found in most books on electromagnetic fields and vector analysis. For example, many of the references cited in Chapter 3 are relevant in understanding vectors and vector analysis, as are [4–8].

A.2 DEFINITIONS OF SCALAR AND VECTOR FIELDS

A *field* may be defined as a mathematical function that describes a physical quantity at all points in space. It may be a function of both space and time. Generally, we encounter two types of fields: scalar fields and vector fields.

A.2.1 Scalar Fields

A scalar field is completely specified by one parameter (i.e., number) for each point in space. Examples of scalar fields include temperature distribution $T(x, y, z)$ at a point $P(x, y, z)$ in space and electrostatic potential or speed at any point P.

Note that all scalar quantities may vary from one point to another—that is, they may be variable in space as well as in time.

A.2.2 Vector Fields

In order to completely specify the physical quantity of a vector field, we need both a number to represent its magnitude and a number to specify its direction. Thus all vector quantities have two dimensions—magnitude and direction—and both must be specified for a complete description of the vector quantity.

Examples of vector fields include temperature gradient, velocity, and electric or magnetic field intensity. Again, the magnitude and direction of a vector may change in space as well as in time.

A.3 VECTOR ALGEBRA

A.3.1 Definitions

Vectors

A vector quantity is usually denoted by either arrow notation ($\overrightarrow{F}$) or a boldface letter ($\mathbf{F}$). Its parameter dependence is sometimes made explicit as $\mathbf{F}(\mathbf{r}, t)$, where $\mathbf{r}$ is the position vector of the point P at which the vector is defined.

The graphical representation of a vector $\mathbf{F}$ defined at the point P is shown in Figure A.1a, where the length of the arrow equals the magnitude of $\mathbf{F} = |\mathbf{F}|$ and the direction of the arrow equals the direction of $\mathbf{F}$. Note that a specific coordinate system is required to identify the point P at which $\mathbf{F}$ is defined,

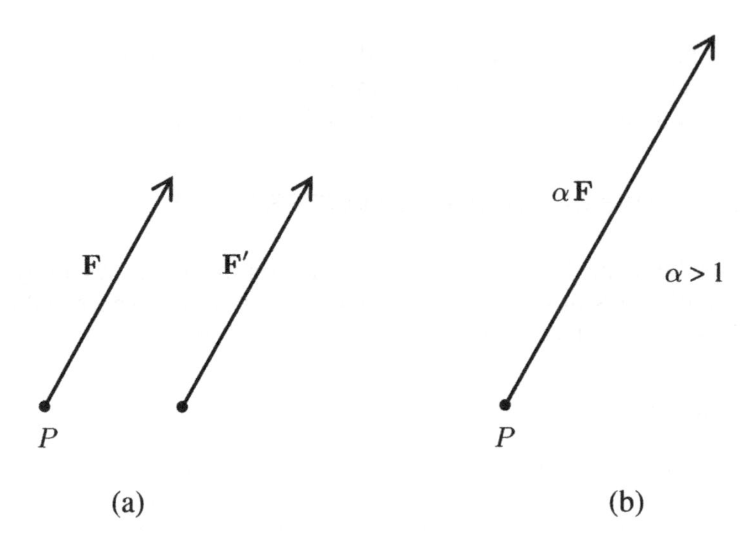

Figure A.1 Representations of vectors: (*a*) Vector $\mathbf{F}$ defined at a point P and one of its parallel vectors $\mathbf{F}'$; (*b*) Vector $\mathbf{F}$ multiplied by a positive scalar constant α.

and note that for a positive constant α, $\alpha\mathbf{F}$ signifies a vector of amplitude $\alpha|\mathbf{F}|$ in the same direction of $\mathbf{F}$, as shown in Figure A.1*b*.

Constant Vector

A *constant vector* or a *vector function* is one whose magnitude and direction are both constant at all points of interest.

Parallel Vectors

Two vectors whose directions are the same are *parallel vectors*; for example, in Figure A.1*a*, $\mathbf{F}$ and $\mathbf{F}'$ are parallel. If $\mathbf{F} \parallel \mathbf{F}'$ and $|\mathbf{F}| = |\mathbf{F}'|$, then they are *equal*. Thus equal vectors are in the same direction and have the same magnitude.

Negative Vectors

The *negative vector* of a vector $\mathbf{F}$ has the same magnitude as that of $\mathbf{F}$ but runs in the opposite direction, as shown in Figure A.2. The negative of a vector $\mathbf{F}$ is denoted by $-\mathbf{F}$.

Null Vector

A null vector is a vector with a magnitude of zero that runs in an unspecified direction.

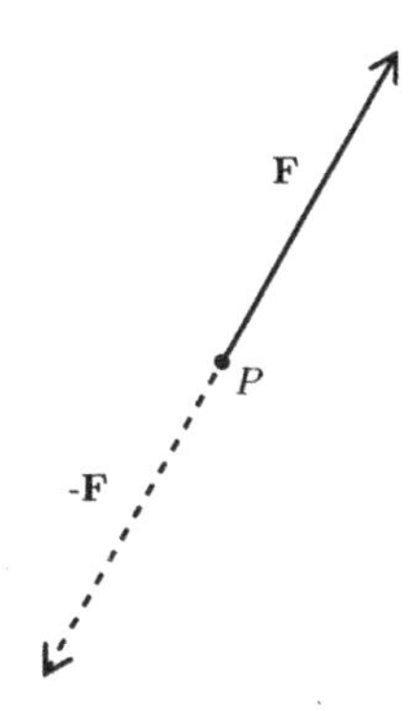

Figure A.2 Vector **F** at P and its negative $-\mathbf{F}$.

A.3.2 Addition and Subtraction of Vectors

Addition and subtraction concepts apply to vectors of the same dimension. The addition of two vectors **A** and **B** follows the parallelogram law of geometry or the triangular (head-to-tail) rule, as shown in Figure A.3.

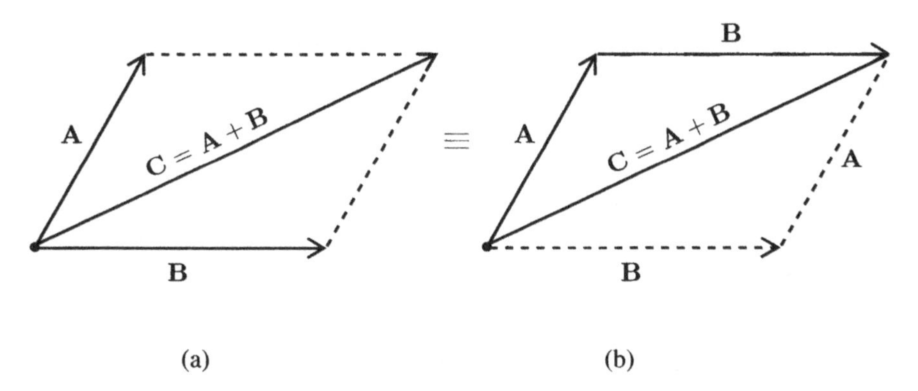

Figure A.3 Addition of two vectors **A** and **B**. (*a*) Parallelogram law; (*b*) triangular or head-to-tail rule.

The addition of two vectors **A** and **B** produces another vector **C**, such that

$$\mathbf{A} + \mathbf{B} = \mathbf{C}. \tag{A.1}$$

It follows from the triangular rule (Figure A.3*b*) that

$$\mathbf{A} + \mathbf{B} = \mathbf{C} = \mathbf{B} + \mathbf{A}. \tag{A.2}$$

Therefore vector addition is commutative; that is, the order in which the addition is performed is unimportant.

It is simple to show that

$$(\mathbf{A} + \mathbf{B}) + \mathbf{C} = \mathbf{A}(\mathbf{B} + \mathbf{C}) = \mathbf{C}(\mathbf{A} + \mathbf{B}), \tag{A.3}$$

which indicates that vector addition satisfies the associative law.

For a positive α, it can be shown that

$$\alpha(\mathbf{A} + \mathbf{B}) = \alpha\mathbf{A} + \alpha\mathbf{B}, \tag{A.4}$$

meaning the vector's definition satisfies the distributive law.

Subtraction of a vector $\mathbf{B}$ from another vector $\mathbf{A}$ produces a third vector $\mathbf{D}$ such that

$$\mathbf{D} = \mathbf{A} - \mathbf{B}. \tag{A.5}$$

If we interpret $-\mathbf{B}$ as a vector that has the same magnitude as that of $\mathbf{B}$ but is of opposite direction, then the operation (A.5) may be considered as the addition of $\mathbf{A}$ and $-\mathbf{B}$, that is,

$$\mathbf{D} = \mathbf{A} + (-\mathbf{B}). \tag{A.6}$$

A geometric interpretation of the subtraction of two vectors is given in Figure A.4

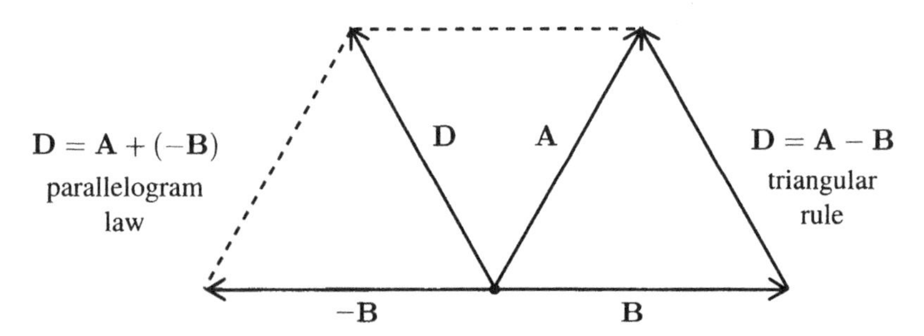

Figure A.4 Geometric interpretation of subtraction of $\mathbf{B}$ from $\mathbf{A}$.

Two vectors are equal if their difference is zero (i.e., a null vector), that is,

$$\mathbf{A} = \mathbf{B}, \text{ if } \mathbf{A} - \mathbf{B} = \mathbf{D} \equiv 0. \tag{A.7}$$

Applications

In electromagnetic problems we frequently add and subtract a number of similar vectors defined at a point in space.

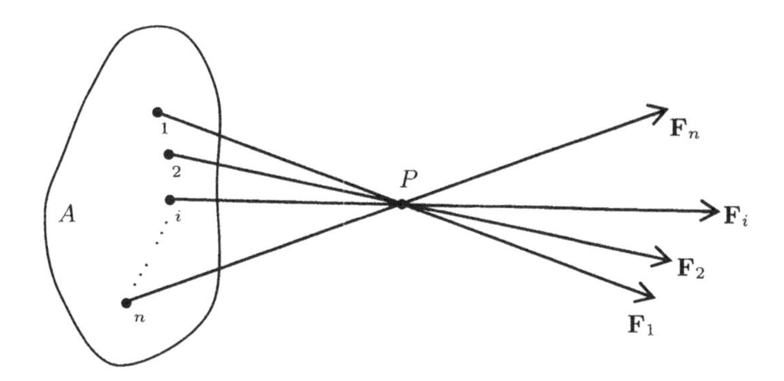

Figure A.5 A number of point sources confined in a region A of space. The fields $\mathbf{F}_1, \mathbf{F}_2, \ldots, \mathbf{F}_i, \ldots, \mathbf{F}_n$ are produced at a point P.

Figure A.5 represents a number of point sources (e.g., charges) $1, 2, \ldots n$ located in a region of space $\mathbf{A}$, producing fields $\mathbf{F}_1, \mathbf{F}_2, \ldots, \mathbf{F}_n$ at the field point P. The total field $\mathbf{E}_\mathrm{T}$ is then represented by

$$\mathbf{E}_\mathrm{T}(P) = \sum_{i=1}^{n} \mathbf{F}_i. \tag{A.8}$$

A.3.3 Multiplication of a Vector by a Scalar Quantity

Multiplication of a vector $\mathbf{A}$ by a dimensionless scalar α yields another vector $\mathbf{B}$, defined as

$$\mathbf{B} = \alpha \mathbf{A}. \tag{A.9}$$

This vector can be interpreted in one of two ways:

1. When α is positive, then the magnitude of $\mathbf{B}$ is α times the magnitude of $\mathbf{A}$, and the direction of $\mathbf{B}$ is the same as the direction of $\mathbf{A}$

2. When α is negative, then the magnitude of $\mathbf{B}$is $|\alpha|$ times the magnitude of $\mathbf{A}$, and the direction of $\mathbf{B}$ is opposite that of $\mathbf{A}$.

Multiplication of two or more vectors will be described later.

A.3.4 Unit Vectors

A *unit vector* is one whose magnitude is unity. It is often found convenient to express a vector in terms of a unit vector specified in its own direction, as

$$\mathbf{A} = \hat{a} A \text{ (or } \mathbf{A} = \hat{a}|\mathbf{A}|).$$

Thus, by definition,

$$\hat{a} = \frac{\mathbf{A}}{A} \text{ or } \frac{\mathbf{A}}{|\mathbf{A}|}, \tag{A.10}$$

where $|\mathbf{A}|$ or A is the magnitude (always positive) of the vector $\mathbf{A}$, and $\hat{a}$ is a vector of magnitude of unity in the direction of $\mathbf{A}$. Note that a unit vector is not necessarily a constant vector.

A.3.5 Vector Displacement and Components of a Vector

The parallelogram rule of addition of vectors considered earlier suggests that any vector can be considered as made up of component values; for example, in Figure A.3a, $\mathbf{C}$ is made up of $\mathbf{A}$ and $\mathbf{B}$.

However, this choice of component vectors is arbitrary. We will see that any vector can be considered to be made up of components (component vectors) associated with a specific coordinate system. These components are defined along the coordinate axes of a given coordinate system. That is, the components of a given vector correspond to the projections of the vector upon the axes of that coordinate system. Obviously the components of a vector depend on the coordinate system used to describe the vector.

Consider a point P in space identified by a set of coordinates (x, y, z) in a right handed rectangular coordinate system shown in Figure A.6.

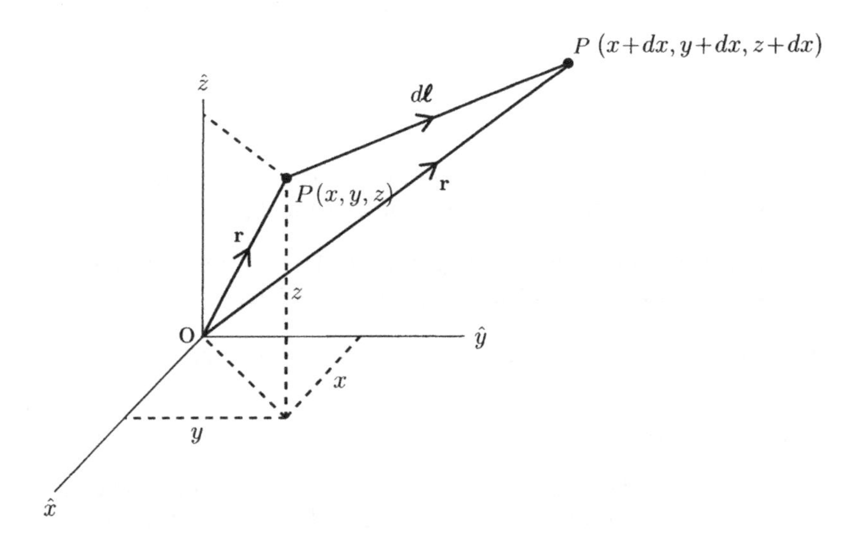

Figure A.6 Right-handed rectangular coordinate system.

Consider another point P' identified by $(x + dx,\ y + dx,\ z + dx)$. It is known from geometry that the distances PP' and OP are given by

$$
\begin{aligned}
PP' &= \left[(dx)^2 + (dy)^2 + (dz)^2\right]^{1/2} \\
OP &= \left[(x)^2 + (y)^2 + (z)^2\right]^{1/2}.
\end{aligned}
\tag{A.11}
$$

We now define the *position vector* of P:

$$\text{position vector of } P = \overrightarrow{OP} = \mathbf{r} = x\hat{x} + y\hat{y} + z\hat{z}, \tag{A.12}$$

where $\hat{x}$, $\hat{y}$, $\hat{z}$ are unit vectors along the x, y, and z axes, respectively. Thus

$$|\overrightarrow{OP}| = |\mathbf{r}| = r = \left(x^2 + y^2 + z^2\right)^{1/2}. \tag{A.13}$$

The vector displacement $\overrightarrow{PP'} = \overrightarrow{OP'} - \overrightarrow{OP} = d\boldsymbol{\ell}$ is a frequently used quantity.
>From the analogy with two position vectors of a point P in space and by assuming that the component of a vector defined at any point behaves like the coordinates of the point, we obtain for any vector $\mathbf{A}$ the following algebraic expression:

$$\mathbf{A} = A_x\hat{x} + A_y\hat{y} + A_z\hat{z}, \tag{A.14}$$

where A_x, A_y, A_z are the scalar components of $\mathbf{A}$ in the $\hat{x}, \hat{y}, \hat{z}$ directions, respectively, and $A_x\hat{x}$, $A_y\hat{y}$, $A_z\hat{z}$ are the vector components of $\mathbf{A}$ in the three directions $\hat{x}, \hat{y}, \hat{z}$, with

$$|\mathbf{A}| = A = \left[A_x{}^2 + A_y{}^2 + A_z{}^2\right]^{1/2}. \tag{A.15}$$

Observe from Figure A.6 that

$$\overrightarrow{PP'} = \overrightarrow{OP'} - \overrightarrow{OP} = d\boldsymbol{\ell}$$

element of vector displacement. Thus

$$d\boldsymbol{\ell} = \mathbf{r}' - \mathbf{r} = dx\,\hat{x} + dy\,\hat{y} + dz\,\hat{z} \tag{A.16}$$

with $|d\boldsymbol{\ell}| = d\ell = \left[(dx)^2 + (dy)^2 + (dz)^2\right]^{1/2}$, as before.
Consider a vector $\mathbf{A}$ defined at O (i.e. $\overrightarrow{OP} = \mathbf{A}$) as shown in Figure A.7. The algebraic expression in terms of its components is given by (A.14). It is convenient to identify the vector by its orientation with respect to three orthogonal axes, namely by the angles α_x, α_y, and α_z (shown in Figure A.7).

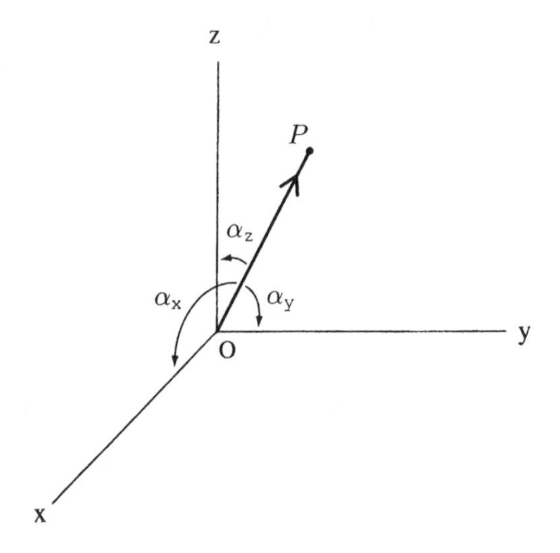

Figure A.7 Vector **A** and its direction with respect to the coordinate axes.

The direction cosines of the vector **A** are defined by

$$\begin{aligned} \cos\alpha_x &= \frac{A_x}{A} \\ \cos\alpha_y &= \frac{A_y}{A} \\ \cos\alpha_z &= \frac{A_z}{A}\,, \end{aligned} \tag{A.17}$$

and thus

$$\cos^2\alpha_x + \cos^2\alpha_y + \cos^2\alpha_z = 1. \tag{A.18}$$

Example A.1

In a rectangular coordinate system two points in space are specified as $P_1(-1, 1, 1)$ and $P_2(1, -1, 1)$. Determine the following: (1) the vector $\overrightarrow{P_1\,P_2}$ and the scalar and vector components of $\overrightarrow{P_1\,P_2}$; (2) the unit vector in the direction of $\overrightarrow{P_1\,P_2}$; (3) the direction cosines of the unit vector in (2) that are the same as those of $\overrightarrow{P_1\,P_2}$; (4) the angles between $\overrightarrow{P_1\,P_2}$ (or the unit vector $\hat{\overrightarrow{P_1\,P_2}}$) and the coordinate axes x, y, and z.

Solution

For an arbitrary origin the geometry of P_1 and P_2 and other parameters are as shown in Figure A.8.

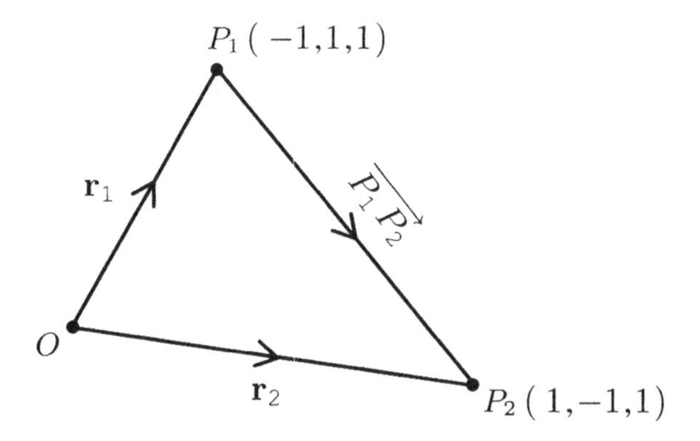

Figure A.8 Representation for the points P_1 and P_2 in space with respect to an origin O.

(1)

$$\begin{aligned} \overrightarrow{P_1\,P_2} = ? \quad \overrightarrow{OP_1} &= \mathbf{r}_1 = -1\hat{x} + 1\hat{y} + 1\hat{z} \\ \overrightarrow{OP_2} &= \mathbf{r}_2 = 1\hat{x} - 1\hat{y} + 1\hat{z}, \\ \overrightarrow{P_1\,P_2} &= \mathbf{r}_2 - \mathbf{r}_1 = (1+1)\hat{x} + (-1-1)\hat{y} + (1-1)\hat{z} \\ &= 2\hat{x} - 2\hat{y} + 0\hat{z}. \end{aligned}$$

The scalar components of $\overrightarrow{P_1\,P_2}$ are

$$\overrightarrow{P_1\,P_2}|_x = 2, \quad \overrightarrow{P_1\,P_2}|_y = -2, \quad \overrightarrow{P_1\,P_2}|_z = 0.$$

The vector components of $\mathbf{P}_1\mathbf{P}_2$ are

$$\begin{array}{rl} 2\hat{x} & \text{in the direction of } \hat{x} \\ -2\hat{y} & \text{in the direction of } \hat{y} \\ 0\hat{z} & \text{in the direction of } \hat{z}. \end{array}$$

(2) The unit vector $P_1\hat{P}_2$ in the direction of $\mathbf{P}_1\mathbf{P}_2$ is

$$\overrightarrow{P_1\hat{P}_2} = \frac{\overrightarrow{P_1 P_2}}{|\overrightarrow{P_1 P_2}|} = \frac{2\hat{x} - 2\hat{y} + 0\hat{z}}{[2^2 + (-2)^2 + 0^2]^{\frac{4}{2}}}$$
$$= \frac{\hat{x} - \hat{y}}{\sqrt{2}}.$$

(3) The direction cosines of $\overrightarrow{P_1\hat{P}_2}$ or $\overrightarrow{P_1 P_2}$ are

$$\cos\alpha_x = \frac{\overrightarrow{P_1\hat{P}_2}|_x}{|\overrightarrow{P_1\hat{P}_2}|} = \frac{\overrightarrow{P_1 P_2}|_x}{|\overrightarrow{P_1 P_2}|} = \frac{1}{\sqrt{2}}$$

$$\cos\alpha_y = \frac{\overrightarrow{P_1\hat{P}_2}|_y}{|\overrightarrow{P_1\hat{P}_2}|} = \frac{\overrightarrow{P_1 P_2}|_y}{|\overrightarrow{P_1 P_2}|} = -\frac{1}{\sqrt{2}}$$

$$\cos\alpha_z = \frac{\overrightarrow{P_1\hat{P}_2}|_z}{|\overrightarrow{P_1\hat{P}_2}|} = \frac{\overrightarrow{P_1 P_2}|_z}{|\overrightarrow{P_1 P_2}|} = 0.$$

(4) $\alpha_x = 45°,\ \alpha_y = -45°,\ \alpha_z = \pi/2.$

Example A.2

Case 1. Consider two points $P_1\,(x_1, y_1, z_1)$ and $P_2\,(x_2, y_2, z_2)$ with respect to a coordinate system having origin at O. Find the position vectors of P_1, P_2, and the vector $\overrightarrow{P_1 P_2}$ identified in Figure A.9*a*.

Case 2. Points P' and P are identified as in Figure A.9*b*. Determine $\mathbf{R}$ and $\hat{R}$.

Solution

Case 1.

$$\overrightarrow{OP_1} = \mathbf{r}_1 = x_1\,\hat{x} + y_1\,\hat{y} + z_1\,\hat{z}$$
$$\overrightarrow{OP_2} = \mathbf{r}_2 = x_2\,\hat{x} + y_2\,\hat{y} + z_2\,\hat{z}$$
$$\mathbf{P}_1\mathbf{P}_2 = \overrightarrow{OP_2} - \overrightarrow{OP_1} = \mathbf{r}_2 - \mathbf{r}_1$$
$$= (x_2 - x_1)\,\hat{x} + (y_2 - y_1)\,\hat{y} + (z_2 - z_1)\,\hat{z}$$
$$\therefore\quad \overrightarrow{P_1\hat{P}_2} = \frac{\overrightarrow{P_{12}}}{|\overrightarrow{P_{12}}|} = \frac{(x_2 - x_1)\,\hat{x} + (y_2 - y_1)\,\hat{y} + (z_2 - z_1)\,\hat{z}}{\left[(x_2 - x_1)^2 + (y_2 - y_1)^2 + (z_2 - z_1)^2\right]^{1/2}}$$

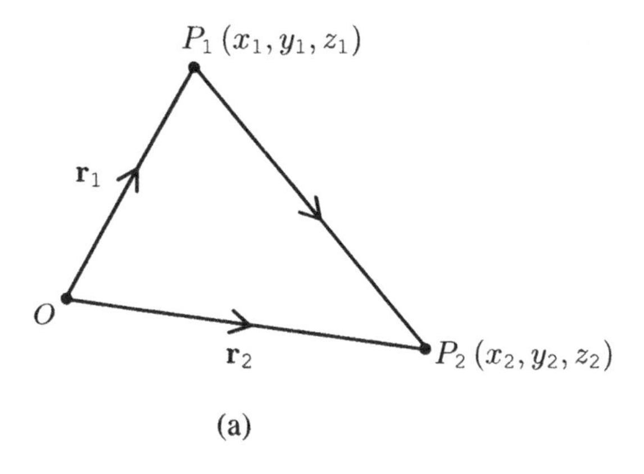

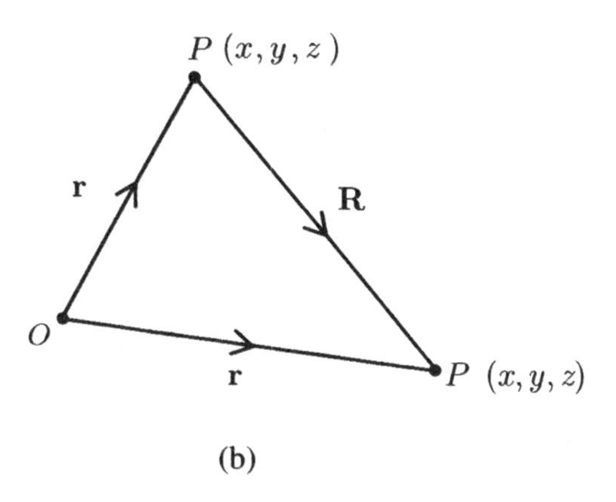

Figure A.9 (*a*) Representation of two points P' and P in space with respect to an arbitrary origin O. (*b*) Representation of points P' and P with respect to an arbitrary origin O.

which represents the unit vector directed from P_1 to P_2.

Case 2. Frequently we identify $P(x, y, z)$ and $P'(x', y', z')$ as the field and source points, respectively (Figure A.9*b*):

$$\begin{aligned} \mathbf{r}' &= \overrightarrow{OP'} \text{ identifies the source point} \\ \mathbf{r} &= \overrightarrow{OP} \text{ identifies the field point} \\ \mathbf{R} &= \text{vector directed from the source to the field point.} \end{aligned}$$

We now have

$$\mathbf{R} = \mathbf{r} - \mathbf{r}' = (x - x')\,\hat{x} + (y - y')\,\hat{y} + (z - z')\,\hat{z} \tag{A.19a}$$

$$\hat{R} = \frac{\mathbf{R}}{R} = \frac{(x - x')\,\hat{x} + (y - y')\,\hat{y} + (z - z')\,\hat{z}}{[(x - x')^2 + (y - y')^2 + (z - z')^2]^{1/2}} \tag{A.19b}$$

Expressions (A.19a) and (A.19b) are extensively used in various field problems.

A.4 VECTOR SURFACE ELEMENT

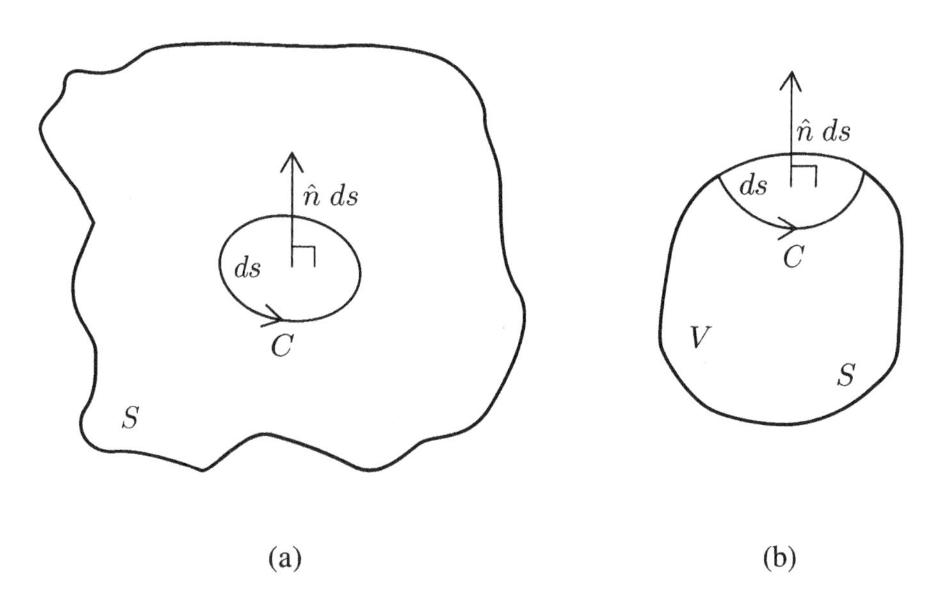

Figure A.10 Vector representation of elementary surfaces ds where ds is part of an open surface S.

Consider an elementary (open) surface of arbitrary shape of area ds shown in Figure A.10a. A vector surface element $d\mathbf{s}$ is represented by a vector whose magnitude is equal to the surface area ds and whose direction is specified by the unit positive normal (vector) of the surface. When ds is part of an open surface, as in Figure A.10a, positive $\hat{n}$ is taken in the direction that a right-hand screw would advance when the screw is placed perpendicular to the surface and is turned along the periphery or contour C such that the area ds is always on the left-hand side.

For a closed surface, $\hat{n}$ is always directed away from the volume enclosed by the surfaces, of which ds is a part of as in Figure A.10b.

A.5 PRODUCT OF VECTORS

There are two types of vector multiplication, the scalar or dot product and the vector or cross product of two vectors.

A.5.1 Dot Product of Two Vectors

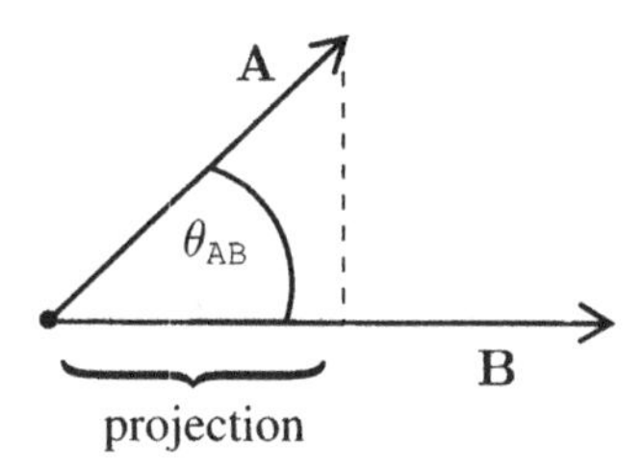

Figure A.11 Dot product of the vectors $\mathbf{A}$ and $\mathbf{B}$ in terms of the projection of $\mathbf{A}$ onto $\mathbf{B}$.

The *dot* (or *scalar*) *product* of two vectors $\mathbf{A}$, $\mathbf{B}$ is denoted by $\mathbf{A} \cdot \mathbf{B}$ and is a scalar quantity defined as

$$\mathbf{A} \cdot \mathbf{B} = AB \cos \theta_{AB}, \tag{A.20}$$

where θ_{AB} is the (smaller) included angle between $\mathbf{A}$ and $\mathbf{B}$, and the process is geometrically shown in Figure A.11.

Observe that (A.20) indicates that the result of the dot product of $\mathbf{A}$ and $\mathbf{B}$ is a scalar quantity, hence the name. Note that $\mathbf{A} \cdot \mathbf{B}$ may be interpreted as the algebraic product of the magnitudes of one vector (e.g., $\mathbf{B}$) and the projection of $\mathbf{A}$ on $\mathbf{B}$, and vice versa. It follows from the definition that $\mathbf{A} \cdot \mathbf{B} = \mathbf{B} \cdot \mathbf{A}$, meaning it obeys the commutative law (i.e., order is unimportant). Also it can be shown that dot product obeys the distributive law $\mathbf{A} \cdot (\mathbf{B} + \mathbf{C}) = \mathbf{A} \cdot \mathbf{B} + \mathbf{A} \cdot \mathbf{C}$.

Proof of the Distributive Law

Take $\mathbf{A}, \mathbf{B}, \mathbf{C}$ as shown in Figure A.12

$$\begin{aligned} \mathbf{A} \cdot (\mathbf{B} + \mathbf{C}) &= |\mathbf{A}| \cdot \text{ Projection of } (\mathbf{B} + \mathbf{C}) \text{ on } \mathbf{A} \\ &= |\mathbf{A}| OQ. \end{aligned}$$

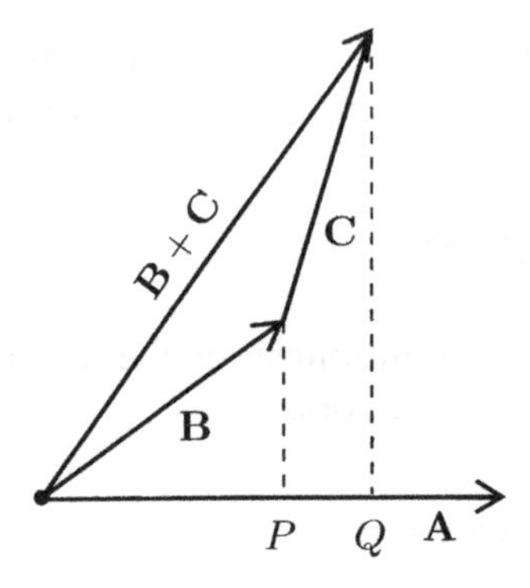

Figure A.12 Three vectors arranged to form a triangle in space.

Now

$$\begin{aligned}
\mathbf{A}\cdot\mathbf{B} &= |\mathbf{A}|\,OP = \mathbf{B}\cdot\mathbf{A} \\
\mathbf{A}\cdot\mathbf{C} &= \mathbf{C}\cdot\mathbf{A} = |\mathbf{A}|\,PQ \\
\therefore \quad \mathbf{A}\cdot\mathbf{B} + \mathbf{A}\cdot\mathbf{C} &= |\mathbf{A}|(OP + PQ) = |\mathbf{A}|\,OQ \\
&= \mathbf{A}\cdot(\mathbf{B} + \mathbf{C}).
\end{aligned}$$

No more than two vectors can be dot multiplied. (The associative law does not apply to the dot product.)

Applications

Scalar products, particularly in the integral form, are frequently encountered in physical problems.

1. Work done by a force $\mathbf{F}$ in moving a body of unit mass along a path ab is (Figure A.13)

$$W = \int_a^b \mathrm{dw} = \int_a^b \mathbf{F}\cdot d\mathbf{l}. \tag{A.21}$$

2. Given a volume current density $\mathbf{J}$ in a region as shown, the current I through a surface S is (Figure A.14)

$$I = \int_S \mathbf{J}\cdot d\mathbf{s}. \tag{A.22}$$

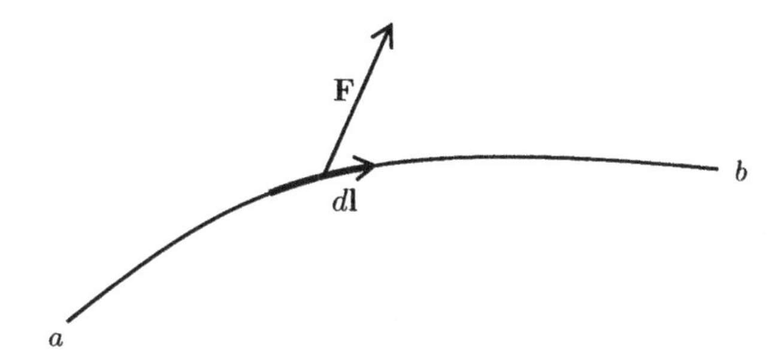

Figure A.13 Line integral of $\mathbf{F}$ along the path $\overline{ab}$.

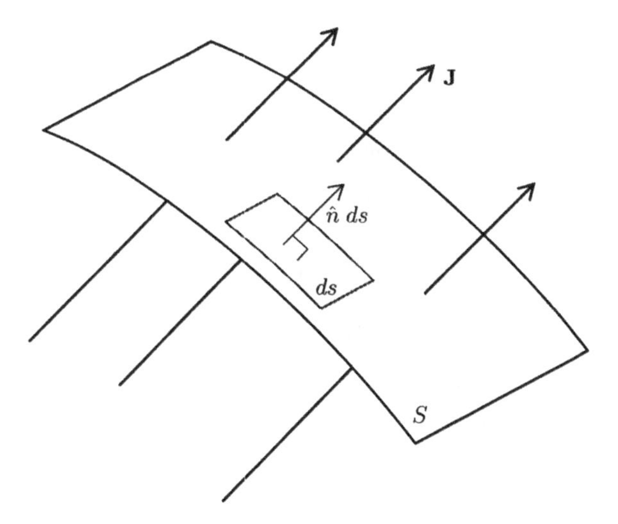

Figure A.14 Current I through the surface S is the integral of $\mathbf{J}$ through a surface S.

Algebraic Expressions

It follows from the definition of the dot product of two vectors $\mathbf{A}, \mathbf{B}$ that (1) if $\theta_{AB} = \pi/2$, then $\mathbf{A} \cdot \mathbf{B} \equiv 0$ (that is, the dot product vanishes for two orthogonal vectors); and (2) if $\theta_{AB} = 0$ (that is, when $\mathbf{A}$ is parallel to $\mathbf{B}$), then $\mathbf{A} \cdot \mathbf{B} = |\mathbf{A}|\,|\mathbf{B}|$. Hence $\mathbf{A} \cdot \mathbf{A} = |\mathbf{A}|^2$, a relation used to determine the magnitude of a vector:

$$|\mathbf{A}| = (\mathbf{A} \cdot \mathbf{A})^{1/2}, \tag{A.23}$$

provided that we know how to evaluate $\mathbf{A} \cdot \mathbf{A}$.

We will illustrate the general procedure by using the rectangular system of coordinates when $(\hat{x}, \hat{y}, \hat{z})$ are the orthogonal set of unit vectors. By definition, it can be shown that

$$\begin{aligned} \hat{x} \cdot \hat{x} = \hat{y} \cdot \hat{y} = \hat{z} \cdot \hat{z} = 1 \\ \text{and} \quad \hat{x} \cdot \hat{y} = \hat{y} \cdot \hat{z} = \hat{z} \cdot \hat{x} = 0. \end{aligned} \tag{A.24}$$

Take the two vectors $\mathbf{A}$ and $\mathbf{B}$ expressed as

$$\begin{aligned} \mathbf{A} &= \hat{x}A_x + \hat{y}A_y + \hat{z}A_z \\ \mathbf{B} &= \hat{x}B_x + \hat{y}B_y + \hat{z}B_z. \end{aligned} \tag{A.25}$$

Then, using (A.23) and (A.24), it could be shown that

$$\mathbf{A} \cdot \mathbf{B} = A_x B_x + A_y B_y + A_z B_z, \tag{A.26}$$

which gives the algebraic expression for $\mathbf{A} \cdot \mathbf{B}$ in terms of the components of the vectors.

From (A.26) it follows that with $\mathbf{A} = \mathbf{B}$

$$\begin{aligned} \mathbf{A} \cdot \mathbf{A} = A^2 &= A_x^2 + A_y^2 + A_z^2 \\ A &= \left[A_x^2 + A_y^2 + A_z^2\right]^{1/2}. \end{aligned} \tag{A.27}$$

Since the scalar product of two vectors $\mathbf{A}, \mathbf{B}$ is independent of the coordinate system used, (A.26) can be generalized as

$$\begin{aligned} \mathbf{A} \cdot \mathbf{B} &= A_x B_x + A_y B_y + A_z B_z \\ &= A_\rho B_\rho + A_\phi B_\phi + A_z B_z \\ &= A_r B_r + A_\theta B_\theta + A_\phi B_\phi, \end{aligned} \tag{A.28}$$

where (A_ρ, A_ϕ, A_z) and (A_r, A_θ, A_ϕ) are the cylindrical and spherical components of $\mathbf{A}$ and $\mathbf{B}$, respectively.

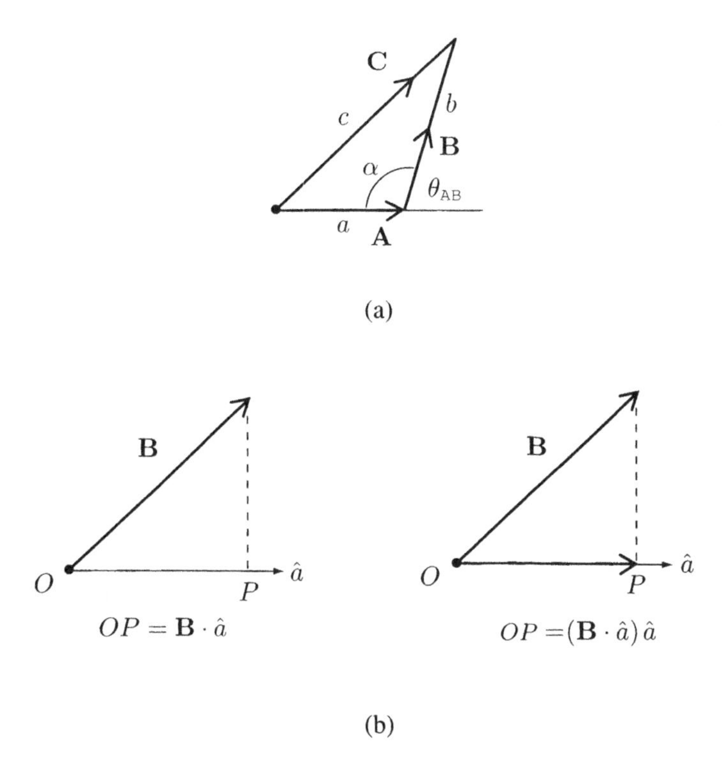

Figure A.15 Vector $\mathbf{A}, \mathbf{B}, \mathbf{C}$ forming a triangle. (*a*) Three vectors forming a triangle in space. (*b*) $OP = \mathbf{B} \cdot \hat{a}$, $\overline{OP} = (\mathbf{B} \cdot \hat{a})\, \hat{a}$.

Example A.3

- **Law of Cosines**
  Given three vectors $\mathbf{A}, \mathbf{B}, \mathbf{C}$, as shown the Figure A.15, with $c^2 = a^2 + b^2 - 2ab\,$cc where (a, b, c) are the magnitude of the vectors $\mathbf{A}, \mathbf{B}, \mathbf{C}$ respectively and α is the angle included between $\mathbf{A}, \mathbf{B}$, prove the law of cosines using the concept of dot product.

Solution

From Figure A.15, $\mathbf{C} = \mathbf{A} + \mathbf{B}$.

$$
\begin{aligned}
\therefore \quad c^2 &= \mathbf{C} \cdot \mathbf{C} = (\mathbf{A} + \mathbf{B}) \cdot (\mathbf{A} + \mathbf{B}) \\
&= \mathbf{A} \cdot \mathbf{A} + \mathbf{B} \cdot \mathbf{B} + 2\mathbf{A} \cdot \mathbf{B} \\
&= |\mathbf{A}|^2 + |\mathbf{B}|^2 + 2|\mathbf{A}||\mathbf{B}| \cos \theta_{AB} \\
&= a^2 + b^2 + 2ab \cos(\pi - \alpha) \\
&= a^2 + b^2 - 2ab \cos \alpha.
\end{aligned}
$$

- **Dot Product as Projection of a Vector, Geometric Representation**

The dot product is frequently used to determine the component of a vector in a given direction. For example,

$$
\begin{aligned}
OP = \mathbf{B} \cdot \hat{a} &= \text{scalar component of } \mathbf{B} \text{ in the direction of } \hat{a} \\
(\mathbf{B} \cdot \hat{a}) &= \text{vector component of } \mathbf{B} \text{ in the direction of } \hat{a}.
\end{aligned}
$$

A.5.2 The Cross Product of Two Vectors

The *cross* (or *vector*) *product* of two given vectors $\mathbf{A}$ and $\mathbf{B}$ is denoted by $\mathbf{A} \times \mathbf{B}$ (read as "A cross B") and is a vector quantity defined as

$$\mathbf{A} \times \mathbf{B} = \hat{n} |\mathbf{A}||\mathbf{B}| \sin \theta_{AB}, \tag{A.29}$$

where $\hat{n}$ denotes a unit vector perpendicular to both $\overline{A}$ and $\overline{B}$, and is pointed to the right-hand screw advance direction where we turn from $\mathbf{A}$ (the first-named vector) to $\mathbf{B}$ through the smaller included angle θ_{AB} between the two vectors, as in Figure A.16.
By definition,

$$
\begin{aligned}
|\overline{A} \times \overline{B}| &= \text{area of the parallelogram having adjacent sides } \overline{A}, \overline{B}. \\
&= 2 \times \text{area of the triangle } OPQ.
\end{aligned}
$$

Note that from the definition it follows that

$$\mathbf{A} \times \mathbf{B} = -(\mathbf{B} \times \mathbf{A}).$$

This means that the cross product does not obey commutative law, or in other words, the order in which the product is carried out is significant.

The following properties hold for the vector product:

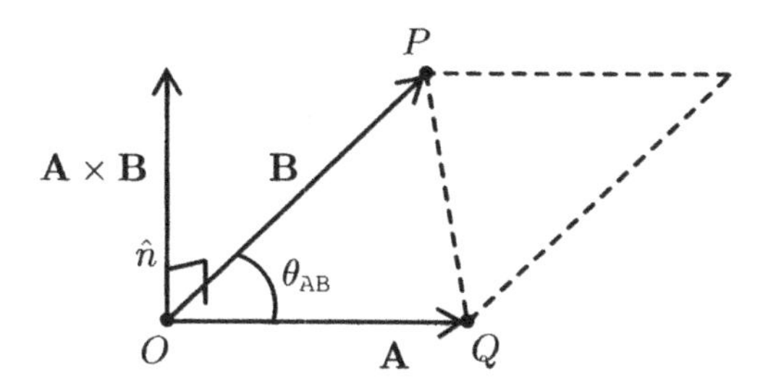

Figure A.16 Geometrical interpretation of the vector product of two vectors **A** and **B** with included angle θ_{AB}.

1. $\mathbf{B} \times \mathbf{A} \neq \mathbf{A} \times \mathbf{B}$, not commutative
2. $\mathbf{A} \times (\mathbf{B} \times \mathbf{C}) \neq (\mathbf{A} \times \mathbf{B}) \times \mathbf{C}$, not associative
3. $\mathbf{A} \times (\mathbf{B} \times \mathbf{C}) = \mathbf{A} \times \mathbf{B} + \mathbf{A} \times \mathbf{C}$, is distributive.

Algebraic Expressions

It follows from (A.29) that

$$\mathbf{A} \times \mathbf{B} = 0 \text{ if } \theta_{AB} = 0, \text{ i.e., when } \mathbf{A} \parallel \mathbf{B}$$

$$\mathbf{A} \times \mathbf{B} = \hat{n}AB \text{ if } \theta_{AB} = \frac{\pi}{2}, \text{ i.e., when } \mathbf{A} \perp \mathbf{B}.$$

Using the above, it can be shown that for rectangular systems of coordinates

$$\hat{x} \times \hat{x} = \hat{y} \times \hat{y} = \hat{z} \times \hat{z} \equiv 0. \tag{A.30}$$

$$\hat{x} \times \hat{y} = \hat{z} : \hat{y} \times \hat{z} = \hat{x} : \hat{z} \times \hat{x} = \hat{y}. \tag{A.31}$$

Equation (A.31) defines $(\hat{x}, \hat{y}, \hat{z})$ as a right-handed triad of orthogonal unit vectors serving as the base vectors of the coordinate system. Using the distributive law plus (A.30) and (A.31), it can be shown that

$$\begin{aligned}\mathbf{A} \times \mathbf{B} &= (A_x\hat{x} + A_y\hat{y} + A_z\hat{z}) \times (B_x\hat{x} + B_y\hat{y} + B_z\hat{z}) \\ &= (A_yB_z - A_zB_y)\hat{x} + (A_zB_x - A_xB_z)\hat{y} + (A_xB_y - A_yB_x)\hat{z} \\ &= \begin{vmatrix} \hat{x} & \hat{y} & \hat{z} \\ A_x & A_y & A_z \\ B_x & B_y & B_z \end{vmatrix}.\end{aligned} \tag{A.32}$$

In cylindrical and spherical coordinate systems it can be shown that

$$\mathbf{A} \times \mathbf{B} = \begin{vmatrix} \rho & \phi & z \\ A_\rho & A_\phi & A_z \\ B_\rho & B_\phi & B_z \end{vmatrix} = \begin{vmatrix} r & \theta & \phi \\ A_r & A_\theta & A_\phi \\ B_r & B_\theta & B_\phi \end{vmatrix}. \tag{A.33}$$

Applications

1. Force on a current element $I\,d\boldsymbol{\ell}$ in a $\mathbf{B}$ field, as shown in Figure A.17*a*, is

$$d\mathbf{F} = I\,d\boldsymbol{\ell} \times \mathbf{B}. \tag{A.34a}$$

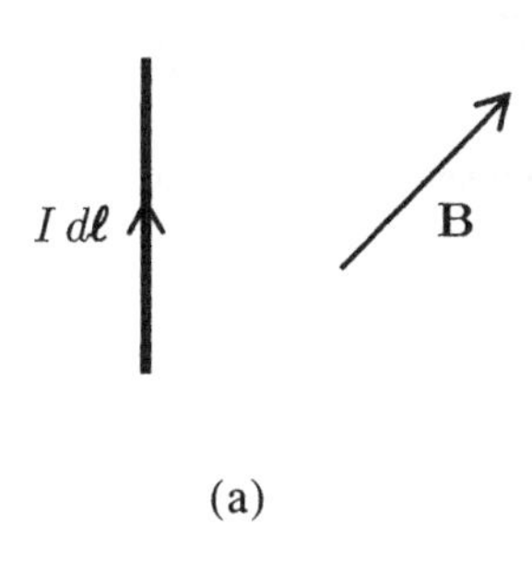

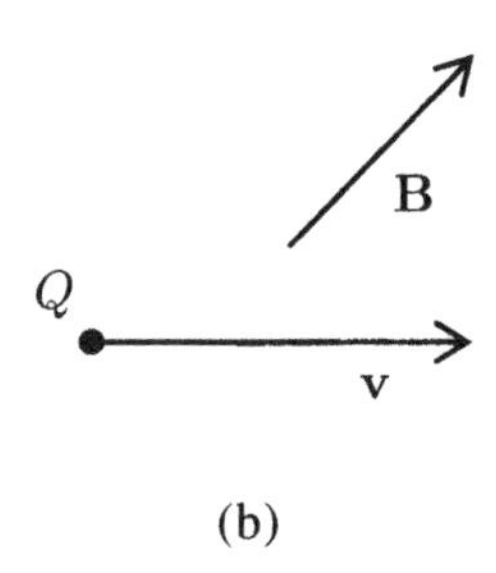

Figure A.17 (*a*) Current element $I\,d\boldsymbol{\ell}$ in a magnetic field $\mathbf{B}$. (*b*) Electric charge Q in free space moving with uniform velocity $\mathbf{v}$ in a magnetic field $\mathbf{B}$.

2. Force on a charge Q moving with velocity $\mathbf{v}$ in a $\mathbf{B}$ field, as shown in Figure A.17*b*, is

$$\mathbf{F}_Q = Q(\mathbf{v} \times \mathbf{B}). \tag{A.34b}$$

A.5.3 Product of Three Vectors

Given three vectors $\mathbf{A}, \mathbf{B}, \mathbf{C}$.

1. $\mathbf{A} \times \mathbf{B} \times \mathbf{C}$ must necessarily be defined as $(\mathbf{A} \times \mathbf{B}) \times \mathbf{C}$ or $\mathbf{A} \times (\mathbf{B} \times \mathbf{C})$; no new definition needed, and one can proceed as before.

2. $(\mathbf{A} \cdot \mathbf{B})\,\mathbf{C}$ is nothing more than the product (scalar) of $(\mathbf{A} \cdot \mathbf{B})$ times $\mathbf{C}$, that is, a vector in the direction of $\mathbf{C}$ with appropriate magnitude.

Similarly $(\mathbf{B} \cdot \mathbf{C})\,\mathbf{A}$, $(\mathbf{C} \cdot \mathbf{A})\,\mathbf{B}$, etc., can be interpreted.

3. **The Triple Scalar Product**

For $\mathbf{A} \cdot \mathbf{B} \times \mathbf{C}$, obviously the cross product is to be carried out first and the end product is a scalar, hence the name.

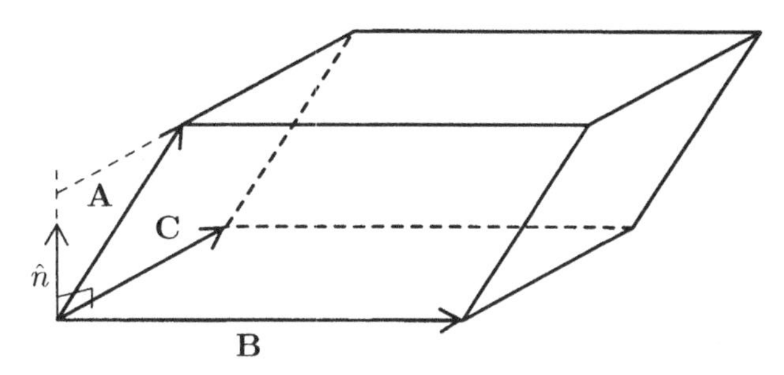

Figure A.18 $\mathbf{A}, \mathbf{B}$, and $\mathbf{C}$ representing the adjacent sides of a parallelepiped.

Choose $\mathbf{A}, \mathbf{B}, \mathbf{C}$ as the adjacent sides of a parallelepiped (Figure A.18). Then

$$\begin{aligned}\mathbf{A} \cdot \mathbf{B} \times \mathbf{C} &= (\mathbf{A} \cdot \hat{n})\ (\text{area of the base parallelogram with sides } \mathbf{B}, \mathbf{C}) \\ &= [\text{height of the parallelepiped with sides } (\mathbf{A}, \mathbf{B}, \mathbf{C})] \times \text{base} \\ &= \text{volume of the parallelepiped with adjacent sides } \mathbf{A}, \mathbf{B}, \mathbf{C}).\end{aligned} \tag{A.35}$$

Similarly

$$\begin{aligned}\therefore \mathbf{A} \cdot \mathbf{B} \times \mathbf{C} &= \mathbf{B} \cdot \mathbf{C} \times \mathbf{A} = \mathbf{C} \cdot \mathbf{A} \times \mathbf{B} \\ &= \text{volume of the parallelepiped} \\ &\quad \text{with adjacent sides } \mathbf{A}, \mathbf{B}, \mathbf{C}.\end{aligned} \tag{A.36}$$

Expression (A.36) is often referred to as the *cyclical rule*.

It can be shown that

$$\mathbf{A} \cdot (\mathbf{B} \times \mathbf{C}) = \begin{vmatrix} A_x & A_y & A_z \\ B_x & B_y & B_z \\ C_x & C_y & C_z \end{vmatrix}. \tag{A.37}$$

4. **Vector Triple Product**

$$\mathbf{A} \times (\mathbf{B} \times \mathbf{C}) = \mathbf{B}(\mathbf{A} \cdot \mathbf{C}) - \mathbf{C}(\mathbf{A} \cdot \mathbf{C}), \tag{A.38}$$

where the bracket on the left-hand side is important. Equation (A.38) can be proved by using algebraic expressions.

Division

$\mathbf{A}/\alpha$ is defined when $\alpha \neq 0$.

$\mathbf{A}/\mathbf{B}$ is not defined unless $\mathbf{A} \parallel \mathbf{B}$.

Example A.4

Geometric proof of the property $\mathbf{A} \times (\mathbf{B} \times \mathbf{C}) = \mathbf{A} \times \mathbf{B} + \mathbf{A} \times \mathbf{C}$.

Solution

Consider a prism of sides $\mathbf{A}, \mathbf{B}, \mathbf{C}$ and $\mathbf{B} + \mathbf{C}$. As shown in Figure A.19. Now we use the theorem that the total vector sum of a closed surface $\equiv 0$

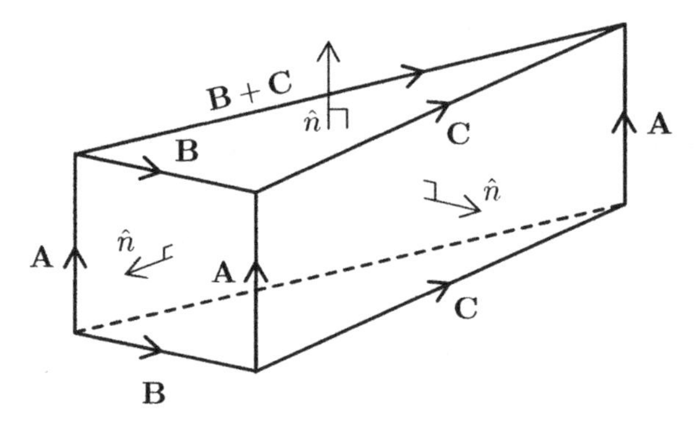

Figure A.19 Three vectors $\mathbf{A}, \mathbf{B}, \mathbf{C}$ forming a closed geometrical shape.

(Note: $\hat{n}$ outward normal, the closed vector surface enclosing the prism):

$$(\mathbf{B} \times \mathbf{A}) + \frac{1}{2}(\mathbf{B} \times \mathbf{C}) + \frac{1}{2}((\mathbf{C} \times \mathbf{B}) + \mathbf{C} \times \mathbf{A} + \mathbf{A} \times (\mathbf{B} \times \mathbf{C}) \equiv 0$$

$$\therefore \quad \mathbf{A} \times (\mathbf{B} \times \mathbf{C}) = (\mathbf{A} \times \mathbf{B}) + \mathbf{A} \times \mathbf{C}.$$

A.6 COORDINATE SYSTEMS

Choice of a specific coordinate system depends on the geometry of the problem under consideration. Various systems are in use. Most common are the right-handed, rectangular, circular cylindrical, and spherical systems.

Detailed discussion of the coordinate systems are available in various references cited. Here we discuss the basic outlines.

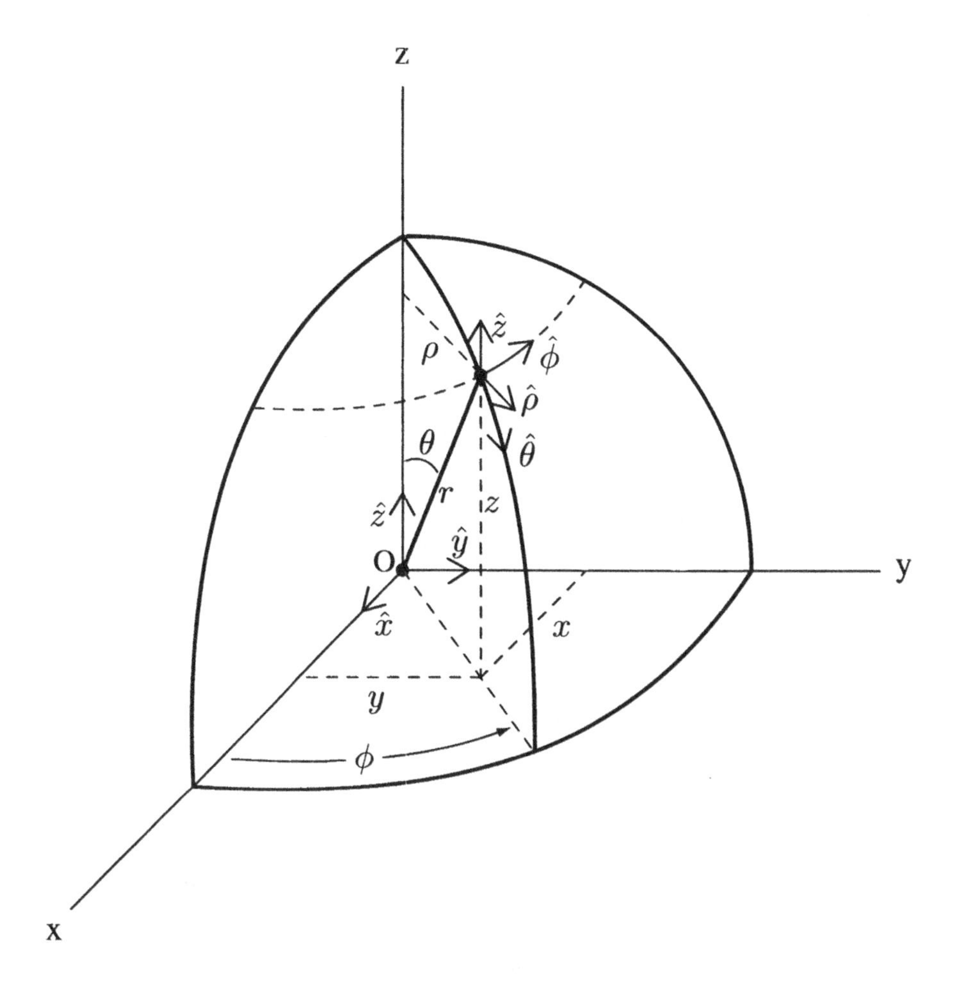

Figure A.20 Rectangular (or cartesian), cylindrical, and spherical coordinate systems, including their basic unit vectors.

A.6.1 Three Basic Coordinate Systems

The representation of a point P in space in the three coordinate systems is given in the compact diagram Figure A.20. The location of any point P in space is specified by a combination of three basic space variables measured in the three orthogonal directions. Any vector can be expressed in terms of its three basic vector components in the three orthogonal directions (or base vector directions). Note that in each system the three unit vectors are orthogonal to each other and they form a positive triad.

A.6.2 Space Variables and Base Vectors

Space Variables

The basic space variables and base vectors for the three systems of interest are as given in Table A.1.

Table A.1 Three Coordinate Systems and Their Associated Coordinates and Base Vectors

Coordinate System	Space Variables	Base Vectors
Rectangular	x, y, z	$\hat{x}, \hat{y}, \hat{z}$
Circular cylindrical	ρ, ϕ, z	$\hat{\rho}, \hat{\phi}, \hat{z}$
Spherical	r, θ, ϕ	$\hat{r}, \hat{\theta}, \hat{\phi}$

Note with reference to Figure A.20, any point P in space is identified by the intersection of the three coordinate surfaces passing through the point P. Thus the coordinates of P are (x, y, z), (ρ, ϕ, z), and (r, θ, ϕ) in the three systems.

The coordinate surfaces in the three systems are:

$$\left.\begin{aligned} x &= \text{constant} \\ y &= \text{constant} \\ z &= \text{constant} \end{aligned}\right\} \text{rectangular plane surfaces.}$$

$$\begin{aligned} \rho = \text{constant, cylindrical surface, } \phi &= \text{constant, plane surface,} \\ z &= \text{constant, plane surface.} \end{aligned}$$

$$\begin{aligned} r &= \text{constant, spherical surface} \\ \theta &= \text{constant, conical surface} \\ \phi &= \text{constant, plane surface.} \end{aligned}$$

The ranges of the space variables are

$$\begin{aligned} -\infty < x, y, z < \infty \\ 0 \leq \rho, r \leq \infty \\ 0 < \phi < 2\pi \\ 0 \leq \theta \leq \pi. \end{aligned} \tag{A.39}$$

The following expressions give the relationships between the space variables in the three systems:

Rectangular to cylindrical systems, and vice versa,

$$\begin{aligned} x &= \rho\cos\phi & \rho &= \left(x^2+y^2\right)^{1/2} \geq 0 \\ y &= \rho\sin\phi & \phi &= \tan^{-1}\frac{y}{x} \\ z &= z & z &= z. \end{aligned} \tag{A.40}$$

Rectangular to spherical systems, and vice versa,

$$\begin{aligned} x &= r\sin\theta\cos\phi & r &= (x^2+y^2+z^2)^{\frac{1}{2}} \geq 0 \\ y &= r\sin\theta\sin\phi & \theta &= \tan^{-1}\frac{\left(x^2+y^2\right)^{\frac{1}{2}}}{z} \\ z &= r\cos\theta & \phi &= \tan^{-1}\frac{y}{x}. \end{aligned} \tag{A.41}$$

Cylindrical to spherical systems, and vice versa,

$$\begin{aligned} \rho &= r\sin\theta & r &= (\rho^2+z^2)^{\frac{1}{2}} \geq 0 \\ \phi &= \phi & \phi &= \phi = \tan^{-1}\frac{y}{x} \\ z &= r\cos\theta & \theta &= \tan^{-1}\frac{\rho}{z} = \tan^{-1}\frac{\left(x^2+y^2\right)^{\frac{1}{2}}}{z}. \end{aligned} \tag{A.42}$$

Base Vectors

In each system the base vectors consist of three unit vectors along the three orthogonal coordinate directions (Figure A.20), and in each system they form a positive triad of unit vectors. In other words, the three unit vectors of each set are orthogonal to each other, so they form a positive triad. In the rectangular, cylindrical, and spherical system of coordinates they are as follows:

Base vectors	Interrelations	
$\hat{x},\ \hat{y},\ \hat{z}$	$\hat{x}\times\hat{y}=\hat{z},\ \hat{y}\times\hat{z}=\hat{x},\ \hat{z}\times\hat{x}=\hat{y}$	
	$\hat{x}\cdot\hat{y}=\hat{y}\cdot\hat{z}=\hat{z}\cdot\hat{x}=0$	(A.43a)
$\hat{\rho},\ \hat{\phi},\ \hat{z}$	$\hat{\rho}\times\hat{\phi}=\hat{z},\ \hat{\phi}\times\hat{z}=\hat{\rho},\ \hat{z}\times\hat{\rho}=\hat{\phi}$	
	$\hat{\rho}\cdot\hat{\phi}=\hat{\phi}\cdot\hat{z}=\hat{z}\cdot\hat{\rho}=0$	(A.43b)
$\hat{r},\ \hat{\theta},\ \hat{\phi}$	$\hat{r}\times\hat{\theta}=\hat{\phi},\ \hat{\theta}\times\hat{\phi}=\hat{r},\ \hat{\phi}\times\hat{r}=\hat{\theta}$	
	$\hat{r}\cdot\hat{\theta}=\hat{\theta}\cdot\hat{\phi}=\hat{\phi}\cdot\hat{r}=0.$	(A.43c)

Note that among the base vectors given above, only $\hat{x},\hat{y},\hat{z}$ are constant vectors; the others are not.

By using the appropriate base vectors, one can express any vector $\mathbf{A}$ in terms of its components in the three coordinate systems, as

$$\begin{aligned}\mathbf{A} &= A_x\,\hat{x} + A_y\,\hat{y} + A_z\,\hat{z}\\ &= A_\rho\,\hat{\rho} + A_\phi\,\hat{\phi} + A_z\,\hat{z}\\ &= A_r\,\hat{r} + A_\theta\,\hat{\theta} + A_\phi\,\hat{\phi}.\end{aligned} \tag{A.44}$$

A.7 ELEMENTARY DIFFERENTIAL RELATIONS

A.7.1 Rectangular System

Consider two points $P(x, y, z)$ and $P'(x+dx, y+dy, z+dz)$. The differential length (vector displacement) is

$$d\boldsymbol{\ell} = \overrightarrow{PP'} = dx\,\hat{x} + dy\,\hat{y} + dz\,\hat{z}. \tag{A.45}$$

The elementary vector surface is

$$\begin{aligned}\mathrm{ds} &= ds_x\,\hat{x} + ds_y\,\hat{y} + ds_z\,\hat{z}\\ &= dydz\,\hat{x} + dzdx\,\hat{y} + dxdy\,\hat{z}.\end{aligned} \tag{A.46}$$

The elementary (differential) volume is

$$\begin{aligned}d\mathrm{v} &= (dx\,\hat{x})\cdot(dy\,\hat{y}\times dz\,\hat{z})\\ &= (dy\,\hat{y})\cdot(dz\,\hat{z}\times dx\,\hat{x})\\ &= (dz\,\hat{z})\cdot(dx\,\hat{x}\times dy\,\hat{z})\\ &= dx\,dy\,dz\\ &\quad \text{(property of triple scalar product).}\end{aligned} \tag{A.47}$$

A.7.2 Cylindrical and Spherical Systems

Note that $ds, d\mathrm{v}$ can be obtained once the expression for $d\boldsymbol{\ell}$ is shown. By inspection we now write the corresponding quantities in the other systems.

Cylindrical system $(\rho, \phi, z), (\hat{\rho}, \hat{\phi}, \hat{z})$:

$$\begin{aligned}d\boldsymbol{\ell} &= d\rho\,\hat{\rho} + \rho\,d\phi\hat{\phi} + dz\,\hat{z}\\ &= ds_\rho\,\hat{\rho} + ds_\phi\,\hat{\phi} + ds_z\,\hat{z}\end{aligned} \tag{A.48}$$

$$\mathrm{ds} = \rho\,d\phi dz\hat{\rho} + dz\,d\rho\,\hat{\phi} + \rho\,d\ell\,d\phi\,\hat{z} \tag{A.49}$$

$$d\mathrm{v} = \rho\,d\rho\,d\phi\,dz. \tag{A.50}$$

Spherical system $f(r, \theta, \phi), (\hat{r}, \hat{\theta}, \hat{\phi})$:

$$d\boldsymbol{\ell} = dr\,\hat{r} + r\,d\theta\,\hat{\theta} + r\sin\theta\,d\phi\hat{\phi} \tag{A.51}$$

$$\begin{aligned} d\mathrm{s} &= ds_r\hat{r} + ds_\theta\,\hat{\theta} + ds_\phi\,\hat{\phi} \\ &= r^2\sin\theta\,d\theta\,d\phi\hat{r} + r\,\sin\theta\,dr\,d\phi\hat{\theta} + r\,dr\,d\phi\,\hat{\phi} \end{aligned} \tag{A.52}$$

$$d\mathrm{v} = r^2\,\sin\theta\,dr\,d\theta\,d\phi. \tag{A.53}$$

A.8 TRANSFORMATION OF UNIT VECTORS

We have seen that any given vector can be expressed in a specific coordinate system in terms of its components in three orthogonal directions. Frequently one needs to express the vectors in other coordinate systems. For this we need to know the relationship between the unit vectors of these systems. For example, let

$$\begin{aligned} \mathbf{A} &= A_x\,\hat{x} + A_y\,\hat{y} + A_z\,\hat{z} \\ &= A_\rho\,\hat{\rho} + A_\phi\,\hat{\phi} + A_z\,\hat{z}. \end{aligned} \tag{A.54a}$$

If A_x, A_y, A_z are known, how do we get A_ρ, A_ϕ, A_z from the knowledge of the rectangular components?

We obtain

$$A_\rho = (\mathbf{A}\cdot\hat{\rho}) = A_x(\hat{x}\cdot\hat{\rho}) + A_y(\hat{y}\cdot\hat{\rho}) + A_z(\hat{z}\cdot\hat{\rho}). \tag{A.54b}$$

Thus we need to know the relationship between the unit vectors in the two systems to carry out the dot products above.

The derivation of the transformation relationships can be found in [4]. Tables A.2, A.3, and A.4 give the required relationships.

Example A.5

1. Express $\hat{x}$ in (ρ, ϕ, z) system.

2. Express $\hat{\rho}$ in $(\hat{x}, \hat{y}, \hat{z})$ system.

Solution

1. Use Table A.2. Go down the $\hat{x}$ column and carry along the corresponding $(\hat{\rho}, \hat{\phi}, \hat{z})$ factors: $\hat{x} = \hat{\rho}\,\cos\phi - \hat{\phi}\,\sin\phi - \hat{z}\,0$.

Table A.2 Transformation between Rectangular to Cylindrical Systems, and Vice Versa

	A_x $\hat{x}$	A_y $\hat{y}$	A_z $\hat{z}$
$A_\rho \hat{\rho}$	$\cos\phi$	$\sin\phi$	0
$A_\phi \hat{\phi}$	$-\sin\phi$	$\cos\phi$	0
$A_z \hat{z}$	0	0	1

Table A.3 Transformation between Rectangular and Spherical Systems, and Vice Versa

	A_x $\hat{x}$	A_y $\hat{y}$	A_z $\hat{z}$
$A_r \hat{r}$	$\sin\theta \cos\phi$	$\sin\theta \sin\phi$	$\cos\theta$
$A_\theta \hat{\theta}$	$\cos\theta \cos\phi$	$\cos\theta \sin\phi$	$-\sin\theta$
$A_\phi \hat{\phi}$	$-\sin\phi$	$\cos\phi$	0

Table A.4 Transformation between Cylindrical and Spherical Systems, and Vice Versa

	A_x $\hat{x}$	A_y $\hat{y}$	A_z $\hat{z}$
$A_\rho \hat{\rho}$	$\sin\theta$	$\cos\theta$	0
$A_\phi \hat{\phi}$	0	0	1
$A_z \hat{z}$	$\cos\theta$	$-\sin\theta$	0

2. Use Table A.2. Go along the row $\hat{\rho}$ and carry along the corresponding $(\hat{x}, \hat{y}, \hat{z})$ factors:

$$\begin{aligned}\hat{\rho} &= \hat{x}\,\cos\phi + \hat{y}\,\sin\phi + \hat{z}\,0 \\ &= \hat{x}\frac{x}{\sqrt{x^2+y^2}} + \hat{y}\frac{y}{\sqrt{x^2+y^2}}\,.\end{aligned}$$

Proceed similarly for the others.

A.9 VECTOR CALCULUS

In general, a vector may be a function of both space and time. Let such a function be denoted by $\mathbf{A}(x, y, z, t)$ or $\mathbf{A}(\mathbf{r}, t)$, or simply by $\mathbf{A}(t)$ with the understanding that space variables are implied. Let us consider the time variation of the time derivative of $\mathbf{A}$.

A.9.1 Time Derivative of Vector A

In an interval of time Δt, the vector **A** may change by $\Delta\mathbf{A}$, which in general represents also a change in both magnitude and direction, as shown in Figure A.21.

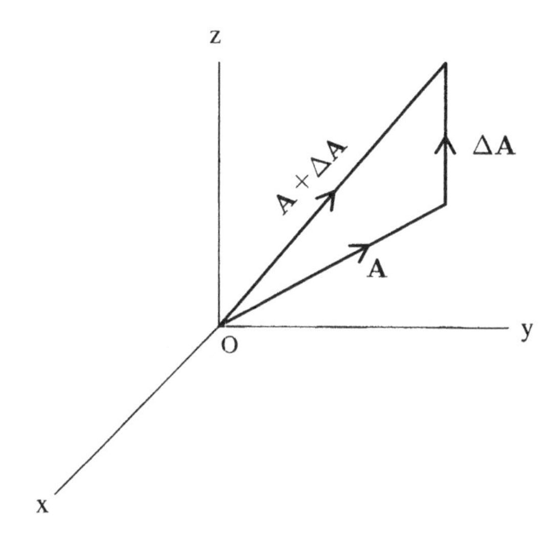

Figure A.21 Vector **A** at time t and Δt.

Thus

$$\begin{aligned} \mathbf{A} &= A_x\,\hat{x} + A_y\,\hat{y} + A_z\,\hat{z} \\ \Delta\mathbf{A} &= \Delta A_x\,\hat{x} + \Delta A_y\,\hat{y} + \Delta A_z\,\hat{z}, \end{aligned} \tag{A.55}$$

where the components are the components of **A** and $\Delta\mathbf{A}$ shown in Figure A.21.

We now use the total derivative of **A** defined by

$$\begin{aligned} \frac{d\mathbf{A}}{dt} &= \lim_{\Delta t\to 0} \frac{\mathbf{A}(t+\Delta t) - \mathbf{A}(t)}{\Delta t} \\ &= \lim_{\Delta t\to 0} \left(\frac{\Delta A_x}{\Delta t}\hat{x} + \frac{\Delta A_y}{\Delta t}\hat{y} + \frac{\Delta A_z}{\Delta t}\hat{z}\right) \\ &= \frac{dA_x}{dt}\hat{x} + \frac{dA_y}{dt}\hat{y} + \frac{dA_z}{dt}\hat{z}. \end{aligned} \tag{A.56}$$

Thus the time derivative of a vector function is equal to the vector sum of the time derivative of its components, and similarly for the derivative with respect to the variables. As a result we have the derivative of a vector with respect to one variable defined as the vector sum of the derivatives of its components with respect to the same variable.

A.9.2 Space Derivatives of a Vector A

Often it is of interest to know certain variations in space of the scalar and vector quantities. Such variations are generally obtained by carrying out three fundamental operations (in space) on the quantity of interest:

Gradient of a scalar function
Divergence of a vector function
Curl (or rotation) of a vector function.

Each of the mathematical operations above implies some physical process. It is, of course, assumed that the vector or scalar functions of interest are physical in nature—meaning they are simple-valued, continuous, and differentiable functions of the coordinates at all points of interest in space. These properties are always true of the physical fields that we will encounter.

In the following sections we will assume that scalar functions $V(x, y, z)$ and vector functions $\mathbf{A}(x, y, z)$ may be time dependent.

A.9.3 Gradient of a Scalar Function

Let us consider two points $P_1(x, y, z)$ and $P_2(x + dx,\ y + dy,\ z + dz)$. We know from calculus that the total change in V from P_1 to P_2 is given by

$$dV = \frac{\partial V}{\partial x} dx + \frac{\partial V}{\partial y} dy + \frac{\partial V}{\partial z} dz. \tag{A.57}$$

If the independent differentials dx, dy, dz are considered as the cartesian components of a differential vector $d\boldsymbol{\ell}$ (or vector displacement) such that

$$d\boldsymbol{\ell} = dx\,\hat{x} + dy\,\hat{y} + dz\,\hat{z}$$

and we define a vector $\mathbf{G}$ such that

$$\mathbf{G} = \frac{\partial V}{dx}\hat{x} + \frac{\partial V}{dy}\hat{y} + \frac{\partial V}{dz}\hat{z} \tag{A.58}$$

we then obtain

$$\begin{aligned} \mathbf{G} \cdot d\boldsymbol{\ell} &= \frac{\partial V}{\partial x}\,dx + \frac{\partial V}{\partial y}\,dy + \frac{\partial V}{\partial z}\,dz \\ &= dV. \end{aligned} \tag{A.59}$$

The vector $\mathbf{G}$ whose components are the rates of change of V with distance along the coordinate axes (directional) is called the *gradient* of the scalar function V. We write it as

$$\mathbf{G} = \text{grad}V = \nabla V, \tag{A.60}$$

where the del operation (in the rectangular system) is defined by

$$\nabla = \hat{x}\frac{\partial}{\partial x} + \hat{y}\frac{\partial}{\partial y} + \hat{z}\frac{\partial}{\partial z} \, . \tag{A.61}$$

Thus we have

$$dV = \mathbf{G} \cdot d\boldsymbol{\ell} = \nabla V \cdot d\boldsymbol{\ell} = \text{grad}V \cdot d\boldsymbol{\ell}. \tag{A.62}$$

The particular form of the gradient $\mathbf{G} = \nabla V$ given in (A.59) is in the rectangular system of coordinates. Assuming $d\boldsymbol{\ell}$ is in the cylindrical and spherical coordinate systems, it can be shown that the gradient of a scalar function V is

$$\mathbf{G} = \nabla V = \frac{\partial V}{\partial \rho}\hat{\rho} + \frac{1}{\rho}\frac{\partial V}{\partial \phi}\hat{\phi} + \frac{\partial V}{\partial z}\hat{z} \tag{A.63}$$

$$\mathbf{G} = \nabla V = \frac{\partial V}{\partial r}\hat{r} + \frac{1}{r}\frac{\partial V}{\partial \theta}\hat{\theta} + \frac{1}{r\,\sin\theta}\frac{\partial V}{\partial \phi}\hat{\phi} \tag{A.64}$$

in cylindrical and spherical systems of coordinates, respectively. Note that the definition of $\mathbf{G}$ implies that it does not depend on the specific coordinate systems used to obtain its value. The gradient operator in cylindrical and spherical coordinate systems are

$$\nabla = \hat{\rho}\frac{\partial}{\partial \rho} + \hat{\phi}\frac{1}{\rho}\frac{\partial}{\partial \phi} + \hat{z}\frac{\partial}{\partial z} \tag{A.65}$$

$$\nabla = \hat{r}\frac{\partial}{\partial r} + \hat{\theta}\frac{1}{r}\frac{\partial}{\partial \theta} + \hat{\phi}\frac{1}{r\sin\theta}\frac{\partial}{\partial \theta} \, . \tag{A.66}$$

Finally, note that the gradient operator operates on a scalar function and yields a vector function.

Two Important Properties of ∇V

The gradient of any scalar function has the following two important properties:

- The direction of ∇V at a point in space is the same as the (unit) normal to the surface $V =$ constant passing through that point, and is pointed in the direction of increasing V (see Figure A.22).

- The magnitude of ∇V (i.e., $|\nabla V|$) at any point is equal to the normal derivative of V at that point (and it gives the maximum rate of increase of the function).

Combining the above, we can write

$$\nabla V = |\nabla V|\hat{n} = \frac{\partial V}{\partial n}\hat{n} \tag{A.67}$$

$$\text{with} \quad \hat{n} = \frac{\nabla V}{|\nabla V|} \,. \tag{A.68}$$

Equation (A.65) is frequently used to determine the positive unit normal to the surface defined by $V =$ constant, shown in Figure A.22

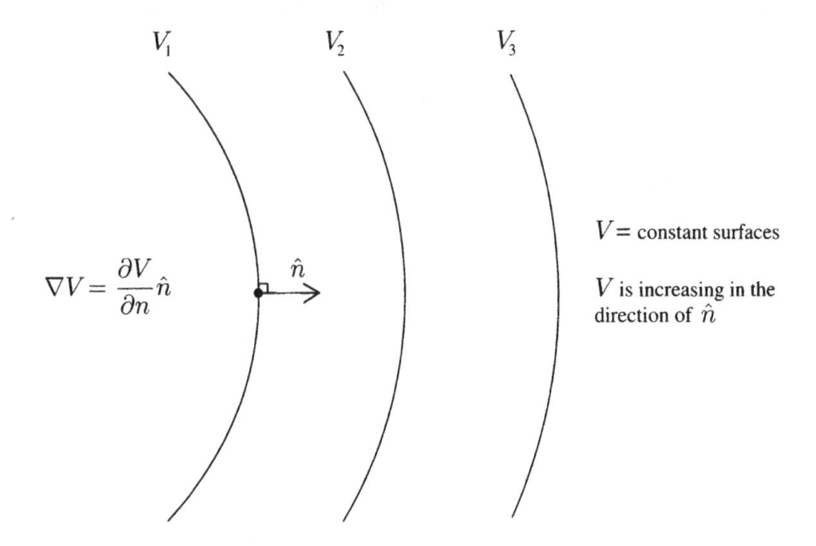

Figure A.22 Diagram illustrating the properties of ∇V.

Directional Derivative

From (A.59) we obtain

$$\mathbf{G} \cdot \hat{\ell} = \frac{dV}{d\ell}$$

meaning the component of $\mathbf{G}$ in the direction $\hat{\ell}$ (i.e. the direction of the displacement $d\boldsymbol{\ell}$) gives the increase of V in that direction. This is also called the *directional derivative* of V in that direction. We consider two points P_1 and P_2 with P_1 on the $V =$ constant surface and P_2 displaced from P_1 by $d\boldsymbol{\ell}$ as shown in Figure A.23.

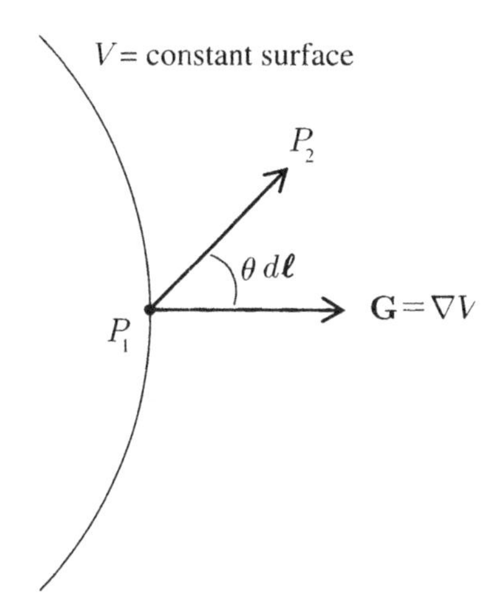

Figure A.23 Typical V = constant surface.

Thus the differential change V from P_1 to P_2 is

$$\begin{aligned} dV &= \nabla V \cdot d\boldsymbol{\ell} \\ &= |\nabla V| \cos\theta \, d\ell \\ \text{or } \frac{dV}{d\ell} &= |\nabla V| \cos\theta \end{aligned} \tag{A.69}$$

for $\theta = 0$, $d\boldsymbol{\ell} = \hat{n}\, dn$. The directional derivative is maximum and is equal to the magnitude of ∇V

$$|\nabla V| = \frac{dV}{dn} \,. \tag{A.70}$$

The gradient operator may be viewed formally as converting a scalar field into a vector field. Certain vector fields can be expressed as the gradient of a fictitious scalar field. This scalar field often plays a significant role in the ensuing mathematical analysis.

An Important Integration Theorem

We know from (A.62) that

$$\nabla V \cdot d\boldsymbol{\ell} = dV.$$

Integrating the above from two points P_1 and P_2 in space we obtain

$$\int_{P_1}^{P_2} \nabla V \cdot d\boldsymbol{\ell} = \int_{P_1}^{P_2} dV = V(P_2) - V(P_1). \tag{A.71}$$

In words, the integral of any vector function $\mathbf{G} = (\nabla V)$ that is the gradient of a scalar function V is independent of the path of the integral.

Example A.6

The vector from the point $P'(x', y', z')$ to another point $P(x, y, z)$ is represented by

$$\mathbf{R} = (x - x')\hat{x} + (y - y')\hat{y} + (z - z')\hat{z}. \tag{A.72}$$

Show that

$$\nabla R = \hat{R}, \quad \nabla\left(\frac{1}{R}\right) = -\frac{\hat{R}}{R^2} \tag{A.73a}$$

$$\nabla' R = -\hat{R}, \quad \nabla'\left(\frac{1}{R}\right) = +\frac{\hat{R}}{R^2}, \tag{A.73b}$$

where

$$\nabla' = \hat{x}\frac{\partial}{\partial x'} + \hat{y}\frac{\partial}{\partial y'} + \hat{z}\frac{\partial}{\partial z'}$$

and the other notations are as explained.

Solution

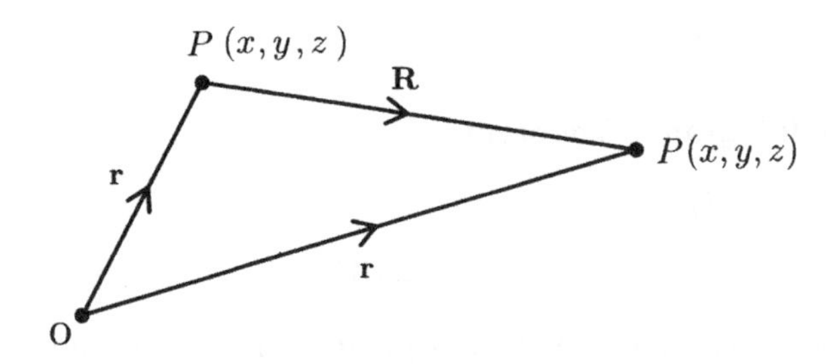

Figure A.24 Geometrical representation of the points P' and P in space with respect to an arbitrary origin O.

The geometry is as shown in Figure A.24. Here, $\mathbf{R} = \mathbf{r} - \mathbf{r}' = (x - x')\hat{x} + (y - y')\hat{y} + (z - z')\hat{z}$. Thus $R = \left[(x - x')^2 + (y - y')^2 + (z - z')^2\right]^{1/2}$.

$$\begin{aligned}
\nabla R &= \frac{\partial R}{\partial x}\hat{x} + \frac{\partial R}{\partial y}\hat{y} + \frac{\partial R}{\partial z}\hat{z} \\
&= \frac{1}{2}\frac{1}{R}2(x - x')\frac{\partial}{\partial x}(x - x')\hat{x} + \frac{1}{2}\frac{1}{R}2(y - y')\frac{\partial}{\partial y}(y - y')\hat{y} \\
&\quad + \frac{1}{R}2(z - Z')\frac{\partial}{\partial z}(z - z')\hat{z} \\
&= \frac{(x - x')\hat{x} + (y - y')\hat{y} + (z - z')\hat{z}}{R} = \frac{\mathbf{R}}{R} = \hat{R} \\
\nabla\left(\frac{1}{R}\right) &= \frac{\partial}{\partial x}\left(\frac{1}{R}\right)\hat{x} + \frac{\partial}{\partial y}\left(\frac{1}{R}\right)\hat{y} + \frac{\partial}{\partial z}\left(\frac{1}{R}\right)\hat{z} \\
&= -\frac{1}{R^2}\left[\frac{\partial r}{\partial x}\hat{x} + \frac{\partial r}{\partial y}\hat{y} + \frac{\partial r}{\partial z}\hat{z}\right] \\
&= -\frac{1}{R^2}\frac{(x - x')\hat{x} + (y - y')\hat{y} + (z - z')\hat{x}}{R} \\
&= -\frac{\mathbf{R}}{R^3} = -\frac{1}{R^2}\frac{\mathbf{R}}{R} = -\frac{\hat{R}}{R^2}.
\end{aligned}$$

Note that

$$\nabla' R = -\nabla R \text{ and } \nabla'\left(\frac{1}{R}\right) = -\nabla\left(\frac{1}{R}\right). \tag{A.74a}$$

Therefore

$$\nabla' R = -\nabla\hat{R} \quad , \nabla'\left(\frac{1}{R}\right) = +\frac{\hat{R}}{R^2}. \tag{A.74b}$$

A.9.4 Flux of a Vector

Given a vector $\mathbf{A}$ in space one can define an elementary vector surface $ds = \hat{n}\, ds$ as shown in Figure A.25. This may be part of a closed (or open) surface S. (Note: $\hat{n}$ is outward normal if ds is part of a closed surface; for an open surface positive normal is defined in the right-hand rule sense.)

Thus for a finite open surface S the total *flux* (of $\mathbf{A}$) through S is by definition

$$\Phi = \int_s \mathbf{A} \cdot ds. \tag{A.75}$$

For a closed surface, the total outward flux is

$$\Phi = \oint_s \mathbf{A} \cdot ds. \tag{A.76}$$

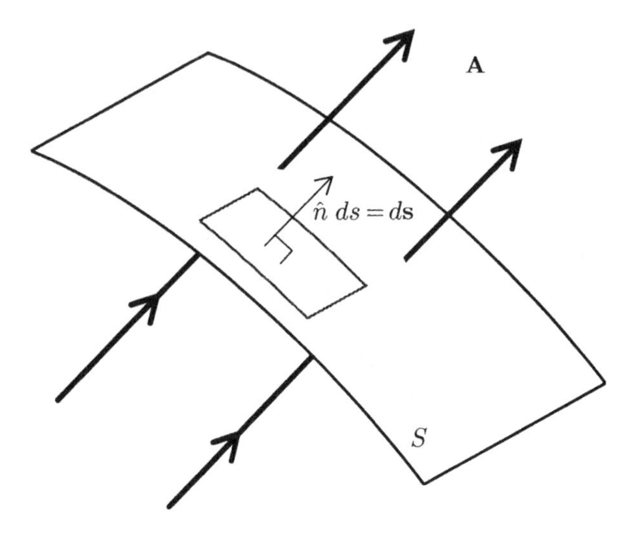

Figure A.25 Flux of **A** through a surface S.

The total outward flux through a closed surface is a measure of a source or sink inside.

A.9.5 Divergence of a Vector A

The *divergence* of a vector **A** at a point P, abbreviated as div $\mathbf{A} = \nabla \cdot \mathbf{A}$, is defined as the net outward flux of **A** per unit volume as the volume ∇V about that point tends to zero. Symbolically (see Figure A.26),

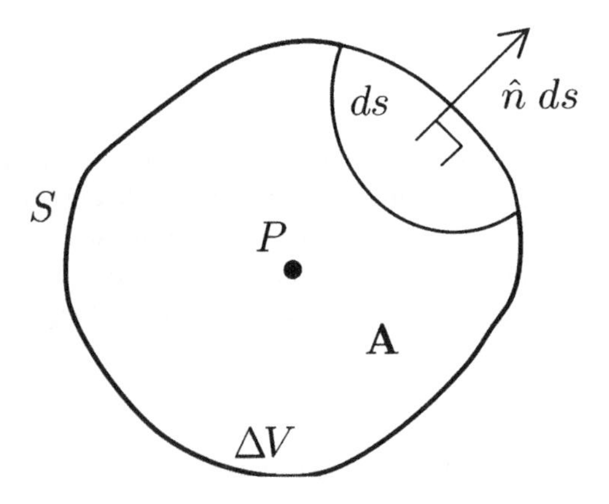

Figure A.26 Diagram for evaluation of the divergence of **A** at a point P. Surface S encloses the volume V.

$$\begin{aligned}\text{div } \mathbf{A} = \nabla \cdot \mathbf{A} &= \lim_{\nabla V \to 0} \oint_S \frac{\mathbf{A} \cdot ds}{\nabla V} \\ &= \lim_{\nabla V \to 0} \sum_m \frac{\mathbf{A} \cdot \nabla \mathbf{S_m}}{\nabla V},\end{aligned} \tag{A.77}$$

where $\nabla \mathbf{S_m}$ denotes the elementary surfaces pertaining to the volume ∇V. By definition, $\nabla \cdot \mathbf{A}$ is a scalar quantity and may vary from point to point. The definition, again, is independent of any coordinate system.

The explicit expression for $\nabla \cdot \mathbf{A}$ in a specific coordinate system may be obtained by assuming an approximate shape for ∇V in that particular system. We will not derive the expression here (it be found in standard textbooks [4–8]); we will just give the expression in the three basic coordinate systems:

$$\text{div } A = \nabla \cdot \mathbf{A} \tag{A.78}$$

$$= \frac{\partial A_x}{\partial x} + \frac{\partial A_y}{\partial y} + \frac{\partial A_z}{\partial z} \tag{A.79}$$

$$= \frac{1}{\rho} \frac{\partial}{\partial \rho} (\rho \, A_\rho) + \frac{1}{\rho} \frac{\partial A_\phi}{\partial_\phi} + \frac{\partial A_z}{\partial_z} \tag{A.80}$$

$$= \frac{1}{r^2} \frac{\partial}{\partial \sigma} (r^2 \, A_r) + \frac{1}{r \sin \theta} \frac{\partial}{\partial \theta} (A_\theta \sin \theta) + \frac{1}{r \sin \theta} \frac{\partial A_\phi}{\partial_\phi}. \tag{A.81}$$

Note that in rectangular systems

$$\begin{aligned}\nabla \cdot \mathbf{A} &= \left(\hat{x} \frac{\partial}{\partial x} + \hat{y} \frac{\partial}{\partial y} + \hat{z} \frac{\partial}{\partial z} \right) \cdot (\hat{x} \, Ax + \hat{y} \, Ay + \hat{z} \, Az) \\ &= \frac{\partial Ax}{\partial x} + \frac{\partial Ay}{\partial y} + \frac{\partial Az}{\partial z}\end{aligned} \tag{A.82}$$

which sometimes interprets the divergence of A as the dot products of ∇ and $\mathbf{A}$. (This is definitely not true in other systems of coordinates; even in the rectangular system it is not logical to interpret it as a dot product [3].)

Divergence Theorem (or Gauss's Theorem)

This is one of the most important theorems of vector analysis. We state it here without proof, which may be found in the references cited. The divergence theorem states that the volume integral of the divergence of a vector field $\mathbf{A}$ taken over any volume V is equal to the surface integral of the normal component of $\mathbf{A}$ taken over the closed surface enclosing the volume V (i.e., the total obtained flux of the vector through the surface enclosing the volume). The mathematical statement of the theorem is

$$\int_V \nabla \cdot \mathbf{A} \, dv = \oint_S \mathbf{A} \cdot ds = \oint_S \mathbf{A} \cdot \hat{n} \, ds, \tag{A.83}$$

where $\mathbf{A}$ is continuous and differentiable in V and on S, and S encloses the volume V.

Example A.7

Show that the vector field $\mathbf{A} = r\hat{r}$ satisfies the divergence theorem $\int_t \nabla \cdot \mathbf{A} d\tau = \oint \mathbf{A} \cdot ds$, where the volume τ is defined by $0 \leq r \leq a,\ \theta \leq 0 \leq \pi/2$, $0 \leq \phi \leq 2\pi$, and S' is the surface enclosing the volume τ. (Figure A.27.)

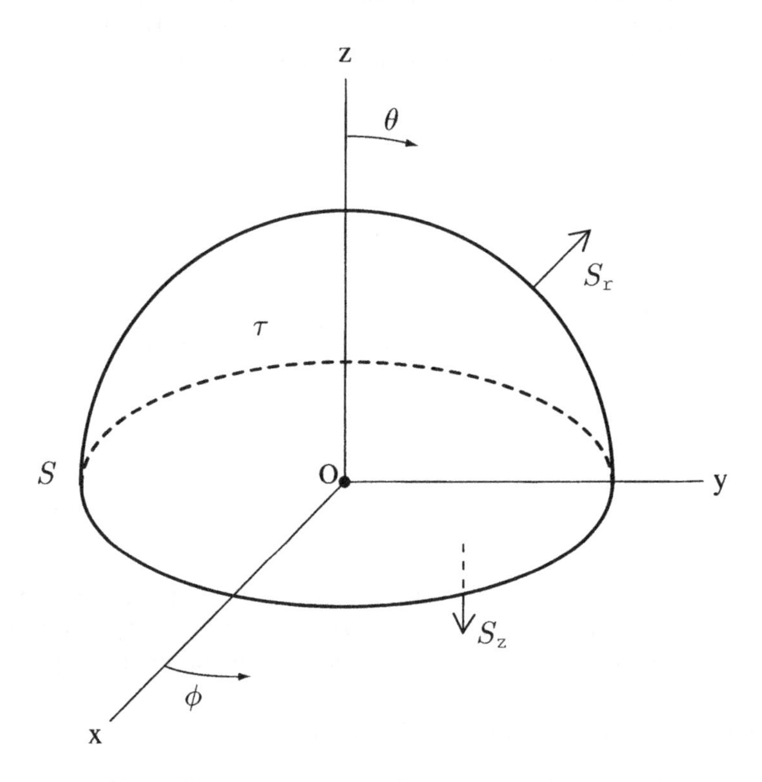

Figure A.27 Volume τ enclosed by surface S.

Solution

Note that the volume τ is the hemispherical volume sketched in Figure A.27. The total surface S enclosing the volume τ consists of the hemispherical surface S_r (outward normal $\hat{n}$) plus the planar surface (bottom surface $z \geq 0$)

in the $-\hat{z}$ direction.

$$
\begin{aligned}
\mathbf{A} &= r\hat{r} \\
\nabla \cdot \mathbf{A} &= \frac{1}{r^2}\frac{\partial}{\partial r}\left(r^2 A_r\right) = \frac{1}{r^2}\frac{\partial}{\partial r}\left(r^3\right) = 3 \\
\int_\tau \nabla \cdot \mathbf{A} d\tau &= \int_0^{2\pi}\int_0^{\frac{\pi}{2}}\int_0^a 3r^2 \sin\theta \, dr \, d\theta \, d\phi \\
&= 3\frac{4\pi a^3}{6} = 2\pi a^3
\end{aligned}
\tag{A.84}
$$

$$
\oint_s \mathbf{A} \cdot ds = \int_{S_r} r\, r^2 \sin\theta \, d\theta \, d\phi|_{r=a} + \int_{S_{-z}} (\rho\,\hat{\rho}) \cdot (\rho \, d\phi \, d\rho)(-\hat{z}).
$$

Note that on the bottom surface

$$
\begin{aligned}
\mathbf{A} = r\hat{r} &= \rho\hat{\rho} \\
&= \int_0^{2\pi}\int_0^{\pi/2} a^3 \sin\theta \, d\theta \, d\phi + 0 \\
&= 2\pi a^3.
\end{aligned}
\tag{A.85}
$$

Thus (A.84) $\equiv$ (A.85), which implies that the given **A** satisfies the divergence theorem where **A** is continuous and differentiable in V and on S and S encloses the volume V.

A.9.6 Curl of a Vector Function

Circulation or Curl of a Vector

The concept of the divergence originated from the translatory motion of fluids. The concept of the curl (or rotation) originated from the circulatory motion of fluids. Circulation of **A** at point P is defined by

$$
\text{circulation of } \mathbf{A} \text{ at } b = \oint_L \mathbf{A} \cdot d\mathbf{l}, \text{ as } L \to 0, \tag{A.86}
$$

where L is the total contour that contains the point P, as in Figure A.28.

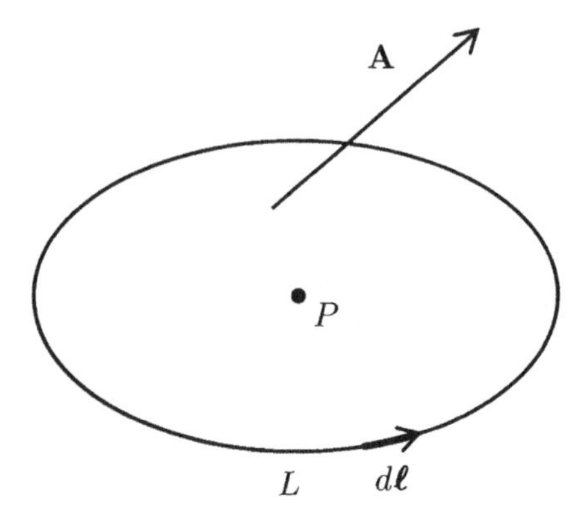

Figure A.28 Diagram defining the curl or circulation of a vector $\mathbf{A}$ at a point P.

General Definition of Curl

The *curl* of a vector function $\mathbf{A}$ at a point P, denoted by curl $\mathbf{A} = \nabla \times \mathbf{A}$, is defined by

$$\text{curl } \mathbf{A} = \nabla \times \mathbf{A} = \lim_{\Delta V \to 0} \sum_m \frac{\boldsymbol{\Delta}\mathbf{S_m} \times \mathbf{A}}{\Delta V}, \tag{A.87}$$

where $\boldsymbol{\Delta}\mathbf{S_m}$, ΔV and the summation have the same significance as those used in the definition of the divergence of a vector function. The only difference is that we have now used the vector product $\boldsymbol{\Delta}\mathbf{S_m} \times \mathbf{A}$ instead of the scalar product $\boldsymbol{\Delta}\mathbf{S_m} \cdot \mathbf{A}$ inverted in the divergence as before, the definition does not depend on any coordinate system, and the result is independent of the shape of ΔV. It is obvious from the definition that $\nabla \times \mathbf{A}$ is a vector function.

Explicit Expression for $\nabla \times \mathbf{A}$

It is simpler to deal with the component of $\nabla \times \mathbf{A}$ in any arbitrary direction. After knowing the three components of $\nabla \times \mathbf{A}$ in the three orthogonal directions of any coordinate system, we can obtain the complete expression for $\nabla \times \mathbf{A}$.

Choose the shape of ΔV to be a flat pillbox of area ΔS and thickness Δt as shown in Figure A.29.

This is permissible since in order to have a unique limiting value of $\nabla \times \mathbf{A}$, defined as before, the shape of ΔV is unimportant. Let the positive unit normal to the top flat surface be $\hat{n}$. The component of $\nabla \times \mathbf{A}$ in the $\hat{n}$ direction is then

$$\hat{n} \cdot \nabla \times \mathbf{A} = \lim_{\Delta V \to 0} \sum_m \frac{\boldsymbol{\Delta}\mathbf{S_M} \times \mathbf{A}}{\Delta V} \quad \text{with } \Delta V = \Delta S\, \Delta \ell. \tag{A.88}$$

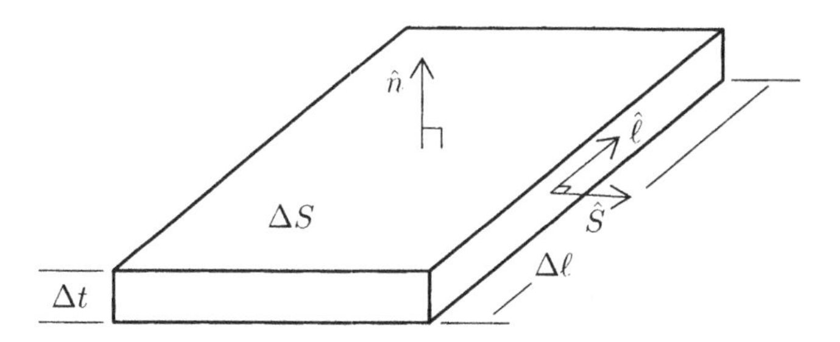

Figure A.29 Pillbox to determine the curl in the rectangular system: $\hat{\ell}$ is the unit vector along $\Delta\boldsymbol{\ell}$ and $\hat{s}$ is the unit vector identifying a surface.

Equation (A.88) can be reduced to

$$\begin{aligned}\hat{n}\cdot\nabla\times\mathbf{A} &= \lim_{\Delta S\to 0}\sum_m \frac{\mathbf{A}\cdot\hat{\ell}\Delta\ell_m}{\Delta S}\\ &= \lim_{\Delta S\to 0}\frac{\oint_L \mathbf{A}\cdot d\boldsymbol{\ell}}{\Delta S},\end{aligned}$$

where ΔS is the area of the loop, enclosed by the loop contour L.

Note that ΔS has its normal in the direction $\hat{n}$ (in which the component is desired) and L is the contour of ΔS; the path of integration follows the right-hand rule with respect to $\hat{n}$. Thus we have

$$\hat{n}\cdot\nabla\times\mathbf{A} = \lim_{\Delta S\to 0}\frac{\oint_L \mathbf{A}\cdot d\boldsymbol{\ell}}{\Delta S} \tag{A.89}$$

which indicates that the $\hat{n}$-component of the curl of a vector at a point P can be interpreted as the closed line integral (or circulation) of that vector function about an infinitesimal closed path on a surface (orientated in the $\hat{n}$ direction) containing that point, on a per unit surface area basis. Circulation per unit area indicates the nature of a vorticity, circulating motion, or curling—hence the name.

To obtain the algebraic expression for $\nabla\times\mathbf{A}$, we orient ds in the three orthogonal systems and obtain the explicit expression. The shape of the area $(d\tau)$ is chosen to correspond to the specific coordinate system. For example, for the rectangular system we choose a rectangular loop and orient it in the $\hat{x},\hat{y},\hat{z}$ directions, etc., meaning $\hat{n}=\hat{x},\hat{y},\hat{z}$.

Thus it can be shown that

$$\nabla\times\mathbf{A} = \begin{vmatrix}\hat{x} & \hat{y} & \hat{z}\\ \dfrac{\partial}{\partial x} & \dfrac{\partial}{\partial y} & \dfrac{\partial}{\partial z}\\ A_x & A_y & A_z\end{vmatrix}. \tag{A.90}$$

Note that in rectangular systems $\nabla \times \mathbf{A}$ may formally be interpreted as the vector product of ∇ and $\mathbf{A}$.

In the cylindrical system,

$$\begin{aligned}\nabla \times \mathbf{A} = & \left(\frac{1}{\rho}\frac{\partial A_z}{\partial \rho} - \frac{\partial A\phi}{\partial z}\right)\hat{\rho} + \left(\frac{\partial A_\rho}{\partial z} - \frac{\partial A_z}{\partial z}\right)\hat{\phi} \\ & + \left(\frac{1}{\rho}\frac{\partial(\rho A\phi)}{\partial \rho} - \frac{1}{\rho}\frac{\partial A\rho}{\partial \rho}\right)\hat{z}\end{aligned} \tag{A.91}$$

and in the spherical system of coordinates

$$\begin{aligned}\nabla \times \mathbf{A} = & \frac{1}{r\,\sin\theta}\left[\frac{\partial A_\phi \sin\theta}{\partial \theta} - \frac{\partial A_\theta}{\partial \phi}\right]\hat{r} + \frac{1}{r}\left[\frac{1}{\sin\theta}\frac{\partial A_r}{\partial \phi} - \frac{\partial(r\,A_\phi)}{\partial r}\right]\hat{\theta} \\ & + \frac{1}{r}\left[\frac{\partial(r\,A_\theta)}{\partial r} - \frac{\partial A_r}{\partial \theta}\right]\hat{\phi}.\end{aligned} \tag{A.92}$$

Stokes's Theorem

Stokes's theorem plays a role as important as Gauss's theorem in vector analysis. It states that when $\mathbf{A}$ is a continuous function of position,

$$\int_S \nabla \times \mathbf{A} \cdot ds = \oint_L \mathbf{A} \cdot d\boldsymbol{\ell}, \tag{A.93}$$

where S is an open surface, L is its condition, and the integration contour is defined in the right-hand sense.

Consider a surface S broken up with elementary surfaces ΔS_i, as in Figure A.30*a*. Apply the definition to the elementary surface ΔS_i:

$$\nabla \times \mathbf{A} \cdot \mathbf{\Delta S_i} \cdot \hat{\mathbf{n}} = \oint \mathbf{A} \cdot \mathbf{d}\boldsymbol{\ell}_\mathbf{i}, \quad \boldsymbol{\Delta}\ell_\mathbf{i} \text{ contour of } \mathbf{\Delta S_i}, \tag{A.94}$$

where $d\boldsymbol{\ell}_i$ is along the perimeter of ΔS_i.

As we evaluate the closed line integral for each $\Delta \ell_i$ and ΔS_i (Figure A.30*b*) and sum it up, some cancellation would occur. We are then left with an integral along the contour of S:

$$\begin{aligned}\sum_i \nabla \times \mathbf{A} \cdot \mathbf{\Delta S_i} &= \sum_\mathrm{i} \oint \mathbf{A} \cdot \mathbf{d}\boldsymbol{\ell}_\mathbf{i} \\ \text{for } i \to \infty : \int_S \nabla \times \mathbf{A} \cdot ds &= \oint_L \mathbf{A} \cdot d\boldsymbol{\ell},\end{aligned} \tag{A.95}$$

where $\oint \mathbf{A} \cdot d\boldsymbol{\ell}$ is the net circulation integral of $\mathbf{A}$ around a closed path; it is a measure of another vector-field property, namely the curling up (or rotation) of the field lines. Hence this is called *curl* or *circulation.*

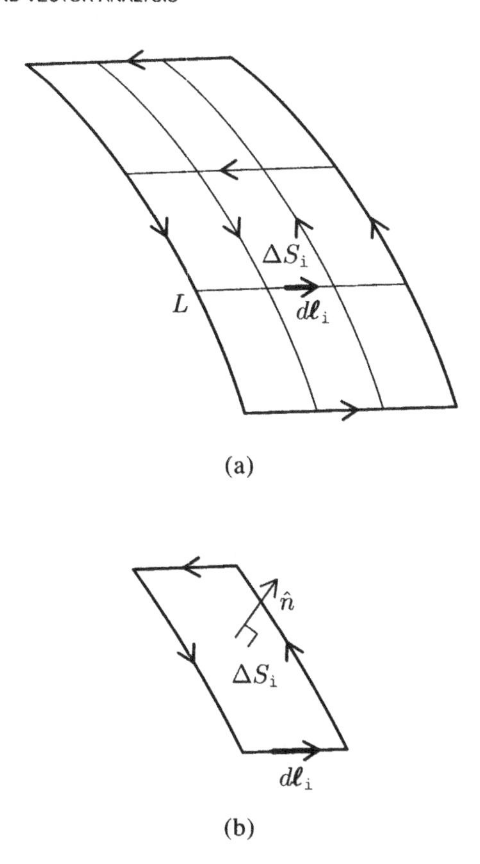

Figure A.30 (*a*) Open surface S with contour L divided into a number of elementary surfaces ΔS_i. (*b*) Elementary surface ΔS_i with contour $\Delta \ell_i$.

Example A.8

Show that the vector function $\mathbf{A} = -y\hat{x} + x\hat{y}$ satisfies Stokes's theorem $\int_S \nabla \times \mathbf{A} \cdot ds = \oint_L \mathbf{A} \cdot d\boldsymbol{\ell}$, where the surface S in the xy plane is defined by $0 \le x \le a,\ 0 \le y \le b$, and L is the appropriate contour of S_1 as shown in Figure A.31.

Solution

Let S be the total surface bounded by the contour L as shown in Figure A.31:

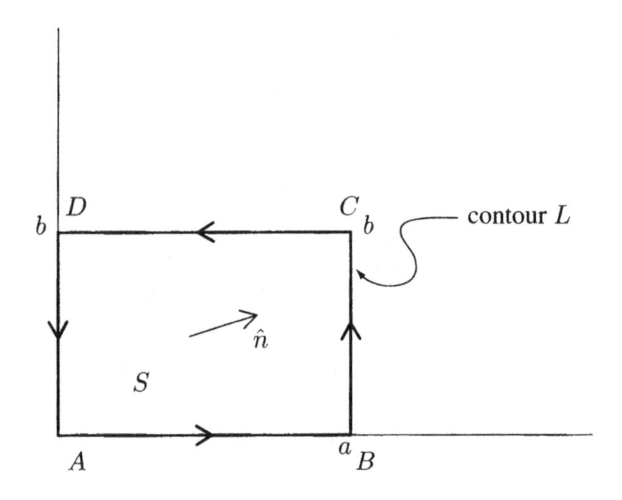

Figure A.31 Contour L surrounding a surface S'.

$$\mathbf{A} = -y\hat{x} + x\hat{y}$$

$$\nabla \times \mathbf{A} = \begin{vmatrix} \hat{x} & \hat{y} & \hat{z} \\ \dfrac{\partial}{\partial x} & \dfrac{\partial}{\partial y} & \dfrac{\partial}{\partial z} \\ -y & x & 0 \end{vmatrix} = 2\hat{z}$$

$$\begin{aligned} \int_S \nabla \times \mathbf{A} \cdot d\mathbf{s} &= \int_S \nabla \times \mathbf{A} \cdot \hat{n}\, ds \\ &= \int_S 2\hat{z} \cdot \hat{z}\, dx\, dy \\ &= 2 \int_0^b \int_0^a dx\, dy = 2ab. \end{aligned} \tag{1}$$

Then,

$$\begin{aligned} \mathbf{A} \cdot d\mathbf{l} &= (-y\hat{x} + x\hat{y}) \cdot (dx\,\hat{x} + dy\,\hat{y} + dz\,\hat{z}) \\ &= -y\,dx + x\,dy \end{aligned}$$

$$\begin{aligned}
\oint_L \mathbf{A} \cdot d\boldsymbol{\ell} &= \int \mathbf{A} \cdot d\boldsymbol{\ell} \\
&= \int_A^B (-y\,dx + x\,dy) + \int_B^C (-y\,dx + x\,dy) \\
&+ \int_C^D (-y\,dx + x\,dy) + \int_D^A (-y\,dx + x\,dy) \\
&= 0 + a\int_0^b dy + \int_a^0 (-b)\,dx + 0 \\
&= ab + ab = 2ab. \qquad (2)
\end{aligned}$$

Therefore (1) ≡ (2), which implies that **A** (given) satisfies Stokes's theorem.

A.10 THE LAPLACIAN $\nabla^2 = \nabla \cdot \nabla$

The divergence of a vector function that corresponds to the gradient of a scalar function often occurs in many mathematical formulations of physical problems.

Let the scalar function be V. Let

$$\begin{aligned}
\mathbf{A} &= \text{grad } V = \nabla V \\
\nabla \cdot \mathbf{A} &= \nabla \cdot \nabla V = \text{div grad} V \qquad (A.96) \\
&= \nabla^2 V.
\end{aligned}$$

The double operator, div grad $\equiv \nabla \cdot \nabla = \nabla^2$ is called the *Laplacian*. It can be shown that

$$\nabla^2 V = \frac{\partial^2 V}{\partial x^2} + \frac{\partial^2 V}{\partial y^2} + \frac{\partial^2 V}{\partial z^2} \qquad (A.97)$$

in the rectangular system, with

$$\nabla^2 = \frac{\partial^2}{\partial x^2} + \frac{\partial^2}{\partial y^2} + \frac{\partial^2}{\partial z^2} \qquad (A.98)$$

only in rectangular systems. In cylindrical systems,

$$\nabla^2 V = \frac{1}{\rho}\frac{\partial}{\partial \rho}\left(\rho \frac{\partial V}{\partial \rho}\right) + \frac{1}{\rho^2}\frac{\partial^2 V}{\partial \phi^2} + \frac{\partial^2 V}{\partial z^2} \qquad (A.99)$$

and in spherical systems of coordinates,

$$\nabla^2 V = \frac{1}{r^2}\frac{\partial}{\partial r}\left(r^2 \frac{\partial V}{\partial r}\right) + \frac{1}{r^2 \sin^2\theta}\frac{\partial}{\partial \theta}\left(\sin\theta \frac{\partial V}{\partial \theta}\right) + \frac{1}{r^2 \sin^2\theta}\frac{\partial^2 V}{\partial \phi^2}. \qquad (A.100)$$

We have seen that the Laplacian ∇^2 operates on a scalar quantity. There are cases where one needs ∇^2 operating on a vector quantity.

Take the vector identity

$$\nabla \times \nabla \times \mathbf{A} = \nabla(\nabla \cdot \mathbf{A}) - \nabla \cdot \nabla \mathbf{A} \tag{A.101}$$

and note that we formally write

$$\nabla \cdot \nabla \mathbf{A} \equiv \nabla^2 \mathbf{A}, \tag{A.102}$$

where ∇^2 is operating on a vector quantity $\mathbf{A}$. Since the quantity $\nabla^2 \mathbf{A}$ is not defined, we define it as

$$\nabla^2 \mathbf{A} \equiv \nabla \cdot \nabla \mathbf{A} \equiv \nabla(\nabla \cdot \mathbf{A}) - \nabla \times \nabla \times \mathbf{A}, \tag{A.103}$$

which is now well defined. It can be shown that only in rectangular systems

$$\nabla^2 \mathbf{A} = \hat{x}\nabla^2 A_x + \hat{y}\nabla^2 A_y + \hat{z}\nabla^2 A_z. \tag{A.104}$$

For other coordinate systems $\nabla^2 \mathbf{A}$ is obtained by carrying and the operations on the right hand side of (A.103) in the desired coordinate systems.

A.11 COMMENTS ON NOTATION

The gradient, divergence, and curl operations on scalar and vector functions are generally denoted by language (literal) and symbolic forms as follows:

	Literal	*Symbolic*
Gradient of ϕ	grad ϕ	$\nabla\phi$
Divergence of $\mathbf{A}$	div $\mathbf{A}$	$\nabla \cdot \mathbf{A}$
Curl $\mathbf{A}$	curl $\mathbf{A}$	$\nabla \times \mathbf{A}$

Here $\mathbf{A}$ and ϕ are the vector and scalar functions and ∇ is the del operator. The literal notations are clear and need no comment.

The symbolic notations were first introduced by J. Willard Gibbs [1], and because of their simplicity were quickly adopted. Most books on electromagnetics and field theory use these symbolic notations. These notations, however, particularly $\nabla \cdot \mathbf{A}$ and $\nabla \times \mathbf{A}$, must be interpreted strictly as representations of the divergence and curl operations and not in the dot- and cross-product sense! For example, $\nabla \cdot \mathbf{A}$ just means the divergence of $\mathbf{A}$ and not the dot product of ∇ and $\mathbf{A}$, although in rectangular systems it yields the correct result. The same applies to $\nabla \times \mathbf{A}$. Even in rectangular systems, Tai [3] pointed out that it is not logical to interpret $\nabla \cdot \mathbf{A}$ and $\nabla \times \mathbf{A}$ as a dot and cross product, respectively. In fact, Tai has suggested the following notation:

Gradient	$\nabla\phi$
Divergence of $\mathbf{A}$	$\nabla\!\!\!\!\cdot\;\mathbf{A}$
Curl of $\mathbf{A}$	$\nabla\!\!\!\!\times\;\mathbf{A}$

It will take some time before Tai's notations are universally accepted. We accept the notations of Tai as the logical ones; however, in view of the extensive use of the $\nabla \cdot \mathbf{F}$ and $\nabla \times \mathbf{F}$ notations, we have retained them with the caveat that they should not be interpreted in their conventional sense.

A.12 SOME USEFUL RELATIONS

$\mathbf{A}, \mathbf{B}$, and $\mathbf{C}$ are three vectors, and ψ and ϕ are two scalar functions assumed to be continuous functions with piecewise continuous first derivatives in V and on S (or on S and the contour boundings) (Figures A.32 and A.33).

A.12.1 Vector Algebra

$$\mathbf{A} \cdot (\mathbf{B} \times \mathbf{C}) = \mathbf{B} \cdot (\mathbf{C} \times \mathbf{A}) = \mathbf{C} \cdot (\mathbf{A} \times \mathbf{B}) \tag{A.105}$$

$$\mathbf{A} \times (\mathbf{B} \times \mathbf{C}) = \mathbf{B}(\mathbf{A} \cdot \mathbf{C}) - \mathbf{C}(\mathbf{A} \cdot \mathbf{B}). \tag{A.106}$$

A.12.2 Vector Identities

$$\nabla(\psi + \phi) = \nabla\psi + \nabla\phi \tag{A.107}$$

$$\nabla \cdot (\mathbf{A} + \mathbf{B}) = \nabla \cdot \mathbf{A} + \nabla \cdot \mathbf{B} \tag{A.108}$$

$$\nabla \times (\mathbf{A} + \mathbf{B}) = \nabla \times \mathbf{A} + \nabla \times \mathbf{B} \tag{A.109}$$

$$\nabla(\psi\phi) = \phi\nabla\psi + \psi\nabla\phi \tag{A.110}$$

$$\nabla \cdot (\phi\mathbf{A}) = \phi\nabla \cdot \mathbf{A} + \nabla\phi \cdot \mathbf{A} \tag{A.111}$$

$$\nabla \times (\phi\mathbf{A}) = \nabla\phi \times \mathbf{A} + \phi\nabla \times \mathbf{A} \tag{A.112}$$

$$\nabla \times \nabla\phi = 0 \tag{A.113}$$

$$\nabla \cdot \nabla \times \mathbf{A} = 0 \tag{A.114}$$

$$\nabla \cdot (\mathbf{A} \times \mathbf{B}) = \mathbf{B} \cdot \nabla \times \mathbf{A} - \mathbf{A} \cdot \nabla\mathbf{B} \tag{A.115}$$

$$\nabla \cdot \nabla \mathbf{A} \equiv \nabla^2 \phi \tag{A.116}$$

$$\nabla \times (\mathbf{A} \times \mathbf{B}) = \mathbf{A}\nabla \cdot \mathbf{B} - \mathbf{B}\nabla \cdot \mathbf{A} + (\mathbf{B} \cdot \nabla)\,\mathbf{A} + (\mathbf{A} \cdot \nabla)\,\mathbf{B} \tag{A.117}$$

$$\nabla (\mathbf{A} \cdot \mathbf{B}) = (\mathbf{A} \cdot \nabla)\,\mathbf{B} + (\mathbf{B} \cdot \nabla)\,\mathbf{A} + \mathbf{A} \times (\nabla \times \mathbf{B}) + \mathbf{B} \times (\nabla \times \mathbf{A})\,. \tag{A.118}$$

A.12.3 Integral Relations

Consider a volume τ bounded by a closed surface S, as shown in Figure A.32.

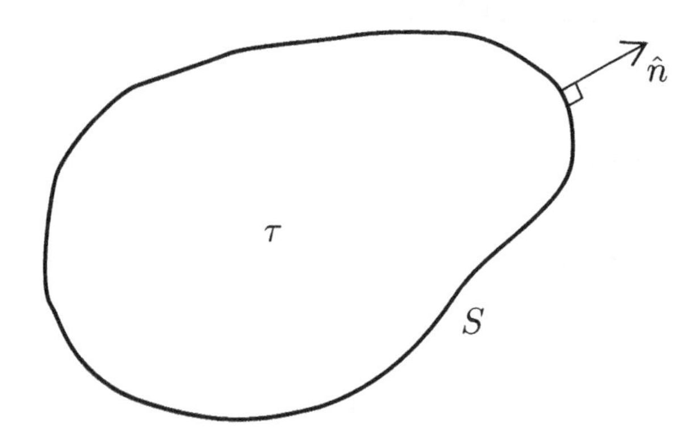

Figure A.32 Volume τ bounded by a closed surface S.

Then

$$\int_\tau \nabla \phi \, d\tau = \oint_S \hat{n} \phi \, ds = \oint_S \phi \, d\mathbf{s} \tag{A.119}$$

$$\int_\tau \nabla \cdot \mathbf{A} \, d\tau = \oint_S \mathbf{A} \cdot \hat{n} \, ds = \oint_S \mathbf{A} \cdot d\mathbf{s}. \tag{A.120}$$

Equation (A.120) is also known as the *divergence theorem* or *Gauss's theorem*:

$$\int_\tau \nabla \times \mathbf{A} \, d\tau = \oint_S \hat{n} \times \mathbf{A} \, ds = - \oint_S \mathbf{A} \times d\mathbf{s}. \tag{A.121}$$

Consider S to be an open (or unclosed) surface bounded by the contour C, as in Figure A.33. Then

$$\int_s \hat{n} \times \nabla \phi \, ds = \oint_C \phi \, d\boldsymbol{\ell} \tag{A.122}$$

$$\int_s \nabla \times \mathbf{A} \cdot \hat{n} \, ds = \oint_C \mathbf{A} \cdot d\boldsymbol{\ell}. \tag{A.123}$$

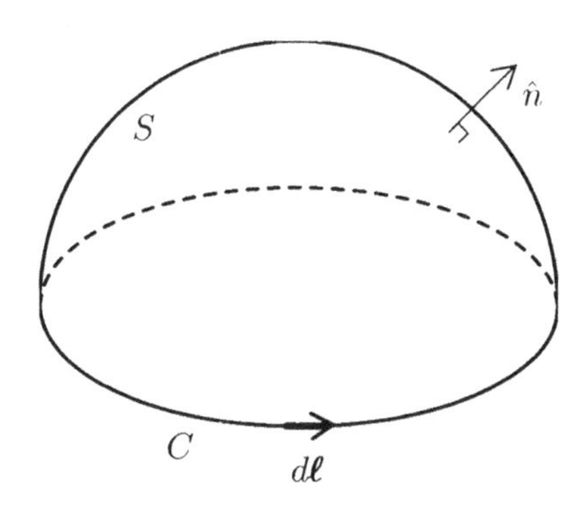

Figure A.33 Open surface S bounded by a contour C.

Equation (A.123) is also known as *Stokes's theorem.* We next give proofs of (A.119), (A.121), and (A.122).

Example A.9

Proof of (A.119), that is, $\int_\tau \nabla\phi\, d\tau = \oint_S \hat{n}\phi\, ds = \oint_S \phi\, d\mathbf{s}$. Note that $d\mathbf{s} = \hat{n}\, ds$ and τ is enclosed by S (see Figure A.10).

Proof

Consider $\phi\mathbf{a}$, where $\mathbf{a}$ is any constant vector and ϕ is a scalar function. Use the vector identity $\nabla\cdot\phi\mathbf{a} = \mathbf{a}\cdot\nabla\phi + \phi\nabla\cdot\mathbf{a} = \mathbf{a}\cdot\nabla\phi$ since $\nabla\cdot\mathbf{a} = 0$. Now apply the divergence theorem to $\phi\mathbf{a}$ and obtain

$$\int_\tau \nabla\cdot\phi\mathbf{a}\, d\tau = \int_\tau \mathbf{a}\cdot\nabla\phi\, d\tau = \oint_S \phi\mathbf{a}\cdot\, d\mathbf{s},$$

that is,

$$\begin{aligned}\mathbf{a}\cdot\int_\tau d\nabla\phi\, d\tau &= \oint_S \phi\mathbf{A}\cdot\, d\mathbf{s} = \oint_s \phi\mathbf{a}\cdot\hat{n}\, ds\\ &= \mathbf{a}\cdot\oint_S \phi\hat{n}\, ds.\end{aligned}$$

Since $\mathbf{a}$ is arbitrary, we have

$$\int_\tau \nabla\phi\, d\tau = \oint_S \phi\hat{n}\, dS.$$

Example A.10

Prove

$$\int_{\tau} \nabla \times \mathbf{A}\, d\tau = \oint_{S} \hat{n} \times \mathbf{A}\, ds.$$

Solution

Take $\mathbf{A} \times \mathbf{a}$ with $\mathbf{a}$ as arbitrary constant vector.

$$\begin{aligned} \nabla \cdot \mathbf{A} \times \mathbf{a} &= \mathbf{a} \cdot \nabla \times \mathbf{A} - \mathbf{A} \cdot \nabla \times \mathbf{a} \\ &= \mathbf{a} \cdot \nabla \times \mathbf{A} \text{ (since } \nabla \times \mathbf{a}) = 0. \end{aligned}$$

Apply the divergence theorem to $\mathbf{A} \times \mathbf{a}$ (i.e., (A.121)):

$$\int_{\tau} (\nabla \cdot \mathbf{A} \times \mathbf{a})\, d\tau = \oint_{S} (\mathbf{A} \times \mathbf{a}) \cdot d\mathbf{s}$$

or

$$\begin{aligned} \int_{\tau} \mathbf{a} \cdot \nabla \times \mathbf{A}\, d\tau &= \oint_{S} (\mathbf{A} \times \mathbf{a}) \cdot \hat{n}\, ds \\ &= \oint_{S} \mathbf{a} \cdot \hat{n} \times \mathbf{A}\, ds \quad \text{(use cyclical rule for triple product).} \end{aligned}$$

Since $\mathbf{a}$ is arbitrary,

$$\int_{\tau} \nabla \times \mathbf{A}\, d\tau = \oint_{S} \hat{n} \times \mathbf{A}\, ds.$$

Example A.11

Prove

$$\int_{S} \hat{n} \times \nabla\phi\, ds = \oint_{C} \phi\, d\boldsymbol{\ell},$$

where S is an open surface with contour C.

Proof

Take a vector $\phi\mathbf{a}$, with $\mathbf{a}$ as a constant but arbitrary vector. Use (A.112): $\nabla \times \phi\mathbf{a} = \nabla\phi \times \mathbf{a} + \phi\nabla \times \mathbf{a} = \nabla\phi \times \mathbf{a}$. Apply Stokes's theorem ((A.123)) to $\phi\mathbf{a}$, that is,

$$\begin{aligned}
\int_S (\nabla \times \phi\mathbf{a}) \cdot d\mathbf{s} &= \int_S (\nabla\phi \times \mathbf{a}) \cdot d\mathbf{s} \\
&= \int_S (\nabla\phi \times \mathbf{a}) \cdot \hat{n}\, ds \\
&= \int_S (\hat{n} \times \nabla\phi) \cdot \mathbf{a}\, ds \\
&= \oint_C \phi\mathbf{a} \cdot d\boldsymbol{\ell}
\end{aligned}$$

$$\therefore \quad \mathbf{a} \cdot \int_S (\hat{n} \times \nabla\phi)\, ds = \mathbf{a} \cdot \oint_C \phi\, d\boldsymbol{\ell}.$$

Since $\mathbf{a}$ is arbitrary,

$$\int_S (\hat{n} \times \nabla\phi)\, ds = \oint_C \phi\, d\boldsymbol{\ell}.$$

REFERENCES

1. J. Willard Gibbs, *Elements of Vector Analysis,* privately printed, New Haven, 1881–1884.
2. J. Willard Gibbs, *The Scientific Papers of J. Willard Gibbs*, Vol. II, Longmans Green, London, 1906. Reprint: Dover, New York, 1961.
3. C.-T. Tai, *Generalized Vectors and Dyadic Analysis: Applied Mathematics in Field Theory,* 2nd ed., IEEE Press, New York, 1992.
4. M. N. O. Sadiku, *Elements of Electromagnetics*, 2nd ed., Harcourt Brace, New York, 1994.
5. J. D. Kraus and D. A. Fleisch, *Electromagnetics,* 5th ed., McGraw-Hill, New York, 1999.
6. F. T. Ulaby, *Applied Electromagnetics*, Prentice Hall, Upper Saddle River, NJ, 1997.
7. U. S. Inan and A. S. Inan, *Electromagnetic Waves,* Prentice Hall, Upper Saddle River, NJ, 2000.
8. C. A. Balanis, *Advanced Engineering Electromagnetics*, Wiley, New York, 1989.

PROBLEMS

A.1 Given two points in space in rectangular system of coordinates as $P_1(1,1,1)$ and $P_2(2,-1,3)$. Determine the following:

(a) The cylindrical and spherical system coordinates (ρ,ϕ,z) and (r,θ,ϕ), respectively, for P_1 and P_2.

(b) The position vectors of P_1 and P_2.

(c) The vector directed from P_1 and P_2, that is, $\mathbf{P_1P_2}$, the direction cosines, scalar, and vector components of $\mathbf{P_1P_2}$.

Answer:

(a) P_1: $(1.414, 45°, 1)$; $(1.732, 57.7°, 45°)$. P_2: $(2.236, 333.4°, 3)$; $(3.742, 36.7°, 3$

(b) $\overrightarrow{OP_1} = \mathbf{r}_1 = \hat{x} + \hat{y} + \hat{z}$; $\overrightarrow{OP_2} = \mathbf{r}_2 = 2\hat{x} - \hat{y} + 3\hat{z}$.

(c) $\overrightarrow{P_1P_2} = \hat{x} - 2\hat{y} + 2\hat{z}$; $\cos\alpha_x = \frac{1}{3}$, $\cos\alpha_y = -\frac{2}{3}$, $\cos\alpha_z = \frac{2}{3}$; $\alpha_x = 70.5°$, $\alpha_y = 131.8°$, $\alpha_z = 48.2°$. Scalar x, y, z components: $1, -2, 2$. Vector x, y, z components: $\hat{x}, -2\hat{y}, 2\hat{z}$.

A.2 Given

$$\mathbf{A} = -2\hat{x} + 3\hat{y} + 4\hat{z}, \quad \mathbf{B} = \hat{x} + \hat{y} + \hat{z}.$$

(a) Find $|\mathbf{A}|$, $|\mathbf{B}|$, the angles made by $\mathbf{A}$ with the $\hat{x}$, $\hat{y}$, and $\hat{z}$ axes.

(b) Determine $\mathbf{A}\cdot\mathbf{B}$ and $(\mathbf{A}\times\mathbf{B})$. Determine the angle between $\mathbf{A}$ and $\mathbf{B}$ by using the definitions of $\mathbf{A}\cdot\mathbf{B}$ and $\mathbf{A}\times\mathbf{B}$. Find the unit vectors normal to the plane of $\mathbf{A}$ and $\mathbf{B}$, and which one is the unit positive normal.

Answer:

(a) $|\mathbf{A}| = 5.39$, $|\mathbf{B}| = 1.73$; angles made by $\mathbf{A}$ with $\hat{x}, \hat{y}, \hat{z}$ axes: $\alpha_x = 111.8°$, $\alpha_y = 56.2°$, $\alpha_z = 42.1°$.

(b) $\mathbf{A}\cdot\mathbf{B} = 5$; $\mathbf{A}\times\mathbf{B} = -\hat{x} + 6\hat{y} - 5\hat{z}$; $\theta_{AB} = 57.5$o. $\hat{n} = \pm(-0.127\hat{x} + 0.762\hat{y} - 0.635\hat{z})$. The positive unit normal is $+\hat{n}$.

A.3 Use the rectangular components of the vectors involved to algebraically prove the following vector identities:

(a) $(\mathbf{A}+\mathbf{B})\cdot\mathbf{C} = \mathbf{A}\cdot\mathbf{C} + \mathbf{B}\cdot\mathbf{C}$

(b) $(\mathbf{A} \times \mathbf{B}) \times \mathbf{C} = \mathbf{A} \times \mathbf{C} + \mathbf{B} \times \mathbf{C}$

(c) $\mathbf{A} \cdot (\mathbf{B} \times \mathbf{C}) = \mathbf{B} \cdot (\mathbf{C} \times \mathbf{A}) = \mathbf{C} \cdot (\mathbf{A} \times \mathbf{B})$

(d) $\mathbf{A} \times (\mathbf{B} \times \mathbf{C}) = \mathbf{B}(\mathbf{A} \cdot \mathbf{C}) - \mathbf{C}(\mathbf{A} \cdot \mathbf{B})$

A.4

(a) Express the vector field $\mathbf{A} = (x + y)\hat{x}$ in cylindrical (ρ, ϕ, z) and spherical (r, θ, ϕ) systems of coordinates.

(b) Express the following two vectors in the cartesian system (x, y, z) of coordinates: $\mathbf{B} = \rho \cos\phi\ \hat{z}$, $\mathbf{C} = r \cos\phi\ \hat{r}$.

Answer:

(a) $\mathbf{A} = \rho\cos\phi(\cos\phi + \sin\phi)\hat{\rho} + \rho\sin\phi(\cos\phi + \sin\phi)\hat{\phi}$.

$\mathbf{A} = (r\sin\theta\cos\phi + r\sin\theta\sin\phi)(\hat{r}\sin\theta\cos\phi + \hat{\theta}\cos\theta\cos\phi - \hat{\phi}\sin\phi)$.

(b $\mathbf{B} = x\hat{x}$.

$\mathbf{C} = (x/\sqrt{x^2 + y^2})(x\hat{x} + y\hat{y} + z\hat{z})$.

A.5 Find the gradient of the following functions:

(a) $\Phi = x^2y^2z^2$

(b) $\Phi = \rho\phi z$

(c) $\Phi = r\theta\phi$.

Answer:

(a) $2xyz(yz\hat{x} + xz\hat{y} + xy\hat{z})$.

(b) $\phi z\hat{\rho} + z\hat{\phi} + \rho\phi\hat{z}$.

(c) $\theta\phi\hat{r} + \phi\hat{\theta} + (\theta/\sin\theta)\hat{\phi}$.

A.6 Given $\mathbf{A} = x\hat{x}$: Determine $\nabla \cdot \mathbf{A}$ in the rectangular system. Transform $\mathbf{A}$ in the cylindrical system of coordinates and evaluate $\nabla \cdot \mathbf{A}$ in that system. Show that the value of $\nabla \cdot \mathbf{A}$ obtained is the same as before.
Answer:
$\nabla \cdot \mathbf{A} = 1$ in both systems.

A.7 Given a scalar function f as $f = x^2 + y^2$, determine the following:

(a) ∇f

(b) Directional derivative of f in the directions $\hat{x}, \hat{y}$, and $(\hat{x} + \hat{y})/\sqrt{2}$

Answer:
(a) $2x\hat{x} + 2y\hat{y}$; **(b)** $2x, 2y, \sqrt{2}(x + y)$.

A.8 Show that the vector identity

$$\nabla \cdot f\mathbf{A} = f\nabla \cdot \mathbf{A} + \mathbf{A} \cdot \nabla f$$

is satisfied by

$$f = x$$
$$\mathbf{A} = x^2 y\hat{x} + xy^2\hat{y}.$$

A.9 Given $\mathbf{A} = y^2 z^2 \hat{x} + z^2 x^2 \hat{y} + x^2 y^2 \hat{z}$, determine $\nabla \times \mathbf{A}$ at the point $P(1, 1, 1)$.

Answer:
$\nabla \times \mathbf{A}|_{(1,1,1)} = 0$.

A.10 Determine the divergence of the following functions:

(a) $\mathbf{A} = xy\hat{x} + yz\hat{y} + zx\hat{z}$

(b) $\mathbf{B} = \sin\phi\hat{\rho} + \cos\phi\hat{\phi} + z^2\hat{z}$

(c) $\mathbf{C} = r\hat{r} + \sin\theta\sin\phi\hat{\theta} - \sin\theta\cos\phi\hat{\phi}$.

Answer:
(a) $y + z + x$; **(b)** $2z$; **(c)** $3 + (2\cos\theta\sin\phi)/r + (\sin\phi)/r$.

A.11 Given $\mathbf{A} = -y\hat{x}$, evaluate $\nabla \times \mathbf{A}$ in the rectangular and cylindrical systems of coordinates.
Answer:
$\nabla \times \mathbf{A} = \hat{z}$ in both systems of coordinates.

A.12 The flux ψ of a vector through a surface s is given by $\psi = \int_s \mathbf{A} \cdot ds$.

(a) If $\mathbf{A} = \hat{r}/r^2$, find ψ through the spherical surface $r = a$

(b) If $\mathbf{A} = \hat{r}/r^2$, find ψ through the surface $z = 1$

(c) If $\mathbf{A} = \hat{\rho}/\rho$, find ψ through the surface $\rho = a,\ 0 \leq z \leq 1$.

Answer:
(a) 4π; **(b)** 2π; **(c)** 2π.

A.13 Determine $\nabla^2\psi$ if ψ equals

(a) $\psi = x^2 + y^2 + z^2$

(b) $\psi = x^2y^2 + y^2z^2 + z^2x^2$

(c) $\psi = x^2y^2z^2$.

Answer:
(a) 8; **(b)** $4\left(x^2 + y^2 + z^2\right)$; **(c)** $2\left(y^2z^2 + z^2x^2 + x^2y^2\right)$.

APPENDIX B

FREQUENCY BAND DESIGNATIONS

Applied Electromagnetics and Electromagnetic Compatibility. By D. L. Sengupta, V. V. Liepa
ISBN 0-471-16549-2

Table B.1 Commonly Used Frequency Band Designations

Name	Frequency Range	Principal Use
Super low frequency (SLF)	30–300 Hz	Power grids
Very low frequency (VLF)	3–30 KHz	Submarine communications, sonar
Low frequency (LF)	30–300 KHz	Radio beacons, navigational aids
Medium frequency (MF)	300–3000 KHz	AM broadcasting, maritime radio, Coast Guard communications, direction finding
High frequency (HF)	3–30 MHz	Shortwave broadcasting, telephone, telegraph, international broadcasting, amateur radio, citizen's band
Very high frequency (VHF)	30–300 MHz	FM broadcasting, TV, air traffic control, police, taxicab, and mobile radio
Ultra high frequency (UHF)	300–3000 MHz	TV, satellite communication, surveillance radar, WiLAN, cellular phones, GPS
Super high frequency (SHF)	3–30 GHz	Airborne radar, microwave links, common carrier land mobile communication, satellite communication
Extremely high frequency (EHF)	30–300 GHz	Radar, automotive radar

Notes: For AM radio, Frequency band = 535–1605 MHz; number of channels = 107; adjacent channel separation = 10 kHz. For FM radio, frequency band = 88–108 MHz; number of channels = 100; adjacent channel separation = 250 KHz.

Table B.2 IEEE Frequency Band Designations

World War II Radar Band Designations	Name	Frequency Range
—	HF	3–30 MHz
—	VHF	30–300 MHz
—	UHF	300–1000 MHz
390–1550 MHz	L	1–2 GHz
1550–3900 MHz	S	2–4 GHz
3.9–6.2 GHz	C	4–8 GHz
6.2–12 GHz	X	8–12 GHz
12.9–18 GHz	Ku	12–18 GHz
18–26.5 GHz	K	18–2 GHz
26.5–40 GHz	Ka	27–40 GHz
	V	40–75 GHz
	W	75–110 GHz
	Millimeter	110–300 GHz

Microwave Bands

Microwave bands cover approximately the UHF to EHF frequencies. The relationship between the World War II radar band designations (old) and the IEEE band designations (new) of radio frequencies are given as shown in Table B.3.

Table B.3 Mobile Telephone Bands

Cellular Phones	
AMPS Uplink:	864–894 MHz
AMPS Downlink:	824–849 MHz
PCS:	1850–1990 MHz
Cordless Telephones	46–49 MHz

Uplink: Mobile Station (MS) to Base Station (BS)
Downlink: Base Station (BS) to Mobile Station (MS)

Each of the VHF and UHF TV channels has a bandwidth of 6 MHz. For each channel, the carrier frequency for the video part of the signal is equal to the lower frequency of the channel band plus 1.25 MHz. The carrier frequency

Table B.4 Television Numbers and Frequencies—VHF Channels

Channel Number	Frequency (MHz)	Channel Number	Frequency (MHz)
2	54–60	8	180–186
3	60–66	9	186–192
4	66–72	10	192–198
5	76–82	11	198–204
6	82–88	12	204–210
7	174–180	13	210–216

for the audio part of the signal is equal to the upper frequency of the channel band 0.25 MHz.

Examples

Channel 4 (VHF):

Frequency band	= 66–72 MHz
Video carrier frequency	= 66 + 1.25 = 67.25 MHz
Audio carrier frequency	= 72 − 0.25 = 71.75 MHz

Channel 15 (UHF):

Frequency band	= 476–482 MHz
Video carrier frequency	= 476 + 1.25 = 477.25 MHz
Audio carrier frequency	= 482 − 0.25 = 481.75 MHz

Table B.5 Television Numbers and Frequencies—UHF Channels

Channel Number	Frequency (MHz)	Channel Number	Frequency (MHz)	Channel Number	Frequency (MHz)
14	470–476	37	608–614	60	746–752
15	476–482	38	614–620	61	752–758
16	482–488	39	620–626	62	758–764
17	488–494	40	626–632	63	764–770
18	494–500	41	632–638	64	770–776
19	500–506	42	638–644	65	776–782
20	506–512	43	644–650	66	782–788
21	512–518	44	650–656	67	788–794
22	518–524	45	656–662	68	794–800
23	524–530	46	662–668	69	800–806
24	530–536	47	668–674	70	806–812
25	536–542	48	674–680	71	812–818
26	542–548	49	680–686	72	818–824
27	548–554	50	686–692	73	824–830
28	554–560	51	692–698	74	830–836
29	560–566	52	698–704	75	836–842
30	566–572	53	704–710	76	842–848
31	572–578	54	710–716	77	848–854
32	578–584	55	716–722	78	854–860
33	584–590	56	722–728	79	860–866
34	590–596	57	728–734	80	866–872
35	596–602	58	734–740	81	872–878
36	602–608	59	740–746	82	878–884
				83	884–890

APPENDIX C

CONSTITUTIVE RELATIONS

Applied Electromagnetics and Electromagnetic Compatibility. By D. L. Sengupta, V. V. Liepa
ISBN 0-471-16549-2

Table C.1 Conductivity of Selected Materials at 20° C

Material	Conductivity σ (S/m)
Aluminum	3.82×10^7
Bismuth	8.70×10^5
Brass (66% Cu, 34% Zn)	2.56×10^7
Copper (annealed)	5.80×10^7
Carbon	3×10^4
Distilled water	$\sim 10^{-4}$
Dry sandy soil	$\sim 10^{-3}$
Fresh water	$\sim 10^{-2}$
Germanium (intrinsic)	2.13
Glass	$\sim 10^{-12}$
Gold	4.10×10^7
Iron	1.03×10^7
Lead	4.57×10^6
Marshy soil	$\sim 10^{-2}$
Mercury (liquid)	1.04×10^6
Mica	$\sim 10^{-15}$
Nickel	1.45×10^7
Platinum	9.52×10^6
Polystyrene	$\sim 10^{-16}$
Quartz (fused)	$\sim 10^{-17}$
Seawater	~ 4
Silicon (intrinsic)	4.35×10^{-4}
Silver	6.17×10^7
Stainless steel	1.11×10^6
Tin	8.77×10^6
Titanium	2.09×10^6
Tungsten	1.82×10^7
Wood	$10^{-11} - 10^{-8}$
Zinc	1.67×10^7

Source: V. S. Inan and A. S. Inan, *Electromagnetic Waves*, Prentice Hall, Upper Saddle River, NJ, 2000, p. APP-27.

Table C.2 Relative Permittivity and Dielectric Strength of Selected Materials

Material (room temp and 1 atm)	ϵ_r	Dielectric Strength (MV/m)
Air	1	~ 3
Bakelite	$\sim$ 4–8	25
Barium titanate (BaTiO3)	1200	7.5
Fused quartz	3.9	1000
Galium arsenide (GaAs)	13.1	~ 40
Glass	$\sim$ 4–9	~ 30
Mica (film)	5.4	200
Polystyrene	2.56	20
Silicon	11.9	~ 30
Paper	1.5–4	15
Rubber	$\sim$ 2.4–3	25

Source: V. S. Inan and A. S. Inan, *Electromagnetic Waves*, Prentice Hall, Upper Saddle River, NJ, 2000, p. APP-26.

Table C.3 Approximate Static Dielectric Constants (Relative Permittivity) of Dielectric Materials

Material	ϵ_r
Air	1.0006
Bakelite	4.8
Ferrite (Fe_2O_3)	12-16
Flint glass	10
Formica	5
Gallium arsenide (GaAs)	13
Lead glass	6
Mica	6
Nylon	3.8
Paraffin	2.1
Plexiglas	3.4
Plywood	2.1
Polyethylene	2.26
Quartz	3.8
Rubber	3
RT/duroid 5870	2.35
Silicon (Si)	12
Soil (dry)	3
Styrofoam	1.03
Teflon	2.1
Water	81

Source: C. A. Balanis, *Advanced Engineering Electromagnetics*, Wiley, New York, 1989, p. 50.

It is convenient to characterize a lossy dielectric medium by a complex dielectric constant $\epsilon_r^c = \epsilon_r' - j\epsilon_0''$, where $\epsilon_r', \epsilon_r''$ are real quantities and both may be frequency dependent, in general. Measured values of ϵ_r^c are usually expressed in terms of ϵ_r' and ϵ_r'' or ϵ_r' and (loss tangent) $\tan\delta = \epsilon_r''/\epsilon_r'$. We will use the former representation in Table C.4.

$$\epsilon_r^c = \frac{\epsilon_c}{\epsilon_0} = \frac{\epsilon' - j\epsilon''}{\epsilon_0} = \epsilon_r' - j\epsilon_r''$$

$$\tan\delta = \frac{\epsilon_c}{\epsilon_0} = \frac{\sigma + j\omega\epsilon''}{\omega\epsilon'} = \frac{\epsilon''}{\epsilon'} = \frac{\epsilon_r''}{\epsilon_r'}$$

Table C.4 Dielectric Properties of Selected Materials

Material	$T^\circ C$	f (GHz)	ϵ_r'	ϵ_r''
Aluminum oxide (Al_2O_2)	25	3	8.79	8.79×10^{-3}
Bakelite	24	3	3.64	190×10^{-3}
Barium titanate ($BaTiO_3$)	26	3	600	180
Concrete (dry)	25	2.45	0.05	0.05
Concrete (wet)	25	2.45	14.5	1.73
Clay soil	25	3	2.27	34×10^{-3}
Fiberglass (Bk174)	24	10	6.64	470×10^{-4}
Milk	20	3	51	30.1
Paper	25	3	2.7	1500×10^{-4}
Potato (78% moisture)	25	3	81	30.8
Plexiglas	27	3	2.6	150×10^{-4}
Polyethylene	25	3	2.55	8.5×10^{-4}
Polystyrene	25	3	2.55	8.5×10^{-4}
Quartz (fused)	25	3	3.78	2.3×10^{-4}
Raw beef	25	2.45	52.4	17.3
Sandy soil	25	3	2.55	160×10^{-4}
Snow (fresh fallen)	-20	3	1.2	3.48×10^{-4}
Snow (hard packed)	-6	3	1.5	1.35×10^{-3}
Teflon	22	3	2.1	3.15×10^{-4}
Water	1.5	3	80.5	2500×10^{-4}
	25	3	76.7	1200×10^{-4}
	85	3	56.5	310×10^{-4}
White rice	24	2.45	3.8	0.8

Source: V. S. Inan and A. S. Inan, *Electromagnetic Waves*, Prentice Hall, Upper Saddle River, NJ, 2000, p. 48.
Note: In practice, it is generally not necessary to distinguish between losses due to σ or $\omega\epsilon''$, since ϵ'' and σ_{eff} (or $\tan\delta$) are often determined by measurement.

Table C.5 Approximate Relative Permeability (μ_r) of Some Materials

Material	μ_r
Diamagnetic	
Bismuth	0.999833
Mercury	0.999963
Silver	0.9999736
Lead	0.9999831
Copper	0.99999906
Water	0.9999912
Hydrogen	~ 1
Paramagnetic	
Air	1.00000037
Aluminum	1.000021
Manganese	1.001
Ferromagnetic	
Cobalt	250
Nickel	600
Soft iron	5000
Silicon iron	70,000

Source: M. N. O. Sadiku, *Elements of Electromagnetics*, 2nd ed., Saunders/Harcourt Brace, New York, 1994, p. 784.

INDEX

Printed and bound by CPI Group (UK) Ltd, Croydon, CR0 4YY

16/04/2025

14658605-0001